Aircraft

Accident

Investigation

Richard H. Wood

Robert W. Sweginnis

Foreword by Jerome Lederer

International Standard Book Number 0-9653706-0-7
(previously ISBN 0-9629524-3-5)
For Sale By: Endeavor Books-Mountain States Litho
133 S. McKinley, Casper, WY 82601
(307) 265-7410 📖 FAX: (307) 237-9521
Toll Free: (888) 324-9303
E-Mail: danh@endeavorbooks.com

Aircraft Accident Investigation

Richard H. Wood
Robert W. Sweginnis

Printed in the United States of America

Endeavor Books-Mountain States Litho
133 S. McKinley, Casper, WY 82601
Phone: (307) 265-7410   Fax: (307) 237-9521
Toll Free: (888) 324-9303
E-Mail: danh@endeavorbooks.com
www.endeavorbooks.com

ISBN: 0-9653706-0-7 (previously ISBN 0-9629524-3-5)

Print History
   8th printing — August, 1998
   9th printing — December, 1998
  10th printing — March, 1999
  11th printing — November, 1999
  12th printing — May, 2000
  13th printing — October, 2000
  14th printing — March, 2001
  15th printing — June, 2001
  16th printing — September, 2001
  17th printing — January, 2002
  18th printing — March, 2002
  19th printing — July, 2002
  20th printing — September, 2002
  21st printing — November, 2002
  22nd printing — February, 2003
  23rd printing — July, 2003

Contents

PART I - THE RULES OF THE GAME

PART II - INVESTIGATION TECHNIQUES

PART III - TECHNOLOGY

PART IV - ANALYSIS, REPORTS AND INVESTIGATION MANAGEMENT

APPENDICES

Acknowledgements

During the preparation of this book, the authors received technical help and advice from a lot of people. While the manuscript was still in draft form, it was submitted to a group of experienced aircraft accident investigators for technical review. Their comments and criticisms were most helpful. They included:

Terry Heaslip. Aircraft Accident Investigation Consultant with AIR, Inc, Ottawa, Canada. A specialist in accident reconstruction.

Gerard M. Bruggink, Dep. Dir. Bureau of Accident Investigation, National Transportation Safety Board (Retired).

Jack Hazlett, Hazlett Associates; an Accident Reconstruction and Safety Education Consultancy. Adjunct Assistant Professor, University of Southern California.

William Waldock, Professor, Embry Riddle Aeronautical University.

Ted Banick, Aviation Safety and Risk Management Consultant specializing in Flight Operations Safety and Aircraft Accident Investigation.

Ira J. Rimson, P.E., FAAFS, FISASI, a Forensic Engineer who has specialized in aircraft accident reconstruction and analysis for 35 years. His principal current interest is in the application of Quality Control methodologies to investigations and their output reports, recommendations and findings. He has been editor of the ISASI "Forum" since 1977.

Olof Fritsch, Retired Chief Accident Investigation and Prevention, ICAO, and past President of the International Society of Air Safety Investigators.

C.O. Miller, Consultant-System Safety. Internationally recognized authority in aviation accident investigation and prevention. Forty years of specific air safety experience in industry, research, university, government and private consulting activities.

While the individual chapters were being written, **Don Clifford**, a Forensic Photographer from Los Angeles, helped with the chapter on photography. **Dr. Robert Alkov** and **E. John Kennedy, MD** reviewed and contributed to the chapter on Human Factors.

The figures in the chapter on Mid Air Collisions were done by **Kenneth S. Wood**.

The sketches in Chapters 34 and 35 were originally drawn by **Thomas P. Garvin** in 1957 when he was in the United States Air Force and attending Flight Safety School. His drawings were originally printed in a government publication and they have appeared in at least a dozen books; usually without credit. It is a pleasure to reprint some of them here with full credit to Tom Garvin and our thanks for his permission to use them.

The cover was painted for this book by **Bernard Oliver**. It depicts the first fatal crash of a heavier-than-air powered aircraft on September 17, 1908. The pilot, Orville Wright, survived with severe injuries. The passenger and observer, Lieutenant Thomas Selfridge, was fatally injured. In the painting, a Boeing 747 rises from the mist of the crash suggesting the progress in aircraft design and safety that has come from investigation of accidents.

Authors

Richard H. Wood has been involved in aviation safety and aircraft accident investigation since 1963. He is a Professional Engineer (Safety), a Certified Safety Professional (CSP), a retired pilot (U.S. Air Force) and a retired Professor of Safety Science at the University of Southern California.

He is presently Director of Aviation Safety Programs for Southern California Safety Institute and an active consultant in both aircraft accident investigation and aviation safety.

He is the author of numerous books, articles and professional papers on aviation safety and aircraft accident investigation. He has investigated over 125 aircraft accidents and lectures on the subject for Southern California Safety Institute.

He currently lives (and writes) in Snohomish, Washington.

Robert W. Sweginnis is an Associate Professor at Embry-Riddle Aeronautical University's Prescott, Arizona campus. He is a Professional Engineer (Safety), a Certified Safety Professional (CSP), and a retired U.S. Air Force command pilot. He has been active in aircraft accident investigation and system safety since 1973. His safety experience spans the military, industry and academe. In addition to teaching at ERAU's Center for Aerospace Safety Education, he has also taught aircraft accident investigation and system safety for the University of Southern California's Institute of Safety and Systems Management, the University of Washington's Engineering Professional Programs, and Southern California Safety Institute.

For fun, he occasionally "slips the surly bonds" and provides unusual attitude recovery training or enters aerobatic competitions.

Although both authors contributed to all chapters, Dick Wood was primarily responsible for Parts I, II, IV and the Appendices. Bob Sweginnis wrote Part III.

Foreword

Jerome Lederer

The authors of *Aircraft Accident Investigation* (Wood and Sweginnis) provide a superb review of the philosophy, reasoning, methods, controversial issues and associated ramifications that are applied in the investigation of the unique and complex challenges posed by aircraft accidents.

Aircraft operations and the resulting accidents are unique because:[1]

●The aircraft is continuously fighting the unrelenting law of gravity which instantaneously takes advantage of any failures or weaknesses in this struggle for survival. Human errors, carelessness and complacency are likely to be more catastrophic in air transportation than in any other means of transportation.

●Unlike surface traffic, an airplane cannot stop to attend to emergencies such as power plant failure, crew incapacitation or structural failure.

●It requires the coordinated cooperative efforts of a greater variety of associated technologies than any other system of transportation--air traffic control, airport management, weather, navaids, flight planning and dispatching, communications, ramp operations, et cetera.

●It is three dimensional, requiring navigation in three dimensions, subject to the variable hazards of the atmosphere, of terrain, and of air traffic threats from every direction. In the next century will be the problem of increased exposure of aircraft occupants to cosmic radiation in sub-orbital operations.

●"Aviation to an even greater extent than the sea is terribly unforgiving of any carelessness, incapacity or neglect."[2]

Despite these formidable hurdles, which *in toto* are not present in any other form of transportation, safety has been achieved in a remarkably short time. Safety is here defined as the achievement of normal life expectancies by those most exposed to the hazards of air transportation--the airline pilots. Since 1955, they can secure life insurance at the same cost as the typist who prepares the insurance policy! Over 1 billion global passengers fly the airlines annually, approaching the excellent fatal accident rate of 1 per million airline flights. Aircraft operated by corporations for business purposes enjoy a similar record--they too are flown by professional pilots. Millions of others fly in general aviation with a higher but steadily improving accident rate.

Much of these successes in safety have been due to lessons learned from the investigation of aircraft accidents by dedicated, patient, objective and thorough engineers, pilots and technicians who are trained in the art and science of aircraft accident investigation. The two very knowledgeable, experienced authors of this text show you how it is done. So read on!

Jerome F. Lederer
March, 1995

---

[1]Adapted from "Safety Science in Aviation," J. Lederer, First World Conference on Safety Science, Cologne, Germany, September, 1990.

[2]This well know quotation is credited to A. G. Lamplugh, British Aviation Insurance Group, 1931.

JEROME LEDERER

Jerry Lederer was born in 1902 in New York City and received his degree in mechanical engineering from New York University in 1925. In 1926, he served as aeronautical engineer for the United States Air Mail Service, the world's first successful system of scheduled air transportation. From 1929 to 1940 he served as chief engineer for pioneer aviation insurance underwriters. In 1940 he was appointed Director of the newly created Safety Bureau of the Civil Aeronautics Board. As Director, he was in charge of all civil aviation safety regulations and the investigation of all civil aviation accidents. The Safety Bureau was the forerunner of today's National Transportation Safety Board.

In 1947, he organized the Flight Safety Foundation and, concurrently, the Cornell-Guggenheim Aviation Safety Research Center. He served as Director of the Flight Safety Foundation until 1967 when, following the deaths of three astronauts in a fire at Kennedy Space Center, NASA appointed him Director, Office of Manned Space Flight Safety. In 1970 he was appointed Director of Safety for all of NASA. He retired in 1972 and has continued a distinguished career of writing and speaking on aviation safety issues.

Jerry is known internationally as "Mr. Aviation Safety" and has received numerous awards and recognitions for his contributions over a career spanning 70 years and most of the history of aviation. Each year, the International Society of Air Safety Investigators recognizes outstanding contributions to technical excellence in aircraft accident investigation with the JEROME F. LEDERER AWARD.

The authors are indebted to him for taking the time to review the manuscript and write this foreword.

Preface

Both of us have investigated a lot of accidents and taught aircraft accident investigation for various institutions for a number of years. One problem we encountered was the lack of a usable textbook on the subject. Those available were either out of date or oriented toward military aircraft accidents.

A little over three years ago, we decided to quit procrastinating and write this book. We recognized that the big problem was going to be controlling the contents. The subject of aircraft accident investigation is too big; there is no end to it. There is more information than can fit between the covers of a single book. During the technical review, the most consistent comment was, "Too much!" The second most consistent comment was, "Not enough!" There is probably some truth to both of them.

An apology. Our individual writing styles are about the same as our classroom lecturing styles-- very informal. Nothing fancy or stilted. We use a lot of personal pronouns--I, you, he, she, his, him, her, and so on. Supposedly these should all be changed to he or she, he/she, (s)he, or rewritten so that everybody becomes they or them. That sounds funny and reads worse. It destroys the flow of the words and, therefore, the intended communication. Besides, both of us feel that it is something of an insult to imply that aircraft accident investigation is not gender-neutral. Of course it is. Investigators can be of either sex as can pilots and anyone else involved in aviation. In context, "he" or "she" means both. We apologize to anyone who takes offense at this use of language.

RHW   RWS

PART I

The Rules of The Game

These first three chapters set the stage for the rest of the book. We start by explaining how an aircraft accident is defined and who, generally, gets involved in the investigation. Chapter 1 also gets into some of the ideas behind accident cause selection and development of safety recommendations.

Chapter 2 covers the organization of the United States National Transportation Safety Board and its procedures for field investigations and public hearings.

Chapter 3 describes Annex 13 (Aircraft Accidents and Incidents) to the Convention on International Civil Aviation and describes how international investigations are conducted.

First, a few comments on accidents in general.

In any industry you can name, accidents are defined in terms of damage or injury. You must damage something or hurt someone in order for it to be called an accident. Aviation is no exception. The basic definitions as developed by the International Civil Aviation Organization (ICAO) are statements of the degree of damage or injury necessary to qualify as an accident. All other countries either use the ICAO definitions or develop their own based on their concepts of damage or injury. While the United States Military does not use the ICAO definitions, it still defines accidents (Class A Mishaps) in terms of damage or injury.

So what's wrong with that?

Nothing in an industry where accidents are common and expected. When you have a lot of accidents, it would be normal to focus your investigative abilities on the ones that do the most harm.

Aviation, though, is a little different. Because of the growth of aviation, the total number of aircraft accidents that occur is increasing. That makes sense; more flying, more accidents.

Actually, though, the <u>rate</u> at which the accidents occur is decreasing. Although there are some ups and downs in the rate charts, the overall trend is downward for any segment of aviation. Even though we are having more accidents (growth) we are getting safer. In many flying organizations, airlines, for example, the expectancy of an accident in any particular year is zero. Mathematically, they are not supposed to have an accident.

This situation, which is unique to aviation, has raised some questions about how we define an accident. So far, the questioning is coming from those professionally involved in aircraft accident prevention or investigation. The argument goes something like this.

A particular event or series of events occur which

involve some degree of risk. Sometimes, the result will be damage or injury and we will call it an accident and investigate it. How about the times when no damage or injury occurs? The risk was still there wasn't it? Shouldn't we investigate those events, too?

This is not exactly a new idea. In the 1930s William Heinrich who is considered to be the father of the safety movement in the United States, developed "Heinrich's Triangle" wherein he showed that industrial events that produced serious injuries occurred on the average of 300 times with no injury at all. The occurrence of damage or injury is random and we should actually be investigating the event based on its risk regardless of whether any damage or injury resulted.

This hasn't happened, although safety professionals have been saying for years that "we ought to be investigating the incidents; not just the accidents." True, but we don't do that.

Why?

Simple. We have too many accidents. They keep us busy enough.

Nevertheless, aviation is at the point where the rate of occurrence of accidents is low and it can afford to look critically at how accidents are defined. Is the near-midair collision no less important than the actual collision? If an airline cargo door comes open on the ground and does no harm, is that not as important as the one that opens in flight and results in destruction of the aircraft? Maybe it's time to stop defining accidents in terms of resulting damage or injury.

That would be a monumental change. It would be the equivalent of investigating two vehicles that almost collided or a worker that almost fell off a ladder. Aviation is probably the only industry that could contemplate such a thing.

Nevertheless, ICAO, with its 1994 revisions to Annex 13, is moving in that direction. Safety professionals in our industry are beginning to look critically at how accidents and incidents are defined and how investigative resources are committed. Change is coming.

But not for a few years. Measurements of damage and injury are going to be with us for a while and these chapters deal with the situation as it exists now.

1

Introduction to Aircraft Accident Investigation

1. INTRODUCTION.

We need to understand the rules. This chapter covers various definitions; discusses findings, causes and recommendations; and identifies the key players in an accident investigation.

2. DEFINITIONS.

This covers the most commonly used aircraft accident definitions and the major variants. Comments are provided where appropriate.

A. AIRCRAFT ACCIDENT. An aircraft accident involves some degree of damage or injury associated with the operation of an aircraft. There are several variations, but the most widely accepted definition is the one developed by the International Civil Aviation Organization (ICAO). This definition does not apply to military and other government aircraft accidents.

ACCIDENT (ICAO). An occurrence associated with the operation of an aircraft which takes place between the time any person boards the aircraft with the intention of flight until such time as all such persons have disembarked, in which:

a. a person is fatally or seriously injured as a result of:
    -being in the aircraft, or

-direct contact with any part of the aircraft, including parts which have become detached from the aircraft, or
    -direct exposure to jet blast.

EXCEPT when the injuries are from natural causes, self inflicted or inflicted by other persons, or when the injuries are to stowaways hiding outside the areas normally available to the passengers and crew; or

b. the aircraft sustains damage or structural failure which:
    -adversely affects the structural strength, performance or flight characteristics of the aircraft, and
    -would normally require major repair or replacement of the affected component.

EXCEPT for engine failure or damage, when the damage is limited to the engine, its cowlings or accessories; or for damage limited to propellers, wing tips, antennas, tires, brakes, fairings, small dents or puncture holes in the aircraft skin; or

c. the aircraft is missing or is completely inaccessible.

SERIOUS INJURY (ICAO). An injury which is sustained by a person in an accident and which:

a. requires hospitalization for more that 48 hours, commencing within seven days from the date the injury was received; or

b. results in a fracture of any bone (except simple fractures of fingers, toes, or nose); or

c. involves lacerations which cause severe hemorrhage, nerve, muscle or tendon damage; or

d. involves injury to any internal organ; or

e. involves second or third degree burns, or any burns affecting more than 5 per cent of the body surface; or

f. involves verified exposure to infectious substances or injurious radiation.

COMMENT: Under this definition, someone must be on the plane with the intention of flying in it when the event occurs. If not, then the event may be an accident, but not an aircraft accident. The definition would include injuries to ground personnel (a propeller strike, for example), but exclude the same injury if it occurred during ground handling or maintenance and there was no one on board who intended to fly the plane. In the case of a helicopter, suppose a mechanic is running the engine (and turning the rotor) and the thing starts flying, the subsequent accident is technically not an aircraft accident unless the mechanic intended to fly it. These apparent inconsistencies occur in any safety discipline where we try to categorize an accident as being either this or that. With a slight change in the scenario, the same identical accident could fit in either category. Fortunately, most aircraft accidents occur during intentional flight and obviously belong in that category. The accidents that aren't so obvious are not numerous enough to seriously affect the statistics.

Other ways to define an aircraft accident involve flight (start of takeoff until end of landing) or power (from the time the first engine is started until the last one is stopped.) All of these methods suffer from the same problem mentioned above. The application of the definition to certain events that occur on the ground may be inconsistent.

B. UNITED STATES DEFINITIONS. The official definition of an aircraft accident as used by the National Transportation Safety Board (NTSB) is nearly identical to the ICAO definition. Here are the differences:

The NTSB uses the term, "Substantial Damage" which is defined as damage or failure which adversely affects the structural strength, performance, or flight characteristics of the aircraft, and which would normally require major repair or replacement of the affected component. Engine failure or damage limited to an engine if only one engine fails or is damaged, bent fairings or cowling, dented skin, small punctured holes in the skin or fabric, ground damage to rotor or propeller blades, and damage to landing gear, wheels, tires, flaps, engine accessories, brakes, or wingtips are not considered "substantial damage."

COMMENT: Both the ICAO and NTSB definitions of damage or substantial damage permit considerable latitude in interpretation. The phrase, "...adversely affects the structural strength, performance or flight characteristics of the aircraft...," invites debate. As a practical matter, the NTSB expects all events involving any significant damage to be reported and they decide which ones should be classified as aircraft accidents. Other ways of defining damage include, "Damage to a major component," (followed by a listing of major components) and cost of repair. The "major component" scheme proved completely unworkable as the definitions could not keep up with technology. The cost-of-repair method is used in the United States military. It has the advantage of being simple and consistent; but the disadvantage of always being slightly out of date due to inflation.

The NTSB definition of Serious Injury is identical to the ICAO definition. The NTSB adds a definition of Fatal Injury as any injury which results in death within 30 days of the accident.

COMMENT: The United States military uses injury definitions developed by the Department of Labor under the Occupational Safety and Health Act. Since the United States is a signatory to the ICAO convention, civil aircraft accident injury definitions will probably remain with the ICAO wording. The differences are minor.

C. CIVIL AND PUBLIC AIRCRAFT. These terms are unique to the United States. In our skies, there are only two kinds of aircraft; civil or public. It is important to understand the difference between them.

PUBLIC AIRCRAFT. This term goes back

to the Air Commerce Act of 1926 and was most recently amended in 1994 with an effective date of April, 1995. Originally, it was the intent of Congress to exclude the Army Air Corps and the Postal Service (the only operators of government aircraft at that time) from the air regulations being established in 1926. This worked fine until sometime after World War II when most government agencies and all states began operating aircraft. All of them were exempt from almost all Federal Aviation Regulations and there was no investigation of their accidents. There was no such thing as a "Public Aircraft Accident."

The definition of a public aircraft (paraphrased) reads like this:

Public Aircraft: An aircraft used only for the United States Government or owned and operated or exclusively leased for at least 90 continuous days (except for the United States Government) by any other government including a state, territory, possession or political subdivision of that government.

It does not include a government-owned aircraft transporting property for commercial purposes or transporting passengers except for:

Crewmembers or other persons whose presence is required to perform a governmental function such as fire fighting, search and rescue, law enforcement, aeronautical research, or biological or geological resource management.

Passengers if the aircraft is operated by the Armed Forces or an intelligence agency of the United States.

CIVIL AIRCRAFT. The definition of a civil aircraft is any aircraft other than a public aircraft.

COMMENT: All U.S. military aircraft are technically classified as public aircraft as are all aircraft belonging to intelligence agencies. All other aircraft (federal, state, county, or city) may or may not be classified as public aircraft depending on how they are being used at the time. The significance of this is that an aircraft not being operated as a public aircraft must, therefore, be a civil aircraft. If so, than all appropriate Federal Aviation Regulations apply and aircraft accidents would be under the jurisdiction of the NTSB. If the aircraft is being operated as a public aircraft, then most FARs do not apply and any accident would be investigated by the NTSB only if there was a pre-existing agreement with them to do so.

This is going to lead to some initial confusion (as of April, 1995) but the long range effect will bring most government (non-military) flying under the appropriate FARs and require investigation of their accidents. The military is excluded, but they were never a problem. Their record of aircraft accident investigation goes back to 1908. There will still be a number of government aircraft operated as "public aircraft." When an accident occurs, it may or may not be investigated.

3. FINDINGS, CAUSES AND RECOMMENDATIONS

A full discussion of these subjects could fill a book. No aspect of aircraft accident investigation creates more controversy than the final selection of findings, causes and recommendations. Supposedly, findings are those factors which were significant to the accident. Causes are those findings which are causal; thus a list of findings should contain all the causes plus some other factors which were merely significant, but not causal. Recommendations, of course, are the suggestions of the investigative body as to what should be done to prevent recurrence of the accident. That seems simple enough. Let's take them one at a time.

A. FINDINGS. Currently, ICAO (Annex 13) does not define findings. It has been suggested to ICAO by the International Society of Air Safety Investigators (ISASI) that "findings" should be defined as "all significant conditions and events, causal and non-causal, found in the investigation." The NTSB does not define "findings" although they use the term. Their use of it would pretty well fit the above definition. The United States Air Force uses the following definition:

The findings are the conclusions of the (investigating body). They are based on the weight of evidence, the (investigators') professional knowledge and their best judgement. They are statements of significant events or conditions leading to the (accident). They are arranged in the order in which they occurred. Though each finding is an essential step in the (accident) sequence, each is not necessarily a cause factor.

This has proved to be a reasonably workable method. Development of findings is a very handy way to account for the investigator's conclusions about the accident. Listed chronologically, the findings become a sequence of events which is the easiest (and perhaps the only) way to describe an accident. Finally, a good list of findings makes selection of causes a lot easier.

Figure 1-1, for example, is a list of findings from a hypothetical (but common) accident.

1	The flight crew was properly certificated and qualified for this flight.
2	After the start of takeoff roll, the Number 2 engine oil pressure transmitter failed resulting in a cockpit indication of loss of all oil pressure to the engine.
3	Following flight manual procedures, the Captain rejected the takeoff at a speed of 120 knots, which was 10 knots below V_1.
4	The Captain reduced thrust on both engines to idle, deployed the speed brakes and applied full braking.
5	Almost immediately after full brake application, Number 2 tire on the right main landing gear failed at a point on the tire sidewall which showed evidence of previous damage.
6	Shortly after the failure of Number 2 tire, the Number 1 tire on the same axle failed due to overload.
7	The aircraft departed the right side of the runway at a speed of approximately 60 knots.
8	The left main gear failed when it struck the cement base of a runway distance marker which extended eight inches above the grade.
9	The aircraft came to rest 120 feet from the edge of the runway. There was substantial damage to the aircraft, but no injuries to the crew or passengers.

Figure 1-1. Sample Sequence of Events and List of Findings

Figure 1-1 is a reasonable list of findings and a useful sequence of events. As we will see, it has some deficiencies.

B. CAUSES. In almost all countries where there is a separate board or body created to investigate accidents, the enabling legislation charges that body with determining the cause or causes of each accident. The term "cause" (or "causes") is not well defined. The ICAO definition (Annex 13) is, "Actions, omissions, events, conditions, or a combination thereof, which led to the accident or incident." There is apparently an assumption that everyone knows what causes are. Wrong! Everyone thinks they know what causes are, but there is little consensus.

In the early 1900s, which was the beginning of the safety movement, the dominant idea was that there was only one cause per accident. In attempting to reduce the findings of an accident to a single cause, one invariably had some findings left over which looked suspiciously like causes. Consider Figure 1-1. Findings 2, 5 and 8 are each causal. Each had to occur in order to produce the final damage. Each was essential to the sequence of events. If any of those three do not occur, there is no accident. If you are forced to select only one cause for that accident, it can only be Finding 8; the raised cement base of the distance marker. Why? Because prior to hitting that cement base, there was no accident. All we had was a couple of blown tires. Our definition of an accident requires a certain amount of damage and that didn't occur (or become inevitable) until our plane hit the cement base!

That illustrates the futility of trying to reduce an accident to a single cause. The next effort in cause analysis was to accept more than one cause, but categorize them as primary, contributing, main, direct, underlying, root, probable, related and so on. At one time, investigators were required to assign percentages to each cause to describe its contribution to the accident. This was illogical from the start. Going back to Figure 1-1 for a moment, if each of the causes (2, 5 and 8) were essential to the accident sequence and if the elimination of any one of them would have prevented the accident, then they are each infinitely and equally important. Attempting to prioritize causes has always been an exercise in futility. It can't be done.

Another problem with causes relates to the manner in which the cause is stated. In our legal system, we have "Proximate Cause," which is defined as, "That which stands next in causation to effect, not necessarily in time or space, but in causal relation." In the accident investigation business, a proximate cause is something like, "Pilot Error," or "Material Failure." If that's as far as it goes, this type of cause is worthless as it does not identify any correctable elements. (Of course pilots make errors. What do you expect us to do about it?)

A more useful statement of cause is one that explains why the error (or failure) occurred and includes something that is correctable. This, we call a root cause. "Pilot error due to lack of training on ____ procedures," would be a root cause. We can correct that. Go back to Figure 1-1 and read finding number 2 again. That's about half way between a proximate cause and a root cause. It tells us what failed, but not why it failed. If we need to act to improve our oil pressure transmitters, that statement of cause is not going to help us much. Besides its correctability, a root cause is frequently related to the overall system or its management.

In general, a statement of cause that does not contain some element of correctability is almost useless. The International Society of Air Safety Investigators (ISASI) has recommended to ICAO that two types of causes be defined as follows:

DESCRIPTIVE CAUSES (What happened). The sequence of conditions and events which, singly or in combination, led to the accident; listed in chronological order.

EXPLANATORY CAUSES (Why the accident happened). The predisposing events and conditions which explain why the descriptive causes existed or occurred. Explanatory causes usually form the basis for safety recommendations.

That sounds logical, but it was not incorporated into the latest version of Annex 13.

One of the simpler methods of cause determination is that used by the United States Air Force. They define Causes as follows:

Causes are those findings which singly or in combination with other causes, resulted in the damage or injury that occurred. A cause is a deficiency the correction, elimination or avoidance of which would likely have prevented or mitigated the mishap damage or significant injuries. A cause is an act, an omission, a condition, or a circumstance, and it either starts or sustains the mishap sequence. A cause may be an element of human or mechanical performance. An environmental condition may be a cause if it was not reasonably avoidable. Findings which sustained the mishap sequence, but were normal to the situation as it developed, are not causes. These are often unavoidable effects of a preceding cause. Apply the "reasonable person" concept when determining the causes. If a person's performance was reasonable, considering the mishap circumstance, do not assign cause. It is not appropriate to expect extraordinary or uniquely superior performance in activities.

If we apply USAF logic to the findings in Figure 1-1, we see that Finding 1 is merely a statement of fact. Finding 2 is a cause. Finding 3 is an effect of 2. Finding 4 merely sustains the accident sequence. Finding 5 is a cause. Finding 6 is an effect of 5. Finding 7 is an effect of 5 and 6. Finding 8 is a cause and Finding 9 just sustains the sequence through completion.

The USAF instructions to the investigation board are to merely identify those findings which are causal; there is no requirement to list them separately. The board is specifically instructed to not prioritize the causes with words like main, primary and so on. This system was initiated in the USAF in 1973 and it has proved to be reasonably simple and quite workable.

The NTSB is charged by law with determining the probable cause or causes of all civil aircraft accidents. Even though the plural causes has been on the books since 1974, for the next 20 years, the NTSB persisted in constructing a single statement of probable cause for each of their major investigations. This flew in the face of all reason, because there was no single probable cause for any aircraft accident. The result was frequently a tortuous one-sentence statement of probable cause containing a hundred or more words and several subordinate clauses; most of which are also causes.

Beginning in 1994, the NTSB changed their format and began listing probable causes (plural).

Interestingly enough, the term "probable cause" has never been defined. As early as 1970 it was suggested to the NTSB to either change the enabling legislation to delete "probable cause" or to define probable cause as a description of the accident event along with its cause and effect relationship.

Actually, the NTSB has adopted a system similar to the USAF for their "brief format" reports. These are computerized summaries of accident and they are the report format for well over 90% of the aircraft accidents in the United States. In the brief format, the NTSB lists the factors (findings) and then lists the numbers of the factors that they consider to be causal.

Another problem that comes up in cause determination is the question of when does an accident stop being an accident? When is it over? Some would argue that it is essentially over as soon as the crash becomes inevitable. Chapter 36 (Crash Survivability) should be eliminated from this book as it has nothing to do with the accident!

That logic actually goes back to the days of "Primary Cause." The official definition of primary cause was, "That event after which the accident became inevitable." Since the accident sequence kept right on going, that led to some ridiculous conclusions about what to do with the subsequent events. They couldn't, obviously, contribute to an accident that had already occurred!

The classic illustration of the problem is the military fighter pilot who suddenly finds himself using his ejection seat. Never mind why. Perhaps the engine quit. Perhaps he was shot down. Whatever the reason, he ejects and discovers that either the seat or the parachute doesn't work. He is fatally injured. Granted, that failure had nothing to do with the accident (or shoot-down), but it certainly caused the pilot's fatal injuries. Where does that fit in the report?

That's a fairly simple one to answer. The precipitating reason for the ejection and the ejection itself can be treated as two separate accidents and (if you like) investigated separately. But what of the airline transport that has an inflight fire and lands safely only to have some or all of the passengers die due to failure of the exit doors, evacuation slides or crash fire rescue response? Those failures didn't have anything to do with the accident (inflight fire) either, but we certainly can't ignore them.

One method is to treat those failures as "Aggravating Causes." They didn't cause the accident, but they made the results worse. That idea works, but it hasn't been received with a lot of enthusiasm. Who needs another set of causes?

Current thinking is to treat the accident as a system in much the same way as we treat the airplane as a system. It takes a lot to make the airplane fly including the plane, the crew, the mechanics, the airport and air traffic control. Likewise, the accident involves everything that was supposed to either prevent it or reduce its severity. That includes the airplane, the crew, the restraint devices, the airport, the fire department and so on. The accident stops when the damage stops and the injuries stop. Everything that contributed to the final damage and injury is part of the accident and should be part of the investigation.

That logic (along with acceptance of multiple cause factors) will cover almost all accidents. It can, of course, be carried too far. One of the authors investigated a gear-up landing which did minimal damage to the plane; mostly the propeller and the boarding step. In moving the plane from the runway, though, a jury-rigged lifting device failed and dropped the plane on its nose. This broke all four engine mounts and did a substantial amount of damage. Question: Is that part of the original accident? Answer: No, the accident was over by that time and we have to draw the line somewhere.

In spite of all these ideas, there is no universal agreement among investigators with any of them; not with ICAO, not with ISASI, not with USAF, not with NTSB, and certainly not with the authors of this book. Anybody can have an opinion and on this subject, everyone is an expert.

The problem, in the opinion of the authors, is that causes are argued emotionally. Anyone involved in the investigation of an accident or even aware of the circumstances forms some opinions on what caused it. These opinions are based largely on the background and experience of the individual and they are not necessarily right or wrong; but they are strongly held. It is very difficult for people with differing backgrounds and experience to agree on causes. Several general conclusions about the process of cause determination are possible.

1. There is no such thing as a single cause aircraft accident. All accidents result from multiple causes.

2. All causes should be identified and listed chronologically. Listing them any other way implies some sort of priority ranking; which is impossible.

3. Arguments about whether a particular finding is or is not a cause are generally a waste of time. It doesn't make that much difference. If it has any causal elements in it, identify it as a cause.

Among the more avant-garde thinkers in the investigation world, there are some strong feelings that causes serve no useful purpose as far as prevention is concerned and we would do better to eliminate them. Australia has adopted this idea. They go directly from "Significant Factors" to "Recommendations." The Australian Bureau of Air Safety Investigation (BASI: an independent government organization similar to the NTSB) is very pleased with that system and intends to continue it.

In theory, they are correct. Causes really don't serve any useful purpose and time spent arguing about them is time wasted. As a practical matter, though, they are not going to go away. All accidents have causes and the public, the news media and most governments insist on knowing what the cause is. Notice the singular of the word "cause" is used in that last sentence, because the public, the news media and most governments don't accept conclusion number one stated above. They want a single cause and they are not going to accept no cause at all. Even in Australia, the news media insists on converting BASI's Significant Factors into a single statement of cause. The debates will continue.

C. RECOMMENDATIONS. The recommendations of the accident investigation board are the most important aspect of the investigation. After all, the whole purpose of investigating the accident in the first place is to do something to prevent the accident from happening again and correct the deficiencies identified in the process.

A persistent problem arises with recommendations. First, the accident investigators are hired for their skills in determining what happened and why. Since fixing the problem may involve some economic trade-offs related to time, cost and feasibility, it is quite possible that the investigators are not the best qualified group to decide what should be done. Second, the investigating board's recommendations get published as part of the report and they are now chiseled in stone as the official recommendations on that accident. No subsequent modification or addition to the recommendations will ever achieve that level of permanence. Even bad recommendations will be on the books forever.

For these reasons, it can be argued that investigators should not make recommendations at all. One of the authors of this book presented a paper to that effect and suggested that recommendations would be best developed by a separate group with a background in how to fix things in specific areas; not how to investigate them.

Nevertheless, the history and tradition of having the investigation board make recommendations is pretty solid. It is not going to change and we certainly would not want to totally ignore the suggestions of the investigation board. What do we do?

First, we should remove the aura of omnipotence presently attached to investigation board recommendations. They are suggestions and nothing more. If there is a better way to correct the problem, no one should feel constrained by the original recommendation. One way to do this would be to remove the investigation board recommendations from the basic report and submit them as a separate document.

Second, the recommendation should be as non-specific as possible as to method (but specific as to objective) to permit maximum flexibility in how the problem is solved.

Third, before the report is published, there should be some sort of technical review to insure that the recommendations are both reasonable and practical.

Earlier it was stated that arguments about causes tend to be emotional arguments. Arguments about recommendations, on the other hand, tend to be logical arguments. To illustrate this, let's go back to Figure 1-1. Now that we have accepted Findings 2, 5 and 8 as causes, let's suppose that we are going to make recommendations in the three areas. One relates to the engine oil pressure transmitter; another to the tires or their inspection; and the third involves the airfield obstruction caused by the cement base of the distance marker. In considering which of those recommendations is likely to have the greatest potential for accident prevention, we would probably pick the engine oil pressure transmitter. We use the engine (and the oil system) each time we fly and failure of the system or its indicators potentially affects that plane every time it flies and every plane that uses that engine. If we can fix that problem, we've done a good days' work.

At the other end, we could say that the least effective recommendation is going to involve the cement base of the runway distance marker. That is only a hazard on one runway at one airport and you have to run off the runway to encounter it. It is worth correcting, but, realistically, we are not going to prevent a lot of accidents by doing so. The recommendation concerning the tires is priority 2 as it falls somewhere between the engine oil system and the runway marker.

Now that's fairly logical and most people would agree with assigning priorities to the recommendations in that manner. That's a far easier argument to win, because people are not emotionally involved with recommendations; only causes. Bottom line: it is easier and much more productive to spend your time debating recommendations than causes.

4. THE KEY PLAYERS

When an aircraft accident occurs, a surprising number of organizations may have some authority or responsibility. It is essential to understand who these groups might be.

A. INTERNATIONAL CIVIL AVIATION ORGANIZATION (ICAO). See Chapter 3. While ICAO itself has no particular authority and rarely participates in investigations, it does set the ground rules on how aircraft accidents involving more than one country (state) will be investigated. Since most countries of the world are signatories to the ICAO convention, the rules established by ICAO Annex 13 (Aircraft Accident and Incident Investigation) will probably prevail.

B. INVESTIGATOR-IN-CHARGE (IIC). This is an ICAO term used almost universally throughout the world to describe the person or group charged with conducting the investigation.

C. NATIONAL TRANSPORTATION SAFETY BOARD (NTSB). See Chapter 2. This is an independent board charged with investigating all civil aircraft accidents in the United States. There are similar independent boards or groups in Canada, England, Australia, New Zealand, Switzerland, Sweden, Denmark and Norway. Other countries may use their equivalent of the Federal Aviation Administration to investigate accidents. Still others may use their judiciary or even their military. In the United States, the NTSB may delegate certain investigations (currently experimental and agricultural aircraft accidents) to the FAA for investigation.

D. FEDERAL AVIATION ADMINISTRATION (FAA). The FAA is the government agency responsible for aviation safety in the United States; not investigation. They will always participate in NTSB investigations as a party (See Chapter 2.) and may conduct the entire investigation if the NTSB has delegated it to them. Their principal areas of concern are violations of Federal Air Regulations and deficiencies in FAA systems or procedures.

E. LOCAL LAW ENFORCEMENT. All land in the world belongs to somebody or some government and someone has jurisdiction over it. That's not completely true. The question of jurisdiction arose when an Air New Zealand DC-10 crashed on Mt. Erebus in Antarctica. Since the only local operating agency was the U.S. Navy, that question bothered the lawyers more than the investigators. Aside from that unusual circumstance, the presence of law enforcement at the scene of an accident is common and expected. Normally, their primary concern is the protection and safety of the public within their jurisdiction. Secondarily, they want to be sure that this was in fact an accident and not a criminal act. Although most accidents look like accidents, the possibility of a criminal act usually cannot be completely ruled out without some investigation. Either way, the local authorities are interested in preserving evidence. See Chapter 8. In addition, the local authorities may find that the accident has generated an enormous traffic jam and their presence is vital to permit emergency response vehicles to get where they need to go.

F. LOCAL FIRE DEPARTMENT. If there is a fire or if there are people to be rescued, the local fire department is usually in charge. It is almost a worldwide tradition that the Fire Captain or Fire Marshall is in charge as long there is a fire to be put out or people to be rescued. When that is complete, he expects to transfer on-scene authority to either law enforcement or government investigators, whichever is appropriate.

G. CORONERS AND MEDICAL EXAMINERS. The Coroner or his equivalent in almost any part of the world has complete authority over any human remains found in the area under his jurisdiction. He is ultimately responsible for identifying the remains; determining the cause of death; and issuing the death certificate. One exception to his authority in the United States is that the NTSB, by law, can require an autopsy of any crewmember or passenger involved in an aircraft accident.

H. MILITARY. The military obviously has complete jurisdiction over accidents occurring on military installations. Off the military installation, jurisdiction reverts to the local law enforcement structure unless the military (United States) can declare the accident scene a National Security Area. That does not usually happen unless the aircraft itself is classified or it is carrying a large amount of classified equipment. Generally, the military expects to conduct the investigation, but comply with all local laws in doing so.

I. INSURANCE CARRIERS. The agency

carrying the hull insurance on the civil aircraft becomes the owner of the wreckage after the hull loss is paid. In the United States, the wreckage is "owned" by the NTSB until they are through with it. It is then released to the owner. If the owner is now the insurance company, they either store it in anticipation of litigation, repair it or salvage it.

J. PROPERTY OWNER. If the aircraft crashed on private property, the property owner retains all of his rights. These are frequently ignored in the aftermath of the accident, but his rights must ultimately be recognized. As an example, the property owner can claim damage to his property as a direct result of the accident. He may also have a second claim for damages as a direct result of the investigation if it was necessary to build a road or cut down trees to gain access to the wreckage. As a rule, it is wise for the investigators to establish anticipated property damage costs with the owner before doing the damage.

K. AIRCRAFT OWNER/OPERATOR. Unless and until the wreckage has been transferred to the insurance company, the owner/operator retains some authority over it and its contents.

L. OCCUPATIONAL SAFETY AND HEALTH ADMINISTRATION (OSHA). In the United States in some accidents (airplane collides with vehicle; driver injured, for example) the injury or fatality is workplace-related and technically falls under the jurisdiction of OSHA

M. ENVIRONMENTAL PROTECTION ADMINISTRATION (EPA). In the United States, if the accident resulted in damage to the environment (fuel spill, for example) EPA is likely to get involved.

N. FEDERAL BUREAU OF INVESTIGATION (FBI). In the United States, the FBI is responsible for investigations involving sabotage to aircraft, air piracy (hijacking) bomb threats or criminal acts involving commercial aircraft.

O. UNITED STATES CUSTOMS SERVICE (USCS). If smuggling is suspected, USCS may have some responsibility.

P. NEWS MEDIA. Finally, the news media may have a certain amount of authority and access to accident scenes. In general, law enforcement officers would prefer to let them in rather than attempt to keep them out. As a general rule, aircraft accidents do not stay newsworthy for very long unless there is some intriguing aspect that will keep the story alive. Attempts by the investigators to deny the news media access and information is one aspect that will keep the story alive. The sooner the reporters are given the known facts and the opportunity to take a few pictures, the sooner they will lose interest in the accident.

5. THE INDEPENDENT INVESTIGATOR.

So far, it looks like most of the key players in an investigation are government employees charged with investigating it. Actually, there are probably more independent investigators at work than there are government investigators. Here, we use the term "independent" to mean any investigator not affiliated with a government agency. Here is where you find independent investigators.

A. AIRFRAME, ENGINE AND COMPONENT MANUFACTURERS. These organizations have a strong interest in the quality and performance of their products and they maintain investigation staffs who either participate in government investigations or conduct independent investigations.

B. AIRCRAFT OPERATORS. Many of the larger aircraft operators maintain aircraft accident investigation staffs. Some of the smaller ones have people trained in investigation techniques, although they are not full time investigators.

C. UNIONS. The major unions (Air Line Pilots Association, Allied Pilots Association, etc.) maintain staffs of well trained and very competent accident investigators.

D. INSURANCE COMPANIES. All of the major aviation insurance underwriters have aircraft accident investigators on their staffs

E. CONSULTANTS. Aircraft accident investigation (or reconstruction, as it is called in the legal arena) is something of a cottage industry in the United States where accident litigation is common. The consultant finds work as an expert for plaintiff or defense attorneys. Occasionally, the consultant gets involved in "public aircraft" accidents which would not be otherwise investigated.

6. THE FUTURE.

As this is being written, the transport ministers

from the European Community nations have met and agreed to an aircraft accident investigation policy. The highlights:

●Mandatory investigations of all accidents and serious incidents.

●Separation of technical and legal investigations.

●A single investigative agency for each EC nation.

●Annual reports from each EC nation.

●Standardized procedures to prevent technical investigations from being hampered by legal inquiries.

In addition to the European Community, we can expect changes involving the former East European nations; the Southeast Asian nations; and, perhaps, African nations. Will this affect the basic ICAO practices? Who knows? Because aviation cuts across so many borders and because aviation safety objectives are universal, there is a strong incentive for any group to develop procedures that can exist in harmony with the rest of the world.

7. SUMMARY.

This chapter sets the stage for what is to come. Now you understand the basic definitions of an aircraft accident and have gained an insight into the principal organizations involved. We have dabbled enough in the findings--causes--recommendations morass to appreciate the difficulties involved in this subject.

This, incidently, will be the last extensive discussion of findings, causes and recommendations in the book. From here on, we are going to deal with the technology of investigation without getting into any more deep discussions on causes. You may want to dog-ear one of the pages of this chapter so you can refer to it when you get involved in an argument on accident causation.

2

United States Investigations

1. INTRODUCTION.

All investigations of civil aircraft accidents in the United States, its territories and possessions, are conducted by or on behalf of the National Transportation Safety Board (NTSB). They are not chartered to investigate public aircraft accidents (see Chapter 1) although this is changing. Fewer government owned or operated aircraft will be eligible for the public aircraft classification. They will be considered civil aircraft and their accidents investigated by the NTSB.

The Board itself is composed of five persons appointed by the President for terms of five years. One of them is appointed Chairman for a term of two years. A Vice-Chairman is likewise appointed for two years. Each appointee (and the Chairman and Vice-Chairman) must be confirmed by the Senate.

A little history might be in order. The Air Commerce Act of 1926 established the requirement to investigate accidents. In 1938, the Civil Aeronautics Act established a three-member Air Safety Board for investigation. In 1940, an amendment created the Civil Aeronautics Board (CAB) Bureau of Safety which was charged with all civil aviation safety regulations and the investigation of accidents. Jerome Lederer (who wrote the Foreword to this book) was its first chairman. The Federal Aviation Act of 1958 created the Federal Aviation Administration (FAA) and relegated the CAB to economic regulation and accident investi-

gation. The NTSB was created in 1966 under the Department of Transportation Act. At that time, the NTSB was part of the Department of Transportation (as was the FAA) and was physically located in the same building as the FAA. In 1974, the Independent Safety Board Act divorced the NTSB from the Department of Transportation and it became a non-regulatory board completely independent of any other federal agency. In 1991, the NTSB moved out of the FAA building into its own quarters. In tracing this history, it can be shown that each significant act or amendment followed a year or so after one or more dramatic aircraft accidents captured the attention of the public and the Congress. Safety professionals will recognize this as the principle of "blood priority" in action.

The NTSB receives its appropriations directly from Congress and provides an annual report to Congress summarizing its activities for the year. It is a multi-modal Board in that it also investigates transportation accidents in the areas of railroad, marine, vehicle and pipeline. The NTSB also has a number of administrative law judges on its staff who handle appeals from adverse decisions of the FAA. As of 1993, the NTSB had approximately 350 personnel including about 50 Air Safety Investigators and an operating budget of 37 million dollars. Aside from its headquarters in Washington, it operates offices in Anchorage, Atlanta, Chicago, Denver, Fort Worth, Los Angeles, Miami, Kansas City, Seattle and Parsippany, New Jersey. See Figure 2-1.

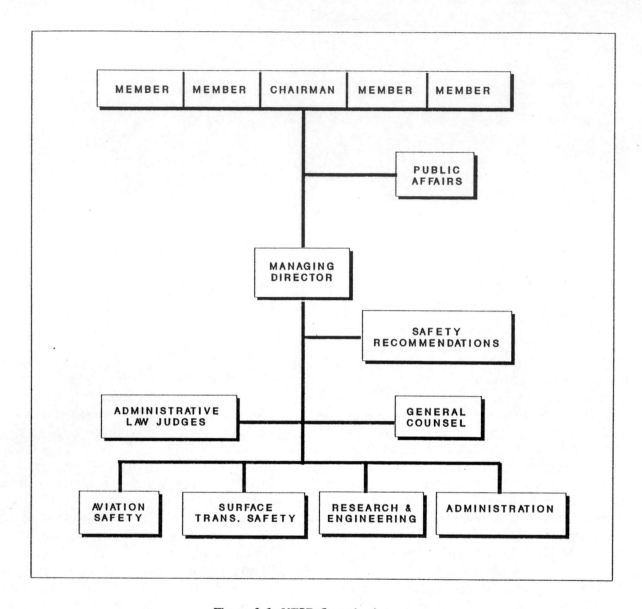

Figure 2-1. NTSB Organization

Although the NTSB deals primarily with accidents, it also investigates incidents of its choosing and conducts special studies on various transportation safety subjects. Most of its recommendations go to the FAA (or the regulatory agency for the transportation system involved), but recently the NTSB has begun targeting specific segments of industry with its recommendations. The recommendations are not mandatory and only have to be answered. Answers are required within 60 days and may consist of acceptance, partial acceptance or rejection. Partial acceptance or rejection require an explanation. The recommendation process has improved over the years and the present rate of acceptance is over 90 percent.

2. LEVEL OF INVESTIGATION.

Considering the number of aircraft accidents occurring in the United States, the NTSB cannot investigate all of them with equal intensity. Here is a list of the investigative levels generally available.

1. Major Investigation. This involves the headquarters "GO" Team and is generally reserved for air carrier accidents and some mid-air collisions. The "GO" team consists of one of the Board members, a senior IIC, a public affairs officer and perhaps a dozen specialists in various investigation areas. These specialists become the Group Chairmen of the groups organized in the field investigation. The "GO" team travels by commercial air and expects to be enroute to the scene within two hours.

2. Major Field Investigation. This is usually headed by an IIC from the headquarters using technical assistance from the headquarters and the Regional and Field Offices.

3. Field Investigation. Usually a single IIC from the Regional or Field Office.

4. Limited Investigation. These are "fender-bender" type accidents wherein the operator submits the report and the NTSB reviews it.

In addition, the NTSB may delegate certain accidents (usually agricultural or home-built) to the FAA.

Regardless of which of these is used, they have certain things in common. The investigator in charge of the case is an NTSB IIC (Investigator-In-Charge) and he has the same authority regardless of the level of investigation. In spite of the fact that the field investigation may have been conducted by the FAA or even the operator, it is still an NTSB investigation and the final determination of probable cause(s) is made by the NTSB.

3. RULES AND REGULATIONS.

The NTSB Rules and Regulations are found in Title 49 Code of Federal Regulations, Parts 800 through 850. Although it would be best to maintain the entire set, the parts directly applicable to aircraft accidents are listed here.

800. Organization and Functions of the Board and Delegations of Authority.

801. Public Availability of Information.

830. Notification and Reporting of Aircraft Accidents or Incidents and Overdue Aircraft, and Preservation of Aircraft Wreckage, Mail, Cargo, and Records.

831. Accident/Incident Investigation Procedures.

845. Rules of Practice in Transportation; Accident/Incident Hearing and Reports.

4. REQUIRED REPORTS.

The NTSB requires that they be immediately notified of all civil aircraft accidents, selected incidents, and overdue aircraft which are believed to have been involved in an accident. The NTSB also requires notification of all public (except military) aircraft accidents or incidents. (See Chapter 1.)

5. ACCIDENTS AND INCIDENTS.

The definition of an accident is found in Chapter 1. The term "incident" is imprecisely defined and depends on who is defining it. ICAO, NTSB, FAA and each U. S. Military Service have different definitions of incidents. ICAO and the NTSB define an incident as:

An occurrence, other than an accident, associated with the operation of an aircraft, which affects or could affect the safety of operations.

ICAO further defines a "Serious Incident" as an incident involving circumstances indicating that an accident nearly occurred.)

The NTSB goes on to list the following occurrences as reportable incidents:

1. Flight control system malfunction or failure.

2. Inability of any required flight crewmember to perform normal flight duties as a result of injury or illness.

3. Failure of structural components of a turbine engine excluding compressor and turbine blades and vanes.

4. In-flight fire.

5. Aircraft collide in flight.

6. Damage to property, other than the aircraft, estimated to exceed $25,000 for repair (including materials and labor) or fair market value in the event of total loss, whichever is less.

7. For large multiengined aircraft (more than 12,500 pounds maximum certificated takeoff weight):

(i) In-flight failure of electrical systems which requires the sustained use of an emergency buss powered by a back-up source such as a battery, auxiliary power unit, or air-driven generator to retain flight control or essential instruments.

(ii) In-flight failure of hydraulic systems that results in sustained reliance on the sole remaining hydraulic or mechanical system for movement of flight control surfaces.

(iii) Sustained loss of the power or thrust produced by two or more engines.

(iv) An evacuation off an aircraft in which an emergency egress system is utilized.

6. REPORTING METHOD.

According to the regulations, reports should be filed with the nearest NTSB Regional Office. As a practical matter, the NTSB is not particular how it is notified. If the FAA is aware of the accident or incident (as they frequently are) notification of the NTSB is automatic through the FAA Command Center in Washington. If the local police know of the accident, the NTSB will eventually be notified, either directly or through the FAA. The ones that the NTSB does not know about are the incidents where no one knows it occurred except the operator who may not know a report is required.

7. INVESTIGATION ORGANIZATION.

As already mentioned, the NTSB adjusts the scope of its investigation to fit the magnitude of the accident. In all cases, the investigator heading the investigation is the IIC. In a large investigation, he establishes as many groups as he needs to cover the accident. In a large air carrier accident, there will typically be groups established for weather, witnesses, propulsion, structures, systems, recorders, air traffic control human performance and so on. Each group is headed by an NTSB investigator who is called the Group Chairman. The other people on the groups are not members of the NTSB staff, but are called, "Representatives of Parties to the Investigation."

8. PARTIES TO THE FIELD INVESTIGATION.

The use of people from the manufacturer and aircraft operator to assist in the investigation is somewhat unique to the United States and Australia. While other countries may permit outside observers or participants under very narrow restrictions; the United States does it as a standard practice. The reason is largely a matter of economics. The NTSB has neither the personnel nor the resources to completely investigate a large accident without considerable help from industry. Even if the NTSB had the personnel, there is no way that all of their people could possibly be current in all aircraft and all technologies.

The parties to the field investigation all have vested and sometimes competing interests and the process has been compared to allowing the mice to investigate the missing cheese. This is an "adversarial" type investigation wherein the truth tends to emerge because none of the adversaries will allow "non-truths" to survive. It should be remembered that the entire legal system in the United States is an adversarial system. This type of investigation works surprisingly well and the relationship between the NTSB and the "parties" who normally participate in investigations is very professional.

One question that frequently arises is the propriety of returning a component (an engine, for example) to the manufacturer's facility for further examination and investigation. This sounds bad, but it isn't. When the component arrives at the manufacturer's facility, it is placed in secure storage until the investigation group (headed by an NTSB Group Chairman) arrives to observe and participate in the manufacturer's investigation. Considering that the manufacturer has the tools, equipment and technical knowledge to disassemble and inspect the component, this usually results in the best possible investigation.

9. DEFINITION OF PARTIES TO THE FIELD INVESTIGATION.

Parties to the Field Investigation are limited to those persons, government agencies, companies, and associations whose employees, functions, activities, or products were involved in the accident or incident and who can provide suitable qualified technical personnel to actively assist in the field investigation.

Typically, parties to a large investigation will include the air carrier, the airframe manufacturer, the engine manufacturer, the pilot or crewmember associations (unions) and the FAA. By law, the FAA is entitled to status as a party to any NTSB investigation. The U.S. Military is entitled to status as a party to any investigation involving both civil and military aircraft; which will be investigated by the NTSB. For all other organizations, status as a party is permissive and it depends primarily on whether the NTSB feels it needs the assistance that the party is offering. In almost all

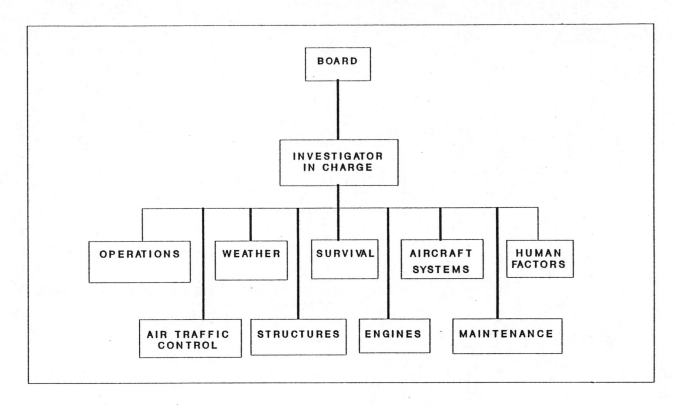

Figure 2-2. Typical NTSB Field Investigation

cases, the IIC makes the final determination.

The process is not automatic. After an accident occurs, an organization wanting status as a party must apply for it to the IIC and be accepted. If accepted, the party may participate subject to the rules and procedures established by the IIC.

Two classes or groups of people are specifically excluded from party status. These are people who represent claimants or insurers which translates into lawyers and insurance investigators. This exclusion has been a matter of some controversy over the years. A fairly strong argument can be made for excluding lawyers as their interest in the accident does not begin until the accident occurs. Furthermore, the NTSB feels that it is in the safety business; not the civil litigation business. The argument for excluding insurance investigators is less solid. The insurance company's interest in the aircraft (or operator) preceded the accident. In many aircraft accidents, the aircraft operator never gets a chance to participate as a party to the field investigation. He is either dead or he doesn't understand the system, or he is not qualified to help. Logically, one would think that the insurance company investigators could legitimately participate in the investigation on behalf of their clients, but that's

not the way it works.

10. INVESTIGATION PROCEDURES.

Each investigation group (and the parties serving on that group) conduct the field investigation and establish the facts on that particular subject. They prepare a factual report which each member of the group approves. These factual reports go into the public docket and are available to the Board. When the field investigation (or public hearing, if held) is complete, the IIC and his GC's prepare an analytical report covering their analysis of the accident and proposed findings. Parties to the field investigation do not participate in the development of this analytical report. This, along with the Group factual reports is forwarded to the Board.

11. PUBLIC HEARING.

The Board may order a Public Hearing as part of an aircraft accident investigation whenever a hearing is deemed necessary in the public interest. This only occurs three or four times a year and usually only for air carrier accidents. The hearing is normally held in the city nearest the scene of the accident and usually occurs two to three months after completion of the

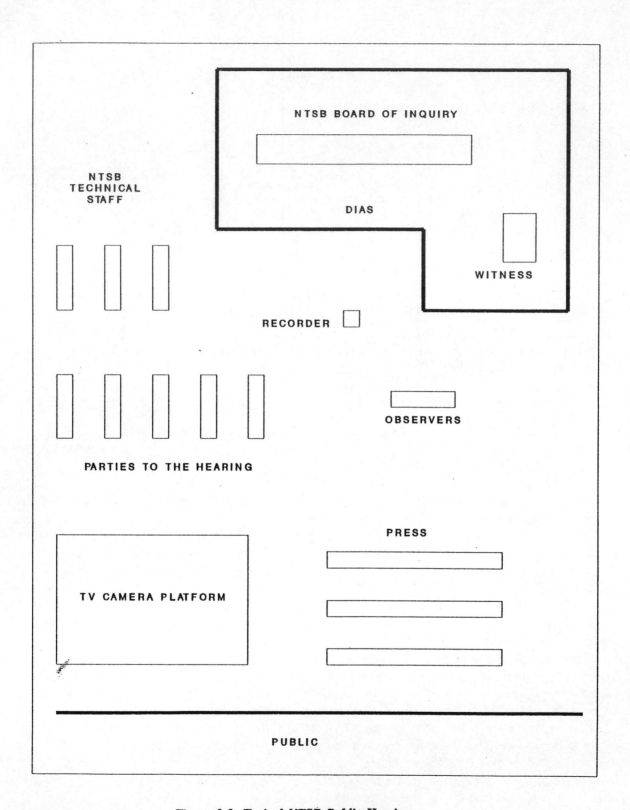

Figure 2-3. Typical NTSB Public Hearing

field investigation.

The hearing is chaired by one of the appointed Board members and conducted by a panel of NTSB staff and senior investigators. The Chairman may designate "Parties to the Hearing" under definitions and ground rules almost identical to those for "Parties to the Field Investigation." If an organization achieved status as a party to the field investigation, it will almost certainly apply for and be accorded status as a party to the hearing.

Witnesses are examined by the Chairman, the Hearing Panel (NTSB) and the Parties to the Hearing. Since this is an administrative hearing, normal rules of evidence do not apply. the transcript of the hearing is forwarded to the Board for consideration along with the factual and analytical reports of the field investigation.

12. PROPOSED FINDINGS.

Any party may propose to the Board the probable cause(s) and safety recommendations that should be developed from the facts and the testimony.

13. TECHNICAL REVIEW AND REPORT ADOPTION.

Although not codified in their regulations, the NTSB has adopted a policy of permitting a technical review of the factual portion of the draft report for technical accuracy prior to publication. There is no technical review of the Board's analysis or findings.

The final report is formally adopted by the Board in open session. This is scheduled and announced in the Federal Register and is called a "Sunshine Meeting." This expression comes from a federal requirement for regulatory agencies to conduct the public's business in public; thereby exposing their activities to sunshine. Spectators at the sunshine meetings can listen, but not participate in the discussion.

If a quorum (three members) of the Board exists, a report can be adopted by a simple majority. A dissenting member (or members) may file a dissenting opinion which is published as part of the report.

14. REPORTS.

Eventually, (usually eight months or longer), the Board will issue a report on the accident. Those accidents which, in the opinion of the Board, warrant such a report, will get a detailed narrative report summarizing the facts, analysis, findings, probable cause(s) and recommendations. There are perhaps a dozen of these reports issued per year. Of the remaining accidents, some of them will be summarized in a short form "Summary Report," and the rest will be logged into the system with no narrative report. For these, there will be a computer worksheet stating the basic facts, factors and probable cause(s).

Regardless of the type of report issued, all of the factual reports developed by the field investigators; all of the technical reports; all of the witness statements and many of the photographs are placed in a file called the "Public Docket." There is only one copy of this, but any person has a right to examine this docket, copy it or have a copy made for him. The IIC determines what reports and corroborative data go into the docket. Other raw data (such as photographs) collected during the field investigation is usually destroyed.

15. PETITIONS FOR RECONSIDERATION.

By regulation, aircraft accident investigations are never closed. After the report is issued, any person having in interest in the accident may petition the Board for reconsideration or modification of the findings. These petitions are normally entertained only if new evidence has been discovered or gross errors have been identified in the analysis.

16. SUMMARY.

The NTSB, like any investigating agency, is the first organization to take a position on an accident and publicly announce their statement of probable cause(s). As was discussed in Chapter 1, this pleases almost no one and the NTSB is under constant criticism from individuals and organizations who see the accident differently. For the reasons stated in Chapter 1, this is unlikely to change.

One almost universal problem confronting the NTSB is the tendency of the news media (and therefore the public) to confuse "cause" with "blame." The NTSB along with ICAO and most other investigative organizations in the western world see themselves in the safety and prevention business; not the blame and punishment business. The NTSB goes to great lengths to exclude blame and punishment ideas from their investigations, but they are only partially successful.

In the United States, a serious aircraft accident almost always spawns civil litigation. The litigants are interested in establishing blame (or lack of it) and receiving compensation, which might be considered a form of civil punishment. This litigation probably started while the NTSB field investigation was still in progress. Thus the litigants develop a strong interest in the NTSB's activities as their report (and the public docket) becomes the only record of the initial investigation. The NTSB's report is unlikely to suit the needs of the litigants because it was not written for that purpose.

To make matters worse, the airframe and engine manufacturers and the operator may have participated in the NTSB investigation as "Parties." If litigation ensues, they are almost certain to be involved and their participation in the NTSB investigation seems to give them an advantage denied other litigants. We must remember that the reason they were permitted to participate was because they could contribute to the investigation and (therefore) the prevention of future accidents. The fact that their participation may have been some help to them in future litigation is hardly their fault.

The basic problem (in the opinion of the authors) is that our society (any society) finds it difficult to agree that the prevention of aircraft accidents in the

long run serves the public interest better than the assessment of blame in individual accidents. Unfortunately, the two goals tend to be incompatible. There are plenty of counter arguments to this point of view and change is unlikely.

We might say that the NTSB (and any other professional aircraft accident investigation organization) has a task that can never be done perfectly and cannot hope to please everyone. Even under the best of circumstances, their report on any accident is almost certain to displease someone. For all of the criticism that it takes, the NTSB is staffed with professional and dedicated investigators who feel strongly about their contribution to the improvement of public transportation safety. They have a worldwide reputation for professionalism and technical excellence.

3

International Investigations

1. INTRODUCTION.

Following World War II, it became apparent that international air transport was going to grow and agreements were needed to insure that aircraft could move freely and safely from the airspace of one country to that of another. Not the least of the problems was the protocol to be followed when an aircraft registered in one country crashed in another country.

In 1944, delegates of 52 nations met in Chicago to discuss these problems. The outcome was the Convention on International Civil Aviation. By April of 1947, the requisite 26 nations had ratified the Chicago Convention and the International Civil Aviation Organization (ICAO) was born. Also in 1947, an agreement was concluded between ICAO and the United Nations to provide for close collaboration to fulfill both the requirements of the ICAO Convention and the United Nations Charter.

As of 1994, 183 nations of the world were signatories to the ICAO Convention. A glance at the titles of the Annexes to the Convention gives some idea of the scope of ICAO activities.

1. Personnel Licensing

2. Rules of the Air

3. Meteorological Service for International

Air Navigation

4. Aeronautical Charts

5. Units of Measurement to be used in Air and Ground Operations

6. Operation of Aircraft

7. Aircraft Nationality and Registration Marks

8. Airworthiness of Aircraft

9. Facilitation (Entry and Departure)

10. Aeronautical Telecommunications

11. Air Traffic Services

12. Search and Rescue

13. Aircraft Accident and Incident Investigation

14. Aerodromes

15. Aeronautical Information Services

16. Environmental Protection

17. Security

18. The Safe Transport of Dangerous Goods

The one that concerns us, of course, is Annex 13; Aircraft Accident and Incident Investigation. A word about terminology. In the United States, we use the word, "state," to mean one of the fifty states of our union. Internationally, the word, "State," means country or nation. For the rest of this chapter, we will use the wording of ICAO Annex 13. Just remember that "State" means "country."

2. INTERNATIONAL DEFINITIONS.

Annex 13 is the source of the definition of an Aircraft Accident quoted in Chapter 1. Beyond that, these are some definitions significant to international investigations.

A. ACCREDITED REPRESENTATIVE. A person designated by a State, on the basis of his qualifications, for the purpose of participating in an investigation conducted by another State.

B. ADVISOR. A person appointed by a State, on the basis of his or her qualifications, for the purpose of assisting its Accredited Representative at an investigation. An Accredited Representative may bring several advisors with him.

C. CAUSES. Actions, omissions, events, conditions, or a combination thereof, which led to the accident or incident.

D. INCIDENT. An occurrence, other than an accident, associated with the operation of an aircraft which affects or could affect the safety of operation.

E. INVESTIGATION. A process conducted for the purpose of accident prevention which includes the gathering and analysis of information, the drawing of conclusions, including the determination of causes and, when appropriate, the making of safety recommendations.

F. INVESTIGATOR-IN-CHARGE. A person charged, on the basis of his or her qualifications, with the responsibility for the organization, conduct and control of an investigation.

G. SERIOUS INCIDENT. An incident involving circumstances indicating that an accident nearly occurred.

Note: The difference between an accident and a serious incident lies only in the result.

H. STATE OF DESIGN. The State having jurisdiction over the organization responsible for the type design.

I. STATE OF MANUFACTURE. The State having jurisdiction over the organization responsible for the final assembly of the aircraft.

J. STATE OF OCCURRENCE. The State in the territory of which an accident or incident occurs.

K. STATE OF THE OPERATOR. The State in which the operator's principle place of business is located or, if there is no such place of business, the operator's permanent residence.

L. STATE OF REGISTRY. The State on whose register the aircraft is entered.

3. INVESTIGATION.

A. STATE OF OCCURRENCE. A fundamental rule is that the State of Occurrence has both the right and the responsibility to conduct investigations of accidents occurring in its territory. It may delegate all or part of the conduct of the investigation to the State of Registry or the State of the Operator. It may also ask for technical assistance from any Contracting State.

The State of Occurrence notifies ICAO and appropriate Contracting States; designates the Investigator-In-Charge; and prepares and issues a report of the accident.

B. ACCIDENTS IN INTERNATIONAL WATERS. When the accident occurs outside the territory of any State, the State of Registry assumes the rights and responsibilities of the State of Occurrence.

C. STATE OF REGISTRY AND STATE OF OPERATOR. When notified of an accident involving one of its aircraft, the State of Registry and the State of Operator (if different) are obliged to furnish the State of Occurrence with any available information regarding the aircraft and flight crew. The States of Registry and Operator are both entitled to appoint an Accredited Representative to participate in the investigation. The States may also appoint advisors to assist the Accredited Representative.

D. STATE OF MANUFACTURE AND STATE OF DESIGN. These States are likewise entitled to appoint an Accredited Representative. In addition, the States of Manufacture and Design may request that the aircraft (wreckage) remain undisturbed pending inspection by their accredited representatives. Other States who may have participated in the manufacture of assembly of the aircraft may request permission to send an Accredited Representative.

E. OTHER STATES. States providing information, facilities or experts to the State conducting the investigation are entitled to appoint Accredited Representatives. States having suffered fatalities to its citizens in the accident may request permission to send an Accredited Representative.

4. PARTICIPATION.

Accredited Representatives (and their advisors) are entitled to:

A. Visit the scene of the accident.

B. Examine the wreckage.

C. Obtain witness information and suggest areas of questioning.

D. Have full access to all relevant evidence as soon as possible.

E. Receive copies of all pertinent documents.

F. Participate in readouts of recorded media.

G. Participate in off-scene investigative activities such as component examinations, technical briefings, tests and simulations.

H. Participate in investigation progress meetings including deliberations related to analysis, findings, causes and safety recommendations.

I. Make submissions in respect of the various elements of the investigation.

5. APPLICABILITY.

With a July, 1994 change, Annex 13 has been expanded to include all aircraft accidents and incidents wherever they occur. This would include domestic accidents not involving two or more Contracting States.

6. OBJECTIVE.

The sole objective of the investigation of an accident or incident is the prevention of accidents and incidents. Annex 13 specifically states that the purpose is not to apportion blame or liability.

7. REPORTS.

The State conducting the investigation is expected to provide a Preliminary Report, an Accident/Incident Data Report, and a Final Report. The Preliminary Report is sent by mail within 30 days of the accident and follows the format specified in the ICAO Accident/Incident Reporting Manual (DOC 9156.) It goes to the States of Registry, Occurrence, Operator, Design, Manufacture, States which provided information to the investigation and, of course, ICAO. The Accident Incident Data Report is only required for accidents involving aircraft weighing over 5700 kg. This report is largely for the ICAO Accident Data Reporting System (ADREP). The Final Report may be in any suitable format, but the one most commonly used is the format suggested in ICAO Annex 13. This is discussed at length in Chapter 38. States which provided Accredited Representatives to the investigation are entitled to receive a copy of the final report.

8. DISCLOSURE OF RECORDS.

Annex 13 urges that certain records not be disclosed unless they are pertinent to the analysis of the accident or incident. These include witness statements, communications between persons involved in the accident, medical information, cockpit voice recording (CVR) transcripts and opinions expressed during analysis. Unfortunately, this is contrary to the laws of many Contracting States. The United States, for example, can withhold personal medical information and non-pertinent CVR transcripts, but witness statements become part of the public record of the investigation.

9. HOW DOES IT ALL WORK?

On balance, pretty well. The quality of the investigation varies with the technical capability of the State conducting the investigation. Some are very sophisticated, but some are not. Those that are not can still conduct a good investigation providing they seek help from States that do have the capability. Within the aircraft accident investigation community, technical help is freely exchanged.

Another variable is the laws of the State conducting the investigation. On November 10, 1994, ICAO published a revised Annex 13 which included resolutions adopted nearly a year earlier. Under the ground rules, each Contracting State has one year to notify ICAO of "Differences." These differences are the exceptions each State must take because of its laws. Eventually, ICAO will publish an amendment to the Annex which will list all of the differences (or exceptions) received from the reporting States.

Some States view civil aircraft accidents as matters to be resolved through their criminal court system. This, of course, leads to an entirely different investigation than one aimed at the prevention of future accidents.

As a practical matter, many States have people who have been formally trained in aircraft accident investigation techniques and have a background in aviation safety. Through the efforts of organizations like the International Society of Air Safety Investigators (ISASI) and the Flight Safety Foundation (FSF), knowledge, techniques and procedures are freely exchanged. The overall quality of international investigations is improving.

To illustrate how this actually works in practice, here is how the investigations of some well-known accidents were actually conducted.

1. Two B747s (Pan Am and KLM) collide on the runway at Tenerife. State of Occurrence is Spain. They designated the IIC. The United States is a State of Registry, Operator and Manufacturer. They designate an Accredited Representative who is always an NTSB investigator. He brings Advisors from Boeing, Pan Am, ALPA, and FAA. Netherlands is also a State of Registry and Operator. They designate an Accredited Representative who brings advisors from KLM.

2. An Air India B747 crashes in the North Atlantic off the coast of Ireland. There is no State of Occurrence. India, as the State of Registry, designates the IIC. The United States, as the State of Manufacture, sends an accredited representative with Advisors from Boeing and FAA. Canada, as a State whose

citizens suffered fatalities in the accident, receives permission to send an Accredited Representative and Advisors from Transport Canada.

10. EXCEPTIONS.

One exception was mentioned in Chapter 1. The Air New Zealand DC10 that crashed in Antarctica generated a few minutes of concern over who would be the State of Occurrence. The answer, of course, was that Antarctica is considered international territory. The investigation was conducted by New Zealand as the State of Registry with an Accredited Representative from the United States and Advisors from the manufacturer and the FAA.

Other exceptions usually involve military aircraft accidents. Among members of NATO, there is a Standard NATO Agreement (STANAG 3531) which deals with aircraft accidents. It permits separate or concurrent investigations by either involved State; joint investigations or investigations by one State with participation from the other. Unless there is some unusual amount of civil damage, the most common resolution is to permit the State operating the aircraft to conduct the investigation using its own procedures.

In situations where one or both States are not members of NATO, there is frequently a Status of Forces Agreement (SOFA) in effect. This is an agreement between two States specifying the rules under which military forces from one State operate in another State. One of the "boiler plate" paragraphs in a SOFA usually deals with aircraft accidents.

In situations where there is no SOFA, the two States may agree to apply NATO STANAG 3531 because it is in writing and it deals with military aircraft accidents. On the other hand, the State of Occurrence may insist on applying ICAO Annex 13. Even though that applies only to civil aircraft, it is also in writing and it is the only one they have signed. What follows is usually a discussion at the diplomatic level and there is no predicting how it will come out.

11. SUMMARY.

Internationally, almost anything is possible. ICAO Annex 13 has served the world well, but it doesn't address every conceivable situation. Fortunately, as long as we stay focused on the basic objective--safety and accident prevention--reasonable people will overcome bureaucracy and get the job done.

PART II

===========================

Investigation Techniques

The next 22 chapters start with the basics of field investigation and deal with each definable portion of the investigation as a separate chapter.

Inevitably, there is a certain amount of overlap and some techniques discussed in one chapter are perfectly usable in some other chapter. Portions of some chapters spill over into Part III on Technology or even Part IV on Analysis and Reports.

In most cases, the chapters are cross-referenced. It is fairly important to read all of them in order to gain an understanding of the complete investigation process.

All of these chapters deal generally with the gathering of data. This accumulation of knowledge represents everything the investigator knows about a particular accident and becomes the basis for analysis and subsequent determination of causes. It is important for all investigators involved in a particular accident to know what data is being collected and how it is being collected.

The difficult part of writing this section of the book was not in deciding what to include; but what to leave out. We have made no attempt to cover everything about a particular aircraft system; only enough to explain how to investigate it. The person wanting to learn more about turbine engines, for example, will have to consult one of the referenced texts on turbine engines. This book is already thick enough.

Another decision was on the depth of investigation. Since this book is meant for the field investiga-

tor, the emphasis is on what can be done in the field; not in the laboratory. Although that's easy to say, it became evident during the writing that the field investigator has to at least understand what can be done in the laboratory if not how to do it. We have tried to include that knowledge where appropriate.

Finally, if there was ever an activity that benefits from the synergistic effects of group participation, it has to be the investigation of aircraft accidents. Investigations by a single investigator are really difficult as there is almost certain to be some aspect of the accident that is beyond the knowledge or experience of a single person. There is a tendency, unfortunately, for the single investigator to either ignore that which he does not understand or to make some assumptions about a particular facet of the accident which simply aren't true. The investigator is essentially a gatherer of facts and the resulting analysis and conclusions can be no better than the facts gathered.

This is a good place to talk about objectivity. Some insist that the investigator should approach the investigation with absolutely no preconceived ideas about what might have been a factor in the accident. This confuses objectivity with ignorance and it is both impossible and impractical. Investigators are hired for their knowledge and experience and asking them to not use it is silly. If the initial information about the accident looks suspiciously like a problem in propulsion, any experienced investigator is going to look at the engine(s) first. There is nothing wrong with that. That is not lack of objectivity. Lack of objectivity occurs when the investigator encounters evidence that the engines were running just fine, but refuses to accept it because it doesn't fit his initial opinions.

There are a lot of investigators out there who commit to a particular point of view too early and spend the rest of the investigation ignoring facts that conflict with that point of view. That's lack of objectivity.

All good investigators share at least three attributes.

A. They are not afraid to be wrong. They will accept facts that are contrary to their present theory.

B. They readily admit that they don't know everything. When they need help, they seek help.

C. They listen to the other investigators. They don't necessarily believe them, but they listen to them.

Those are pretty good traits to acquire.

4

Field Investigations

1. INTRODUCTION.

All aircraft accidents are different, but the accident investigation process doesn't change very much. At its simplest, it is a process of accumulating knowledge about the situation (data acquisition), analyzing that data and developing some conclusions about what happened. This entire process must be somehow documented in a report which becomes the official record of the accident and its investigation.

Initially, almost all efforts are directed toward data acquisition and there is an illusion created that analysis does not start until all data is collected. Likewise, conclusions are not developed until the analysis is complete and the report is not started until the conclusions have been reached.

Actually, this is incorrect. Accidents are investigated "piecemeal" and the report (except for the final conclusions) is composed of a number of smaller reports which fit together. At any given time during the investigation, the activities of data acquisition, analysis, conclusions and report-writing are all occurring simultaneously for different parts of the investigation.

Suppose, for example, that one of the initial unknowns in the accident involves the engine. Was it capable of putting out sufficient power to sustain flight? The engine is investigated (data acquisition) and no evidence is found that suggests the engine was not capable of full power operation. There is, in fact, evidence that the engine was operating at high RPM at impact (analysis.) Engine failure was not a factor in

this accident (conclusion.) At this point, the smart thing to do is to document those facts, analysis and conclusions in the form of a mini-report and move on to something else. This is best done immediately while the investigator's observations and impressions are fresh. If the engine report is written now in the correct report format, it can be easily merged into the final report. If the engine report is not written until the entire investigation is complete, the quality of the final report will suffer. See Chapters 38 and 39 for some other ideas on the report process and managing the investigation.

As a practical matter, the idea of data acquisition, analysis, conclusions and report-writing are principles; not procedures. We need to expand those a little to develop a useful list of procedures. Let's suppose that we have just been notified of an accident and we are going to respond to it immediately and be in charge of the field investigation. How do we proceed?

Obviously, that depends a lot on the circumstances. Let's take it one step at a time.

2. PREPARATION.

It is pointless to head for the accident scene without taking a few preparatory actions first and bringing the right equipment. Most organizations that can anticipate being involved in the investigation of an aircraft accident would be well advised to develop a pre-accident response plan.

A. INITIAL COORDINATION. If it is obvious that you will need assistance, notify the proper agencies. Arrange for transportation and an initial meeting location. If possible, arrange for the wreckage to be secured pending your arrival. Notify appropriate agencies to begin collecting and preserving documents and information relevant to the accident. This could include the FAA, aircraft operator and manufacturer. Assemble the equipment you think you will need.

B. INVESTIGATION EQUIPMENT. There is no standard investigation kit. What you need depends on accident location, weather, type of aircraft and your investigative specialty. Two pretty good rules are:

1. Bring everything you need. Do not depend on someone else to bring equipment for you.

2. Be prepared to carry whatever you bring. Do not depend on anyone else to carry it for you.

It goes without saying that you must bring the rain gear or cold weather clothing appropriate for the weather and location. If you can't keep yourself warm and dry, you are not likely to be of much use at the scene. Beyond that, here are some lists of equipment that you might need depending on what you plan to do during the investigation.

C. PERSONAL SURVIVAL ITEMS. Take care of yourself first. You can't investigate if you're not prepared to survive.

Appropriate severe weather clothing including sturdy boots.

Gloves. Wreckage is sharp! Get a pair of heavy leather gloves or industrial safety gloves.

Hat for sun protection.

Insect repellant.

Small First Aid kit.

Whistle or other signalling device.

Moist towelettes for cleaning your hands.

Disposable latex gloves.

Ear protection - plugs or muffs.

Toilet paper.

Water container - plastic canteen or flask.

Food bars.

D. DIAGRAMMING AND PLOTTING EQUIPMENT. If you expect to make a diagram of the scene, you can do a good job with these items.

Pad of quad-ruled paper.

Navigation plotter with protractor.

100' tape measure.

Ball of string

Compass

E6-B Computer (or electronic equivalent) for wind calculations.

Calculator

Notebook, pencils, marking pens, etc.

Topographical map of the area if available.

E. WITNESS INTERVIEWING EQUIPMENT. If you expect to interview witnesses, bring these.

Tape recorder, tapes, spare batteries.

Hand microphone with switch.

AC power adapter.

Statement forms (if used).

Model aircraft (a generic model made of cardboard which doesn't resemble any particular aircraft will work fine.)

Headphones and foot switch for the tape recorder. These are useful if you plan on transcribing the tapes while in the field.

F. EVIDENCE COLLECTION EQUIPMENT. If you expect to examine or collect evidence samples, you will need these.

Sterile containers. You can usually obtain these locally at a pharmacy.

Magnifying glass.

Small tape measure.

Mirror. A mechanic's mirror with a handle and swivel head is useful.

Flashlight.

Tags, Labels and Markers.

Plastic bags and sealing tape.

G. PHOTOGRAPHIC EQUIPMENT. If you intend to photograph the wreckage, this is the minimum equipment needed for a professional job. See Chapter 6.

35mm SLR camera body.

#1 lens - infinity to macro capability.

#2 lens - wide angle - 24mm preferred.

Electronic flash with extension cord.

Small tripod.

Locking shutter cable release.

Plastic bag and rubber band to protect camera in rain.

Ruler - size reference.

Photo log - notebook.

Spare batteries for camera and flash.

Film.

H. REPORT WRITING AND ADMINISTRATIVE EQUIPMENT.

Accident report forms.

File folders and labels.

Paper

Stapler, paper clips.

Laptop or notebook computer.

I. TECHNICAL DATA.

Parts catalog or illustrated parts breakdown for the aircraft.

Flight manual for the aircraft.

Color photographs of the undamaged aircraft, if available.

Handbook of common aircraft hardware.

Investigation manual and investigation reference material.

J. PERSONAL ITEMS.

Company/agency identification.

Expense record.

Credit cards, blank checks and money.

Passport.

Immunization record.

International Driver's License

K. CARRYING THE EQUIPMENT. You will rarely need all of this for any particular accident. Except for the personal and survival items, you only bring what you expect to use. Actually, all the equipment listed here will fit in a large catalog-style briefcase and a camera bag. It is also handy to bring a light weight backpack for carrying essential items in rugged terrain.

3. INITIAL ACTIONS.

What do you do when you arrive? If it's a big accident, consider these steps.

A. ESTABLISH A BASE OF OPERATIONS. This should be a location near the scene where you can work, store your equipment and communicate with the rest of the world. The NTSB will routinely use a local motel or hotel including meeting rooms and will arrange for extra telephones and FAX machines to be installed.

B. ESTABLISH LIAISON WITH THE LOCAL AUTHORITIES. This includes the police or sheriff, fire department and coroner.

C. ARRANGE FOR SECURITY AND PROTECTION OF THE WRECKAGE.

D. DETERMINE WHAT HAS HAPPENED SO FAR.
How many total people are involved? How many fatalities? What was the cargo? What was done to the wreckage in order to extinguish the fire? Rescue the injured? Remove the bodies?

E. CONDUCT AN ORGANIZATIONAL MEETING.
Find out who you have available to assist. Establish ground rules with respect to investigation and group leadership, wreckage access, news media relations and so on.

F. ESTABLISH SAFETY RULES.
Rule #1 in the aircraft accident investigation business is, "Don't get hurt." Things are bad enough already. We don't need any additional injuries. Unfortunately, wreckage can be very dangerous. Here are some of the dangers you should anticipate.

Chemical Hazards. Fuel, hydraulic fluids, liquid oxygen, hydrazine.

Pressure Vessels. Hydraulic accumulators, Oleo struts, Tires, Fire extinguishers.

Mechanical Hazards. Springs (Ram air turbines, gear doors, drag chute mechanisms.)

Pyrotechnic Hazards. Ejection seats, munitions, survival equipment.

Hygiene Hazards. If there are human remains present in the wreckage, bloodborne pathogens present a serious risk to the investigator. See Chapter 24 for additional information. In the United State, Occupational Safety and Health (OSHA) Standard 1910.1030 deals with bloodborne pathogens and is applicable to the scene of aircraft accidents. See Bibliography.

Miscellaneous Hazards. Radioactivity (instruments, avionics, flight control balance weights), fumes, dusts, and vapors resulting from composite materials or cargo.

Investigators should understand the potential hazards of the wreckage. If a known hazard is present such as munitions, hydrazine or a hazardous cargo, then initial steps should be to remove or neutralize that hazard before the investigation proceeds.

G. CONDUCT AN INITIAL WALK-THROUGH OF THE WRECKAGE.
This is valuable even for investigators who won't spend much time with the wreckage. It provides perspective on the accident and facilitates future discussion about it.

H. TAKE INITIAL PHOTOGRAPHS.

I. COLLECT PERISHABLE EVIDENCE.
Here's a list of potentially perishable items.

Fuel samples. Unless you can absolutely discount fuel contamination as a factor (mid-air collision, perhaps) the rule of thumb is to get fuel samples. Ideally, samples should be collected and stored in dark, sterile containers. Bulk fuel suppliers usually have sampling equipment and containers. If you have to do it yourself, use sterile containers and get samples from tank drains; from a fuel line or from the tank itself. Take samples from as many points as possible. If you suspect local contamination of the samples, get a sample of the surrounding soil and submit that for analysis along with the fuel. Label the containers and set them aside until you determine whether fuel contamination is a likely problem or not. If the accident fits typical fuel contamination scenarios (loss of power immediately after takeoff) get samples from the truck or tank used to fuel the aircraft and consider sampling other aircraft fueled from that source. For a full qualitative analysis, laboratories usually want a gallon or more of fuel. That's usually a little optimistic. Get what you can.

Oil and hydraulic fluid samples. These are not as perishable as fuel and they can usually be obtained later. If you are in the sampling business, though, you might as well get them all. Follow the sterile container procedures specified for fuel samples.

Loose papers, maps and charts. Collect these before they blow away.

Evidence of icing. Was structural ice a problem? If it is cold enough, maybe it is still there. Talk with the first people on the scene. Same with reciprocating engine carburetor ice. If it's a power loss-type accident and it is still cold, check the carburetor for ice. This might also apply to turbine engine fuel control units. See Chapter 32.

Runway condition. For takeoff and landing

accidents, the exact condition of the runway at the time of the accident may be important. Likewise, any runway skid marks that can be associated with the accident are perishable.

Switch positions and instrument readings. These are perishable because people will move the switches and change the readings. Get full photographic coverage of the cockpit before anyone is allowed in it.

Control surface and trim tab positions. These will change (or be changed) during the course of the investigation.

Flight data recorders and cockpit voice recorders. These are not exactly perishable, but they should be located and retrieved early as they will always require laboratory analysis to be of any use to the investigators.

Ground scars. These are somewhat perishable, particularly if you intend to do some calculations on impact angle and deceleration distance. Measure and photograph them.

Other Perishables. This could include anything that is likely to be moved, damaged or destroyed before it can be investigated.

J. INVENTORY THE WRECKAGE. Make sure you have all of it. This sounds simple, but it is not. If you are missing a major chunk of airplane, that is likely to be significant to the investigation and you are not going to know it if you haven't accounted for it. Also, this may be your first opportunity to spot something that should <u>not</u> be in the wreckage such as a tool or a part of another aircraft.

K. BEGIN A WRECKAGE DIAGRAM. If you need one, now is the time to start one. See chapter 7 for more ideas on this.

L. DEVELOP A PLAN for the remainder of the investigation. By now, it should be obvious how the investigation will progress. Here are some ideas.

What is the immediate problem?

Human remains? If this is the problem, you are not going to get deeply into the technical investigation until you have taken care of the human remains problem. As a general rule, it is important to

identify where the remains were found as this can significantly aid the pathologist in remains identification. Thus the investigator with his ability to diagram the wreckage can be a big help to the pathologist. This may be the first priority of the investigation.

Wreckage recovery? Sometimes, the wreckage is inaccessible. It is either underwater or strewn across an unclimbable piece of mountain. What do we do now?

Underwater wreckage. If you don't know where the wreckage is, locating it can be very difficult and expensive. The cost of locating and recovering it must be balanced against the potential knowledge gained from it. Sometimes, it's not worth it. If you know where it is, the cost of recovering it depends on the depth. As a rule, you will never get all of it. You must decide what parts of the wreckage you would really like to have and concentrate on those. This will involve teaching the salvors what those parts look like and how they should be recovered. Engines, for example, have built-in lifting points where slings or hooks can be attached. As part of the underwater wreckage recovery, you should also decide what you are going to do with it after you get it. As a rule, wreckage recovered from seawater should be washed thoroughly with fresh water and sprayed with a light moisture-inhibiting oil to reduce corrosion. Wreckage recovered from fresh water should be air-blown dry. Oil spray should not be used unless absolutely necessary. If the recovered part is small enough, the best way to preserve it is to keep it submerged in fresh water during transport to the laboratory.

Inaccessible wreckage. Sometimes, the only wreckage you recover is that which can be brought back by the climbing team or the helicopter. If so, that team should have been briefed extensively on what you would really like recovered from the wreckage.

What will be the general direction of the investigation?

Field investigation only. We can do it all with what we have available in the field.

Cursory field investigation followed by laboratory analysis. We'll document the wreckage as it crashed, but we plan to take the engines and other key components back to the shop for detailed examination.

Reconstruction. We need to reassemble all or portions of this aircraft to trace fire or structural failure patterns. We will document the location of the wreckage as it crashed, but move all of it to a hangar or open space where we can reconstruct it.

How will we (the investigators) work on it? Set up a work schedule for transportation, daily meetings, press briefings, etc. Develop a time line reflecting how you think the investigation ought to progress (See Chapter 39).

4. SUMMARY.

Here are some pretty good rules for the initial stages of the investigation.

Don't get hurt.

Bring what you need.

Carry what you bring.

Get organized.

Do the important things first.

Document everything you do.

Keep your hands in your pockets. Don't pick up anything until you know what you are going to do with it.

5

Wreckage Distribution

1. INTRODUCTION.

Although all aircraft accidents are different, there are certain elements that are common to aircraft impacts and wreckage distribution. By organizing those common elements, we can predict and explain most impact situations. Some, for various reasons, will defy any rational explanation.

The two questions that usually need to be answered are:

      1. What was the attitude of the aircraft at impact?

      2. Was this impact survivable?

2. INFLUENCE OF AIRCRAFT VELOCITY AND IMPACT ANGLE.

Crash dynamics and wreckage distribution is influenced primarily by aircraft velocity and impact angle. The velocity of the aircraft generally determines the degree of breakup and destruction of the wreckage. The impact angle determines how the wreckage will be distributed. We will deal with each of these separately.

3. INFLUENCE OF TERRAIN.

As we will see, we can cope with the upslope or downslope impact. Where we run into trouble is with the nature of the terrain itself. Obviously, there is a big difference in the crash dynamics of an airplane crashing on a frozen wheat field and another one going into a forest at the same angle and velocity. Theoreti-

cally, we could look at an impact as a problem in deceleration and energy absorption and (for example) calculate impact velocity as a function of the distance various parts of the aircraft traveled. Unfortunately, all of those calculations require some estimate of the retarding effect of the terrain (not to mention terrain angle) and there just isn't any accurate way to deal with that.

4. WATER IMPACTS.

These deserve a brief mention. What happens when a plane hits the water?

Actually, it is primarily a function of velocity. If the plane is going fast enough, the water is a solid object as far as the plane is concerned. It comes apart in about as many pieces as it would if it had crashed on land. At the other extreme, a plane that was deliberately ditched will probably sink to the bottom in one piece.

In a water impact, there is an impact angle, but it is related to the wave action on the surface of the water. If the water was absolutely calm, then the impact angle may be determined by examination of the wreckage. If there was heavy wave action, impact angle becomes meaningless as you don't know whether the plane impacted on the front or backside of a wave.

What is the significance of the distribution of wreckage on the bottom of the lake, ocean, or river? In general, the wreckage distribution doesn't mean much unless it resulted from an inflight breakup. The actual distribution is influenced by depth, currents and the tendency of various parts to float (or be slightly buoyant) or to sink in different fashions. Some parts of the plane will sink like a rock. Others will sink like a dinner plate. Try it and you'll see the difference.

One rule that does hold up fairly well is the dispersed part rule (see Chapter 9.) A portion of the airplane or its contents found at a significant distance from the main wreckage probably came off inflight.

5. LAND IMPACTS.

All aircraft land impacts can be categorized in five different types.

A. HIGH VELOCITY, HIGH ANGLE. This is the "Smoking Hole" type of impact. The aircraft crashed at a very high angle and a very high velocity. Depending on the terrain (of course) the aircraft dug a deep hole and most of the fuselage followed the nose into the hole. The dirt that was originally in the hole had to go somewhere. It splashes out of the hole and forms a rim around the crater. If the impact angle was exactly vertical, which it almost never is, the rim of dirt around the crater would be symmetrical. Since the impact was at some angle other than vertical, there will be more dirt piled up in the direction the plane was going--if that makes any sense.

The front parts of the aircraft will probably stay in the crater and be covered with dirt. Other parts of the aircraft will actually be pushed out of the hole by the following wreckage and will be distributed in random fashion around the crater. Again, the bulk of this distribution will occur on the side in the direction the plane was going.

While the actual distribution of the wreckage is meaningless, the gross distribution of it is a clue as to the impact heading.

B. HIGH VELOCITY, LOW ANGLE. This wreckage is spread across a lot of real estate--perhaps a mile or more. There is an initial impact scar and, because of the velocity, the aircraft started to come apart immediately. What happens next is a little unpredictable, but remember that you are no longer dealing with an intact airplane; only portions of it.

As a rule, the wreckage will be distributed in a fan-shaped pattern from the impact point. The heaviest portions of the wreckage will travel farthest. If the engines were ripped from the nacelles or cowlings, it is not unusual to find them down at the end of the wreckage trail. The engines have the most mass and therefore the most energy and will travel furthest. If the engines do not separate from the aircraft, the parts of the wreckage containing the engines usually goes a long way.

This distribution pattern may be affected slightly by the local winds. Light weight parts may be blown slightly downwind. You can always trust the engines, though. They don't get blown around much by the wind and they are affected less by the terrain than the rest of the wreckage. A line from the impact point to the engine(s) is probably close to the impact heading.

C. LOW VELOCITY, HIGH ANGLE. This plane crashed at a high angle, but it wasn't going very fast. This is typical of a lot of general aviation accidents. The impact crater will be shallow and the plane will be largely intact. There won't be a lot of wreckage strewn around the crater.

In many general aviation aircraft, the aft part of the fuselage connecting the empennage to the cabin is not particularly strong. Its sole purpose is just that; to connect the tail to the rest of the airplane. Since the tail has quite a bit of mass, it is not unusual for the tail to keep going when the front of the plane comes to a stop. Depending on the impact angle (and the geometry of the plane) the tail either goes over the top or under the bottom of the rest of the plane; buckling the aft fuselage in the process. This over or under phenomenon is a pretty good clue as to whether the impact angle was greater or less than (nominally) 45°.

D. LOW VELOCITY, LOW ANGLE. This plane hit the ground at a fairly low angle, bounced, hit again and perhaps hit a third or fourth time. In the process, it may have shed wings and engines, but it still looks pretty much like an airplane. There is none of the mass destruction seen in high velocity impacts. This is also typical of general aviation impacts, particularly those associated with landing approaches.

Each of the impact scars and the distance between them can be used to calculate velocity and vertical and horizontal G forces felt during the impact sequence.

E. STALL/SPIN IMPACTS. These are essentially low velocity high angle impacts and are described in detail in Chapter 27. Keep in mind that the plane was not in control and may not have impacted nose first. If a spin was involved, it may be possible to determine the direction and the nature of the spin.

6. DETERMINING IMPACT ATTITUDE.

First, check the ground scar. Pitch, bank and yaw may be measurable from the evidence on the ground. Next, look at the front end of the wreckage. In low velocity impacts, the impact attitude is frequently captured by the nose of the aircraft. This doesn't work very well in a high velocity impact, although there may be alternatives. In one high velocity, high angle impact, all of the flight instrument "cans" in the instrument panel were bent upward and sideways in an identical manner. Since the instrument panel was mounted vertically in the airplane and the structure ahead of it was mostly empty baggage compartment, the direction of the force that bent the instrument cans was correlated with pitch and bank attitude at impact.

Third choice is the attitude indicators in the cockpit. If the damage was severe enough to capture the gyro gimbals, (See Chapter 14) these instruments may provide a reliable indication of attitude.

7. DETERMINATION OF VELOCITY.

Start with the airspeed indicator in the cockpit. This is a geared instrument (Chapter 14) and it will frequently capture the impact velocity if it was severe enough. Next, find out if anyone happened to have a video tape of the accident. That happens just often enough to be worth checking. If you find one, calculation of velocity is a matter of timing the plane's progress past known points in the background.

Check the configuration of the aircraft. Assume that the lowest possible velocity would be the aircraft's stall speed for that weight and the highest possible velocity would be the limiting speed for that configuration. The correct velocity would be somewhere in the between those two. Pick the correct speed for the situation (approach speed, normal cruise) and run your calculations for all three numbers.

Check the ATC radar plot and the A/G communications tapes. You can use these to help establish a velocity prior to the impact and estimate the impact velocity.

If, of course, you are dealing with an aircraft with a flight data recorder, no problem. It records velocity. (Chapter 18.) Regardless of the availability of recording devices, it is always a good idea to verify velocity

calculations with all available methods.

8. DETERMINING IMPACT ANGLE.

We need to understand that impact angle is the flight path angle of the aircraft adjusted for the pitch attitude (or deck angle) of the aircraft with respect to its flight path. If the terrain is not level, that also has to be considered. One way to deal with this is to consider level flight to be zero degrees; a climbing flight path to be negative (-) and a descending flight path to be positive (+). Level terrain is likewise considered to be zero degrees while an upslope is positive (+) and a downslope is negative (-). Thus impact angle becomes the algebraic sum of flight path angle and terrain angle. That's OK as long as the deck angle of the aircraft is the same as its flight path angle. This is probably true (or close enough) for most impacts, but not (say) for an aircraft that is mushing toward the ground in a nose-high stall. Deck angle is largely a function of airspeed and can be estimated if the airspeed is known. If the deck angle is significant, then a nose-up deck angle should be treated as negative (-) with respect to the flight path angle and a nose-down deck angle should be treated as positive (+). If we stick to that convention, we can just add all the angles algebraicly and come out with impact angle, which is what the plane actually feels when it hits the ground. Some examples of this are shown in Figures 5-1 through 5-3.

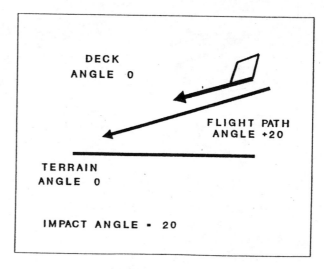

Figure 5-1
Impact on Level Terrain

Figure 5-1 shows the simple situation where the airplane impacts on level terrain and deck angle and flight path angle are the same. In this case, impact angle equals flight path angle.

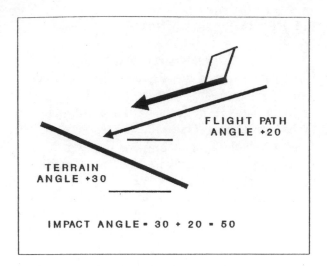

Figure 5-2.
Impact on Rising Terrain

In Figure 5-2, the airplane is impacting on rising terrain. The impact angle is the sum of flight path angle and terrain angle.

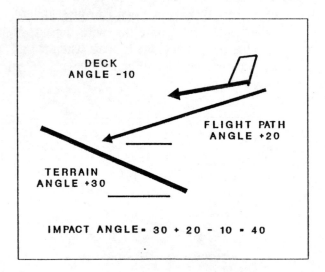

Figure 5-3.
Rising Terrain and Nose-high Deck Angle

Figure 5-3 shows how to cope with terrain angle, deck angle and flight path angle. Using the convention suggested earlier, the impact angle will be the algebraic sum of the other angles. In this case, 40°.

Suppose a plane in a climb after takeoff impacts a steep hillside. As far as the impact is concerned, this would be the same as if the plane dove into level terrain at a high angle. Both would appear to us as low velocity high angle impacts. Likewise, a plane in a steep dive could impact on the downslope of a hill and the result would resemble a high velocity impact. Thus we always have to account for these angular differences.

Impact angle may be captured in the first ground scar. The slope of the scar would be equal to impact angle. Knowing that, we can factor in terrain angle to obtain flight path angle.

Another way to obtain impact angle is to look for something (a tree, perhaps) that the plane hit prior to impact. By measuring the distance from the impact crater and the height of the broken branches, we can calculate impact angle.

A third way is to look at the front end of the wreckage. In low velocity impacts, the impact angle may be captured by the damage to the nose. In some general aviation aircraft accidents, the leading edge of the wing can provide a good indication of impact angle.

9. CALCULATION OF IMPACT FORCES.

Now that we have impact velocity and impact angle, we can make an assessment of the forces involved and whether or not the impact was survivable. See also Chapter 36.

Basically, we need to resolve the impact velocity into horizontal and vertical velocity. This is easy to do by constructing a right triangle where the impact velocity vector is the hypotenuse and the impact angle is the adjacent angle. See Figure 5-4.

The vertical and horizontal velocities may be calculated with the following formulas.

$$V_V = V \sin\alpha \qquad (1)$$

$$V_H = V \cos\alpha \qquad (2)$$

Where:
    V = Impact Velocity
    V_H = Horizontal Velocity
    V_V = Vertical Velocity
    α = Impact Angle

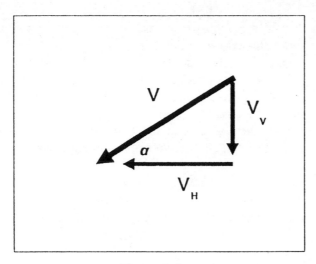

Figure 5-4.
Resolution of Impact Velocity

Occasionally, you get lucky and you can derive the vertical and horizontal velocities directly. Suppose, for example, the aircraft hit wires shortly after takeoff and crashed immediately thereafter. You can assume that the aircraft's vertical velocity was essentially zero at the time it hit the wires. The vertical velocity at impact can be calculated by applying acceleration due to gravity (32 feet per second[2]) to the height of the wires. We can also calculate the time it took to fall that distance. Using that time and the distance of the impact horizontally from the wires, we can calculate the velocity the aircraft must have had in order to travel that distance in that time. Now we have the two velocities we really need, and very accurately at that. If we really want to know impact velocity, it is the square root of the sum of the squares of the other two velocities.

Anyway, now we've got vertical and horizontal velocities at impact. Now we need to know the distance over which the aircraft decelerated from that velocity to zero. That equates to G force in the vertical and horizontal directions.

The deceleration distances are composed of the distances that the aircraft plowed into the ground plus the distance that the structure of the aircraft collapsed during the deceleration process. Think of it this way. Suppose we drop a ripe tomato onto a sand bed. We can calculate the vertical velocity at impact easily enough by knowing the drop height. If we want to know the deceleration distance; the distance through which the tomato went from impact velocity to zero; we would measure the depth of the divot in the sand and the dent in the tomato and add them together.

Same with airplanes. The deceleration distance is the depth (or length) of the impact scar plus the vertical or horizontal "crush" distance of the fuselage structure.

There are a couple of problems with this. First, while the vertical velocity at impact may have gone to zero, the horizontal velocity probably did not. As the aircraft left the first impact, it was still traveling forward at some velocity. This can be calculated and that's why it is important to examine and measure all the impact scars. When we use V in the formula for horizontal velocity, it would be more technically correct to use V_{enter} minus V_{exit} for each of the impacts.

The other problem is accuracy. The aircraft structure crush distance is an estimate. The way around this is to run the calculations for crush distances above and below the estimate and see what difference it makes.

Now we have the velocities and the deceleration distances we need and we are ready to go to Chapter 36 and calculate vertical and horizontal G forces. Once we have the G forces in both the vertical and horizontal direction and the duration of the forces (which we calculate at the same time), we can consult tables that will tell us whether that force in that direction for that duration was reasonably survivable or not. That's how its done.

10. SUMMARY.

There are a lot of variables that affect wreckage distribution. We have only covered two of them to any depth; velocity and impact angle.

While terrain angle can be factored in to aid in the calculation of forces, we haven't said much about the effect of terrain angle on distribution. Obviously, the distribution of the wreckage impacting a steep slope is going to be downhill. It would be normal to find the heaviest parts at the bottom regardless of the direction of impact.

We mentioned the nature of the terrain as being a factor. Sometimes, this makes calculations of forces almost impossible. There is, for example, no way to accurately calculate the forces involved when a plane crashes into the tops of trees. Likewise, there is no good way to predict exactly how the wreckage will be distributed after such an impact.

When a plane hits something prior to impact, that

sometimes influences the impact forces. A large plane hitting the top branches of a tree is probably not significantly affected. That might not be true of a general aviation aircraft. Wire strikes or wire impacts are likewise difficult to deal with. The retarding effect of the wire is almost impossible to calculate.

For all of these difficulties, the distribution of the wreckage can sometimes tell us quite a bit about the dynamics of the impact. The ground scars are important and should be diagrammed and measured before they are destroyed during the investigation process. Likewise, the damage to the aircraft can tell us a lot about aircraft attitude and impact angle.

One caution. It is easy to get carried away with the mathematics of this business and forget that post-impact measurements are never precisely accurate and crash decelerations are never uniform. Mathematics, being an exact science, allows us to take these imprecise measurements and calculate something out to several decimal places; forgetting that the accuracy of the result can be no better than the data used.

It is important sometimes to stand back and look at the overall wreckage distribution and make sure that our conclusions make sense and none of them violate any of the basic laws of physics.

6

Accident Photography

1. INTRODUCTION.

Perhaps the most useful tool the investigator can bring to the scene is a camera--and the knowledge of how to use it. There is no better tool to document the situation as it exists now, or the situation as it changes due to movement or disassembly of the wreckage. The camera can permanently record the fracture patterns; the dents, scratches and gouges; and the instrument readings. It can compare correct and incorrect or damaged and undamaged. It can show the view seen by a witness. It can record documents. There is almost no limit to its usefulness.

As a technology, photography (or "imaging" as we are learning to call it) is advancing at about the same speed as aviation. Anything written about it is probably obsolete at the time of publication.

If we are not there already, we are moving toward a form of digital imagery where images are stored digitally on a computer disk or memory chip where they can be played back and manipulated on a computer. In the accident investigation business, this computer manipulation capability is a mixed blessing as it can raise evidentiary questions. Is, for example, the resulting product a true depiction of what was seen?

From the computer, the image can be printed on paper in high resolution color. We can do about the same thing with a video camera. From the digital video images, we can select the one we want, transfer it electronically to a different storage medium and print it in color.

Video cameras that can do the picture-taking end of this are already available. They are highly portable and, with a macro lens capability, they can do a remarkably good job of technical photography. Digital image still cameras are also available. They look a lot like 35mm SLR cameras and have a similar lens system, but the image is stored on a disk; not on film. What's going to happen to the good old film camera?

It will be with us for a while since the quality of digital imagery has not yet achieved film quality and equipment needed to produce good digital imagery is not readily available. The shopping malls are not yet full of 1-hour digital image processing stores. While video cameras have their uses, our present requirements are for pictures that can be enlarged and printed in multiple copies that will fit in the written accident report. Including a video cassette in the report is not practical and the process of making high quality prints from standard video tapes still needs some work. It seems likely that digital imaging will really become practical about the same time we decide to give up "hard copy" written reports and do them all electronically. That's coming, too.

In the meantime, most of us are going to use a regular camera and film to document the accident. This chapter deals primarily with that technology, although many of the basic techniques would apply to any "imaging" system.

2. SOME PHOTOGRAPHIC BASICS.

While all this digital activity has been going on, the film camera makers have not been idle. It is becoming very difficult to buy a plain vanilla camera that doesn't automatically do everything for you. While all this automatic technology is nice, you need to keep it in perspective and decide whether you really want it or need it.

There are two types of photographers out there; tourists and professionals. While their needs overlap somewhat, the professional is frequently working at the fringes of the camera's capability where the tourist dare not and need not tread. The tourist seldom worries about precise focus or depth of field, because he never gets close enough to anything to encounter those problems. The tourist is generally unconcerned about the placement or the output of the electronic flash. The built-in flash on the camera works fine for him.

Aircraft accident photography is a form of technical photography that is closer to professional than tourist. Unfortunately, the world is full of investigators who don't understand this and act more like tourists.

Consider this. The quality of the resulting pictures really depends on just four elements.

A. THE FILM. While all films are good, some are better than others for technical work. In general, the bigger the film size (format), the better the results.

B. THE LENS SYSTEM. The quality of the lens system determines how sharp the resulting pictures are going to be.

C. THE AMOUNT OF LIGHT ILLUMINATING THE SUBJECT. Controlling this means being able to cope with the existing light and control the light you are adding with electronic flash or floodlight.

D. THE AMOUNT OF LIGHT ALLOWED ONTO THE FILM. This is called exposure and each type (speed) of film needs to see a specific amount of light. It gets this light through a combination of size of opening (aperture) and amount of time the opening is left open (shutter speed). Since the film doesn't care how it gets the light, there are an infinite number of combinations of shutter speed and aperture (f-stop) that will each produce exactly the same amount of light. We care, however, because aperture affects depth of field and shutter speed affects our ability to hold the camera still. Therefore, if you contemplate buying a camera system for technical work, buy one that has a manual mode that can override the automatic features. You'll need it.

Remember also that you are buying a camera body and a lens system. Although 35mm cameras are usually packaged with a 50mm lens, you don't have to buy it that way. Decide which lenses you want first; then buy a camera body that will accept those lenses.

3. BASIC EQUIPMENT.

As a rule, the bigger the negative, the better the print. Unfortunately, bigger negatives means bigger and bulkier cameras. In aircraft accident photography, your camera equipment has to be portable as you are seldom able to drive to and park at the scene of the accident. The 35mm Single Lens Reflex (SLR) emerges as the best compromise considering size of negative and portability of camera. The 35mm negative will enlarge to 30" x 40" with reasonable quality and maybe larger if the picture was taken with a fairly slow film and a good lens properly focused. Equally important, the SLR has a through-the-lens focusing system. For you computer buffs, this is a WYSIWYG system. Twin lens reflex and view finder cameras lack this advantage.

The "normal" lens for a 35mm SLR is considered to be a 50mm lens. This is a perfectly good lens for tourist activities, but, for technical work, it won't let you get closer to anything than about two feet. That's not good enough. This will produce an image size ratio of about 1:7 (photographic size to actual size.) An ideal lens is one with a macro capability that will allow you to focus at about three inches. This produces a 1:1 or 1:2 size ratio and is excellent for technical work. This same lens will handle pictures at infinity, which means that you can do almost all of your photography with it. There is seldom any need to change lenses or add extension tubes. Today, while camera bodies have gone automatic, lenses have gone zoomish. It is hard to find a lens that doesn't zoom from telephoto to macro to wide angle to whatever.

Keep this in mind. The zoom capability is another

convenience. It doesn't make the picture any better. It may in fact make it slightly worse (when compared to an equivalent fixed focal length lens) because the zoom thing sometimes adds extra lens elements (more glass) between the subject and the film. In aircraft accident photography, the telephoto lens has almost no value. You will need to take good close-ups, so get a lens specifically designed for that purpose. A zoom-macro is very difficult to use for technical close-ups. A wide angle lens does have some value. Get one that is really wide; perhaps 24mm or 28mm. A zoom wide angle is not that wide.

Electronic flash. You will absolutely need one of these. Here, bells and whistles do make a difference. The ability to aim and control light output and have the flash unit measure the output and automatically turn itself off is worth the price. The difference in price between a cheap flash unit and a top of the line model is not that great. When you are buying the flash, pick up a PC extension cord so you can take the flash off the camera and hold it in some other location. You'll need this.

The best type of flash unit is one where the light measuring device (thyristor) stays on the camera shoe while the flash itself can be unhooked from the measuring device and held at any distance or angle. This means that light output is always being measured at the film plane where it counts.

Another very handy type of flash is called a ring light. This is a circular flash unit that screws into the filter holder on the front of the lens. This was originally developed for dentists who wanted to take pictures inside someone's mouth but couldn't get both the camera and the flash in there at the same time. It produces nice evenly lighted shadowless pictures at closeup distances down to about six inches.

You need something to hold the camera steady while you shoot at slow speeds or, perhaps, want exact duplicate pictures with and without signs, captions or arrows. A good professional tripod with a quick-release fitting for the camera is really handy; unless you have to carry it a mile or so to the wreckage site. Then it's not handy at all. A small telescoping tripod will work, but it doesn't provide perfect support for the camera. Tabletop tripods can also be used. In a pinch, a beanbag or a bag of rice can be used. You can also buy a camera clamp that screws into the camera's tripod socket and clamps to any stationary object.

Here's a list of the minimum equipment you need to do a professional job. See also Chapter 4.

A. 35mm SLR camera body.

B. Macro lens with 1:1 capability (1st lens)

C. Wide angle lens - 24mm (2nd lens)

D. Electronic flash with extension cord

E. Shutter cable release with locking capability

F. Portable tripod

G. Spare batteries for flash and camera.

H. Plastic bag, tie and rubber band to protect camera.

I. Film

If you have money left over, go for the ring light. If you want to buy some insurance, look for a used camera body of the same make and model series as the one you own. Most of the mechanical things that can fail are in the camera body; never the lens. This gives you a backup and it allows you to load and shoot two types of film by just switching bodies.

4. IDENTIFICATION OF PICTURES.

Tourists never identify their pictures as they take them. Half the fun of being a tourist is getting the pictures back from the processor and trying to figure out where and when each was taken.

Professional photographers aren't very good at identifying them either. Most of their work involves several pictures of the same item, so picture identification is not much of a problem.

Aircraft accident photographers absolutely must identify their pictures as they take them. It is not unusual to shoot five or six rolls of film in a day. That's over 100 pictures and many of them will be medium or close-ups which, by themselves, are unidentifiable. Furthermore, aircraft accident investigation is a slow process. It may be months between the time the pictures are taken and the time they are needed for the report. It may be years before the accident gets into litigation. One of the authors of this book has testified in court as to the identification of wreckage pictures he took nine years earlier!

You can buy a new back for your camera (called a data back) which will add a number or a date or a time to your negatives. Unfortunately, this does not solve your basic problem which is to identify the subject of the picture. You have no choice but to write down or record what the picture is of and there is no automatic device you can buy that will do it for you.

Actually, identifying the pictures is fairly easy. First you need a system for identifying the film roll. The best way to do this is to use the first shot of each roll to photograph an identification board. This is a small erasable card that has your name and address; the identification of the accident, the date and the film roll number.

Next, you need a method of identifying each individual frame. Keep in mind that the film manufacturer has already numbered the frames for you. He has put little numbers (1, 1A, 2, 2A and so on) on the edge of the film where it won't show in the print. There is really no need to buy an expensive gadget that numbers something that is already numbered. All you really have to do is record, frame by frame, the subject of the picture. The lazy way to do this is to use a small pocket tape recorder, but there is nothing wrong with a notebook and a pencil. This photo log then becomes your caption sheet for that roll of film. Your pictures are forever and ever identified.

5. FILM.

A. COLOR OR BLACK AND WHITE? In some cases, document photography for example, black and white works a little better; but not much. Color film is preferred in almost all accident photography situations. Today, the cost of color film and paper is actually less than the cost of black and white because of the high silver content of B&W films and papers. Where color gets expensive is in the custom printing of large enlargements. More on that later. In some situations (color changes in materials due to exposure to heat, for example) color is absolutely essential. Remember that you can always print a color negative in black and white--if that's what you want--but it's very hard to print a black and white negative in color.

B. SLIDE FILM OR PRINT FILM? What do you want the final result to be? A slide or a print? If you want prints, shoot print film in the first place.

Granted, you can make a print from a slide (or vice versa), but you have to go through an intermediate step to get there and the result will never be as good as the original slide or print.

Another factor to keep in mind is that slides are a one-step process. What you shoot is what you get. If there is an error in exposure or lighting, that error will show up on the slide and there is nothing you can do about it. Prints, on the other hand, are a two-step process. First you get a negative and from that you make a positive or a print. While all errors made will show up on the negative, there is a lot that can be done to eliminate those errors in the printing process. This, in fact, is done routinely by automatic film processing machinery. When the tourist brings in a roll of poorly exposed pictures, the machinery compensates for the lousy negatives and produces good prints. Since the tourist never looks at the negatives, he has no idea what a poor photographer he really is.

C. FAST FILM OR SLOW FILM? Fast film permits more latitude. You can goof up the exposure more and still get a good picture. Fast film can also be used under low light conditions. Slow film requires more attention to lighting and exposure, but it enlarges better. Any film should enlarge satisfactorily to (say) 8 X 10. Beyond that, the grain in the fast film starts to show up. As the film speed gets faster, so does the film's sensitivity to airport X-Ray machines.

Since this is technical photography and we are professional investigators, exposure and lighting shouldn't bother us. If we need more light, we add it with our electronic flash. If we are concerned about exposure, we take multiple pictures at different exposures to cover our bets. Since we want to be able to enlarge the negative to perhaps 30 X 40, we shoot a slow film in the ASA 100-200 (DIN 21-24) range. Professional photographers frequently use film that is even slower than that. The problem with most professional grade films is that they have to be kept refrigerated. An ASA 100 film is a good compromise. It is readily available and relatively insensitive to airport X-Ray machines.

6. PHOTOGRAPHIC PRIORITIES.

You have two choices. You can drape your camera around your neck and snap pictures of anything that looks interesting as you wander through the wreckage. This is what we call the tourist technique.

On the other hand, you can park the camera bag at some convenient place while you do your wandering and plan what pictures you want. Then you retrieve the camera bag and take them all at once. This works out a lot better.

Consider these ideas. You don't have to take the pictures in the order you intend to look at them. You may, for example, want some distant or medium shots to put a close-up in context (see "Hollywood Technique"), but you don't have to shoot them in that order. The easy way is to shoot all your distant and medium shots first as those can be taken with a hand-held camera and no extra equipment. After you've done that, then take your close-ups. For those, you will need the tripod, the flash and the cable release. (Most professional photographers have mastered the technique of bracing their elbows against their chest, thereby imitating a tripod, and squeezing the shutter release as though it were a gun trigger so they don't need a tripod or a cable release. They look funny, but they get good results because they have a lot of practice with that technique. The rest of us will merely look funny without the good results.) The idea here is to save yourself time by not jumping back and forth between the two types of photography.

Also, there is (or should be) a logical order in which aircraft wreckage should be photographed. Here is a list that will work for most accidents.

A. PERISHABLE EVIDENCE. These are pictures of things that are likely to change or disappear altogether if not photographed immediately. Included are pictures of the crash aftermath or rescue in progress; medical documentation pictures, cockpit instrument panel readings, switch positions and flight control positions.

B. AERIAL VIEWS. If possible, get some aerial views early. The appearance of the wreckage from the air will change rapidly as investigators move through it.

C. OVERVIEWS OF THE SCENE. One suggestion is to photograph the wreckage from the eight points of the compass. That, obviously, only works if the wreckage is fairly compact. If it is spread out over a long distance, try a series of overlapping pictures along the wreckage trail. The prints can be edge-matched and a montage created.

D. SIGNIFICANT SCENE ELEMENTS.

Try to photographically establish the terrain gradient, if any. Photograph the initial impact scars in such a manner as to allow future analysis of crater size and depth. Photograph anything hit by the aircraft prior to impact.

E. WRECKAGE INVENTORY. You will almost always want to inventory the wreckage to make sure all the major parts are there. The camera is a very handy inventory tool.

F. CLOSE-UPS. You've taken the easy pictures. Now get the tripod and flash and get set up for close-ups. Where possible, it is always easier to set the camera up once and bring the parts to it. If you are photographing individual parts as they are removed from a component, for example, it is worthwhile to acquire a table or even a copy stand where the camera can be mounted vertically and the parts positioned beneath it. Most tripods with swivel heads can be used to mount the camera vertically. To create a bland background for small parts, buy a roll of freezer paper. Use the dull side and just unroll some more of it whenever it gets dirty. When using a light background like this with a small dark object on it, the camera light meter will tend to meter the background and the object will appear too dark. You can correct this by overexposing one stop (bigger opening; smaller f-stop number). Another trick is to hold the camera close to the subject where the light meter can't see much of the background. Now see what f-stop the meter recommends. Put the camera back on the tripod and use that setting. The background will be overexposed, but who cares? Incidently, this same logic applies to what are called black-on-black photographs. If you are trying to photograph an aircraft tire or perhaps a section of carbon fiber composite, the camera will almost always underexpose it. Here, you can safely open the lens two full f-stops over that suggested by the meter. If this bothers you, shoot pictures at both the metered f-stop and metered plus two and see which you like best. This is called bracketing the picture and it is how a professional assures himself that he will always get at least one picture that is correctly exposed.

It is not always possible, of course, to bring the parts to the camera. Field setups for close-ups take time and there are no shortcuts.

G. DOCUMENTS. Remember that you can use the camera to copy documents that you otherwise cannot retain or copy. This could include licenses or logbooks. It could even include a map or chart on someone's wall. One neat use of a camera is to take a close-up photograph of the crash area on a large scale topographic map. By enlarging the negative, you create a very large scale diagram of the area.

H. WITNESS VIEWS. It may be important to document the view a witness had of the accident. Remember that the witness has very wide angle eyes. Use a tripod and the montage technique to duplicate his view.

I. EXEMPLARS. An exemplar is a model or a pattern for the real thing. Sometimes it is hard to tell from a wreckage photograph what the part or component is supposed to look like. In some investigations, it is worth a roll of film to have pictures of an identical undamaged aircraft for comparison.

7. BASIC TECHNIQUES.

A. CLOSE-UPS. The two main problems with close-ups are focus and depth of field. It is hard to focus precisely at extreme close-up ranges and even harder to maintain the focus while you trip the shutter. This means that you ought to seriously consider a tripod and a cable release. See comments above under "Priorities." Once you get the camera focused, at least it will stay that way while you take the picture.

There are really three ways to focus a camera. 1. Rotate the barrel of the lens. 2. Set the lens where you want it and move the whole camera back and forth. 3. Set the lens and move the object back and forth in front of it. For some reason, rotating the barrel of a macro lens is the most difficult. Try extending the lens to maximum close-up and moving either the camera or the object.

Depth of field is a pesky problem as it is critical at close-up ranges and easy to overlook. Depth of field is the distance, front to back, that things will be in focus in the picture. At extreme (1:1) close-ups, depth of field may be only a fraction of an inch. It is possible, for example, to focus on the filament of a small light bulb and be unable to get both ends of the filament in focus.

If possible, try to get everything that is important in the same plane as the film. You would shoot a document, for example, with the paper perpendicular to the lens and parallel to the film. That way, there is no depth of field problem. If you shoot the paper from an angle, you will probably be unable to focus on both ends of it. If you are forced to shoot objects that are not all in the same plane, the solution of choice is to close the lens down to the smallest possible aperture (largest f-stop number.) The reason is that a large aperture produces a very poor depth of field, while a small one improves it significantly. If you think about how aperture and shutter speed are related, you realize that in order get the same amount of light through the small hole, you must leave the hole open longer, which means slow shutter speed. This, in turn, means tripod, because your ability to hold the camera still drops off rapidly at speeds slower than about 1/60th of a second. Perhaps, in order to shoot at the smallest aperture, you will have to leave the shutter open for a full second. So what? The film doesn't care and neither should you if you have a tripod.

You should realize that most good camera lenses have what is called a 'preview button' somewhere on the barrel. This is to allow you to see exactly what depth of field you are going to get. Remember that when you set the aperture, nothing actually happens. The diaphragm stays wide open so you can see to focus. When you trip the shutter release, the viewfinder mirror snaps up (so you can't see what happens next), the diaphragm closes down to the preset position, the shutter opens and the flash fires. All of those items now happen in reverse order and you are suddenly able to see through the lens again. It all happens so fast that many (tourist) photographers squint through the viewfinder thinking that they are actually going to see something. Frequently they close the other eye and thus have no idea whether the flash fired or not. If you insist on looking through the viewfinder as you trip the shutter, practice keeping both eyes open. That way, you'll see the flash. Back to the preview button, this manually closes the diaphragm to the selected setting so you can see what the film is going to see. If you use the preview button, you will find a flashlight useful as it gets dark with the diaphragm closed and you can't see the subject very well. Sometimes a flashlight is a big help in focusing even with the lens wide open. One professional trick is to have someone hold a penlight at the point of focus and aim it back into the lens. this gives you something to focus on.

If getting the camera perpendicular to the subject and closing down the lens doesn't produce the depth of field you need, your only alternative is to back up a little. You're too close.

The proper setup for a good close-up is camera on a tripod; flash (if needed) detached from the camera and aimed at the subject; and a cable release used to trip the shutter. If you are using flash, incidently, you don't really need the cable release. The flash will record the scene faster than you can spoil it by moving the camera.

B. AERIAL PHOTOGRAPHY. The 35mm SLR will do a reasonably good job on aerial photography. Some suggestions:

1. Try and fly in something that has an open window or door. If that's not possible, hold the camera as close as possible to the window without touching it. This will minimize reflections from the window.

2. Don't touch the camera to any part of the aircraft or helicopter. It will pick up vibration and make the picture look fuzzy. The human body acts as a very good shock absorber to dampen out vibrations.

3. Arrange for something of known dimensions to be visible in the wreckage scene. This will permit measurements to be made on the aerial photographs.

4. Try and get at least one picture as close to vertical as possible. Shoot obliques from several directions, particularly the direction of flight.

5. When shooting aerials, always take two shots in rapid succession for each picture. This creates a stereo pair which can be viewed three-dimensionally using a stereo viewer. Hills, valleys and contours can be seen in this manner.

6. As part of your photo log, note the altitude for each picture taken.

C. FILL-IN FLASH. Remember that film can only cope with one set of circumstances. If you have a subject with both bright spots and shadows, you can expose for one or the other, but not both. If you want both, your only choice is to expose for the bright spots and throw some extra light into the dark spots to make them bright, too. This is called Fill-in Flash and it is a technique you must master. Unfortunately, the actual technique varies slightly with the capability of your camera and your electronic flash. The instructions came with the flash. Find them and read them.

D. NIGHT PHOTOGRAPHY. This is to be avoided. There is no way you can duplicate the sun as a source of light. The light output from your flash drops off as the square of the distance, which means that you're not going to do very well on scene overviews with a single flash of your flash. Unfortunately, there are times when you have no choice. Either the wreckage is going to be gone by daylight or the sun is not scheduled to rise tomorrow. You're going to have to take your pictures now.

There is a neat trick called "painting with light" which is easy and yields consistently good results. The idea is to set the camera on a tripod with the shutter locked open. Then, with the flash completely disconnected from the camera, literally paint the scene with light by firing the flash with the test button, moving to a new location and firing it again. Here are the procedures.

1. Camera on tripod, prefocused. Shutter set on the "B" setting (which means that it will be open as long as the picture-taking button is held down.) F-stop set for the anticipated flash-to-subject distance; probably 12-15 feet. Locking cable release attached.

2. Push the cable release (opening the shutter) and lock it.

3. Move in front of the camera and, starting at one side of the scene, fire the flash at the scene. Move to a new location and fire it again as soon as it recharges. Try to stay about 12-15 feet from the wreckage and hold the flash high and at arm's length when you fire it.

4. Return to the camera and close the shutter by unlocking the cable release.

You might want to try this two or three times with different amounts of "painting," but the only really critical item is that the camera must be held absolutely still (tripod.) The film will record whatever is illuminated. It will not record anything that is not illuminated, which explains why you won't appear in the picture even though you are in front of the camera. If your body is between the flash and the lens, your body will appear as a shadow. That's the reason for holding the flash high and away from your body. Any other lights in the scene will appear as bright spots on the negative, but they usually don't detract from the picture. Other people who are milling around in the scene may appear as fuzzy images unless they stand absolutely still, which means they aren't doing much.

E. FLASH PHOTOGRAPHY. By the time you get to the wreckage, it may have been moved to a hangar and all of your pictures will need additional light. If this is going to be a major photographic effort, you might seriously consider bringing or renting a pair of photoflood lights and some telescoping light stands. These will provide better light than your flash and you can shoot pictures as though you were in daylight. If you are going to use flash, a second flash unit on its own stand with a remote firing (slave) gadget can improve your lighting significantly. If you only have a single flash, remember that the camera-mounted flash tends to bounce light off a shiny subject right back into the camera lens. You see this very frequently in pictures of instrument panels. The photographer forgets that the instrument panel is mostly glass and he gets a beautiful picture of his own flash. If you suspect that reflection may be a problem, detach the flash from the camera and aim it at the subject at a 45 degree angle. This justifies another comment about aiming things. If you put the camera on a tripod, there is nothing going on in the viewfinder that you can watch. Keep your eye out of the camera and concentrate on aiming the flash correctly. The flash beam is actually pretty narrow and it is easy to aim it incorrectly.

When you are using a single flash, particularly off camera, remember that you are not only adding light, you are creating shadows which only the film will see. You won't see them until you process the film. For this reason, you should always take duplicate shots with the flash aimed from a different position. One of them should give you what you want, but you won't know which until you process the film.

8. TRICKS OF THE TRADE.

Here are some simple techniques that can improve your aircraft accident photography.

A. SHADOWS. Human eyes are three-dimensional devices. We see in 3-D and we take it for granted. The camera, however, is a two-dimensional device. Things that appear to have depth to us come out flat in the picture. This is particularly true of things like scratches, wrinkles, dents and gouges. They just don't look right in a picture. You can improve that somewhat by creating shadows with the sun, another light source or your flash. The shadows create

the illusion of depth.

B. MIRRORS. Sometimes you just can't get the camera where it needs to be to take the picture you want. Try using a mirror. A lady's square compact mirror works pretty well, particularly if you can stick the mirror where you need it. Put a roll of styrofoam mounting tape in your bag along with a small spring clamp.

C. PHOTO MONTAGE. Our normal peripheral vision is nearly 180 degrees and we can increase it with a slight turn of our head. The widest wide angle lens you can buy won't come close to that. If you are trying to illustrate what a witness might have seen or you want to photograph an entire scene close enough to see what's in it, try the montage technique. Put the camera on a tripod and swing the camera through the scene with about a 1/3 overlap of each picture. Good tripods have a built-in leveling device and a protractor to help you do this accurately. Actually, you can skip the tripod and get fairly good results with a hand-held camera. When you get the prints back, edge match them and do a little creative cutting and pasting to get a long, skinny picture. Re-photograph this and print it.

D. MODEL AIRCRAFT. Occasionally you will want to photograph one or more aircraft models to illustrate something about the accident. First decide on the background. A background of gray or blue velour cloth will work fine. Get two sticks. Wrap them with some of the background material and glue or staple it. Mount the models on the sticks by beveling each stick and gluing it to the bottom of a model. Hang the background cloth on the wall and have two assistants "fly" the models with the sticks. When photographed, the sticks disappear and you can take a lot of model pictures in a short time.

E. HOLLYWOOD TECHNIQUE. This refers to the standard Hollywood practice of never showing the audience a close-up without preceding it with a distant shot and a medium shot to put the close-up in context. This is a really good technique for aircraft accident photography. The close-ups are what tell the story, but they are so close that no one can relate them to the aircraft as a whole or even that component. When you know you will need a close-up, always take a medium and a distant shot so you can lead the viewer into the close-up and help him understand it.

F. HEIGHT. Sometimes, it would be nice if you had a twelve foot ladder and could get a picture looking down on the wreckage. Actually, you can do this if you have a long pole (anything), a camera clamp and a camera with an automatic timer. Clamp the camera to the pole and pre-focus it. Use the length of the pole plus your height with your arm extended as the minimum distance. Set the exposure for the prevailing conditions. Set the automatic timer and hold the pole up in the air to get your picture. The hard part of this is aiming the camera correctly. If you clamp the camera to the pole in such a manner that the pole is just out of sight when you look through the viewfinder, then the pole itself becomes the bottom edge of your picture. Just stand where you want the bottom of the picture to be and hold the pole vertically. If you have a wide angle lens, this is a good time to use it.

G. PHOTOGRAMMETRY. It is possible to scale dimensions from a photograph providing there is something in the picture that provides a size reference. An obvious example would be a ruler or measuring scale included in the scene. Even without a ruler, an object of known dimensions can be used to establish perspective and distances. This technique is called photogrammetry and is useful when there is nothing left of the scene except the photograph of it. If a need for this technique can be anticipated, a simple method is to construct a perspective grid reference consisting of four pieces of cardboard measuring one foot square each. Tape these together to form a two foot square and outline the edges and diagonals of all four pieces with black tape. With this grid photographed in the foreground of the picture, a photogrammetist can determine any dimension in the picture.

9. EQUIPMENT PROTECTION.

You have some expensive equipment and you don't have much control over the climate or the weather conditions of the accident scene. Here are some suggestions on how to protect your equipment.

A. LENS PROTECTION. This is technical photography under adverse conditions and the most vulnerable equipment you carry are your lenses. Whenever you buy a lens, buy a Skylight filter to go on it. This filter has almost no effect on the resulting pictures, but it protects the lens. If the filter gets scratched, replace it. This also eliminates any need to carry lens cleaning equipment. If you never take the filter off, the lens never gets dirty. The filter gets dirty and you need to clean it occasionally. If you don't have any lens tissue, an old photographer's trick is to use a dollar bill. American money has a high rag content, is lint-free and doesn't scratch. If you want to impress people, use a five dollar bill.

B. RAIN. This is the most common problem and the easiest to solve. Carry some plastic bags, ties and rubber bands. Place the camera in the plastic bag and tie off the opening. Put the rubber band around the lens and rip open the bag in front of the lens. Tuck the edges of the ripped-open bag under the rubber band and you're all set. You can work all the camera controls through the bag. What's the only thing that gets wet? The filter, of course, see paragraph A above.

C. HEAT AND COLD. Temperature extremes are not bad in themselves. The problem is with the change in temperature when you bring the cold camera into the warm room or take the warm film to the cold wreckage. As much as possible, you would like to maintain your equipment and film at a fairly even temperature. One inexpensive way to do this is to buy a styrofoam beverage cooler that will hold one or two six-packs. Keep your camera, flash and film in there and only take them out to use them.

The two major problems with severe cold are camera lubrication and battery life. If you know you are going into severe cold, it is possible to have your camera specially lubricated with a silicon lubricant. Regarding batteries, there is not much you can do about them except bring spares and keep them in your pocket where they'll stay warm. If you own a plain vanilla camera body, the only thing the battery does is run the internal light meter. You can do without a battery completely if you own a selenium light meter which does not need a battery.

Severe heat can effect both your camera and film. At some point, the glue or bonding material fastening the lens elements melts and you've just ruined a lens. Unfortunately, the heat doesn't have to be just ambient heat. Automobile interiors and trunks can act like ovens and expose your equipment to higher temperatures than you are feeling. Don't store your equipment in closed automobiles. The best carrying case in these conditions is an insulated case with reflective surfaces.

D. FILM. Your film will last a lot longer if you keep it refrigerated or even frozen. If you do this, allow the film to warm up to room temperature before you open the film canister. If you allow warm air to enter the film cartridge when its cold, the moisture will condense on the film and you'll get spots. If you are working in freezing conditions, it is OK to use cold film providing you remember to seal it back in its canister before you let it warm up. If you are using cold film, advance and rewind it carefully. It is brittle when cold and you don't need broken film.

At the hot end of the temperature scale, at some point heat will melt the gelatinous layer of the film and you will be out of business. You must keep the film below that temperature.

10. PROCESSING.

Unless you used slide film, there are two steps in processing the film. First, the film has to be developed. These days, this is pretty much automatic and the 1-hour photo places can do a good a job. There is some risk here, as they can also do a poor job by using old chemicals and dirty machinery. You need to inquire into their quality control before you give them your film.

Next, the negatives need to be enlarged and printed. Here, the 1-hour photo place will do a reasonable job with their automatic machinery, but they are not the place to go for custom work. If all you want is a set of quick prints, use the 1-hour place. If you want serious enlargements for the report or legal work, take your negatives to a professional processing lab.

Earlier, it was mentioned that color photography is not appreciably more expensive than black and white photography. That's true as long as you keep the enlarging and printing under control. Typically, you shot perhaps a hundred pictures of your wreckage. Of those hundred shots, many of them were duplicates with slight differences in exposure or lighting. Many of them were taken on speculation and turn out to be of no value to the investigation. The secret to controlling costs in this business is to never enlarge and print a picture that you don't need.

A professional photographer works primarily with negatives. The rest of us need to look at a positive to see what the picture is supposed to look like. One of the easiest ways to do this is to have the negatives printed on proof sheets. For 35mm work, a proof sheet is a contact print of up to 30 negatives on an 8 x 10 print. The one in the upper left corner is, guess what, the roll identification picture. The rest of the

negatives are in order with the edge numbers visible. These correlate with your photo log. This is an easy way to examine an entire roll of film at once and pick the ones you really want to enlarge. It is also an easy way to file your proof sheets, negatives and captions in a way that is easy to retrieve. Caution. An 8 x 10 proof sheet will only hold 30 negatives, so shoot 24 exposure rolls.

Where do you get proof sheets? From professional processing labs. Some 1-hour photo places may have the capability. Ask.

11. PRACTICE.

Perhaps the biggest difference between a tourist and a professional is that the tourist is unwilling to waste an entire roll of film just to find out if the camera (or flash) really works. The professional always runs a test roll on any new piece of equipment before he takes it out on a job. If you want to add the "painting with light" scheme to your bag of tricks, waste a roll of film (and an evening) finding out exactly how it works. If you are not clear on "fill-in flash," practice. Don't expect to learn how it's done by photographing a real accident.

When you are practicing, either use slide film, which will capture all your errors, or proof sheets, which will capture most of them. A proof sheet has to be exposed for the average density of all the negatives and there is no correction of exposure for individual negatives. While you are practicing, keep track of what you did. If you don't, you'll have to buy another roll of film as you won't know what to correct.

12. SUMMARY.

Aircraft accident photography is fairly demanding. Most of the time, you only get one opportunity to get the picture. By the time you find out that the results are no good, it is too late to recreate the accident scene. For this reason, you need to have dependable equipment and be very familiar with its capabilities. You need to have a plan for covering the wreckage photographically and you need an system for keeping track of what you are doing. Photography itself really isn't very difficult and you should have no trouble taking better pictures than those normally found in aircraft accident reports. Happy shooting!

7

Accident Diagrams

1. INTRODUCTION.

It is common practice to prepare one or more diagrams for inclusion in the written report. These are the types of diagrams usually found.

A. FLIGHT PATH DIAGRAMS. These depict the flight path of the aircraft prior to impact. They are a useful method of correlating and displaying a lot of information about the accident derived from several different sources. See Chapter 37 for the use of a flight path diagram as an analytical technique.

B. WRECKAGE OR SCENE DIAGRAMS. These diagrams plot the location of the impact and the subsequent distribution of the wreckage. Some of the information typically found on wreckage diagrams include:

Location references (roads, buildings, runways, etc.)

Direction Reference

Scale

Elevations including contours

Impact scars

Impact heading

Human remains location

Final position of major aircraft components

Burn areas

Damage to buildings or structures

Location of eye witnesses

C. TECHNICAL DIAGRAMS. These could be anything from a cabin seating chart to a schematic of a hydraulic system. Usually, these are copied from aircraft technical manuals and overlaid with appropriate accident information.

D. FLIGHT DEPICTIONS. Based on the investigation, it may be possible to accurately depict an aircraft's flight attitude or, perhaps, the geometry of a mid-air collision. In the past, these were created by graphic artists under the supervision of the investigators. Now, they are more likely to be produced on a computer. Theoretically, if digital information from the flight data recorder is available as input to the computer, the accuracy of the result should be high. This is true up to a point. The same computer technology that allows us to display digital data in a number of different ways also allows us to manipulate it and display it in the way most favorable to our point of view. It is possible, for example, for two competing sides in an accident litigation case to use the same data to produce significantly different depictions. All depictions of this type should include information as to how they were created.

2. PURPOSE OF DIAGRAMS.

Diagrams tend to be labor intensive. They may look nice in the report, but the time spent in preparing them sometimes outweighs their usefulness. Any diagram should serve a purpose. If it doesn't, you probably don't need it. Here are the common uses of diagrams.

A. ANALYSIS. A good scene or wreckage diagram is essential to most calculations involving crash dynamics and survivability. Likewise, a diagram showing burn areas is helpful in inflight fire analysis. If inflight structural failure is suspect, a wreckage diagram may be essential.

B. LOCATION AND INVENTORY. If the wreckage is spread out over a large area, a "map" of the scene can be very useful. If the wreckage must be moved, a scene diagram may be the only accurate record of how the wreckage was distributed. If remains identification is a problem, a diagram of where the remains were found is an essential first step and is very helpful to the pathologist.

C. CORRELATION OF INFORMATION. A flight path diagram can be very helpful in depicting information from various sources such as the FDR, CVR, ATC radar, A/G communications, and so on. A plot of this information on the same diagram with it all correctly referenced by time and location enables the investigator to see how a number of different events are related to each other.

D. EXPLANATION. Sometimes, the investigators don't really need a diagram, but the report reader does. Time spent creating the diagram may be time saved in trying to describe the situation without a diagram.

3. DIAGRAMMING TECHNIQUES.

If you decide that you need a diagram, your next decision involves how good the diagram should be. Do you want it to be pretty, accurate or useful? If you want it to be useful, meaning that you need it as soon as possible while the field investigation is still in progress, it is probably not going to be either pretty or accurate. If, on the other hand, you are concerned with resolving claims resulting from the accident, you would probably vote for accuracy and accept the necessary delays in preparing the diagram. If your sole purpose is to make the report look good and assist the reader, vote for pretty and hire a professional surveyor or graphic artist to do it for you.

Let's talk about accuracy for a minute. In diagramming wreckage, for example, you really don't need to know exactly where each piece of wreckage is on the face of the earth. What you need is relative accuracy. You need to know where each piece of wreckage is in relation to the other pieces. How accurately do you need to know this? Are we talking surveying equipment accuracy? Tape measure accuracy? Pace-it-off-and-count-the-steps accuracy? You probably don't need surveying-type accuracy, although some surveying equipment (a laser range finder, for example) could be useful if long distances are involved. A tape measure is useful in laying out the basic diagramming scheme. If that is done with reasonable accuracy, then pacing off the distances to the wreckage parts is probably good enough. Regarding angles and headings, you can usually get within five degrees with simple equipment and that should be good enough for most calculations.

If you would like to prepare the diagram today and have it ready for distribution to the other investigators tomorrow, you are going to have to use some fairly simple methods and sacrifice some accuracy. The rest of this chapter describes how to do that.

4. FLIGHT PATH DIAGRAMMING TECHNIQUES.

Most flight path diagrams are drawn on a map. Unless the flight path covers a large area or it is essential to show navaids and aeronautical features, aeronautical charts usually do not make good flight path diagrams. The scale is too small and they don't show contours or man-made cultural features. In the United States, the chart used most often is called a 7.5 minute Topographic chart published by the Department of Interior, Geological Survey. This has a scale of 1:24,000 and it shows contours, elevations and cultural features very accurately. These charts are available for the entire United States and can be ordered from the U.S. Geological Survey in either Denver, Colorado or Reston, Virginia. An easier (and quicker) way to get the charts is to call local engineering blueprint services and find out who stocks the charts locally. Since these charts are widely used by contractors and developers, someone in every population center stocks them for that area. Other countries have similar charting systems.

Other sources of large scale maps include city and county offices and sporting goods stores. In some areas, the maps used by campers and hunters are the best available. In some states, the State Forest Service may have excellent charts of forest areas. If the accident occurred in the water, but near land, accurate charts should be available from marine supply stores, harbor offices or the U.S. Coast Guard. If the accident occurred on or near an airport, there is almost certainly a scale diagram or aerial photograph of the airport available.

Once you obtain a chart, it is probably best to overlay it with tracing paper and plot your diagram on that. This allows corrections to be made and the final product can be transferred to the map itself.

5. WRECKAGE DIAGRAMMING TECHNIQUES.

Creating a diagram of an aircraft accident scene looks pretty hopeless, particularly if you expect to have it done by tomorrow. Actually, you can diagram a fairly large scene in a few hours if you are organized and stick to simple techniques.

First, look for simple solutions. A good vertical aerial photograph of the scene just might serve as a diagram all by itself, particularly if you have included a size reference in the scene which will permit measurements. In any case, the photograph will be useful in improving the accuracy of whatever diagram you draw yourself. If your wreckage is spread out over a quarter of a mile or more, start with a topographic map of the area. The scale of this map is too small to be used as a scene diagram, but you can use a macro lens to photograph the portion of the map where the wreckage is and enlarge the results. You can also copy that portion of the map on a photocopier with an enlarging capability and repetitively enlarge it to a size that suits your needs. This way, you get an accurate large scale map of the area and all you need to do is plot the wreckage on it. If your wreckage is located on or near some man-made feature (airport, city, residential area) don't waste your time drawing roads and runways. Someone has already done that for you and you'll save time by plotting your wreckage on an existing chart or diagram.

Second, go for relative accuracy rather than absolute accuracy. No one really cares about the latitude and longitude of the left wing. They are more likely to be interested in its relationship to the right wing.

Third, keep it simple. You are going to need some help and you haven't got time to teach surveying techniques.

Fourth, pick the simplest possible method. Here are several.

A. GRID SYSTEM. This works for a fairly compact accident and it is useful for wreckage that, due to terrain or vegetation, is very difficult to move through easily. It can also be used for underwater wreckage, although it is difficult to imagine what purpose an underwater diagram would serve. Basically, you overlay the wreckage with a grid of colored tape or rope. Each grid square is numbered and the wreckage within each square can be sketched with reasonable accuracy.

B. POLAR SYSTEM. This is also useful for a compact wreckage area. Plant a stake or pole in the center of the wreckage and extend lines in various compass directions from the center stake. Mark each line with distances and plot each major piece of wreckage as a direction and distance from the center stake. This, incidently, is the method commonly used to plot underwater wreckage with the salvage ship as the center.

C. SINGLE POINT SYSTEM. This is identical to the polar system except that the stake is at the edge of the wreckage or perhaps at the impact point or a referenced cultural feature. The lines are extended through the wreckage in a pie-shaped fashion and the wreckage is plotted as a direction and distance from the stake.

D. STRAIGHT LINE SYSTEM. This is by far the most commonly used and (fortunately) the simplest of all the systems. Because of its usefulness, it will be described in some detail. This method is depicted in Figure 7-1.

1. Establish a point of reference that can be later located on a map. If you can't do this, consider using a portable GPS (Global Positioning Satellite) receiver or a portable Loran-C receiver.

2. Decide where you are going to start your stake line. It would be helpful if it started from the point of reference, but it doesn't have to. Just locate the stake line to the point of reference by direction and distance. It would also be helpful if the stake line started at the point of impact ran through the wreckage

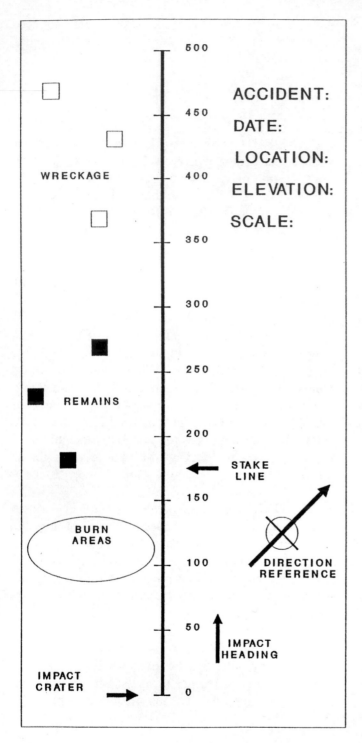

Figure 7-1.
Basic Requirements - Straight Line Method

every 20 meters.

4. Measure the compass heading of the stake line.

5. With a straight line established through the wreckage, you are now ready to locate each major piece of wreckage as a distance down the line and a distance right or left of the line and perpendicular to it. Distances between stakes can be estimated within four or five feet and distances from the line to the item of wreckage can be paced off with reasonable accuracy. Remember, you are trying to locate things correctly in relationship to each other. Absolute accuracy is not necessary.

6. This is a group activity and the more people involved, the quicker it can be done. Assemble all available investigators and assign each an area of the wreckage. It doesn't matter if they overlap slightly. Explain the measuring system described above and ask them to write down each major item of wreckage in their area with the identification of the wreckage item followed by distance down the stake line, R or L, and distance perpendicular to the line. As an example, LMLG 125 L 30 would mean that the left main landing gear was found 125 feet down the line and 30 feet left of it. This is easy to do and the most time-consuming part is the actual identification of the piece of wreckage. For this reason, mechanics who know the parts of the aircraft may be more useful than engineers who don't.

7. Collect the notes from all participants and begin plotting. Start with a straight line scaled with the same distances as your stake line. Plot each item of wreckage in its correct location relative to the line based on the notes of your participants.

8. Now you've got the wreckage correctly plotted relative to itself and the chosen reference point. If your stake line did not start at the impact point, draw a new line from the impact point through the wreckage on the impact heading. The easiest and probably most accurate way to do this is to just connect the impact point and the farthest piece of heavy wreckage. Measure the angular difference between this line and your arbitrary stake line which you measured with a compass in step 4 above. This is the impact compass heading, or very close to it. You can refine this with variation and deviation and obtain an accurate aircraft heading.

on the impact heading. This is also not necessary. Start the stake line at any convenient point and run it though the wreckage so that you can see from one stake to the next.

3. Using a tape measure, put stakes every 50 feet and mark each with the distance from the starting point. If you are using the metric system, put stakes

9. You can now erase the stake line and even pull up the stakes, if you want to. You have the wreckage correctly referenced to the impact line and that's what you really want.

10. Make any necessary adjustments to your diagram based on aerial photographs.

11. Add a scale, elevation information, geographical location notes, date of accident and so on.

12. Print and distribute your diagram. You're done and it didn't take long at all.

6. EQUIPMENT.

A. LINEAR MEASURING EQUIPMENT. Your basic measuring tool is a 100 foot tape measure. Cloth is preferred over steel as there is no chance that it will conduct electricity as you run it through the wreckage. Steel lasts longer. For long distances, an optical range finder can be useful. These are available through surveying equipment supply stores. For really long distances, a laser range finder is handy.

B. VERTICAL ANGLE MEASURING EQUIPMENT. There are a variety of instruments available that will measure vertical angles. These are used by surveyors and people in the lumbering and forestry industries and they tend to be expensive. An easier solution is to use an air navigation plotter; the one with a ruler on one edge and a protractor on the other. Put a string through the protractor hole and tie a weight to the other end of the string. As you sight along the ruler edge of the protractor, the string is giving you angular measurement on the protractor scale.

C. HORIZONTAL ANGLE MEASURING EQUIPMENT. This is called a magnetic compass. A military-type lensatic compass will provide desired accuracy.

D. PLOTTING EQUIPMENT. Graph paper or quad paper with a grid already printed makes plotting easier. Use a decimal grid system (ten to the inch, for example) printed in light blue. When you print your diagram, the grid will not reproduce. If you use tracing paper (with or without grids) available at blueprint supply stores, you can draw the diagram in pencil instead of ink. This makes corrections a lot easier. Other plotting equipment is a ruler or a straight edge and a pair of dividers. A scaled engineering ruler is helpful, but not essential.

7. SPECIAL TECHNIQUES.

A. SLOPE OF TERRAIN. Pick a friend about your height and have him take one end of the tape measure up the hill. Use the plotter-and-string trick to sight from you to him. This gives you the average terrain angle over a hundred foot distance.

B. ANGLE OF IMPACT CRATER. Lay a straight edge down the side of the crater that you want to measure. Hold the ruler of the plotter on the straight edge and read angle from the string on the protractor.

C. DEPTH OF IMPACT CRATER. Extend a straight edge over or across the crater and measure depth down from the straight edge.

D. HEIGHT OF TREES, WIRES OR POLES.

Method 1. If there are shadows, put a stake in the ground and measure the height of the stake and the length of its shadow. Measure also, the shadow of the power pole. This is a simple problem in proportionality. The height of the pole will equal the product of the length of the pole's shadow times the height of the stake divided by the length of the stake's shadow.

Method 2. Measure the angle from the ground to the top of the pole (tree, wire) at a known distance from the base. Use the plotter-and-string trick to get the angle. Measure the distance with the tape measure and adjust it to compensate for your height as you stood to measure the angle. The height of the object will be the distance from the base times the tangent of the measured angle. If you want to do this without resorting to the trig tables, just find the spot where the angle to the top is 45 degrees. Then the distance from the base equals the height of the object.

Method 3. If you can stand directly beneath an object, measure the height directly with an optical range finder. This works well with power lines.

8. SUMMARY.

Some aircraft accidents just don't need wreckage diagrams. In a mid-air collision, for example, the accident occurred at some point in space. The actual distribution of the wreckage as it fluttered to earth is random and won't tell us much. A wreckage diagram might be useful for legal purposes, but it is unlikely to be of any value to the accident investigators. On the other hand, a flight path diagram of a mid-air collision could be very useful.

In general, don't make diagrams without reason. Even the simple ones take a certain amount of time and effort. If you do decide to make one, keep it simple and useful to you, the investigator.

8

Fire Investigation

1.INTRODUCTION.

Fire can either be the cause of an aircraft accident or result from it. As a cause, it is fairly rare considering all causes of all aircraft accidents. Modern aircraft design practices separate, as much as possible, flammables from sources of ignition and provide for containment or extinguishment of fires that do occur.

Nevertheless, fires can occur in engines, engine bays, cockpits, cabins, cargo holds, wheel wells, and fuel tanks. If the fire is not contained or extinguished and the aircraft crashes, a post-impact fire invariably results which destroys considerable evidence of the nature and origin of the in-flight fire.

While the inflight fire may be rare, the post impact fire is not. Once the fuel tanks of the aircraft are ruptured, the resulting fuel mist is almost certain to encounter one or more ignition sources in the form of hot engine parts, sparks or electrical arcs. Even when there was no inflight fire (or reason to suspect inflight fire) the post impact fire can destroy a lot of evidence related to the aircraft structure or systems. Thus knowledge of how materials behave in the presence of fire is useful to the investigator even though the fire itself is not suspected as a cause.

This chapter will deal with some basic concepts about fire chemistry, the behavior of aircraft fluids and materials, the difference between inflight and post impact fires and the characteristics of inflight explosions.

2. DEFINITIONS.

The following terms and definitions will be used throughout this chapter.

A. FIRE. This is a collective term for an oxidation reaction producing heat and light. There are several types of fires.

B. DIFFUSION FLAME OR OPEN FLAME. A rapid oxidation reaction with the production of heat and light. A gas flame or a candle flame is termed an open flame. So is the burning of residual fuel following the initial "fire ball" during an aircraft impact.

C. DEFLAGRATION. Subsonic gaseous combustion resulting in intense heat and light and (possibly) a low level shock wave. Most aircraft impact "fireballs" are technically deflagrations.

D. DETONATION. A supersonic combustion process occurring in a confined or open space characterized by a shock wave preceding the flame front.

E. EXPLOSION. Detonation within a confined space resulting in rapid build-up of pressure and rupture of the confining vessel. Explosions may be further categorized as either mechanical or chemical. A mechanical explosion involves the rupture of the confining vessel due to a combination of internal overpressure and loss of vessel integrity. A chemical explosion involves a chemical reaction resulting in catastrophic overpressure and subsequent vessel rupture.

F. FLASH POINT. This is the lowest temperature at which a material will produce a flammable vapor. It is a measure of the volatility of the material.

G. AUTO-IGNITION TEMPERATURE. This is also called Ignition Temperature or Autogenous Ignition Temperature. It is the temperature at which the material will ignite on its own without any outside source of ignition.

H. FLAMMABILITY LIMITS. These are generally listed as the upper and lower flammability or explosive limits. These describe the highest and lowest concentrations of a fuel in air by volume percent which will sustain combustion. In other words, a fuel-air mixture below the lower limit is too lean to burn while a mixture above the upper limit is too rich to burn. This is of little consequence in a post impact fire, because all possible combinations of fuel-air mixtures will be present. In considering inflight fires, though, the upper and lower limits may be useful as they vary with temperature and altitude. Thus, for an inflight fire to occur, the aircraft must be operating in a temperature/altitude regime where a combustible fuel-air mixture can exist. See Figure 8-1.

I. FLASHOVER. This term is used to describe the situation where an area or its contents is heated to above its autoignition temperature, but does not ignite due to a shortage of oxygen. When the area is ventilated (oxygen added) the area and its contents ignite simultaneously, sometimes with explosive force.

3. FIRE CHEMISTRY.

Fire is essentially an oxidation reaction. In order for fire to occur, four conditions must exist.

1. Combustible material

2. Oxidizer

3. Ignition

4. Enough heat or energy to sustain the reaction.

A. COMBUSTIBLE MATERIAL. The flammable liquids (fuel, hydraulic fluid) used on aircraft do not burn as liquids. Their vapors burn. Thus the fire chemistry involving a liquid is essentially a gaseous reaction and the characteristics of the liquid itself are of little consequence. Since combustion takes place in the vapors above the surface of the liquid (or in a mist formed of the liquid) the key characteristic of a liquid is its tendency to form flammable vapors. ("Vapors" refer to the gaseous form of a substance. "Mists" refer to suspended liquid droplets of a substance.) This is a function of the liquid's volatility and temperature. A highly volatile liquid such as AvGas produces flammable vapors at a very low temperature. Jet-A, on the other hand, is much less volatile and requires elevated temperatures to produce flammable vapors. See Figure 8-2. Flammable vapors can also be created in the form of a mist. This is basically what happens in a jet engine when fuel is sprayed into the combustion chamber in a fine mist. It is also what happens when a leak or rupture occurs in a hydraulic system. The fluid under a pressure of 3000 psi can spray out in a fine mist and form a flammable vapor. A misting fluid can generally be ignited at a lower temperature than the fluid's vapor flash point.

Solid materials used on the aircraft may burn, char or melt. Some may burn or melt depending on physical composition. A steel forging, for example, would melt before it burned; but a pile of thin steel shavings would probably burn before it melted. If the material burns, the burning takes place on the surface of the material only.

In the process of charring or burning, some cabin materials may produce a thick black smoke which impedes both breathing and vision and, therefore, escape. Others, in the process of burning or charring, may decompose into other elements or compounds. An organic material (wool, for example) will usually decompose into a compound containing hydrogen cyanide. A material containing a carbon molecule is almost certain to produce carbon monoxide as a byproduct of combustion. Products of decomposition may have different flash points than the original material and may contribute to flashover.

During the investigation of fires involving solids, it is well to remember that almost any substance found on an airplane will react, somehow, to heat. The nature of the reaction and the temperature at which the reaction occurs is a function of the material itself.

B. OXIDIZER. Since air is 20% oxygen, ordinary air is sufficient to support most fires. If a fire occurs in flight and the fire is exposed to the slipstream, oxygen is added and the fire will burn faster and

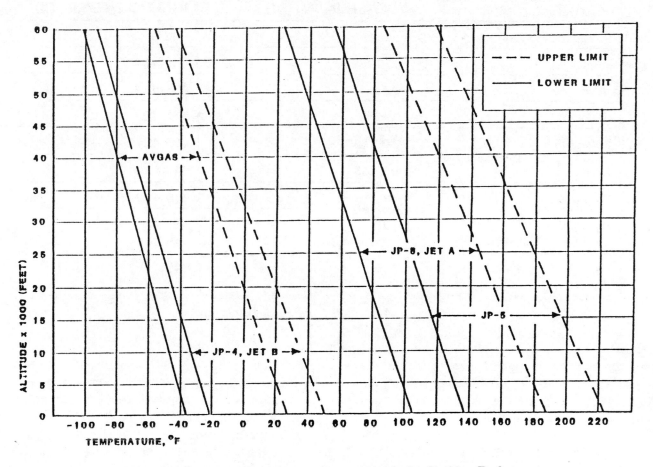

Figure 8-1. Combustible Limits Versus Altitude for Various Fuels

hotter.

Although the percentage of oxygen in air is constant at any altitude, the partial pressure is not. This is reduced as atmospheric pressure is reduced and may be thought of as a reduction in the quantity of oxygen available. At some point, depending on the volatility of the fuel and the temperature, the quantity of oxygen available becomes insufficient to support the oxidation reaction. This may be thought of as a problem in fuel-oxygen mixture or, more properly, upper and lower flammability limits. In theory, an exactly correct mixture of fuel vapor and oxygen would be called a stoichiometric mixture and it would result in a complete and perfect reaction of the fuel and oxygen molecules. There would be no byproducts in the form of smoke. Surrounding this perfect mixture is a range of flammability defined by the upper and lower flammability limits. Above the upper limit, the mixture is too rich to burn. Below the lower limit, it is too lean. Jet-A, for example, cannot form a flammable mixture at certain altitudes and temperatures (See Figure 8-1) because the available oxygen will always form a mixture with the Jet-A vapor that is too lean.

Thus if the fuel, flight altitude and temperature are known, the investigator can make a determination whether an inflight fire was theoretically possible. This does not apply, of course, to fuel that is misted in the combustion chamber (or cylinders) and supported by compressed air (oxygen) through the air intake. That's why the engines keep running even at high altitude. This theory also does not apply to fires originating within the pressurized compartment or an oxygen-rich area; or if the fire starts inside the plane and is then exposed to the slipstream where additional oxygen can be continuously added.

Another factor not taken into account in Figure 8-1 is ignition dwell time. This is the time flammable vapors must be in contact with an ignition source before fire results. This is a function of absolute pressure and varies greatly with altitude.

C. IGNITION. In order for a fire to ignite, the ignition source must first raise the temperature of the combustible vapors (or material) in its immediate vicinity to the ignition temperature of the material. Sparks from the aluminum alloys, for example, are

FLUID	FLASH POINT (°f)	IGNITION TEMP. (°f)
Aviation Gasoline	-45	825-960
JP-4	-20 to -30	435-484
JP-5	147-150	435-484
JP-8	115	435-484
Jet A/A1	105-140	435-484
Kerosene	95-145	440-480
Engine Oil	437	440-480
Hydraulic Fluid (Petroleum-Based)	195	437
Hydraulic Fluid (Synthetic)	320	945

Figure 8-2. Fluid Flash Points

generally incapable of igniting turbine fuel vapors; they are not hot enough. Likewise, it is pointless to hypothesize a bleed air duct leak as an ignition source if the temperature of the bleed air is not above the ignition temperature of the fuel.

D. HEAT OR ENERGY TO SUSTAIN THE REACTION. If the ignition process provides this energy, the fire will be self-sustaining. If not, the fire will go out when the source of ignition is removed. Most synthetic hydraulic fluids ("Skydrol" and Mil-83282) exhibit this characteristic. They will burn, but only in the presence of continuous ignition; assuming, of course, that the fluid has not been heated to above its auto-ignition temperature. Petroleum based hydraulic fluids will continue to burn after the ignition source is removed.

Once properly ignited (and in the presence of sufficient heat or energy) the fire will continue until one of four events occur:

1. The combustible material is consumed or removed. An example would be an engine fire wherein the fuel is shut off at the firewall or tank. The fire burns the fuel remaining in the lines--and then goes out.

2. The oxidizing agent concentration is lowered to below that necessary to support combustion. This can occur in a closed area such as a cargo hold or a passenger cabin. A fire starts and consumes enough of the oxygen so that the remaining mixture is too fuel-rich to burn. At this point, the fire goes out. This led to the early design philosophy on the need for fire extinguishing systems in aircraft cargo holds. The thinking was that any fire would rapidly consume the available oxygen and extinguish itself. Unfortunately, this does not hold true for very large cargo holds or very intense fires that burn through the aircraft skin and obtain additional oxygen from the slipstream. It also doesn't work if the cargo itself contains oxidizers.

What sometimes occurs, though, is that the fire, while it is consuming oxygen, raises the temperature of the surrounding materials to above their auto-ignition temperature. Auto-ignition does not occur until the cabin or cargo hold is opened and oxygen is added. Now we have a violent situation called, "Flashover" which is the instant ignition of all materials which have been heated above their auto-ignition temperatures.

3. The combustible material is cooled to below its ignition temperature. This seldom occurs by itself. Once heated, the material stays heated until it is consumed. This is one of the basic methods of fire fighting, i.e. lower the temperature of the material by cooling it with an extinguishent.

4. The fire is chemically inhibited. The fire extinguishing chemical Halon works by chemically terminating the reaction between the fuel and oxygen.

4. LEVEL OF BURNING REACTION.

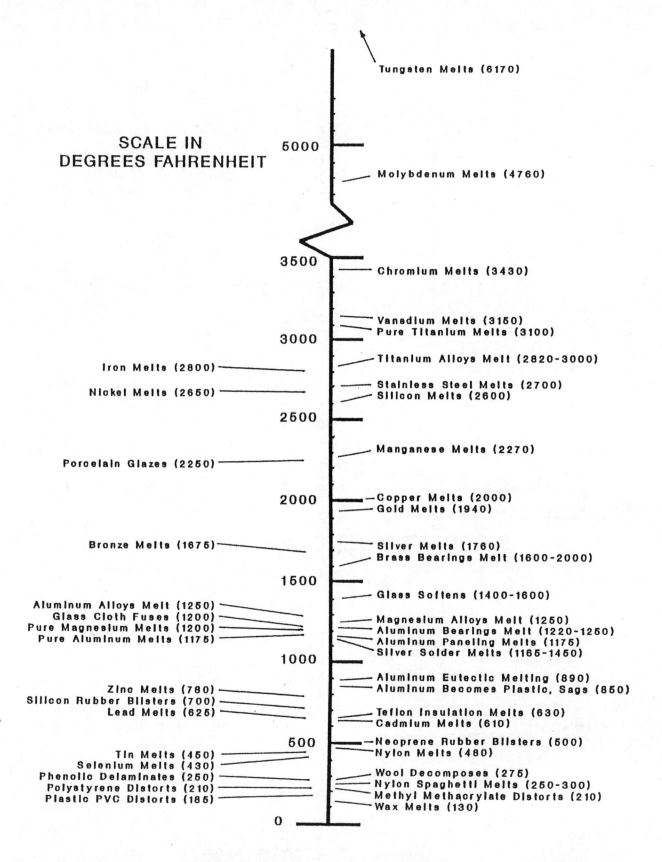

Figure 8-3. Melting Points of Aircraft Materials

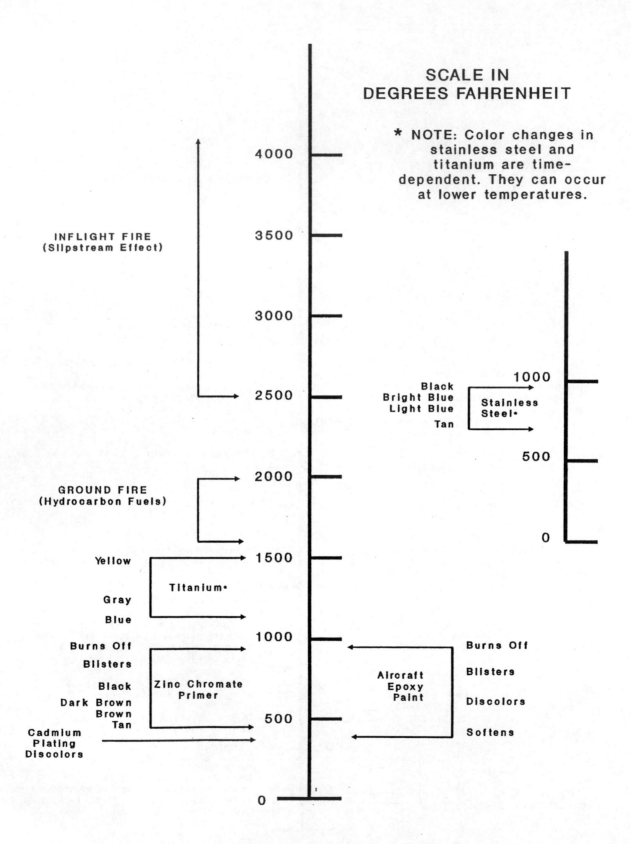

Figure 8-4. Useful Temperature Ranges

There are several different types or levels of burning. These should be understood as the terms tend to be used interchangeably. The term, "explosion." for example, may mean something entirely different to a witness than it does to a fire specialist. These terms are defined earlier in this chapter.

A. DIFFUSION FLAME OR OPEN FLAME. This is the lowest level of burning reaction and is analogous to a candle flame.

B. DEFLAGRATION. Most "fireballs" seen immediately after an aircraft impacts are deflagrations.

C. DETONATION. This is the third level of burning reaction and differs from an explosion only in that it is unconfined.

D. EXPLOSION. This is a form of detonation occurring in a confined space and may be either mechanical or chemical. The extensive damage is due largely to overpressure and a supersonic shock wave.

5. CHARACTERISTICS OF AIRCRAFT FLUIDS AND MATERIALS.

A. FLUIDS. Characteristics of common aircraft fluids are listed in Figure 8-2. The flash point is the lowest temperature at which the fluid will produce a flammable vapor. The auto-ignition temperature is the temperature at which the fluid vapors will ignite (assuming sufficient oxygen) without any outside source of ignition. Fluid flash points are plotted against a temperature scale in Figure 8-2. Flammability limits of common aircraft fuels are shown in Figure 8-1.

B. AIRCRAFT MATERIALS. The melting points of metals and materials commonly used in aircraft are shown in Figure 8-3. The behavior characteristics of some aircraft materials and some useful temperature ranges are plotted in Figure 8-4.

C. COMPOSITE MATERIALS. Technically, any non-homogenous material could be called a composite material. The principal composites currently used in aircraft construction are fiberglass or carbon fiber. These may be used alone, in combination with each other, or sandwiched around a metallic or non-metallic core.

When exposed to fire, fiberglass will melt at around 1200 degrees f. The reaction of a carbon fiber composite depends on the resin in which the fibers are imbedded. The fibers, being pure carbon, aren't going to decompose any further. The resin will melt which liberates the fibers or reduces their structural integrity. The temperature at which the resin melts varies with the resin; a characteristic frequently considered proprietary among manufacturers. Most resins will "burn out" at 1100 degrees f. or below.

D. ALUMINUM ALLOYS. Most aircraft metal structure is about 95% pure aluminum alloyed with copper or zinc and small amounts of other elements. The behavior of the structure in a fire depends on the alloying elements, the configuration (heavy forged or cast structure vs thin paneling), the temperature and time of exposure, and the amount of stress on the structure.

Initial heating. The principal result of exposure to heat is loss of strength. This is a function of time and can occur rapidly at a high temperature or slowly at a low temperature. If some assumptions about time of exposure can be made, then it is possible to test the hardness of the alloy and calculate the temperature as a function of loss of hardness over that specified for the alloy.

Eutectic melting. The is the lowest melting temperature of any of the alloying metals. At this temperature, an interesting phenomena called the "broomstraw effect" occurs if the aluminum part is highly stressed. In this case, the failed area will show delamination along grain boundaries and would resemble the fibers one would see in a "green stick" fracture. This is considered highly indicative of inflight fire, the assumption being that the heating occurred inflight and the stress occurred at impact. This can also occur if the part is under high stress as it is normally used in the aircraft and then heated. Thus a "broom straw" fracture is not a 100% guarantee of inflight fire. The eutectic melting temperature of aluminum alloys is approximately 890 degrees f.

Melting. Aluminum alloys will become plastic around 850 degrees f. and begin to sag. This is most commonly seen in post impact fires and can produce a false appearance of structural bending. At about 1175 degrees f., the aluminum alloy will melt completely and the molten aluminum will either follow gravity or the slipstream. The slipstream will reduce molten aluminum to very small droplets which may impinge on the aircraft structure. Droplets of aluminum formed by gravity will be much larger.

6. SOURCES OF FUEL.

A. AIRCRAFT FUEL. This is the most obvious source. Possible locations would be the fuel tanks, tank vent system, fuel lines and engines. At impact, fuel from ruptured tanks will form a mist that can be ignited by friction sparks, engine exhaust or engine hot section parts. Fuel vented or released in flight has the appearance of white smoke.

B. OIL. Aircraft engine oil is not a common source of fuel for a fire inasmuch as it is confined to the engine and separated from the rest of the aircraft by a firewall. Oil leaked or vented into the exhaust gases will produce a grayish-white smoke. Oil ignited within the engine will produce a black smoke.

C. HYDRAULIC FLUIDS. All hydraulic fluids currently in use can be ignited and will sustain combustion if the temperature is high enough. If the source of the hydraulic fluid is a leak in the system, then the fluid will be atomized under pressure and the resulting mist can be ignited at temperatures well below the flash point of the fluid. This is the basic scenario for a wheel fire wherein the brakes and wheels have been overheated and there is a leak in the hydraulic brake system.

D. BATTERY GASES. All aircraft batteries outgas hydrogen under certain circumstances. If the battery is improperly vented, hydrogen can accumulate in the battery compartment and be ignited, usually by an electrical arc. The resulting fire is likely to be a low-order detonation and will probably consume all the available hydrogen. The amount of damage done will depend on how much hydrogen was accumulated and how much it was confined before ignition. There will be no byproducts in the form of smoke or soot because there was no carbon present in the reaction.

E. CARGO. If the fire originates in the fuselage, the aircraft cargo is always suspect. Cargo can contain both fuel and oxidizers.

F. WASTE MATERIAL. The waste materials generated in aircraft lavatories or galley operations must also be considered as possible sources of fuel.

7. SOURCES OF IGNITION.

The ignition of a fire depends on the flammability of the fuel and the temperature or energy level of the ignition source. Possible ignition sources are listed below with comments where appropriate.

1. Engine hot section parts.

2. Engine exhaust.

3. Electrical arc.

4. Overheated equipment.

5. Bleed air system

6. Static discharge. Static electricity is generated by the contact and separation of dissimilar materials. An aircraft flying through the air will generate static electricity. Fuel flowing through a hose or sloshing in a tank may generate static electricity depending on the electrical characteristics of the fuel. If the difference in the electrical potential becomes great enough, static discharge may occur in the form of an electrical arc at the instant of separation of the materials. This discharge process varies with temperature and humidity and is not entirely predictable. Likewise, the energy level of the discharge is unpredictable and it may or may not ignite flammable vapors. It has been found that some types of foam installed in fuel tanks for fire protection reasons have a high electrical resistance and can generate a static discharge which can ignite flammable vapors in the tank. This has resulted in fuel tank fires (probably mini-low order detonations) which, because of the foam, were localized to the immediate area of the discharge and quickly extinguished.

7. Lightning. Normally, lightning striking an aircraft does not cause significant damage. The lightning strikes an extremity (nose, wing tip); passes through the metal structure; and exits another extremity. If the structure is properly bonded and the bond is maintained, there will be no arc as the stroke passes through. If the structure is not properly bonded, the stroke will arc across that area. If the arc occurs in the presence of flammable vapors, usually a fuel tank or fuel vent system, a fire can occur. The point at which this occurs is seldom the point at which the stroke enters or exits the aircraft. This is helpful to the investigator as the evidence of the lightning strike will not necessarily be destroyed by the fire.

The mark left by a lightning strike on metal structure can be small and difficult to find; sometimes no more than a pin hole. Composite structures are a different story. Since the composite is non-conductive (except for lightning diverter strips or a layer of wire mesh on the composite), the damage tends to be much more obvious.

8. Hot Brakes or Wheels. If overheated, the brake/wheel temperature will probably be hot enough to ignite a spray of hydraulic fluid. If the conditions that lead to the overheated brakes are known (time, gross weight, stopping distance, etc.) it may be possible to calculate maximum temperatures achieved. Overheated aircraft wheels usually do not reach their maximum temperature until approximately 15 minutes after they are stopped. Actually, the brakes are already at their maximum temperature, but it takes time to transfer this heat to the wheels and the air in the tires.

9. Friction Sparks. The energy in a friction spark depends on the metal involved. Tests have indicated that aluminum or aluminum alloys sparking on concrete will not produce sparks of sufficient energy to ignite fuel vapors. The sparks from stainless steel, magnesium and titanium will ignite fuel vapors.

10. Aircraft Heaters.

11. Auxiliary Power Units.

12. Inflight Galleys/Ovens/Hot Cups

13. Smoking Materials. These have been the ignition source of several fires on commercial aircraft, particularly in the waste containers of lavatories.

Sometimes the source of ignition is obvious, but frequently it is not. Locating the source is best done by first locating the point of origin of the fire. For an inflight fire exposed to the slipstream, the point of origin is likely to be at or near the damaged area furthest forward on the aircraft. If the inflight fire was contained (not exposed to the slipstream), the point of origin is at or near the lowest point of the fire damage. If the fire is in a ventilated area, but not exposed to the slipstream, the point of origin will be upstream toward the source of the ventilation. In a classic fire pattern, the fire will progress upward from the point of origin in a V-shape and the smoke will initially accumulate on the ceiling and move downward along the walls or sides of the aircraft. Unfortunately, these patterns are frequently destroyed in the post-impact fire.

8. INFLIGHT FIRE VS POST IMPACT FIRE.

The key question in most aircraft fire investigations is, "Was there an inflight fire?" This can be a difficult question to answer because there is almost always a post impact fire which tends to mask or destroy the evidence of the inflight fire.

Unless the wreckage is totally consumed by the post impact fire, there should be some evidence of an inflight fire if one existed. Caution is advised because in any aircraft fire there will always be isolated clues that individually could be interpreted as substantiating inflight fire. Single clues are not good enough. A pattern of consistent evidence will be necessary to support an argument for inflight fire.

A. INDIRECT VS DIRECT EVIDENCE. Indirect evidence is evidence that strongly points to an inflight fire because of the circumstances of the accident or the statements of the flight crew or witnesses. It is generally easier to examine the indirect evidence first. Direct evidence consists of positive indications in the wreckage that inflight fire existed. Direct evidence can be broken down into inflight fire effects, ground fire effects, crash dynamics, and impact effects.

B. INDIRECT EVIDENCE. Consider all known statements including radio transmissions by the flight crew. Did they know or suspect there was an inflight fire? Review all eye witness statements. Review the circumstances of the accident. Consider what the flight crew would have done if they had suspected an inflight fire. Would they have actuated an extinguishing system? Dumped cabin pressure? Donned smoke masks? Deactivated electrical circuits? The procedures vary from aircraft to aircraft, but the actions of the flight crew are sometimes easier to verify than direct clues to the fire itself.

C. DIRECT EVIDENCE. This will be examined under four major headings: Inflight Fire Effects, Ground Fire Effects, Crash Dynamics, and Impact Effects.

1. Inflight Fire Effects. If a fire occurs inflight and is contained by the aircraft structure, it will be indistinguishable from a ground or post impact fire unless there is some internal forced ventilation system that changes the characteristics of the fire. Most inflight fires, though, eventually burn through the structure and are exposed to the slipstream. This adds oxygen to the fire and creates two significant clues.

The temperature of the fire will be raised substantially. A ground fire of pooled hydrocarbon fuel burns in the 2000 degree f. range. An inflight fire of the same fuel with added oxygen will burn in excess of 3000 degrees f. This is significant because the inflight fire is hot enough to melt materials that could not be melted in a ground fire. The melting points of common aircraft materials are provided in Figure 8-3. With this information and a knowledge of aircraft structure, it is possible to plot temperature contours in the fire areas and determine if the temperatures encountered were within or in excess of ground fire range.

The products of combustion in an inflight fire will follow the slipstream. The origin of the fire will appear as a point source expanding in a cone or V-shaped pattern aft with the slipstream. Soot and molten metal will adhere to anything in the path of the slipstream. (Caution: Soot will not adhere to surfaces heated above 700 degrees f.) The molten metal may fuse to the downstream part and create the appearance of an elliptical shaped splash of metal melted into the structure. The fusion process requires that the structure be hotter than the molten metal which impinged on it. This is seldom seen except in the turbine and exhaust sections of a jet engine. It is more likely that the molten metal was hotter than the structure. This will result in adhesion which can be recognized as grain-sized bits of metal stuck to the downstream structure. At first glance, this may resemble dirt, but it can be detected by rubbing your hand over the surface. The difference between fusion and adhesion is similar to the difference between clear ice and rime ice. One is smooth; the other is rough.

Sometimes the post impact fire will mask this flow pattern. Then it may be possible to find "shadows" in the slipstream path. When soot encounters an obstruction in its aft flow, such as a rivet head, lap joint, antenna, boarding step, etc., it may leave a clean area or "shadow" free of soot on the downstream side of the obstruction. This is a good clue as to the flow pattern.

Ground Fire Effects. A post impact or ground fire, as opposed to an inflight fire, has different characteristics. First, it burns at a lower temperature. Second, the flown pattern of the products of combustion is up as modified by the local wind. The flow pattern of molten metal is down. (Caution: "Up" and "Down" are relative terms. The wreckage will not necessarily be right side up when it comes to rest.

Also, the wreckage could create a local "chimney" effect which could substantially raise the temperature of the fire.) A ground fire will normally start at a point and progress upward in a V-shaped pattern. If it is fuel fed, however, the base of the fire tends to spread out to cover the pooled fuel. This usually destroys the evidence of origin and the V-shaped pattern. Nevertheless, the downward flow and puddling of the molten metal should be a positive indication of ground fire.

2. Crash Dynamics. High speed photography of low angle impacts indicate that the fire sequence goes something like this. The fuel tanks are ruptured at impact and the fuel exits toward the rear in the form of a mist. As the aircraft slows down, the mist of fuel catches up with it and is ignited by hot engine parts. The result is a deflagration (although no one calls it that) with a fire ball, a low pressure shock wave, and heavy clouds of black smoke and soot. Anything passing through the fireball or smoke cloud at this point is instantly sooted. The aircraft comes apart and the various pieces come to rest. Those containing fuel continue to burn (open flame) and create a burn area surrounding that portion of the wreckage. High angle impact scenarios are probably similar except that everything happens faster and is confined to a smaller area.

Logic would suggest that all portions of the wreckage that came to rest in a burn area should show evidence of fire. Wreckage that is not in a burn area should not be burned. If it is, that would suggest that the burning must have occurred inflight before impact. This is true if adjacent parts and equipment for the same area also show fire damage. Be careful here, because it is possible that the part picked up evidence of fire damage (soot) as it passed through the initial impact fire ball and then landed in a non-burn area. Soot and smoke damage does not necessarily equate to inflight fire damage. Remember that we need a consistent pattern of evidence; not just a single clue.

3. Impact Effects. In examining a piece of wreckage damaged by fire, the key question is, "Which came first? The fire damage or the impact damage?"

Crumpled Parts. If evidence of fire is found inside the folds of crumpled metal, this would suggest that the fire damage occurred before impact. In a high intensity ground fire, of course, fire damage can occur throughout the part, crumpled or not.

Fracture Edges. If exposure to fire damage occurred before impact, the edges of parts fractured in the impact ought to be clean and free of soot. This can be misleading as the fracture might have occurred during initial impact and the fire exposure when the part came to rest in a burn area.

Scratches. A clean scratch on top of soot suggests inflight fire. Caution: fire exposure can occur during initial impact and scratches can occur during subsequent tumbling.

Protected Parts. If a part is normally covered by another part (a lap joint, for example) the protected area should show fire damage only if it was exposed to fire after the protection was removed during impact. Caution: if the part was painted, the protected area may only have been primed or not painted at all. In this case, the appearance of the protected area will be different and may mislead the investigator.

Buried Parts. The portion of the wreckage that is buried in the ground at the impact site should not be exposed to post impact fire. If the buried portion shows fire damage, it must have occurred prior to impact.

Rivet Holes. Metal structures frequently fail at a rivet line. If the part was exposed to fire before failure, the area under the rivet heads and the edges of the rivet holes should be clean.

Mud and Soot. In theory, the mud should be on top of the soot in an inflight fire and soot on top of the mud in a post impact fire. This can be misleading as the part can pick up soot in the initial fire ball and mud later when it comes to rest.

Molten Metal. In a post impact fire, the flow of molten metal will always be down, and probably in large drops or puddles. Inflight molten metal flow is likely to be finely dispersed and should follow the slipstream.

Consistent Fire Pattern. If a part is suspected of receiving inflight fire damage, there ought to be a consistent fire pattern across adjacent parts regardless of whether they were found in a burn area or not. Furthermore, if inflight fire is suspected in a particular part of the airplane, then all aircraft components and equipment in that area should show evidence of exposure to fire. It is not logical to hypothesize fire in an equipment bay, for example, and then find items of equipment from that bay that show no evidence of heat; much less fire.

9. SUMMARY OF FIRE INVESTIGATION TECHNIQUES.

It bears repeating that a single clue does not prove inflight fire. Isolated clues supporting inflight fire can be found in any aircraft wreckage that burned on impact. What counts is a consistent pattern of evidence.

10. INVESTIGATION OF EXPLOSIONS.

A. GENERAL. An explosion is a form of detonation which takes place in a confined space and creates large overpressures. An explosion occurring in the air space of a fuel tank would be termed a low-order detonation and the velocities produced would be something less than 10,000 feet per second. An explosive compound such as nitroglycerin, PETN, RDX, etc. will produce velocities in excess of 20,000 fps and would be termed high-order detonations.

In dealing with aircraft fuels, a low-order detonation (or explosion) is always a possibility if the flammable vapors are confined as they might be in a tank or a vent system. The results of this type of explosion is a function of where it happens to occur and not the material involved. The investigation techniques are those already discussed for inflight fire. This section will deal with high-order detonations or explosions which are distinctive due to the high velocities involved.

There is nothing on any airplane as built that will produce a high-order detonation. If one occurred, it occurred in something (munitions, explosive device, etc.) that was carried by the plane or placed on board it. Any ordnance carried by military aircraft is obviously suspect. If it wasn't carrying any (or the ordnance is accounted for) then the explosion must have been caused by some explosive device placed on the aircraft and is quite possibly sabotage.

If sabotage is suspected, then the investigation begins to follow a different path. While still technically an accident, it is an accident resulting from a deliberate criminal act; not an accidental set of circumstances. On a world-wide basis, there have been over 60 documented cases of bombs on aircraft causing inflight explosions. There have been another dozen or so accidents wherein an inflight explosion was suspected but never proven. It is, unfortunately, a possibility. If inflight explosion is strongly suspected, then the character of the investigation shifts more toward the criminal side and away from the safety side. Emphasis on proper documentation and control of evidence now becomes paramount and new forensic skills will be needed to assist in the investigation.

The results of an inflight explosion depend, of course, on the size of the explosive device and its location. In general, aircraft structure survives internal explosions fairly well. The theory behind surviving an explosion is based on venting the pressure and containing the fragments. If a bomb is placed in a baggage compartment or cabin near the skin of the airplane, it will blow a hole in the fuselage (which releases the pressure) and its fragments will be largely contained by the rest of the baggage or surrounding cabin seats. The hole itself will probably not progress to structural failure, as fuselage construction methods tend to contain this type of damage. There are several cases on the books where a bomb has blown a hole in the fuselage measuring several square feet and the plane has landed without further incident. The affect of the explosion on systems or aerodynamics might be more severe. A powerful explosive could result in structural failure if it exploded in the vicinity of a manufacturing joint and the overpressure was not rapidly vented as the explosion did not occur near the skin of the aircraft. This approximately describes the situation involving the Pan American B747 over Lockerbie, Scotland. At this writing, there is research among manufacturers and government agencies to determine if better design of fuselages and cargo containers could contribute to better explosive survivability.

In any event, the results of a high-order inflight explosion are distinctive and the investigator should always be on the lookout for significant evidence. The remainder of this section deals with evidence that the field investigator might reasonably encounter and recognize.

B. CHARACTERISTICS OF HIGH-ORDER EXPLOSIONS. First, explosions are omni-directional. The velocities involved are much greater than the velocity of the aircraft and the damage radiates about equally in every direction from the source. This is different from a fire which follows the slipstream or progresses upward. Second, explosions create an "opening up" effect on adjacent paneling and structure. Most impact damage is from the outside in. Explosive damage is from the inside out. Unlike fire, explosions force their way through the structure by overpressure; not melting.

Explosions create unusual damage to heavy structure. This would be observed as bending damage in directions not consistent with impact loads. In addition, explosions result in fragmentation of lighter structures. These are brittle failures in normally ductile materials. Due to the overpressures, they fail before they have a chance to yield. In some cases, the heat of the explosive reaction melts the edge of the fracture creating a smooth or rolled edge appearance.

Explosions result in high speed penetrations of adjacent objects and structure by metal fragments. If explosion is suspected, all human remains, seat cushions and cargo should be x-rayed and examined for fragment penetration. Typically, the hot fragment will also leave a burned, melted or charred area at the point of penetration. If the fragment hits metal structure, it may leave a rimmed crater caused by the instant melting and splash-back of material. These indications are almost certain to be the results of a high-order explosion. These will not be produced by impact forces.

If some of the above indications are noted in the field, it is time to consult with forensic and laboratory experts for confirmation. In general, a laboratory can, through examination with a scanning electron microscope (SEM), observe surface pitting, erosion, and grain structure changes unique to explosions and not observable by the field investigator. Also, there are chemical methods of detecting and identifying explosive materials and by-products.

11. SUMMARY.

The investigation of aircraft fire and explosions is difficult, but it follows a logical pattern. Explosions leave characteristic evidence that could not be produced any other way except by explosion. Although post-impact fires may mask and destroy a lot of evidence, it is unlikely that all evidence of explosion or inflight fire will be completely destroyed.

Caution. You can always find single elements of evidence which will support any conclusion you care to draw. There must be a consistent pattern of evidence to support an argument for or against inflight fire or explosion. The conclusions reached must be consistent with the basic laws of chemistry and physics.

9

Structural Investigation

1. INTRODUCTION.

Structural failure, in one form or another, is a significant factor in many aircraft accident investigations. To get philosophical for a moment, structural failure in itself should never be cited as a cause of an accident. Things don't just sit there and break. They are either under designed, incorrectly manufactured, improperly installed or maintained, inadequately inspected, or over stressed as they are used. All of those failures are human failures. Thus the finding of, "failure of the whoozis," is something of a cop-out.

So much for philosophy. The fact is that parts do fail and, as investigators, we must have an organized method of identifying those failures and dealing with them. This chapter complements Chapters 34 and 35 and, without getting too technical, attempts to apply some logic to this problem.

2. TYPES OF STRUCTURAL FAILURES.

It is helpful to categorize structural failures along the following lines.

A. OVERSTRESS. THE PART SHOULD HAVE FAILED. Here, we put more stress on the part than it was designed to withstand. We can do this in three ways.

PILOT INDUCED OVERSTRESS.

Aerobatics.

Over reaction to turbulence or controlled

flight departure.

Improper recovery techniques or excessive landing loads.

Any other operation outside of the aircraft's structural envelope.

WEATHER INDUCED OVERSTRESS.

Excessive gust loading (turbulence.)

Wind shear.

WAKE TURBULENCE INDUCED OVERSTRESS.

Downwash.

Wing tip vortices.

B. UNDERSTRESS. THE PART SHOULD NOT HAVE FAILED. Here, the part failed at a stress level supposedly within its operating capacity. Here is how we can do that.

FAULTY MANUFACTURE. The part did not meet the design specifications.

FAULTY REPAIR OR MODIFICATION. Sometime during the service life of the part, its strength characteristics were altered by repair or modification.

REDUCTION OF LOAD BEARING

CAPACITY. Over time, metal parts may corrode or develop fatigue cracks. The result of either of these is that the part can no longer sustain the specified load.

CORROSION

METAL FATIGUE

C. AERODYNAMIC OVERSTRESS. Here, a perfectly good part, was locally stressed beyond its capability for purely aerodynamic reasons. Technically, these are called aeroelastic phenomena.

FLUTTER or DIVERGENCE

CONTROL REVERSAL

3. APPEARANCE OF STRUCTURAL FAILURES.

This section may seem a little dull, maybe even self-evident, but it makes a couple of points that many investigators miss. Let's suppose that we have a metal part that is loaded in tension. A fatigue crack is developing in the part and (as it will) the crack is propagating into the cross section of the part on a line perpendicular to the principle tension stress. This hypothetical situation is illustrated in 9-1.

The net effect of this crack is to reduce the cross sectional area of the part and, therefore, its ultimate strength. As long as the remaining cross sectional area of the part is adequate to support the applied load, we are OK. If the remaining cross sectional area of the part can no longer support the applied load, it fails. We understand that.

Now let's suppose that the fatigue crack exists, but it has not gone far enough to lead to failure. Unfortunately, that plane crashes and this part fails during the impact.

Question. Where does it fail?

Answer. At the fatigue crack, of course. Where else?

Question. Just by looking at the fracture, how can you tell the difference between the failure that occurred inflight and the one that resulted from the impact?

Answer. Well, uh, I'm not sure.

Let me help you out. You can't. There is no difference. This little discussion points out two things worth remembering about structural failure.

First, all structural failures are ultimately overstress failures. There is no such thing as a pure fatigue (or corrosion) failure. A fatigue crack never makes it all the way through the part. What happens is that there is so little left of the part that it fails under its normal load; but it fails in overstress, not fatigue. The same is true of parts that fail during the impact sequence of an aircraft accident. They all fail in overstress. Furthermore, if there is a weakness present (fatigue or corrosion) that's where they are going to fail.

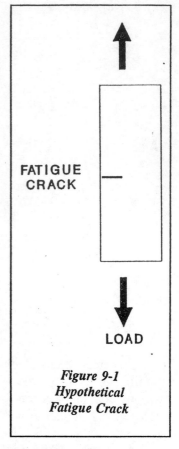

Figure 9-1
Hypothetical Fatigue Crack

Second, all aircraft flying (assuming metal structures) have fatigue cracks and most have some areas of corrosion. That's a given. You can take that to the bank and borrow money on it. Most of these are of little consequence because they are not significant to the load-bearing capacity of the structure and they will be identified and corrected during the aircraft maintenance process. That's how the system works.

Nevertheless, when the plane hits the ground, that's where the parts will fail. Thus the investigator will always find evidence of fatigue or perhaps corrosion if he looks hard enough. It's there! Finding it does not necessarily mean he has solved the accident!

The big question is always, "Did this have anything to do with this accident?"

That can be really tough to answer, because the appearance of the fracture is the same regardless of whether it failed inflight or during impact. Here are a couple of questions you might ask.

A. WAS THE MANNER OF FAILURE CONSISTENT WITH THE WAY THIS PART WAS STRESSED INFLIGHT? If the part is normally loaded in tension, but it failed in bending or torsion, you can pretty well chalk that one up as resulting from the impact. If it happened inflight, it should be consistent with the manner in which the part is normally loaded inflight.

B. IF THIS PART DID FAIL INFLIGHT, WOULD THAT EXPLAIN THIS ACCIDENT? If the answer to this is, "No," go look at some other part. You are wasting your time.

4. COMPOSITES.

So far, we have been talking about metal structures. Composites behave differently. Technically, a composite is any non-homogenous material, which could be anything from plywood to cement. The composite most commonly found in structural applications on aircraft is called carbon fiber reinforced plastic. This may be found alone or sandwiched around a metallic or non-metallic honeycomb structure.

Composites don't develop fatigue cracks; they develop delaminations, which can be hard to find. When they fail, they don't fail in the classic ductile/brittle manner of metal structures. They delaminate. Sometimes the covering sheet of composite may delaminate from the underlying honeycomb structure due to accumulated moisture freezing at altitude.

This can be confusing to the investigator as there is little or no evidence of exactly what happened to make the part fail in that manner. The composite has no memory and does not retain evidence of load as a metal structure does. Here's a couple of suggestions.

At some point, the composite is probably fastened to something or connected to a metal structure using bolts, flanges, brackets, etc. Look closely at these items as they have a memory and may retain evidence of the load on the composite.

All composite surfaces are painted. While the composite itself may not retain evidence of load, the paint might. Examine the edges of the paint at the fracture and look for signs of bending in the direction of the applied load.

If the composite sandwiches a metallic honeycomb structure, the deformation of the honeycomb may indicate the manner of loading of the composite.

Carbon fiber composites are moisture sensitive which sometimes accounts for the delaminations. The moisture content of the composite can be determined by laboratory examination providing the part was sealed during packaging so that it did not absorb additional moisture during transit.

A word of caution about carbon fiber composites. The fibers in these composites are held together by a resin which has a fairly low melting temperature. When the structure is exposed to fire, the resin melts and the carbon fibers (which don't melt) are liberated. This creates some problems. First, breathing the carbon fibers which are floating around in the air is not healthy. Second, handling the structure without protection is like handling raw fiberglass--maybe worse.

One quick solution is to spray the composite structure with something that will recapture the carbon fibers and hold them. Shellac, lacquer or even sprayed paint will work. This is strictly a safety solution. If the structure is critical to the investigation, this is not a good idea. The best thing to do here is to get a large roll of plastic sheeting and some tape and wrap the structure in it.

5. PRINCIPAL EVIDENCE OF INFLIGHT STRUCTURAL FAILURE.

The best possible evidence that the plane came apart inflight is to find pieces of it at some distance from the main wreckage. Being smart enough to go look for those pieces involves knowing that they are missing in the first place. More than one investigator has spent a few days milling around in the wreckage not realizing that he didn't have it all.

Unless you can absolutely rule out structural failure because of the circumstances of the accident, you would be well advised to inventory the wreckage to make sure it's all there. Discovering a week or so later that a key piece of structure is missing is a little embarrassing.

There are a couple of ways to do this. One easy way is to have a plan and profile diagram of the aircraft and just shade in the parts as you find them. Another is to use the TESTED acronym, which has been used by NTSB investigators over the years.

T stands for Tips. Locate the tips of the wings and empennage. Some call these "the corners." The logic here is that if the right wingtip (say) is with the wreckage, the rest of the right wing must be there also.

E stands for Engines and includes propellers and all propeller blades.

S is for Surfaces. Locate all of the primary and secondary control surfaces.

T is for Tail. Make sure you have the entire empennage.

E stands for External devices including landing gear, tanks, pods and so on.

D is for Doors including hatches, canopies and windows.

There. If you've gone through that drill, you can feel reasonably confident that any missing pieces were not essential to flight.

Keep in mind that having all the pieces in one pile does not necessarily mean that there wasn't an inflight failure. It is entirely possible that a horizontal stabilizer (say) did fail in flight, but did not depart the aircraft. Here, you have to examine the failure and analyze whether it occurred before impact or resulted from it. Let's suppose, for example, that a wing is bent up. If the impact scarring is not continuous across the bend, one could argue that the wing was bent up before the impact. If the bend occurred at a point on the wing where the stresses are highest (just outboard of an engine nacelle, for example), then the case for inflight failure is looking even better.

6. LOCATING MISSING PARTS.

Logically, the place to start looking for missing parts is back along the flight path of the aircraft if you know it. Depending on the terrain and location, you might try advertising for it. Give the news media a press release stating that you are missing a certain part and if anyone finds it, please call this number. This is likely to get you a lot of junk which was never part of an aircraft; or maybe was part of some other aircraft. It may also get you the part you are looking for.

Other methods that have been used with some success are aerial search and infrared photography.

The metal part, hidden by the vegetation, will be either warmer or cooler than the surrounding land and may show up on the film.

Another method which has been used with some success is called trajectory analysis.

Anything that falls from a plane will follow a fairly predictable trajectory to earth. The elements that influence that trajectory are initial altitude, airspeed, direction, wind, mass of the object and its flat plate drag characteristics.

Well, you say, we don't have enough information to do much with that.

Perhaps you do. Estimating the mass of a part is fairly easy. The drag characteristics of various shapes have already been calculated and are in the engineering handbooks. Your part will match one of those shapes. If you know where this thing happened in space, which you may know from the FDR or ATC radar data, You've got the altitude, airspeed and direction. Wind is a matter of meteorological analysis. If you don't know where it happened in space, you may be able to determine it if you have enough wreckage that did fall from the plane. All you have to do is work the trajectory calculations backwards. You know where the pieces landed, If you calculate the trajectory curves for enough parts, you may find that the curves tend to intersect at a point in space. Assuming that's where it all happened, you can calculate a new trajectory curve for your missing part and go there and pick it up!

Has this ever been done? Yes it has. There are several papers on the technique available in the literature and even some computer software. Is it accurate? Not very, but the accuracy improves with the number of trajectories calculated. Obviously, the trajectory of a part that has some mass and is fairly symmetrical is more predictable than a part that "flies" or "floats" in the air. Nevertheless, there are many cases on the books where this method was able to put the investigators within a quarter of a mile of a missing part, which is not bad at all.

During the Pan American B747 Lockerbie investigation, the British turned this technique into a science. They had wreckage scattered over a large portion of Scotland. They had very good information from the FDR and ATC Radar and they also had good meteorological information on the upper air winds. They wrote a computer program that would calculate trajectories for the wreckage they had and predict where other wreckage would be found.

7. RECONSTRUCTION.

The reconstruction of aircraft wreckage is a technique pioneered by the United Kingdom Aircraft Accidents Investigation Branch. They got their start with an early British Comet accident and were able to show where a fatigue crack in the fuselage originated and how it progressed. More recently, they reconstructed the front end of the Pan American B747 that crashed near Lockerbie, Scotland. This allowed them to accurately determine the location of the bomb in the forward cargo compartment and show how the force of the explosion had destroyed the aircraft.

Reconstruction is a useful technique for both structural and fire investigation. See Chapter 8 for fire investigation. It can also be used examine systems. See Chapter 13. It is rarely necessary to reconstruct the entire plane; just the part that interests you.

If you decide to reconstruct part of the plane, there are a few things you might consider.

First, this is going to take time. There is no quick way to do it.

Second, it is going to take up a lot of room. You need to pick your location wisely. You want to be protected from the weather and you don't want to move the wreckage more than once.

Third, you will have to move the wreckage to the reconstruction site. You can save yourself a lot of time by identifying and tagging parts before you move them.

Fourth, the easiest time to prepare the reconstruction site is before the wreckage arrives.

There are two basic types of reconstruction; two dimensional and three dimensional.

A. TWO DIMENSIONAL RECONSTRUCTION. This is by far the easiest method and the one that should be used if at all possible. Basically, the parts of whatever it is that you are examining, system or structure, are laid out on the floor in their correct relationship to each other.

This can even be done with a section of a fuse-lage. Imagine, for example, what a fuselage would look like if it were sliced open along the top and unrolled. It might look something like the diagram in Figure 9-2, which represents the forward fuselage of a business jet.

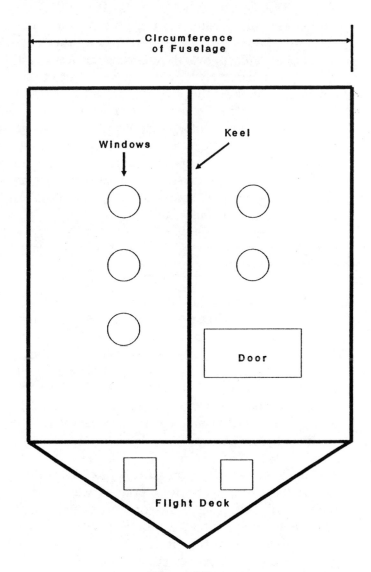

Figure 9-2
Sample Two Dimensional Layout of a Fuselage Section

The first thing to do is to calculate the dimensions of the layout and locate the key parts as shown in Figure 9-2. Next, using spray paint, tape or chalk, draw that diagram on the floor of the reconstruction site. (One investigator uses a shoe black compound meant to be applied to the soles of shoes. It is water soluble and comes with a handy application brush.) Now it is fairly easy to layout the wreckage on the diagram and get it right the first time. The fact that the layout is flat instead of round is a minor problem.

This general procedure works well for any two

dimensional reconstruction. Establish the dimensions of the layout and mark the floor with a diagram to help locate the parts.

B. THREE DIMENSIONAL RECONSTRUC-TION. This is really a lot of work. If you decide to go this way, hire a carpenter and give him the dimensions of the section you are going to reconstruct. He is going to build plywood bulkheads connected with formers and covered with chicken wire. Once he has done that, you are going to hang the wreckage pieces on the chicken wire frame work and make it look like an airplane. A couple of suggestions. Build the framework on a pallet that you can move around with a forklift. Better still, put wheels under it so you can roll it around. If possible, build it in sections that are easier to handle. Build access to the inside so you can examine the results internally.

8. SELECTION OF PARTS FOR LABORATORY EXAMINATION.

This can get tricky. The problem is that the confirmation of metal fatigue or stress corrosion or whatever must come from a metallurgical laboratory. The selection of the part to be taken to the laboratory in the first place, though, is usually done by the investigator. Thus the investigator gets into the metallurgical business whether he likes it or not. The eyeball method of examining fractures leaves a lot to be desired.

Here is a logical approach that may help. First, look at the overall fracture in terms of whether there is a single fracture zone or two or more distinct zones. If there is only a single zone, then only one thing happened to the part as it failed. The chances are pretty good that it is an overstress failure, which is the common type of failure found in aircraft wreckage. This is not likely to be significant unless you believe it happened inflight.

If, on the other hand, you can see more than one distinct zone on the fracture, then at least two things happened to the part during the course of its failure. Examine the line separating the zones. If it is sharp and distinct, then whatever happened to it was mech-

chanical. Metal fatigue will leave this type of indication. So will stress corrosion, although not all metallurgists would agree with this. To the naked eyeball, fatigue and stress corrosion look a lot alike. If the line separating the zones is fuzzy, meaning that it is hard to tell where one zone stops and the other starts, then whatever happened to the part was chemical; most likely corrosion. Any of these multi-zoned fractures could be significant.

Finally, you ask yourself the key question. If any of these fractures occurred in flight, would they have produced this accident? If the answer is, "Yes," those parts go to the lab.

A word of caution. Resist the urge to fit broken parts back together. The metallurgist is going to examine the fractures under a scanning electron microscope and he can tell the difference between a "virgin" fracture and one that has been smeared by being reassembled. If you do that, you are destroying evidence.

9. INVESTIGATIVE STEPS.

If you suspect inflight structural failure, this is the general approach to the investigation.

1. Diagram the wreckage. See Chapter 7.

2. Inventory the wreckage.

3. Locate any missing parts.

4. Reconstruct the portion of the wreckage where the failure occurred.

5. Analyze the reconstructed portion to determine origin and progression of the failure.

10. SUMMARY.

In degree of difficulty, investigating inflight structural failures ranks right up there with investigating fires. Nevertheless, there are some logical steps and methods that can keep you moving in the right direction.

10

Reciprocating Engines

1. INTRODUCTION.

Compared to turbine engines, recips are quite difficult to investigate. First, they always show evidence of rotation as that is their normal wear pattern. Second, there is nothing on the recip that consistently captures evidence of what was happening at impact. That is why so much attention is paid to the propeller. It provides at least an indication of what was going on. See Chapter 11.

2. BASIC STEPS.

Step one in a reciprocating engine investigation is to assemble everything that is known so far about the accident. This includes witness statements, radio transmissions and the basic circumstances of the accident. Is there any positive evidence that the engine was running--or not running? Did anyone hear it? Be cautious in accepting at face value statements about what the engine sounded like. Statements like, "The engine was cutting in and out," or, "The engine was running rough," need verification. An idling engine can sound like that. Do the accident circumstances fit the common scenarios for engine failure-type accidents? Engine failure on approach or landing, for example, is not a common accident cause as the engine(s) are already being operated at low power and the aircraft is already in a nose low descending attitude. If loss of power occurred, it is usually manageable. On the other hand, a takeoff accident is much more likely to be related to engine power. Here, full engine power is needed and a sudden loss of power can be difficult to manage.

Second, determine what you really need to know about the engine. Was it completely stopped (propeller feathered or stopped?) Was it turning at something less than full power? Was it turning at something close to full power? Let's approach this logically.

3. COMPLETE ENGINE FAILURE OR INFLIGHT SHUTDOWN.

If the propeller was feathered, the engine was not rotating at impact and the feathering occurred at some point prior to impact. The pilot either deliberately shutdown the engine and feathered the propeller due to some cockpit indication or the engine failed and the propeller feathered itself because an auto-feather circuit was installed and armed. Given that this all occurred inflight with the engine above 600-800 RPM, examination of the propeller should tell us which of these occurred.

If we conclude that the pilot deliberately shutdown the engine and feathered the propeller due to cockpit indications, we ought to attempt to determine what those indications were. We've got several cockpit instruments and switches to help us. See Chapter 14 for instrument investigation techniques.

If the engine merely failed (not deliberately shut down), then we are not likely to find much evidence of the cause in the cockpit. In these situations, a large percentage of engine failures are related to fuel; or lack of it. We should start with a routine check of the fuel system.

A. WAS THERE FUEL ON BOARD? We can investigate this indirectly by considering fuel consumption and the duration of flight since last refueling and directly by examining the tanks if they have not been destroyed. If significant quantities of fuel were present at the impact, the vegetation in the vicinity of the impact may be burned and the unburned fuel that soaked into the ground can be chemically detected for some time after the accident. The first people on the scene (crash-rescue personnel, perhaps) may have noticed a odor of unburned fuel in the air. If there was no fuel, of course, none of these clues will be present and there probably wasn't any post-crash fire either.

B. WAS THE FUEL THE CORRECT TYPE? Fueling a reciprocating engined aircraft with turbine fuel is a known producer of power loss on takeoff. Since the turbine fuel has a higher specific gravity, it goes to the bottom of the tank and is fed to the engine very early. If you suspect turbine fuel, try the paper test. Place a few drops of the fuel on a sheet of white paper. Avgas will evaporate completely while turbine fuel will leave a small grease spot that you can see by holding the paper up to a light.

C. WAS THE FUEL FREE OF CONTAMINANTS? Realistically, about all you can do in the field is to check for water and gross sediments in the fuel. Any accurate assessment of fuel quality takes place in a laboratory.

D. COULD THE FUEL GET TO THE ENGINE? This involves some sub-choices.

Was the fuel selector on a tank that had fuel in it?

Was there any obstruction in the selected tank that would prevent the fuel from leaving it? Strange things have been found in fuel tanks.

Were the fuel pumps working? You might have one or more fuel pumps in the tank; another mounted on the engine; and a boost pump for the fuel system. Each of these pumps can be checked for presence of fuel and (if the impact damage was severe enough) evidence of rotation. See also Chapter 13.

Were the fuel filters clean? If fuel filters become clogged, most of them are designed to bypass

fuel around the filter in order to avoid starving the engine. Evidence that fuel has been bypassed can usually be found in the filter itself.

E. DID THE FUEL ACTUALLY GET TO THE ENGINE? Check engine fuel lines and the carburetor for the presence of fuel.

F. WAS THE ENGINE GETTING AIR? This is fairly easy to check. The induction system should be clear through the air filter to the throat of the carburetor.

G. WAS THE ENGINE GETTING IGNITION? Since the recip has a dual ignition system which works independently of the aircraft electrical system and two spark plugs per cylinder, total loss of ignition is fairly rare. Checking the position of the magneto switch in the cockpit is not an absolutely positive clue. First, the mag switch can move during the impact sequence. Second, if the pilot deliberately shut down the engine, he most likely turned the mag switch off. Check the mag switch anyway. In some engines, both magnetos are driven off a single shaft. Check to see if this shaft has sheared.

4. INTERNAL ENGINE FAILURE.

No luck, huh? OK, now we're down to the possibility of massive internal damage to the engine that just made it quit running. If possible, you might try turning the engine over by hand. The recip is a rugged piece of machinery and it frequently survives an impact and can still be rotated. If it turns without any weird noises, there is probably no internal damage serious enough to keep it from running.

5. ENGINE DID NOT FAIL, BUT WAS NOT PRODUCING FULL POWER.

There might be several reasons for power loss.

A. INDUCTION SYSTEM ICE. This is more commonly called, "Carburetor Ice," but the term, "Induction System Ice," is a little more accurate as it includes ice that can form in the filters and bends in the induction system as well as the ice that can form in a float-type carburetor. Positive evidence of carburetor ice is rarely found as it can (and generally does) form at ambient temperatures well above freezing. By the time you get there, it has melted.

The mechanism of induction system icing is fairly

simple. As the air goes through the induction system, it goes through a series of narrowing passageways culminating in the carburetor. These provide a venturi effect and, in accordance with the universal gas law, the velocity of the air is increased and its pressure and temperature are both decreased. The venturi effect can lower the air temperature about 5 degrees C. The evaporation of fuel in the fuel-air mixture can lower the temperature up to 40 degrees C. If the temperature is lowered below freezing and the air contains a certain amount of water, it forms instant ice which may block further air flow or impede the operation of the butterfly valve in the carburetor. With no air flow, the engine quits and the plane crashes (or lands prematurely.) Before the investigator shows up, the ice melts and the engine runs just fine. Now what do you do?

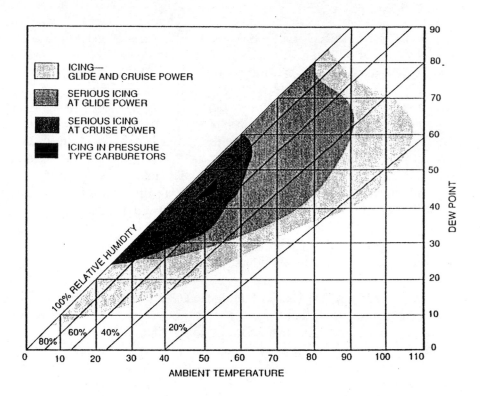

Figure 10-1. Carburetor Icing Probability Chart

You might check the carb heat control in the cockpit to see if the pilot thought he had carb ice. This isn't a positive clue since the carb heat control can move during impact. Most determinations of carb ice are made inferentially. There is nothing else wrong with the engine and atmospheric conditions and probable power settings were conducive to carburetor (or induction system) ice.

The term "conducive to carb ice" is not well defined. It depends primarily on the ambient temperature and dew point, but it is also affected by the type of engine and its installation on the aircraft. There is no universal chart that depicts when carb ice will and will not form. the chart in Figure 10-1 was derived from one developed by Transport Canada (Canadian FAA) showing a curve within which conditions might be considered "conducive" to the formation of carburetor ice. Engine manufacturers have more accurate information for each of their engines.

B. IGNITION SYSTEM FAILURE. It is possible to remove a magneto in the field and spin it by hand to determine if it will produce a spark. Before you do this, carefully note the position of the timing

marks on the engine. Once you remove the magneto, you won't be able to tell if the engine timing was correctly set if you didn't note the marks. Checking the mag in the field, of course, does not tell you much about how well the magneto was working. The magneto and the entire wiring harness can be easily checked by a repair station using a magneto tester.

C. SPARK PLUG FAILURE. Keep in mind that the loss of power from the failure of a single plug would be barely noticeable, if at all. It must be something more serious than that. Examination of the plugs can tell a trained engine mechanic a lot about how the engine was running. Before you remove any plugs, remember that you've got two per cylinder and it would be nice to know which hole each plug came from. Obtain a plug rack with numbered holes or tag each plug before you remove it.

D. CYLINDER FAILURE. If the engine can still be rotated, it is possible to remove one plug from a cylinder and, by closing the plug hole with your thumb, make a rough check for compression as the engine is rotated by turning the prop. If there is something seriously wrong in the cylinder, you will get no compression. It is also possible to examine the cylinder and valve faces with a borescope through the spark plug hole.

E. LUBRICATION SYSTEM FAILURE. Loss or partial loss of oil can result in the engine being shut down or operated at less than full power. Check first for the presence of oil in the system. Start with the oil tank and check the lines and the oil pump. Next examine the oil filters and sump drains for the presence of contaminants. It is possible to test a sample of the oil spectrographically for wearmetal content, although this is not particularly accurate unless there is an oil analysis history on that engine. You will find metallic content in the oil as that is the wear pattern of the reciprocating engine. Finally, remove a rockerbox cover and look for the presence of oil. If it got that far, it was probably in the cylinders, too.

E. TIMING FAILURE. Timing refers to the relationship between what the engine crankshaft is doing and what the valve camshaft is doing. If the engine is not properly timed, it is not going to run right, but it is difficult to see how that, in itself, would cause an accident. An aircraft with a badly mis-timed engine never makes it off the ground.

Nevertheless, it is possible to make a rough check of engine timing in the field providing the engine can still be rotated by turning the prop. This procedure will work on most engines.

First, rotate the engine so that the piston of the number one cylinder is at top dead center (TDC) on the compression stroke. This occurs when the TDC scribe mark on the rear of the starter ring gear is aligned with the scribe mark on the top of the crankcase.

Next, remove the rocker box cover for number two cylinder so that you can see the valve rocker arms. Both valves should be closed. Rock the propeller back and forth slightly and note that each valve (intake and exhaust) alternately comes open. If this does not happen, something is wrong with the timing and the engine will not produce full power.

F. TURBOCHARGERS. A word about turbochargers and superchargers. Both do about the same thing, but they do it differently. Both increase the efficiency of the engine at altitude by compressing the air that goes into the induction system. The supercharger uses a compressor that is geared to the crankshaft and the power needed to run the compressor comes at the expense of the power output of the engine. Thus superchargers are not used until they are needed and their benefit exceeds their cost. Turbochargers, on the other hand, use a turbine placed in the exhaust path to turn the compressor. This improvement in performance is essentially free although increased back pressure does cost some power. The turbocharger is available and can be used throughout flight except for takeoff. Misused, it can lead to overboosted engines and loss of power or even engine failure.

Many aircraft with turbocharged engines have automatic control systems which prevents engine overboost. Some have manual systems wherein the pilot must position the controls correctly or risk serious engine damage or failure. In these aircraft, the clue would be the position of the turbo control in the cockpit.

7. NOW WHAT?

Still a mystery, huh? OK, stand back and take an overall look at the engine. Do you see any signs of obvious mechanical damage? How about the exhaust manifold? That frequently breaks off or crumples during impact. If it is bent and crumpled (ductile behavior), the manifold was probably hot at impact. If it fractured (brittle behavior) it may have been cold. Do you see any signs of fire that seem to emanate from a point? A cracked fuel pump housing, for example, might not be detectable in the field, but the fire pattern resulting from it might be obvious if you back up a little bit.

8. SUMMARY.

Don't get discouraged. We said recip investigation was difficult. Your next step is to remove the engine from the wreckage and take it to a facility where it can be thoroughly torn down and examined. You have done all you can realistically do in the field and, if you have been careful, you have not destroyed any evidence or compromised any future investigation.

11

Propellers

1. INTRODUCTION.

Propellers are common to both reciprocating engines and turbine engines (turboprops). An examination of the damage to the propeller can sometimes be very useful in determining what the engine was doing at the time of impact.

Before we get into this, it should be pointed out that it is not possible to completely explain all of the bends and twists that can occur to a propeller during impact. There are too many variables including RPM, airspeed, attitude and the nature of the surface or soil where the propeller hit. There is a lot of misinformation in the literature on how propellers behave and there is a rumor that the clever investigator can accurately assess engine power output by the appearance of the propeller. It can't be done, and we will try to explain why.

2. EVIDENCE OF ROTATION.

You ought to be able to examine a propeller and determine whether it was rotating or not at impact. What are the clues?

A. Blades bent opposite the direction of rotation.

B. Chordwise scratches on the front side of the blades. It is almost impossible to produce a scratch that is exactly perpendicular to the edges of the blade unless the blade was turning at the time.

C. Similar curling or bending at the tips of all blades. It is also impossible to damage the tips of all blades in a similar manner unless it was turning at the time.

D. Dings and dents to the leading edge of the blades.

E. Torsional damage to the prop shaft or attachment fittings.

Before you get excited over this, remember that the propeller in all probability was turning at impact. Even if the engine failed or was shut down, the propeller will windmill at an RPM high enough to produce these indications of rotation. The exceptions to this are:

A. The propeller was feathered. If this occurred, the propeller will, of course, show no signs of rotation. More on this later.

B. The propeller was not feathered, but was completely stopped due to either internal failure (seizure) of the engine or aerodynamic stall of the propeller. If the engine seized internally; no problem. There will be plenty of evidence of that. The aerodynamic stall theory involves shutting down the engine and then slowing the plane down to the point where the propeller stops windmilling. As a practical matter, this is very difficult to do. It is generally necessary to hold the plane near a full stall while waiting for the propeller to stop.

So. If we are dealing with a non-feathering propeller (almost all single engine aircraft), we would expect it to be rotating at impact. Finding evidence of rotation doesn't tell us much. Finding absolutely no evidence of rotation should lead us to suspect massive internal engine failure.

3. PROPELLER DAMAGE AND RPM.

What can the propeller damage tell us about engine power output? Not much unless we have propeller strike marks that will allow us to calculate RPM. More on this later.

There is a propeller tip-bending phenomena which may give us a rough idea of RPM related to velocity. Many investigation texts will suggest that if the prop tips are bent backward, the RPM was low. If they are bent forward, the RPM was high. This is very misleading.

What actually happens is that the tips of the blades as they strike the ground may bend either forward or backward depending on the relationship between RPM and forward velocity. This is a simple exercise in forces. The prop blade is not straight, but is twisted forward at a blade pitch angle. If the RPM is high compared to the forward velocity, then the dominant force tending to bend the blade is the blade pitch angle and it tends to curl the end of the blade forward. On the other hand, if the RPM is low compared to the forward velocity, then the dominate force on the blade comes from the forward velocity. This, of course, tends to curl the end of the blade backward. Thus this blade tip curling phenomena is really a function of the relationship between RPM and forward velocity. It is not a direct measurement of RPM. The RPM may be high, but if forward velocity is also high, the blade tips are likely to curl backward. If the tips are curled forward, you can be sure that the prop RPM was not only high in relation to the forward velocity, but the propeller was being driven under positive power from the engine.

Some other comments about this phenomenon. First, it occurs only at the blade tips and it appears as a curling starting with the leading edge corner of the tip. A blade bent at mid-span, either forward or backward, is not an indication of high or low RPM. Second, it occurs on all blades. If only one blade is bent, it was caused by something else; not rotation. Third, this only occurs at relatively low angles of impact; perhaps five degrees or less. It is seen most frequently on landing accidents where the angle of impact was very low. The classic use of this phenom-

ena is following gear-up landings. If the pilot had no idea that the gear was up until he heard that horrible screeching sound, then the blade tips will be bent back; the engine was near idle power. If, on the other hand, the pilot realized at the last moment that there were no rollers under the plane and he shoved the throttle full forward to go around, the blade tips will be bent forward. This may give you a smug feeling of satisfaction in knowing what the pilot was thinking about during the last seconds before touchdown, but it doesn't tell you much about why he failed to lower the landing gear in the first place.

Back to the point about impact angle, If the angle of impact was high, then the blades, tips and all, are going to bend backward regardless of RPM. In low angle impacts, this tip-bending thing is, at best, an indication of RPM relative to velocity. That's about all you can say for it.

4. CONSTANT SPEED PROPELLERS.

The idea behind a constant speed propeller is that it will maintain constant engine (or propeller) RPM while adjusting to power demands by changing blade pitch angle. This means that the expected post-impact evidence would confirm the RPM setting of the propeller--which is not very helpful. If we could determine blade pitch angle, we could correlate that with power demanded for the constant RPM.

This is not something that can be done easily in the field. First, most constant speed propellers use engine oil as a hydraulic fluid to force either the propeller dome or a piston within the dome outward to change the blade angle (pitch) of all blades towards the low pitch position. When the aircraft crashes, engine oil pressure is lost and the oil in the prop dome either leaks out due to damage or bleeds back into the oil system. When this occurs, there is nothing holding the blades in any particular pitch angle and they are free to move (twist) in any direction. Thus the investigator who assumes that the way the prop is found is the way it was at impact is asking for trouble. Unless there is severe damage to the propeller hub, it is possible that the blades can be twisted by hand to any position and will be twisted to a new position each time the prop is moved. Knowing this, it would be a good idea, before moving the prop, to put a scribe mark on each blade shank and hub so that the blades can always be returned to their "as found" position. This may be of no importance, but we don't know that yet.

The situation is not hopeless. Depending on the manufacturer, each prop blade has shims or areas on the barrel which tend to capture impact "witness" marks. Determining this is a three-step process.

1. Decide which blade hit first. This is the one with the most leading edge damage. Look for evidence on this one and ignore the other blades.

2. Have the propeller disassembled by a specialist familiar with that propeller and knowledgeable of the method for locating witness marks that would indicate blade pitch at impact.

3. Convert blade pitch angle to engine power for the assumed RPM of the propeller at impact. This can be done using charts in the propeller or engine maintenance handbook.

5. FEATHERING PROPELLERS.

Full feathering propellers are usually found only on multi-engine aircraft and single engine turbine powered (turbo prop) aircraft. Except for some agricultural aircraft, there is not much reason to feather a single reciprocating engine prop. The differences between a constant speed prop on a single engine aircraft and the same propeller installed on a twin are minor. Externally, they look the same. Internally, the feathering prop has a spring which will drive the prop blades into the feather position and centrifugal latches on each blade which will not allow the blades to go past the low pitch stop unless the prop is turning at (nominally) 600-800 RPM. Above this RPM, the latches are retracted (centrifugal force) and the blades can get past the stop into the feather position. Below this RPM, the blades cannot get past the low pitch stop. Without this feature, every time the plane was parked and the engines shut down, the props would automatically feather themselves (due to the feathering spring) as the oil bled back into the engine system. This makes the engine hard to start and unfeather the next time you want to use the plane.

What happens during an impact is that prop dome oil is lost (or bled back) and the prop blades tend to twist themselves toward the feathered position due to the feathering spring. Thus it is normal to find the prop in the high pitch position after an accident. If you want to know where it really was, see the procedures listed above.

A problem occurs because the high pitch position

and the full feathered position look a lot alike. It is difficult to tell the difference, particularly if the prop is damaged. The feathered position is measured with a special protractor at a specified distance from the hub. If the blade is bent, this is hard to do.

This has resulted in a lot of incorrect determinations that the prop was feathered; therefore the engine failed or was shut down. The only positive way to determine if the prop was actually feathered is to disassemble it (qualified prop technician) and determine whether the blades are on the low pitch stops or have gone past them into the feather position. If the propeller was, in fact, feathered; then it occurred inflight with the engine operating (perhaps windmilling) at 600-800 RPM. Contrary to some opinions, a propeller cannot feather itself at zero RPM after impact due to the feathering spring.

An exception to the above is a propeller which is not directly geared to the engine. A Hartzell prop on a Pratt-Whitney Canada PT-6 is an example. This is a turbine engine and the power turbine drives the propeller turbine in much the same way as an automobile engine drives the transmission gears in a fluid drive transmission. In this case, there are no centrifugal latches on the blades and the prop will feather itself if oil pressure is lost. The plane is normally parked with the propellers feathered.

6. PROPELLER SLASH MARKS.

What? You say you've found slash marks made by the propeller as the aircraft impacted? The Investigation Gods are surely smiling on you, for now you will be able to calculate propeller RPM in terms of velocity or vice-versa. There is a nice linear mathematical relationship involving the distance between the slash marks; the number of blades on the propeller (usually known); the RPM of the propeller; and the forward velocity of the aircraft. This relationship has been used very successfully during investigation of not only ground impact accidents, but mid-air collisions where prop strikes from one aircraft were measurable on the other aircraft. Helicopter rotor blade strikes have been used to assist in helicopter investigations. The basic formulas are listed below in formulas (1) through (4).

$$RPM = \frac{V_{kts} \times 101.3}{D \times N} \quad \textbf{(1)}$$

$$RPM = \frac{V_{mph} \times 88}{D \times N} \qquad (2)$$

$$V_{kts} = \frac{RPM \times D \times N}{101.3} \qquad (3)$$

$$V_{mph} = \frac{RPM \times D \times N}{88} \qquad (4)$$

Where: D = Distance in feet between slash marks.
   N = Number of blades on propeller.

NOTE: These formulae assume that the engine drives the propeller directly and engine RPM and propeller RPM are the same. If there is reduction gearing, multiply the distance (D) and the number of blades (N) by the prop-to-engine reduction ratio (R) in all formulas.

Wait a minute. You must know either RPM or velocity in order to calculate the other and usually you don't know either one! What good are these?

Let's think this thing through. First, if we are dealing with a constant speed propeller, we know RPM; or at least we can start by assuming that the engine was operating at the correct RPM and calculate velocity from that. Second, we can run this calculation for a variety of circumstances. Regarding velocity, we can take the lowest possible velocity for that gross weight (stall speed) and the highest possible for that configuration and calculate the upper and lower limits of RPM. Furthermore, we can calculate the effect of an assumed change of RPM on velocity or, conversely, the effect of an assumed change in velocity on RPM. Let's try this on an actual case and see how it comes out.

Our accident involves a general aviation aircraft with a reciprocating engine. It crashes shortly after takeoff and all occupants are fatally injured. At the scene, we discover some propeller slash marks just short of the main impact crater. What can we do with these?

Let's measure the distance between the first two marks. Subsequent marks are inaccurate because the prop is slowing down and bending. To make it easy, let's say the distance is two feet. Now the prop has three blades and if it is turning at full RPM as it should be just after takeoff, it should be turning at 2800 RPM. Using equation (4), we get a velocity of 191 MPH which is impossible. The plane couldn't go that fast on its best day. Let's assume a normal climb speed of 80 MPH. Using equation (2), we get an RPM of 1173, which is way low. We can simplify our equation somewhat and say that for a blade-to-blade slash distance of two feet and a three bladed propeller, RPM will always be 14.6 times any assumed velocity. We can even plot a simple graph showing how RPM varies with velocity and state with confidence that for any reasonable velocity between stall and maximum, this engine was not producing full power at impact. This, of course, does not tell us what the engine was doing before impact or why it was doing it. Nevertheless, this is a useful technique.

If our accident involved a twin engine aircraft and we had slash marks for both engines, we would reasonably expect the distances between the slash marks to be the same for either engine if they are both running at the same RPM. If the distances are significantly longer for one engine, we can say without going further that this engine was turning at a lower RPM than the other one. Let's assume that the other engine was operating correctly. We can use the correct RPM (and the slash mark distances for that engine to calculate airspeed. We can then take that airspeed over to the other engine (it probably didn't change much from engine to engine) and calculate RPM using that engine's slash marks. That ought to be pretty accurate.

7. SUMMARY.

There is a lot that can be learned from propellers. First we should be able to tell if the prop was rotating at impact. Next, it is possible to determine blade pitch, although this is an activity to be done in a repair facility. If we are dealing with a feathering prop, we should be able to tell whether it was feathered or not, but this will usually require disassembly. If we are lucky enough to find propeller slash marks, we may be able to make a reasonably accurate determination of either RPM or velocity.

12

Turbine Engines

1. INTRODUCTION.

In many ways, modern turbine engines are easier to investigate than reciprocating engines. The recip's normal wear pattern always provides internal evidence of rotation regardless of what it was doing at the time of the accident. Moreover, the recip internal components tend to be heavy chunks of metal which do not readily pick up impact evidence.

The turbine engine, on the other hand, does not develop a wear pattern in normal use. If everything is working well, the only metal-to-metal contact points are the bearings and the accessory drive gears. If there is rotational evidence elsewhere, then the engine was either turning at impact or there was something wrong with it.

2. TURBINE ENGINE THEORY AND TERMINOLOGY.

This is the wrong text for a complete treatise on turbine engine theory and principles. There are plenty of references available to deal with that subject. We will limit ourselves to just enough theory to explain the investigation techniques to be covered later.

A. TURBOJET. This was the original uncomplicated jet engine. Air came in the intake and was compressed by a set of rotating blades in the compressor section. Most smaller turbine engines use centrifugal compressors which are efficient, but heavy. Larger engines use axial compressors which are a series of rotating blades (called stages) which compress the air axially as it passes through the engine. Between each

compression stage is a set of fixed vanes called stators which increase the efficiency of the compression process. Some engines have both axial and centrifugal compressors. One of the normal by-products of compression is heat. All turbine engines take advantage of this by tapping off high pressure/high temperature air from the compressor. This is called bleed air and it is used for various purposes on the aircraft including anti-icing and cabin heating and pressurization. All turbine engines have compressors and there are some common features that the investigator should remember.

1. The compressor is supposed to be "cool (relatively) and clean". There should not be any evidence of combustion, melting or sooting in the compressor section.

2. The compressor is the first moving part of the engine encountered by anything entering it. If some foreign object enters the engine, it will hit the compressor first.

3. In most pure jet aircraft, the compressor is at the front of the engine. It will impact the ground first and receive the most damage. This is not necessarily true of turboprop or turboshaft engines. The PWC PT-6, for example, has the compressor at the rear and the turbine in the front where it is close to the propeller.

To continue, the compressed air leaves the compressor and enters the combustion section. The combustion chamber may be either a single annular chamber or a series of interconnected chambers or "cans." Since the heat of combustion comes close to a temperature that can cause the materials used in the combustion cans to fail or deteriorate, most of the air will bypass the cans and be used to cool them. Some of the air will enter the cans where it is mixed with a spray of fuel at the front end. Once the fuel-air mixture is ignited during engine start, combustion is continuous. From this point on, the engine is "hot and dirty," and it should always show evidence of combustion and high temperature.

The process of combustion raises the temperature and pressure of the gasses. The pressure difference between the combustion gasses and ambient causes acceleration of the gasses. These gasses leave the combustion section and enter the turbine section. As they pass through, the gasses are used to turn one or more turbine wheels which are connected directly to the compressor.   The gasses continue through the turbine section and exit through the tailpipe or exhaust. The reaction to this volume of gas exiting the engine is that the engine moves in the opposite direction--and thrusts the plane forward.

The turbojet engine as described has a single shaft and (probably) four bearings. Three of the bearings carry the radial load of the shaft and one is a thrust bearing which carries the forward thrust (axial) load of the engine. Since the rotating shaft represents a source of mechanical energy, this is tapped off through a series of gears and the power is used to run various accessories, typically fuel pumps, hydraulic pumps and generators.

This is the original basic jet engine at its simplest. There were no fancy fuel controls or variable geometry gadgets installed and the principal investigation technique consisted of looking at the remains and judging power or thrust as a function of rotational damage. As we will see, this was never a particularly accurate method.

B. MULTI-SPOOL OR COMPOUND TURBO-JETS. This was the first significant variation on the basic turbojet. Engineers discovered that by adding one or more additional compressor sections, efficiency could be increased. Each additional compressor section required an additional turbine section to turn it. To make this work, each compressor is connected to its turbine with its own shaft and bearings. Here, terminology can get a little confusing. The first compressor is called the low pressure (LP) compressor and is connected to the LP turbine which is the one at the end

of the engine furthest from the combustion section. If it is a twin spool engine, the next compressor is the high pressure (HP) compressor. This is connected to the HP turbine which is the first turbine or the one closest to the combustion section. If it is a three-spool engine, things are renamed slightly. The middle compressor becomes the intermediate pressure (IP) compressor connected to the middle (IP) turbine. OK so far. Some manufacturers and users number the compressors N1, N2 and N3 which are connected (in order) to turbines T3, T2 and T1. Confusion sets in when you realize that the N2 compressor might be either the IP or HP compressor depending on how many spools there are in that engine. Figure 12-1 is a schematic of a three spool turbofan. Note that the N1 (LP) compressor is the fan itself and it is connected to its own three stage turbine section. The N2 (IP) compressor is a seven stage compressor connected to a single turbine. The N3 (HP) compressor has six stages and is likewise connected to a single turbine. Note how the three shafts are concentric, but not connected to each other. This engine has eight bearings which are not shown on the schematic. We could use this schematic to illustrate almost any turbine engine. If we remove the fan and its turbines, it becomes a twin spool turbojet. If we remove another compressor and its turbine it becomes a basic single spool turbojet.

Keep in mind that there is no direct mechanical connection between any of the engine spools. Any of them can turn independently of the others.

From the investigator's point of view, this twin (or triple) spooling complicates the investigation. Typically, the LP compressor will hit first and absorb the brunt of the impact and the damage. Since the HP compressor (if there are only two) is not directly connected to the LP compressor, it does not feel the same damage. It is entirely possible, for example, for the LP compressor to show heavy rotational damage indicating high RPM while the HP compressor shows no damage at all indicating low RPM. The net result of this is to make thrust determination as a function of rotational damage almost impossible on a twin spool engine.

Aside from the fact that there are more compressors and turbines, more shafts and more bearings, the multi-spool or compound engine is really no more complicated than the basic turbojet.

C. FANJETS. The next major change of signifi-

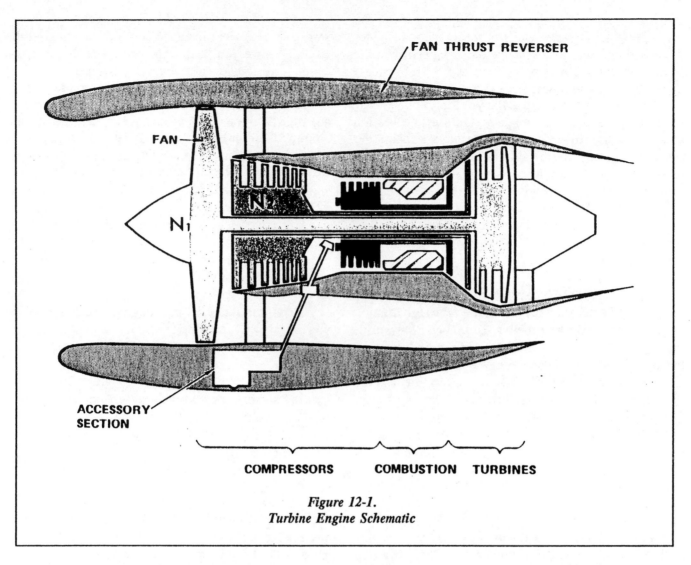

Figure 12-1.
Turbine Engine Schematic

cance was to add a large fan on either the front or the rear of the turbojet. Most of them are on the front. This fan behaved somewhat like a propeller and provided a better distribution of power for subsonic and transonic engines and a substantial increase in power. Most of the air passing through the fan bypassed the core engine, hence the name "bypass engine." A fanjet was termed either high or low bypass based on the ratio of air bypassed to air processed through the core engine. In early fanjets, the first stage compressor was merely replaced with a stage containing larger blades which bypassed air outside of the main engine casing. Later models featured fans which were huge in comparison with the rest of the engine (high bypass) and had their own shaft and turbine wheels. Consult Figure 12-1 to see what this looks like.

From an investigation point of view, this changed the post-impact appearance of the engine significantly. The fan tended to absorb all of the impact energy and protect the rest of the engine. Thus it was not unusual for an engine to show no external damage at all except for the fan section. Additionally, the fan tended to behave more like a propeller than a compressor stage. Depending on the impact circumstances, fan tip bending could be observed which could be related to aircraft velocity and engine RPM. See Chapter 11 on propellers.

Internally, the fanjet could be either a simple or compound engine with no significant differences from a basic turbojet.

D. TURBOPROP. Turboprop engines differ from turbojet engines in that they do not depend on reaction or thrust created by the moving gases. The energy in the gases exiting the engine is used to turn an additional turbine (called a power turbine) instead of creating thrust. The power turbine is connected to the propeller through a series of reduction gears. The engine turns at essentially constant RPM and thrust/power is controlled by varying the pitch of the propeller blades. Because this is not a reaction engine, there is no requirement to have the compressor in the front and the exhaust pointed to the rear or sides. Some turboprop engines, such as the PWC PT-6, have the air intake at the rear of the engine, move the air forward through the compressor section, combustion section, and turbine section and finally exhaust the remaining gasses out the side of the engine. This puts the power turbine up next to the propeller and has a lot of engineering advantages. Nevertheless, the engine still contains the same basic sections as a simple turbojet. The principle difference is that the impact damage will be different as the compressor is no longer in front.

E. TURBOSHAFTS. A turboshaft engine is any jet engine that does not produce thrust through reaction and does not drive a propeller. In the aviation world, this means helicopters. The big difference between a turboprop and an turboshaft engine is that the turboshaft engine does not have to be mounted in line with the direction of flight or the main rotor or anything else. The engineer has considerable latitude in how the engine is actually installed. This, of course, will affect the impact dynamics of the engine. Other than that, there is no essential difference between a turboprop and a turboshaft engine; or, for that matter, a turbojet engine.

That might be a good summary sentence for this part. No matter how it is mounted or what it drives; a jet is a jet is a jet. It has a compression section, a combustion section and a turbine section and the investigation of all of them is about the same.

4. FIELD INVESTIGATION LIMITATIONS.

If the engine needs to be disassembled as part of the investigation, it is almost always best to take the engine to an engine facility where there are hoists, mounting stands, tools and good lighting. Taking a turbine engine apart in the field just isn't practical.

There are, however, some basic techniques that can be used by the field investigator. While these won't always provide the final answer, they may give the investigator a pretty good idea of whether the engine contributed significantly to the accident. This chapter is limited to what the investigator can reasonably do in the field.

5. PREPARATION.

If possible, you should know what engines (by serial number) were installed at which position on the aircraft before you begin the investigation. Since the engines are the heaviest components on the airplane, it is not unusual for them to separate from the aircraft at impact and travel a good distance down the wreckage trail. If the engines are no longer attached to the aircraft, there are three ways to tell which engine is which: serial numbers, mounting hardware and accessories. Unless you are very familiar with how the engine is mounted and what accessories it drives, identification by serial number is the easiest. Keep in mind that if the engine is of modular construction, each module will have a serial number. It is also possible that some of the accessories (such a generators) will be serially numbered.

Bring the manuals and diagrams that will allow you to identify the major engine components and accessories. The parts catalog will also be helpful.

Bring a flashlight as you will want to look as far into the compressor and turbine sections as you can. Modern engines are built with numerous borescope ports to allow internal inspection. This makes the borescope a useful tool for the investigator although lugging one into the field is not always practical.

Review everything that is known about the accident. This includes the basic facts of the accident, any known air-to-ground transmissions, and witness statements. As you investigate, pay attention to what the other investigators are learning. If there was a FDR or CVR, the solution to your engine investigation may be on the recorder.

6. INVESTIGATION PROCEDURES.

Locate all the engine components and examine the engine casing for gross evidence of failure. You are looking for obvious indications of mechanical failure of the engine itself. This would include holes that were either burned or punched through the casing from the inside. If you find such a hole, it probably occurred before impact. Locate the cowling (which should have a matching hole) and start looking for engine components that would have been inside the hole. A shift in a burner can will create a blow-torch effect and rapidly burn through the engine casing. Likewise, a failed blade or rotor segment in either the compressor or turbine section will try to exit the engine sideways through the casing. If not contained, this will leave an obvious hole in the casing.

Next, examine both the compressor and turbine section for signs of interference with the casing itself or with the stators between the rotor disks. Although the tolerances are tight, the rotating parts of the engine aren't supposed to touch anything. Actually, new engines may have knife-edge seals at the tips of the compressor and stator vanes which are meant to wear in during initial engine runs. This is normal and does not resemble rotational scoring on the casing or wear of the blade tips. If there is rotational interference with the casing, the blade tips will be ground down and there will be heavy rotational scoring on the inside of the casing. This will occur if something has caused the rotor shaft to shift radially. When looking for this type of damage, keep in mind that some rotational damage will occur in almost all accidents. A small turbine engine may rotate at 40,000 rpm or more and the large ones turn at least 10,000 rpm. Even though the impact may last only a few seconds, the engine can still make several revolutions during that period.

If the rotor shaft shifts axially, the compressor rotor blades will begin contacting the adjacent stator vanes. As they touch, they will tend to sharpen each other and the edges of the mating blades and vanes will become razor sharp. You can feel this with your fingers and it is a good clue that rotor shift has occurred while the engine was still turning.

The most likely (but not the only) cause of rotor shift is failure or displacement of a bearing. We'll discuss bearings later. In the case of a turbine engine, the bearings themselves are rarely visible during the field investigation. If they are, the engine has suffered massive destruction and the bearing parts (races, retainers and rollers) may be widely scattered. Collect all of these pieces for analysis.

Among all of the causes of bearing failure, the most likely is lack of lubrication. Even if you can't get to the bearing itself, you can check the oil system. You can also analyze an oil sample spectrographically. If a bearing was wearing excessively, the oil will contain evidence in the form of metallic elements. Other checks of the lubrication system include checking for oil quantity and examining the engine for any obvious evidence of oil leaks. You might also check the oil filter and any chip detectors. If you are going to remove the chip detectors, you should be prepared to collect an oil sample from that point. If you find chips or "fuzz" on a chip detector, leave it alone. The detector can be analyzed to determine if the metallic particles were sufficient to activate the chip detector

warning light in the cockpit. Remember that you can also examine the cockpit oil pressure and temperature instruments and their associated warning lights. See the appropriate chapters.

By now you should see that there is a certain logic that the field investigator follows in evaluating engine damage. "I do not see any holes or burn-throughs, so I don't think there was a violent inflight failure of the rotating parts. I do see evidence of rotational interference with the case or stator vanes; therefore we may be dealing with a rotor shift and possibly a bearing failure. Since I can't examine the bearings, I think I will check the lubrication system."

So far, we've looked at the turbine engine for evidence of some mechanical failure that preceded the impact. What else can we do?

Usually, there are two big questions that need to be answered: What power or thrust was the engine producing at impact? Was the engine operating in a normal temperature range?

7. POWER DETERMINATION.

In the early days, the power at impact was judged by examining the rotational damage to the engine. Lots of rotational damage meant high power. Little or no rotational damage meant little or no power.

Theoretically, when the rotating compressor section of the engine hits the ground, the compressor blades should bend opposite the direction of rotation and they should all be bent alike. Depending on the composition of the blade, some types may snap off. Even then, evidence of initial bending should be seen at the remaining blade root. This consistent bending opposite the direction of rotation would certainly be evidence that the compressor section was turning at impact. Conversely, if the blades are not bent or bent in different directions, that is most likely impact damage; not rotational damage. Consider the case where the engine has seized and is not turning at all. When the engine impacts, the ends of the blades may leave blade-shaped marks on the inside of the compressor casing where the impact forces were felt. These are called "knife marks" and they are a pretty good clue that nothing was turning at impact. If something was turning, you would expect rotational scoring or gouging on the inside of the case.

When an engine goes into the water, different

indications can occur. If the engine was windmilling or not turning at all, the water will act against the compressor blades and any rotational damage will appear as blade-bending in the direction of normal rotation instead of opposite to it. If the engine was running under power, this will not happen. Another type of water-related damage results from the hydrodynamic effect of water rushing into the intake. If the engine was running, this is frequently more than the compressor casing can handle and it splits open. This is sometimes called "butterflying" and is simply a failure in hoop stress resulting from the impact.

In a ground impact, it is not unusual for the engine to ingest quantities of dirt and debris. The distance this debris travels through the engine is taken as an indication of RPM. If it wasn't running, it won't go very far. If it was turning at high RPM, the debris may make it all the way to the turbine section in certain types of engines.

Depending on the engine, other evidence of rotational damage may be seen. Transitional casings may show evidence of torsion. Bolt holes or lightening holes may be elongated with the short axis of the elongation perpendicular to the shaft (bolt hole) or perpendicular to the helical damage lines indicating torsion (lightening hole). The engine shaft itself or its splines may show torsional damage.

Thus examination of rotational damage (or lack of it) can give you a pretty good idea of whether the engine was turning or not and a rough idea of whether it was turning fast or slow. Trying to translate that into thrust is where we get in trouble.

In a turbine engine, the relationship between RPM and thrust is not linear. To keep the math easy, let's suppose that our engine will rotate at a maximum of 10,000 RPM. That's 100% RPM and 100% thrust. If the engine is running at 80% (8,000) RPM, it is producing less than 50% of its thrust. In flight (depending on the engine and conditions) the engine will idle at about 65% and windmill in the 25-35% range. Thus if the engine is completely shut down and producing nothing but drag, it will still be rotating at 2,500 RPM or more when it hits the ground. That can do a lot of rotational damage, but it is still zero thrust. If you think you can judge high power by the amount of rotational damage, you are saying that you can tell the difference between 8,000 and 10,000 RPM just by looking at the damage. Good luck!

Keep in mind that in a twin spool engine or a high bypass turbofan, the damage indications are even less useful as a determinant of power.

Fortunately, as engines got more complex, so did their control mechanisms. The engineers (probably inadvertently) provided the investigator with several ways of determining power.

A. VARIABLE GEOMETRY ENGINES. In order to increase the efficiency of the compressor section, modern engines may have different ways of controlling or adjusting the airflow. The most common are variable inlet guide vanes and variable stators. Some engines may have mechanisms that can change the size and shape of the intake. Military engines with afterburners have mechanisms that can control the size of the tailpipe.

Inlet guide vanes (IGVs) are airfoil-shaped vanes in the engine intake which twist to vary the airflow into the engine. Variable stators are a method of twisting each of the stator vanes between each compressor stage to control the airflow through the engine. Both of these devices move as a function of RPM and are directly related to it. Thus if you know IGV or variable stator position, you can determine RPM and therefore thrust.

The mechanism that makes this happen may be either hydraulic or pneumatic. If hydraulic, fuel is usually used as the hydraulic fluid. Either system consists of one or more actuators, some circumferential push-pull rods or bands and some bell cranks attached to the IGVs or stator vanes. The position of the vanes or blades can be measured with a protractor, but an easier way is to measure the extension of the actuator. By consulting a chart in the engine maintenance manual, this extension can be translated into RPM.

The hydraulic mechanism is pretty solid and usually does not change position during impact. The pneumatic mechanism is not nearly as solid and may change depending on the direction of the impact forces. If this is a concern, then the same investigative techniques you would use for any actuator may be appropriate. See Chapter 13.

B. FUEL CONTROL UNIT. Modern fuel control units can contain a wealth of information about what was being demanded of the engine and what it was actually doing. In most cases, it takes a fuel control expert to analyze it.

C. BLEED AIR VALVE. On some engines, there is a bleed air valve (or strap) which will dump excess compressor air automatically to improve engine acceleration characteristics. This valve automatically closes as the engine reaches a speed where it can process all of the air. While this does not provide an exact indication of power, knowing whether the valve was open or closed at impact can tell you the operating range of the engine.

D. FUEL FLOW. If we knew the engine fuel flow, we could calculate engine thrust fairly accurately. One way is to analyze the fuel flow indicator(s) in the cockpit. Another is to analyze the fuel flow transmitter itself. The flow of the fuel moves a mechanical vane which is spring-loaded to the no-flow position. The vane is calibrated and its position is transmitted to the cockpit as fuel flow. In a severe impact, the vane may leave witness or capture marks on the inside of the transmitter. This position can be duplicated on an operational fuel flow transmitter and the fuel flow determined.

E. COCKPIT INDICATIONS. In the cockpit, there may be evidence available from the engine power levers (throttles) and the engine instruments. The position of the power levers tends to be unreliable as impact forces generally move them full forward, although there have been cases where a severe impact collapsed the throttle quadrant around the throttles and generated a witness mark on the quadrant. If the power levers are someplace other than full forward, perhaps their position can be correlated with evidence on the engines themselves. See Chapter 14 for examination of engine instruments.

8. TEMPERATURE DETERMINATION.

Let's turn to the second big question. Was the engine operating in a normal temperature range? We have already examined the engine for gross mechanical failure and didn't find any. We've made an assessment of engine thrust at impact. Now we should check for evidence that the engine was operating at higher than normal temperature.

In normal operation, the engine gets pretty warm from the combustion section to the tailpipe. Over time, the outside of the engine will show evidence of constant exposure to high temperatures. Certain metals will pick up a discoloration after exposure to temperatures. The point is that we should have some idea what the engine normally looks like before we identify that

appearance as overtemperature.

Actually, the best clues to operating temperature are going to be found in the turbine section. If the engine is running too hot, things in the combustion section are going to start melting. If the engine gets really hot, the turbine blades can melt very rapidly. Examine the turbine blades. If all the blades on any turbine wheel are jagged and deformed, that's melting and that wheel has been exposed to severe overtemperature. If the blades on a wheel are intact, but have longitudinal cracks at the ends, that wheel was also exposed to overtemperature and the blades were just getting weak.

If the blades are intact, but coated with a metallic grit like very rough emery paper, something has melted down somewhere forward of the turbine and impinged on the turbine blades. This will occur (logically) on the leading edge and front face of the blades and can be felt with your fingers. This is called "metalizing." If the turbine blades are intact (not melted) and have not picked up any metalization, the engine was probably operating within its normal temperature range.

Knowledge of how molten metal behaves can be useful. In theory, molten metal in a turbine engine usually comes from the combustion section. There have been cases where a titanium fire in the compressor section due to blade interference (friction heating) with the titanium casing created molten metal. Metal particles scraped from a different type of compressor casing due to blade interference will become molten as they pass through the combustion section. The molten metal is first noticeable when it impinges on something in the turbine section. What happens next depends on the relative temperatures involved. If the surface that the molten metal hits is hotter than the molten metal, fusion takes place. This is possible because many of the components of the combustion section are aluminum alloy which melts at a lower temperature than the operating temperature of the turbine section. When fusion occurs, the molten metal melds smoothly into the surface of the other metal and becomes part of it. It looks like it was painted on and it cannot be scrapped off with a fingernail or a pocket knife. If, on the other hand, the molten metal is hotter than the surface it hits, then adhesion takes place. Adhesion produces the rough, gritty appearance where the specks of molten metal are merely stuck to the other metal and have not become part of it. These particles can be scrapped off with a knife blade. The significance of all this is that knowing whether the process involved fusion or adhesion and knowing the type of metal involved can lead to conclusions about the actual operating temperature of the engine.

Although not readily noticeable in the field, another indication of overtemperature is a metallurgical phenomenon called "creep." This is the permanent elongation of a metal part due to a combination of stress and high temperature. Since these conditions pretty well describe the normal environment of a turbine blade, they are more susceptible to creep. In operation, the turbine blade "grows" slightly, but, if the temperature is kept within the normal operating range, it returns to its correct size when the engine cools down. Excessive temperatures accelerate the creep process dramatically. The clues that this has occurred are indications of blade rub on the turbine casing or shroud; "necking" of the center of the blade; and permanent elongation of the blade. While you may see evidence of blade rub in the field, confirmation of creep will probably require laboratory analysis. Keep in mind that rotor shift or possibly impact damage could also cause turbine blade rub. If tensile failure of a turbine blade occurs, the fracture surface will tend to be granular in appearance as small fissures may open in the metal near the fracture surface.

A final method of checking for overtemperature is to examine the engine's Time/Temperature or Events History recorder if it has one. Generally, these recorders are designed to let the maintenance and engineering people know how often the engine has been operated at various temperature levels and how often the engine has been cycled through various RPM levels. While this isn't exactly what the investigator needs, it can be helpful. Some engines do not have these recorders. Some have them permanently installed and some have them selectively installed at the whim of the maintenance folks. You have to ask.

9. COMMON TURBINE ENGINE PROBLEMS.

A. FOREIGN OBJECT DAMAGE. Foreign Object Damage (FOD) is easily the most common source of damage to turbine engines. During engine certification, the manufacturer must demonstrate that the engine can ingest a bird of a certain weight and a certain amount of ice without losing significant power. This does not mean that the engine can handle birds and ice without damage. It is likely that the ingestion of any solid object will produce some amount of damage to the compressor section.

In an aircraft accident, FOD may occur as a result of impact and the resulting damage may be of no significance to the accident. Differentiating between the rotational damage due to impact and pre-existing FOD damage can be difficult.

In general, rotational damage during ground impact is most severe at the front of the engine and tends to become less severe as you move back through the various compressor stages. Eventually, you may reach compressor stages which felt no rotational forces at all and are undamaged. This is particularly common in a twin spool engine where the N2 compressor keeps turning independently while the N1 compressor is soaking up the impact forces.

FOD damage tends to produce just the opposite indications. The early compressor stages may be relatively undamaged. The foreign object may sneak through the first stage and only take a small chunk out of one blade. The second stage now sees two foreign objects; the third sees four objects and so on. As you go back into the compressor section, FOD damage gets worse and worse due to this multiplier effect. If the N2 compressor is still spinning, it,too, will be damaged. Eventually, the engine comes to a stop because there are so many blades and chunks of metal jamming the system that it just can't turn anymore. This process is called "log jamming" and once the engine stops, there is no more damage beyond that point.

Key points: rotational damage due to ground impact starts bad and gets better. FOD damage starts good and gets worse.

If you suspect FOD, pay close attention to the blades of the early stages. A metal object may hit a blade hard enough to leave an imprint on it and you may be able to confirm FOD and the nature of the object that caused it. Microscopic traces of objects causing the damage can be spectrographically analyzed to determine the composition of the object.

The three principal sources of inflight FOD are birds, ice and metal parts from the aircraft itself. Birds are not a problem for the investigator as they leave evidence. Ice can come from three principal sources. Tail mounted engines on some aircraft can ingest the ice formed from a leak in a lavatory drain system. This is called "blue ice" because the chemicals used in lavatory systems are blue. It has caused severe engine damage and the best evidence may be stains on the fuselage aft of the lavatory drain. Tail mounted engines are also susceptible to structural ice shed from the aircraft wings as they flex during takeoff or to ice removed by the deicing boots after excessive ice build up. This has also caused severe engine damage and resulted in accidents. Evidence is based largely on the accident history of that type aircraft and atmospheric conditions conducive to the formation of undetected structural ice. The third source of ingested ice would be ice formed on the intake or cowling lips during flight. Most aircraft have anti-icing systems to preclude this. The investigation might focus on the anti-icing system to determine if it was on or working.

If the ingestion of a metal object is suspected, the investigation should focus on the intake, cowling and aircraft structure forward of the engine. This is in addition to a close examination of the compressor blades for imprint marks.

B. VOLCANIC ASH INGESTION. This is a special form of FOD that deserves discussion. Volcanic ash is a dust that is almost pure silicon and has the consistency of well-sifted flour. When it gets into the engine, it is converted to glass in the combustion section and the turbine guide vanes, nozzles, blades and cooling passages are literally glazed with this molten glass. As the turbine can no longer efficiently drive the compressor, the engine shuts itself down. In the old days, this was called a "flame-out" which is a fairly accurate description of what happens when the engine ingests volcanic ash. So far, most of the engines failing in this manner have been restarted once the plane descended to airstart altitude.

If this is suspected as a factor in the accident (due to the presence of a volcanic ash cloud), the investigation is fairly simple. The glazed coating in the turbine section should be readily visible. In addition, the ash will do other damage, particularly to the aircraft transparencies (windows.) They will be scoured to the point where you cannot see through them.

C. COMPRESSOR STALL. The causes and mechanism of turbine engine compressor stalls are beyond the scope of this text. In almost all cases, a severe compressor stall stagnates or stops the air flow. An aerodynamic stall of the turbine can cause back pressure in the combustion chamber resulting in reverse flow of the gases and stall of the compressor. In this case, witnesses have reported seeing fire coming out of the front of the engine. Since the compressor is supposed to be clean and cool, this

produces the best evidence that a severe compressor stall has occurred. The back sides of the rear compressor blades will be coated with soot and, in some cases, the rear compressor blades may show signs of melting or even adhesion of molten metal particles. One of the significant effects of severe compressor stall is disruption of the air flow and over temperature of the turbine section.

D. ACCESSORY FAILURE. Most turbine engine accessory drive systems incorporate shear points or disconnects to decouple the engine from the accessories or the constant speed drive for the generator. If the drive is sheared or disconnected, this may have occurred prior to impact and may be a clue as to the engine's operating condition. The direction of rotational shear failure may offer a clue as to what caused the failure. In some cases, the engine drives accessories that are essential to its own operation. If the engine driven fuel pump fails for any reason, the engine is going to quit. It is fairly easy to examine the fuel pump and its drive to determine if it was running at impact. As a rule, any rotating device (pump or generator) should show evidence of rotation if it was turning at impact and sufficiently damaged by the impact forces.

E. THRUST REVERSER FAILURE. Failure of a thrust reverser to operate properly on landing does not usually cause an accident. If it does, it is fairly easy to investigate as there is usually a flightcrew member still alive to explain what happened. Unwanted deployment of a thrust reverser during takeoff or in flight is much more serious and has caused at least one accident. While some aircraft and engines are designed to allow the use of reverse thrust in flight as a drag device, most are not. None of them are designed to permit thrust reversal while the engine is at high power.

Thrust reversers may reverse either the fan exhaust (cold reverser) or the core engine exhaust (hot reverser) or both. If only a fan reverser is installed, there will usually be a spoiler mechanism for the core exhaust so that it won't negate the effect of the fan reverser.

The reversing devices (clamshells, buckets, shields or spoilers) are usually pneumatically operated using bleed air as a source of power, although they could be hydraulically operated. Depending on the mechanism, the high pressure air may operate a pneumatic ram which moves the device directly or an air motor which drives a jackscrew and traveling nut. For investigation purposes, the pneumatic ram may be treated as a hydraulic actuator. If the mechanism is a jackscrew, the position of the traveling nut correlates directly with the position of the reversing device. Although the jackscrew mechanism has no tendency to change position due to the impact it could continue to turn if it was in transit during impact.

If the reversing devices are obviously deployed and they shouldn't be, the best investigative technique is to examine the surrounding impact damage and determine whether the damage occurred before or after the reversers were deployed. If, for example, areas normally protected by the clamshells in the closed position are damaged--and the clamshells do not have corresponding damage--you can make a pretty good case for in flight reverser deployment.

Additionally, you can check the cockpit controls. Due to the safety features on the system, the reverse thrust select levers should not be in the reverse position unless someone put them there. If they were put there, they should remain there throughout impact. If the levers are not in the reverse position, but the reversers are clearly deployed, then we may have inflight deployment uncommanded and unwanted by the flightcrew. Now, the best approach is to treat the entire thrust reverser system as a separate system and investigate it as such. Thrust reversers are usually considered part of the airframe and not part of the engine.

9. BEARING FAILURE.

We have discussed bearings earlier and some of the indirect indications of bearing failures. As was mentioned, you don't often see the engine bearings in the field. If you do, though, collect and save all the pieces. There is a lot that can be done with them in the laboratory.

We mentioned that the principle cause of bearing problems is lack of lubrication. That's true, but the evidence of that lack of lubrication can take several forms. The bearing rollers or balls may appear burned, glazed or, in some cases, welded to the bearing retainer. Also, there are other bearing failure modes. Laboratory examination will be required.

10. SUMMARY OF FIELD EXAMINATION TECHNIQUES.

Field examination of a turbine engine follows a fairly standard protocol. If internal examination is needed, it is always best to remove the engine from the wreckage and take it to a facility where it can be disassembled.

A. Identify and account for all the major components of the engine.

B. Locate and recover any engine-installed recording devices.

C. Check the external appearance of the engine. Look for gross evidence of mechanical failure or overtemperature.

D. Obtain fluid samples, particularly the engine oil.

E. Examine the fuel and oil filters.

F. Examine the chip detectors if installed. Preserve any chips or "fuzz" for analysis along with the detectors themselves.

G. If possible, use a borescope to examine the engine internally.

H. Examine the engine mechanisms such as IGVs, variable stators, fuel controls, etc. for evidence of power output.

I. Examine the turbine section for evidence of overtemperature operation.

J. Examine the accessory drive train for condition and continuity.

K. Examine the accessories for condition and operation.

11. SUMMARY.

As pointed out at the beginning of this chapter, turbine engines are probably easier to investigate than reciprocating engines. Much can be done by visually examining the external appearance of the engine and what can be seen of the compressor and turbine sections. By applying a little logic you can usually spot evidence that the engine was obviously operating abnormally. Although there are differences in engines made by different manufacturers, each turbine engine must have the same basic components.

13

Aircraft Systems

1. INTRODUCTION.

It is not possible to completely cover all aircraft systems and their variants in a book of any reasonable size. What we can do is describe a systematic approach to the investigation of any system and discuss some of the specific techniques used for the investigation of specific system hardware.

2. GENERAL PROCEDURES.

Aircraft system investigation can be difficult, particularly if you don't know how the system works or what the components look like. While all aircraft systems are not alike, they all have similarities. If you have a basic understanding of (say) hydraulic systems, you should feel comfortable with any aircraft hydraulic system after a short review of it.

This short review is an essential part of the investigation. Take the time to sit down with the maintenance manual and the schematics and the parts catalog and learn how the system is put together. This may take a couple of hours, but it is time well spent.

Next, bring the documents with you (probably the parts catalog) that will allow you to identify the components and verify that they are the correct part. This part verification is always a good idea in any phase of the investigation.

Finally, be prepared to mark and tag the components as you identify them. There are a lot of them and there is no point in doing this twice.

3. GENERAL SYSTEM KNOWLEDGE.

Aircraft systems may be thought of as methods of transferring power or energy from where it exists to where it is needed--and doing something with it. The basic ways of doing this are mechanical, electrical, hydraulic, pneumatic and fluid transfer. All aircraft systems use one or a combination of these methods. A flight control system, for example, may be both hydraulic and mechanical (hydro-mechanical.) A landing gear system could be electrical or hydraulic. Fuel systems are normally just fluid transfer systems, but, aside from delivering an energy source to the engine, the fuel may also be used as a coolant for oil systems or as a hydraulic medium for engine components.

All of these methods of energy or fluid transfer share some common features. Understanding these features allows us to apply some common techniques and methodically investigate any system.

A. SUPPLY. Somewhere there must be a source of energy or fluid that needs to be moved somewhere else. In the case of a mechanical system, the energy source is usually the pilot through the force he applies to levers, control columns and pedals. This force is transmitted somewhere else where something useful happens. Hydraulic systems must start with a reservoir of fluid. Fuel systems have tanks of fuel. Electrical systems start with electrical energy manufactured by a generator or stored by a battery. Pneumatic systems must have a source of air, either produced by the engines or stored in pressurized containers. A good first step in any systems investigation is to check the supply of whatever it is you are dealing with. If there wasn't any hydraulic fluid in the system, for example, you can pretty well skip the rest of the investigation; you're done.

B. POWER. Something shoves the supply through the system. At the front end, there must be some sort of pump or pressure available to move the supply medium. This is a good second step in system investigation. Check whatever furnishes pressure or power to the system and see if it was working.

C. CONTROL. Most systems can be controlled to some extent from the cockpit. The control often consists of an input signal identifying what is desired and a feedback signal identifying what happened. Both aspects of the control should be examined. A good third step in the investigation is to check the cockpit levers and switches to make sure they were set correctly. It doesn't make much difference how much fuel was on board if the fuel selector was off.

D. PROTECTION. Most aircraft systems incorporate protection devices to prevent the system from destroying itself (pressure regulators, fuses, circuit breakers) or from destroying the airplane (check valves, filters, filter bypass plumbing, emergency reservoirs, emergency pumps, etc.) Check these.

E. DISTRIBUTION. The system medium (fluid, air, electricity or mechanical force) has to be distributed somehow to where it is needed. This is done through electrical wiring, hydraulic and fuel plumbing, cables, pulleys, bellcranks and so on. This is usually the hard part of any systems investigation as the distribution plumbing is frequently destroyed or severely damaged in the accident. It is very hard to put it back together and say with certainty that it either was or was not working. The best approach is to skip this item temporarily and go to the next one.

F. APPLICATION. At the end of the distribution scheme, the system does something useful for us. Electrical systems operate motors, lights or electronic equipment. Hydraulic systems use actuators to move things. Mechanical systems move things through leverage and force. One useful investigative approach is to verify that something on the application end of the system was working correctly. If you can do this, then you have pretty much validated the distribution scheme for that branch of the system. If, for example, you are trying to locate an electrical problem, knowledge that the aircraft's navigation lights were seen by

witnesses (and therefore working) can save you a lot of hours of messy wiring investigation. A second useful approach is to examine the actuating device for the part of the system you are interested in. If the motor or actuator wasn't working correctly, then it really doesn't make any difference whether the distribution plumbing was intact or not.

The whole idea of thinking about systems in this way is to focus the investigation on the easy things and avoid, if at all possible, having to reconstruct the system wiring or plumbing.

4. COMPONENT EXAMINATION.

In any system investigation, you inevitably work your way down to looking at actual system components; motors, actuators, switches, valves and so on. There is pretty good acronym called PIXTOW which can guide you through an orderly examination of almost any component.

1. Photograph it. Get pictures of what it looked like before you messed with it.

2. Index it. This applies to components that rotate or slide or have parts that move. Use a scribe to put a mark across two mating surfaces so, no matter what happens next, you can always restore the moving parts to their "as found" position.

3. X-ray it. Before you take it apart, think about the advantages of industrial x-ray. This is non-destructive and can show you things that you probably won't see if you do take it apart. If you are dealing with a switch, for example, the act of opening it is likely to destroy the evidence of whether the contacts were open or closed.

Up to this point, you haven't done anything to the component that would destroy any evidence. From here on, you are likely to change things and you want to be sure that you have done all the non-destructive things first.

4. Test it. If it is possible to put pressure or electrical current to the component, go ahead. See if it works.

5. Open it. This is what everyone wants to do first. Before you do this, it is a good idea to decide what you expect to find and how you are going to recognize it. These components aren't Christmas

presents and you should pretty well know what's in there before you unwrap it.

6. Write about it. Now is the time to document what you have done, what you learned and what you concluded about that component.

5. MECHANICAL SYSTEMS.

When we want to translate the pilot's control stick, column or pedal movements into control surface movements, we use a combination of levers, pulleys, bellcranks, push-pull rods, torque tubes and cables. All of these things move and they are subject to restriction or blockage by some object that is not supposed to be there and got caught in the machinery. In any wreckage examination, you can pretty well count on finding something (tool, flashlight, rag) that was lost or left in the tummy of the aircraft and was a potential source of obstruction to any nearby mechanical system. When you find these stray objects, examine them closely for evidence that they might have been caught or pinched in the machinery.

6. CABLE SYSTEMS.

Cables are a popular method of transferring mechanical force somewhere else. Some systems are single cables which incorporate a spring or retraction device which brings the cable back to neutral or the zero position. These might be used to open landing gear doors or release gear uplocks. Cables used on flight controls are almost always two-cable systems. One moves the rudder left; the other moves it right. If these are separate cables, one may break and the other may still work--in one direction only, of course. If both are parts of the same cable, a single break disables the entire system. Some cable systems incorporate centering or tensioning springs and, when a cable breaks, the spring may drive the system full travel in the opposite direction. Thus if you notice that a trim tab (for example) is fully deflected for no apparent reason, check for a broken cable. Since the cable can break during the impact sequence, this full travel phenomenon can occur and have nothing to do with the actual accident.

When a cable breaks, it is most likely to do so at either a pulley (due to the extra tension as the cable goes around the pulley) or a fairlead (due to abrasion.) Sometimes, a cable will chafe against an electrical contact and show evidence of burning. Aside from that, in-service cable failures are always tension failures. If the cable shows evidence of some other failure mode, shear for example, then the cable was probably cut or guillotined during impact.

Cables are composed of bundles of individual strands of wire spirally wound. A number of these bundles, usually seven, are then spirally wound to form a cable. Typically, aircraft cable will have from 49 to 133 individual strands of wire. When it fails, the strands fail individually one at a time. Viewed with high magnification, each should show the necking down and cup and cone mode characteristic of ductile tension failure. If a cable did fail in flight, it is likely that some individual strands had already failed and weakened the cable prior to the final failure. These can be detected by rubbing a cloth along the cable. The cloth will snag on the failed strands. Wear gloves while you do this or have a first aid kit handy.

If you are dealing with an inflight structural failure on a plane that has a pulley and cable flight control system, remember that the cable always fails last. When the wing fails, for example, tension on the cable is released, but it still connects the wing to the fuselage. This may cause the wing to swing over or under the fuselage and hit some other part of the plane. Eventually, the cable fails, but in doing so it may leave scratches or "tracks" where it was stretched tightly across the fuselage. When the wreckage is examined, the cables may present a curly or coiled appearance similar to Christmas wrapping ribbon that has been stretched and allowed to coil. The same thing happened here. The cable has been seriously stretched around a corner prior to failure. The cable to look at closely is the one that does not show any signs of stretching. Perhaps it failed first. These clues and evidence of secondary collisions as the wing left the aircraft can help explain some of the things that happen after an inflight structural failure.

Cable systems must be tight (tensioned) to work. They must, in fact, be tensioned to a specific value to work properly. Following a severe impact, most of the cable systems will be loose and not working purely as a result of the impact. If this is the case, there is no practical way to determine if the system was correctly tensioned prior to impact. It is probably a safe assumption that if the plane made it that far, the cable system tension couldn't have been too far off.

If you suspect flight control system failure and the controlling cable systems are loose (as you expected them to be) you might look for reasons for that looseness other than the impact itself. The most obvious would be failure of the cable, which has already been discussed. Less obvious would be failure of any pulley, bellcrank or horn in the system. Failure of any of these will result in instant loose cable and (therefore) system failure.

If you find a failed part in the cable system, you now have the problem of determining whether the failure occurred before the crash or resulted from it. A good first step is to check the part number and verify that it is the correct part. Next, you need to examine the fracture surface itself. Is the manner in which it was loaded at failure (tension, bending, etc.) consistent with the impact forces or the way the part was normally loaded in service? Is there any indication of any pre-existing weakness such as a fatigue crack, corrosion or metallurgical defect? Remember that the correct part correctly installed is supposed to be strong enough to carry the in-service load. If it fails in tension overload (say) that would most likely be a result of the impact forces. If there was a pre-existing weakness, then the failure could have occurred in flight. Keep in mind, though, that it also could have occurred during impact. The final failure mode of any metal part is always an instantaneous one due to overload and there is no difference in the appearance regardless of when it occurred. The determining factor may be an analysis of what would have happened if that failure (and subsequent cable system failure) occurred in flight. Would that explain the accident? If not, go look at something else.

7. HYDRAULIC SYSTEMS.

If you are investigating a hydraulic system, you might as well start with the fluid. The fluid must be clean and the correct type of fluid for that system. The most common fluid in use is a red-colored mineral based fluid called Mil-5606 or some variant of that. Most other hydraulic fluids are synthetics which may or may not be compatible with the red stuff. Mil-83282, a synthetic, is straw-colored and fully compatible with Mil-5606. Other fluids may be purple or blue. Some of them are highly corrosive and can only be used in systems with o-rings and seals specifically designed for that fluid. Indications that the fluid was contaminated (or that the o-rings were destroyed by using the wrong fluid) will show up in the system filters. Hydraulic fluid samples can usually be obtained long after the accident. Consider that any hydraulic actuator is supposed to be full of fluid at both ends. One end is bound to have some left in it.

Because of their color, hydraulic fluid system leaks are fairly easy to locate. Most leaks occur at the connecting fittings. If a leak occurs in the middle of a tubing run, the tube must have been damaged (abraded or cut?) somehow. In checking fittings, an interesting phenomena can occur if the system was exposed to fire or high temperature. The "B" nut, the fitting and the tubing all have different coefficients of thermal expansion. Exposure to fire tends to loosen the nut and make it appear as though it was never tight in the first place. As a rule of thumb, if the "B" nut is merely finger tight (post accident involving fire) it's OK. If it is a quarter turn or more loose, it probably wasn't tight in the first place.

A. HYDRAULIC ACTUATORS. Hydraulic systems operate actuators which translate the hydraulic pressure into linear (or sometimes rotary) motion and moves something. Airplanes use actuators to operate flight control surfaces, flaps, spoilers, air brakes, thrust reversers, landing gear, nose wheel steering systems, propellers and wheel brakes. Here are some general truths to remember about hydraulic actuators.

1. An actuator is an actuator is an actuator. They all work about alike and none of them know what they are connected to. It is not uncommon for the same actuator to be used in several different applications in the same aircraft or in other types of aircraft.

2. Except for propellers and wheel brakes, it takes positive hydraulic pressure on one side of the piston to move the actuator. Mere release of pressure from the other side of the piston normally won't do it. Brakes and props are an exception. When pressure is released on a brake actuator, springs pull it back from the wheel. The hydraulic pressure in a prop dome is counterbalanced by the centrifugal force generated by counterweights. Another exception would be a flight control surface exposed to airloads. When hydraulic pressure is lost, airloads will retract the device (spoilers) or move it to a neutral position. With those exceptions, hydraulic actuators are pretty hard to move except with fluid pressure.

3. There is always a predictable (and usually linear) relationship between the extension of the hydraulic actuator and the position of whatever it is supposed to be moving. Consider an aircraft rudder. If hydraulically actuated, full travel in one direction ought to be full actuator extension, while full travel in the other direction ought to be full actuator retraction. Neutral on the rudder ought to be exactly half exten-

sion on the actuator. This means that when you are dealing with an actuator, you can always tell the position of whatever it was actuating by measuring the extension of the actuator. To help you, the maintenance manuals will (usually) contain a graph plotting actuator extension against position of the surface or device being actuated. The mechanic, after all, must have some method of adjusting the actuator correctly when he installs it. He uses this chart and you can use it to find out the position of the control surface (or whatever) at impact.

Is it possible for the actuator to move due to impact forces? Yes. If this happened, the actuator probably took a pretty good whack and Plan B may work. Plan B is where you take the actuator to a machine shop and physically saw it open, after, of course, doing all the non-destructive things already recommended. Once opened, look for witness or capture marks left by the edge of the piston on the inside of the cylinder. This is where the piston was when it felt the impact forces. Reposition the piston to the marks, measure the actuator extension and proceed as in Plan A.

B. HYDRAULIC PUMPS. Hydraulic pumps are normally either engine-driven or electric. There are a variety of types of pumps that could be installed. If the pump has rotating gears or vanes and it was severely damaged in the impact, it may be possible to determine if it was running or not by opening it and looking for evidence of rotational damage. Most engine driven pumps will be connected to the engine accessory drive by a splined coupling designed to fail if the pump fails. Although this is called a shear coupling, it actually fails in torsion. Check this. If it has not failed, the pump was probably working or at least capable of working. If it has failed and the mating surfaces present a burnished appearance (torsion failure), the pump failed while the engine was still turning. If the coupling failed in some mode other than torsion, it probably occurred during impact when nothing was turning. Pump cavitation describes a situation where air has somehow invaded the hydraulic system. The pump, which was expecting to pump incompressible fluid suddenly encounters compressible air. At those operating pressures, it literally explodes droplets of fluid against the cylinder walls. This produces a dented appearance on a surface that should be smooth. Hydraulic pumps are subject to overheating and sometimes the condition of the paint on the pump can provide a clue as to operating temperature. See Chapter 8. Some manufacturers paint their pumps with a

special paint that is designed to identify overtemperature operation.

C. HYDRAULIC ACCUMULATORS. Hydraulic systems use accumulators to dampen pressure surges and store hydraulic power for emergency use. The accumulator is a sphere or cylinder with an internal diaphragm or piston. The system side of the diaphragm has normal system hydraulic pressure while the other side (called the air side) is charged with air or nitrogen to about half of the normal system pressure. From an investigation point of view, there are two things to remember about accumulators.

1. Check the air charge. If it has zero charge, it could be a result of damage during impact. If it was not properly charged, the hydraulic system would probably still work, but the emergency accumulator power source would not be available.

2. CAUTION! Do not mess with an accumulator unless you are sure the air charge has been bled to zero. A damaged accumulator with a full charge can be lethal.

D. HYDRAULIC SYSTEM FILTERS. Hydraulic fluid needs to be very clean to avoid damage to system components. The human eye cannot see contaminants smaller than about 50 microns, which is not good enough. That means that visual examination of the fluid or the filter won't work unless there is gross contamination present. You will need a contamination test kit, but if you suspect contamination, you might as well take both the fluid sample and the filter to the lab. There, they can also check viscosity, moisture and flash point.

Most hydraulic filters have a bypass feature. If the filter is completely clogged, the increased pressure build-up will open a spring-loaded valve and the fluid will bypass the filter. Thus the system will continue to work--for a while. Some of these filters will incorporate an indicator to tell the mechanic that fluid bypass has occurred.

8. PNEUMATIC SYSTEMS.

When we think of pneumatic systems, we tend to think of a system that uses compressed air to do something mechanical for us. From an investigation point of view, that's a little narrow. We should think of a pneumatic system as any system that uses a compressed gas for something. Now, besides mechanical systems, we can include heat and pressurization systems, anti-icing systems, fire extinguishing systems and oxygen systems. On the front end, we may manufacture the compressed gas (turbine engine bleed air) or store it in a pressurized container (oxygen systems, fire extinguishing systems and emergency air pressure bottles.) On the application end, we may want to use the pressure of the gas (emergency brakes, emergency landing gear extension); its temperature; (cabin heat, anti-ice) or its chemical properties (fire extinguishents, breathing oxygen.) The nice part about this approach is that all of these systems work about alike and all use similar hardware in their distribution systems. As a matter of fact, they have a lot in common with hydraulic systems and many of the investigative techniques already discussed will apply.

A. SYSTEM LEAKS. If the system runs on air or oxygen, leaks are difficult to detect under normal circumstances. Post accident they are very difficult, because the system probably has a lot of leaks after the crash. Engine bleed air systems are potential sources of ignition if the bleed air temperature is high enough. This can usually be determined from the maintenance manuals. Oxygen systems can't start a fire, but they can aggravate an existing one. In passenger transport aircraft, the oxygen lines to the passenger seats are normally empty until the system is activated. Built-in fire extinguishing systems usually incorporate a discharge disc which ruptures to let the mechanic know that discharge has occurred.

B. PNEUMATIC EMERGENCY SYSTEMS. Compressed air is sometimes used to do something that is normally done hydraulically. This is the case where air is used for the emergency braking system or the emergency gear lowering system. In this case, the compressed air is introduced directly into the hydraulic system through a shuttle valve that can allow either air or hydraulic fluid to the actuator. Obviously, if the system failure is downstream of this shuttle valve, neither method will work. Also, the compressed air emergency system won't run the whole system. There are only a certain number of brake applications available.

C. PNEUMATIC SYSTEM FILTERS. Contaminated air is not nearly as serious as contaminated hydraulic fluid and the air filter is not as sophisticated. Check it anyway.

9. FUEL SYSTEMS.

In terms of plumbing and distribution systems, fuel transfer systems have a lot in common with hydraulic and pneumatic systems. The biggest difference is probably that they operate at much lower pressure and can take advantage of larger lines and plumbing. Because the lines are large (and full of liquid) the investigator will sometimes notice a type of failure not often seen in hydraulic or pneumatic systems. A large fuel line full of fuel will tend to behave like a closed cylinder when it is exposed to impact forces. Due to the hydraulic effect of the trapped liquid, the line will tend to rupture in hoop stress, i.e. longitudinally. This is normal for the situation and it does not necessarily mean that the line failed in flight from internal overpressure.

A. FUEL VENT SYSTEMS. Since the purpose of the fuel system is to move the fuel from the tank to the engine(s), we must replace the fuel with air or it won't come out of the tank. This is usually a fairly simple system incorporating a check valve which will allow air in, but won't let anything out. Some fuel systems have a vent valve which will vent fuel if the tank is overfilled or overpressurized due to an increase in temperature after the tank was filled. This may or may not be part of the vent system. If you have fuel in the tank and can't figure out why it didn't get to the engine, check the vent system.

B. FUEL RETURN LINES. All modern fuel systems supply fuel to the engine at a pressure and quantity greater than it can use. The engine uses what it needs at the moment and returns the excess to the tank through a return line system. In some aircraft, this return system uses the same plumbing as the vent system. In this case, a check valve failure means that the fuel will be pumped overboard instead of into the tank. This can contribute to rapid loss of fuel.

C. FUEL PUMPS. Comments on hydraulic pumps also apply here.

D. FUEL SYSTEM CONTAMINANTS. Since we regularly add new fuel from sometimes uncertain sources, fuel contamination is an ever-present problem. The most common contaminant is water. Sometimes it comes with the fuel, but it is also possible for water to form on the inside of a partially filled tank through condensation. The water may either be dissolved in the fuel or suspended (entrained) in it. Inadequate or deteriorated seals on the fuel tank caps of general aviation aircraft is a common source of water. Suspended water is fairly easy to detect with the naked eye as there will be two types of liquid in the test container and an obvious separation between them. Dissolved water can make the fuel appear cloudy. As a general rule, if the cloudiness clears up from the top down, its water. If it clears up from the bottom up, it was air.

Other contaminants are going to be either foreign particles or microbial growth. Foreign particles can be almost anything including rust, dirt, metallic particles or rubber. These appear as sediment and can be detected (sometimes) by swirling a sample of the fuel in a white porcelain bucket. Microbial growth results from micro-organisms living in the fuel-water interface. It appears as a brown or black slime.

Another contaminant, of course, is the wrong type of fuel. This was discussed in Chapter 10.

E. FUEL SYSTEM FILTERS. Most fuel tanks have screens over the outlet to catch the big stuff. There will be other fuel filters in the system, probably closer to the engine. These should be checked. Most fuel filters have a bypass feature and a bypass indicator.

F. FUEL CONSUMPTION. Since fuel is consumed during flight, fuel starvation is sometimes a possibility. If the circumstances of the accident suggest that lack of fuel was a factor, start by looking for evidence of fuel at the scene. If the aircraft had no fuel at all, it probably didn't burn at impact. If the tanks didn't rupture, they should still contain fuel. If they did rupture, there should be evidence of fuel either by smell or chemically burned vegetation. Interviewing the people who were first on the scene may be helpful. If there was fuel present and it soaked into the ground, it can be chemically detected in soil samples long after the accident.

A good second step is to sit down and calculate the amount of fuel that should have been there based on amount serviced, rate of consumption and flight time.

10. ELECTRICAL SYSTEMS.

The investigation of an aircraft electrical system requires a healthy degree of skepticism. Things are not always as they appear. Electrical events happen at the speed of light which, for our purposes, is essentially instantaneous. Aircraft crash at some lower speed and a lot can happen, electrically, between the time the plane first hits the ground and the time it finally comes to a stop. Circuits which were "on" at initial impact are now "off." We now find evidence of shorting or arcing which confuses us because it occurred as the system was damaged during impact; not before.

We can save ourselves a lot of time by learning everything we can about the accident before we attack the wreckage. This would include crewmember and passenger statements, witness statements, air-to-ground communications and FDR/CVR information, if available. Anything that indicates that something electrical was operating correctly prior to impact will validate portions of the system and save us a lot of work. As a rule, evidence in the recordings or statements that something was working is probably more credible than evidence in the wreckage that it was not. True story. A pilot took off in IFR conditions, radioed to the tower that he had complete electrical failure and crashed shortly thereafter with no further transmissions. That single statement told us a lot. First, he didn't have complete electrical failure (obviously) and second, he was flying an aircraft with an AC system, but talking on a radio that required DC power. Third, whatever failure he had was one that gave him positive indications in the cockpit. The fact that he was IFR and crashed so quickly suggested a failure that disabled his primary flight instruments. All this made the investigation a lot easier.

Eventually, you'll run out of easy things to do and you'll be forced to look at the wreckage. One approach is to make sure that electrical power was available on the front end and that electrical devices were working on the rear end.

A. ELECTRICAL TERMINOLOGY. A word on terminology. Most older aircraft and many GA aircraft have DC electrical systems and use generators. When they need AC power, they get it from an inverter. Almost all large modern aircraft have AC systems and use alternators. If they need DC power, they use TR (Transformer-Rectifier) units. The word "alternator" has pretty well dropped out of the industry vocabulary and we call everything a generator, or possibly an "AC generator." It really doesn't make much difference as we all know what we are talking about when we say "generator."

Almost without exception, generators are engine-driven. A DC generator can be coupled directly to the engine, but an AC generator needs a constant speed drive (CSD). This CSD is a potential source of failure.

Most of them have shear couplings or electrical disconnect systems to protect the engine in case of generator or drive overheat. Once disconnected, the CSD normally cannot be reconnected in flight, so the evidence should still be there. Obviously, if the CSD has failed, the generator is not going to produce any power.

Examination of the generator itself is a matter of looking for internal shorting or arcing or mechanical interference between the stator and the rotor. If the generator can still be turned, it might be appropriate to bench test it.

B. EMERGENCY POWER SUPPLY. Most aircraft have some sort of emergency backup for electrical power. This may be only the battery which will supply limited power to certain emergency circuits. Many transport aircraft will have an auxiliary power unit (APU) on board that can generate enough electricity to operate most of the electrical equipment on the aircraft. Some aircraft will have a ram air turbine (RAT) that deploys into the slipstream to generate a limited amount of electrical power. Airplanes with multiple generators, circuits and busses usually have the capability to connect an inoperative bus (due to a failed generator) to an operative one. Knowledge of these emergency systems and investigation of their status will contribute to your knowledge of the system as a whole.

C. ELECTRICAL APPLICATIONS. On the application end of electrical systems, we are dealing with motors, lights and electronic equipment. If the motor was damaged during impact, we may be able to determine if it was rotating by looking for evidence of scoring on the inside of the case. Another potential problem with motors is that AC motors can turn in either direction depending on the current phase. If you suspect this, look for evidence of heat damage. If the motor cooling fan turns backwards, it doesn't cool the motor. Analysis of light bulbs is covered in Chapter 15. From the electrical investigator's point of view, he would like to find a bulb that should have been on and was on. Analysis of electronic equipment depends, of course, on the nature of the equipment. Much of it doesn't leave any evidence of whether it was on or off at impact.

D. FUSES AND CIRCUIT BREAKERS. If nothing else has worked, it's a good idea to check circuit breakers and fuses. Most circuit breakers are heat sensitive, which means that they will open if

exposed to post impact fire. It is also possible for them to pop open as a result of impact forces, so finding an open circuit breaker is not necessarily a big clue, particularly if there is other evidence that circuit was working.

E. ELECTRICAL WIRING. Now you are down to the hard part; investigating the distribution system. Start with the electrical terminals and busses and look for evidence of arcing or shorting. Remember that most modern aircraft use the principle of "control power." For reasons of space, weight and electrical efficiency, the heavy duty wiring that runs motors is not routed through the cockpit. The motor switch in the cockpit is actually connected to some light weight wiring that operates a relay which closes the motor circuit. Thus there are generally two ways to fail an electrical circuit. Fail the primary power circuit or fail the control circuit or its relay.

You should understand a few things about electrical wiring. First, all the wires are numbered and can be identified with reference to the aircraft wiring diagram. This is a big help. Second, you need to be a little suspicious of wiring diagrams. Most electrical wiring harnesses were originally built with extra wires (numbered, of course) to accommodate new equipment and future modifications. If the aircraft has been in service a while, expect confusion. Also, if the aircraft has been in service a while, an easy solution to a previous electrical problem is to just reconnect the equipment to an unused wire pair. More confusion for the investigator.

Another bit of useful knowledge about wires is how they behave when exposed to heat. If the heat was applied externally, the insulation should be discolored or burned, but the wire strands inside should be shiny and bright. If the heat was the result of overheating of the circuit itself, then the inside wire should be discolored. Severe external heat will, of course, also discolor the wire. Here, try examining the wire along its full length. External overheat should be localized. Circuit overheat should exist over the entire length.

When a wire carrying current breaks during impact, there may be a small electrical arc at the instant of failure. If so, you may see evidence ranging from a slight discoloration of the failed tip to a round globule of molten metal where the tip was actually melted by the arc. If you find evidence of arcing, that is a positive indication that there was power in that wire at the time it failed. If you do not find such evidence, you're really not sure. The nature of the failure may have precluded any arcing.

F. ELECTRICAL ANALYSIS. Analyzing electrical faults can be difficult. If you have access to an organization that teaches ground school for that system, perhaps to mechanics, they may have the whole system mocked up on a board so you can see the results of any particular component failure. This is a nifty way to learn the system and learn what happens when various parts don't work. Also, consider that the manufacturer has already done a certain amount of analysis on the system during design. He may have completed a fault tree analysis (FTA) of various failure modes or even a sneak circuit analysis to detect potential wiring problems. See Chapter 37 for more ideas on analysis.

11. MISCELLANEOUS SYSTEMS.

These systems operate like one or more of the basic systems already discussed. This will only cover some of the unique features of these systems.

A. FLAP SYSTEMS. In modern aircraft, these are most likely to be electro-mechanical. The mechanical part is frequently a nut and jackscrew arrangement or a rack and pinion gear. This helps the investigator as these mechanisms have no tendency to move during impact and their position can usually be determined even though the flaps themselves have been destroyed by fire. In some flap systems, it is not necessary to determine the position of every segment as there is a connecting device that physically prevents flap asymmetry.

B. LANDING GEAR SYSTEMS. These may be either hydraulic or electric. The investigation should include the sub-systems that operate the landing gear doors and the gear warning indicators. Also, the landing gear has the squat or WOW (Weight On Wheels) switches. Systems that are supposed to work either in the air or on the ground, but not both, are commonly wired through these switches. Failure of the switch either open or closed can affect the operation of other systems.

All landing gear systems incorporate some emergency method of lowering the gear if the primary system fails. Sometimes this is an alternate source of electrical, hydraulic or pneumatic power. Sometimes it is a system that mechanically releases the gear uplocks and allows it to free fall of its own weight.

Some gear that extend against the slipstream have a spring that forces the gear into the locked position.

When the gear is retracted after takeoff, the wheels are rotating at high RPM. To prevent damage in the wheel well, some aircraft have a snubbing system that brings the wheels to a stop and some have a system that automatically applies the wheel brakes. In addition to providing a place to mount the wheels, the landing gear must also absorb the shock of landing. In most larger aircraft, this is done with an air-oil "oleo" strut where hydraulic oil is forced through an orifice at a controlled rate to absorb touchdown forces. Some aircraft use a trailing beam strut (which is essentially an oleo strut mounted on a lever arm) and others use a steel spring strut.

C. BRAKE SYSTEMS. Aircraft brakes are devices for absorbing mechanical energy and converting it to heat. Most modern aircraft brakes are the single or multiple disk type where pucks or one or more stationary disks (stators) are pressed against one or more rotating disks (rotors.) the amount of pressure is usually a function of the pressure applied to the brake pedals in the cockpit. In multi-place aircraft, the use of one set of pedals may isolate the other set. In some aircraft, both sets may be used simultaneously and the effect is additive; that is the total pressure applied to the brakes will be the sum of the pressures applied to both sets of pedals. Some aircraft may have an emergency brake provision where pressure supplied by hydraulic accumulators or emergency air bottles can be applied to the brakes. A characteristic of emergency brake systems is that there will be a limited number of brake applications available, but that each brake application will be effective until released.

The use of aircraft brakes produces a considerable amount of heat. Most aircraft flight manuals have brake energy limit charts which plot the generation of heat as a function of gross weight, speed, stopping distance, taxi distance and so on. Since maximum temperatures are not generated within the wheel assembly until several minutes after brake application and since the effect of subsequent brake applications is additive, it may be necessary to consider previous taxi distance and time when calculating peak brake temperatures.

The problem of excessively hot brakes can lead to other problems. Modern wheel assemblies have fusible plugs which will melt and release tire pressure before the tire explodes. In some systems, brake heat may melt the o-ring seals in the brake cylinders and create a hydraulic fluid leak. If this occurs, a fire invariably follows as the brakes will also be hot enough to ignite Mil-5606 hydraulic fluid.

The brake system itself is not much different from the one found on most automobiles. There is a master brake cylinder (actuator) for each gear and shuttle valves which apply the pressure at the brake. Unlike other hydraulic systems, springs retract the pucks or disks when pedal pressure is released; thus there is not much to be learned by measuring actuator extension. There is information to be had by examining the pucks or disks for correct assembly and evidence of excessive wear.

D. ANTI-SKID SYSTEMS.

D. ANTI-SKID SYSTEMS. Maximum braking force on the aircraft is achieved at about 10% wheel skid. If the wheel is locked (full skid) the braking force is significantly reduced because particles of rubber scrapped from the tires act as rollers and lubricate the skid. If the wheels are not skidded at all, then maximum braking capability is not being used. Expecting the pilot to keep the wheels in a constant 10% skid is asking a bit much; hence the anti-skid device. Essentially, the device senses the onset of a skid and releases brake pressure to that wheel. As soon as the wheel comes out of the skid, it reapplies brake pressure. It does this for each wheel individually and it is very effective in achieving maximum braking without blowing tires or allowing a full skid. In some aircraft, the system is rigged to release brake pressure to matched pairs of wheels so that there is no loss of directional control. Most systems also have a locked brake protection feature wherein brake pressure cannot be applied to the wheels until they have spun up during touchdown.

These are primarily electrical systems and should be treated as such. The system can usually be turned on or off from the cockpit and some incorporate a self-test, fail-safe feature wherein the system automatically turns itself off if there is a problem and leaves the pilot with unprotected manual brakes. Some systems have a warning light to indicate status. In others, switch position is the only indication.

E. DEICING AND ANTI-ICING SYSTEMS.

E. DEICING AND ANTI-ICING SYSTEMS. The difference between a deicing system and an anti-icing system is that a deicing system can remove ice once it is formed. It must, in fact, have a certain amount of ice formed before it will work effectively. An anti-icing system prevents ice from forming, but doesn't remove it very well once it has formed.

Deicing systems can be either pneumatic or thermal. In pneumatic systems, rubber deicing boots cover the leading edges of the wings and empennage and air tubes within the boots are sequentially inflated and deflated to crack the ice and allow it to blow off. The source of air for this system is either an engine drive air pump (reciprocating engines) or bleed air from the engine compressor (turbine engines.) Treat these as pneumatic systems.

Thermal deicing systems are usually composed of electrically heated elements in the airfoil leading edges. Treat these systems as electrical systems.

Thermal anti-icing systems use hot air usually obtained from the compressor section of a turbine engine or a heat exchanger installed in the exhaust system of a reciprocating engine. A thermal deicing system can be constructed using very hot air which is applied in short cycles so there is no heat damage to the aircraft. Treat these systems as pneumatic systems.

F. PROPELLER ANTI-ICING AND DEICING.

F. PROPELLER ANTI-ICING AND DEICING. These systems are somewhat unique. Prop anti-icing systems dribble anti-icing fluid (commonly isopropyl alcohol) out the hub and onto the leading edges of the propeller blades. Treat this as a fluid transfer system as it has all the parts; tank, pump, lines, valves and so on. Prop deicing systems are electrical using heating elements in each blade. Treat these as electrical systems.

12. SUMMARY.

Aircraft systems investigation can get very complicated and this chapter has barely scratched the surface of the subject. Before you attack the wreckage, take the time to learn the system and develop a plan for how you are going to investigate it.

What do I know about this system already? What do I need to know about this system? What components should I examine? Where are they located? How will I recognize them? How will I measure or test them? What do I expect to learn from them?

Good Questions!

14

Cockpit Instruments

1. INTRODUCTION.

It is possible to derive a lot of useful information from the cockpit of a crashed aircraft. There are two general problems with cockpit instrument examination. At best, the cockpit may tell us what was happening at the instant of impact, while we would really like to know what was going on before impact. Also, as cockpit instruments get more sophisticated, they get harder to examine. Consider your VCR. When power is lost (as it is in a crash) the VCR will go totally blank. If you reapply power, you'll get a blinking "12:00" indication which is not particularly helpful if you are trying to figure out what time it was when the power failed. Inside the VCR, there may be a memory circuit that has "captured" the preset TV channels and you may be able to readout this memory. This is a little like the digitally-tuned radio in your automobile. When you turn on the radio, it automatically brings in the last station it was tuned to. Obviously, there is a memory circuit in there somewhere. Reading it out is not going to be easy and it is not something you can do during the field examination of a crashed aircraft. Progress in the direction of the "Glass Cockpit" and CRT displays has not helped the aircraft accident investigator.

Nevertheless, there is information to be gained in any cockpit and a thorough examination of the instruments and switch positions is part of any aircraft accident investigation.

If the situation looks hopeless, i.e. instruments thoroughly crushed and burned, don't give up. Actually, a heavily damaged instrument is more likely to "capture" the impact reading than one which is lightly damaged. Consider a gyroscopic instrument. If the gimbals of the gyro are completely mashed against it and the gyro is locked in its present position, it may well be an accurate indication of attitude at impact. If the gyro is still free to move, it is not to be trusted.

Treat cockpit instruments, in fact treat the whole cockpit, as perishable evidence. Readings will change. Switch positions will change. Rescue workers who got there first can do a lot of damage to a cockpit. There are some investigators out there who are compulsive "switch-flippers." The only way to defeat them is to have good photographic coverage of the cockpit area and a list of all obvious instrument readings and switch positions.

There are some things you can do in the field, but, with a heavily damaged instrument, you may have to take it to a specialist. If this looks like a possibility, figure out how you are going to do this before you remove the instrument(s) from the panel. Locate a compartmented box and cushioning material so you can put the instrument in a specific labeled place and not get it mixed up with the other instruments. Remember that many instruments are duplicated in multi-engine aircraft and the only way to tell them apart is to know what hole in the panel they came from.

2. TYPES OF INSTRUMENTS.

There are a number of different types of instruments found in aircraft cockpits. A general understanding of these types is helpful as the readout techniques are about the same for each type.

A. GEARED INSTRUMENTS. If the instrument indication is driven by a gear train, it can retain its reading at power failure or impact. Internally, some of the gears have indices so that the instrument can be initially set to zero and calibrated. If the impact damage was severe enough to lock up the gears, their position can be established by counting teeth from the index or by matching the gear position on an undamaged instrument and reading that one. Gear-driven clocks also retain their reading at impact, but there is nothing inside a wind-up clock that will tell you what time it was unless it also has a calendar function. Logically, the clock should read impact time, which may or may not be helpful.

B. ZERO-READING INSTRUMENTS. Some instruments are spring-loaded to return to zero when power is lost, thus a reading of zero does not necessarily mean that the instrument read zero at impact. If you obtain any reading at all from this type of instrument, you should treat it as a minimum. The actual reading was at least that amount or higher.

C. SYNCHRO REPEATERS. Some instruments merely display a reading that is established somewhere else on the airplane. These usually retain their reading in a crash if they are servo motor repeaters. There are three general methods of examination. One is to examine the instrument itself. Another is to examine the remote sensing element. If this is a feedback potentiometer (for example) it may also retain its position and the location of the wiper on the pot can be translated into instrument reading. The third method is to use a synchro tester which is a device used to set and calibrate the two parts of the system in the first place. This is instrument technician work.

D. METER MOVEMENTS. Some instruments are D'Arsonval meter movements or Wheatstone Bridge circuits. They are quite delicate and when power is lost they are free to move to any indication. You are unlikely to obtain any useful information from this type of instrument.

E. TAPE INSTRUMENTS. Some instrument displays are narrow tapes which move vertically past a lubber line. The tapes themselves are delicate, but the sprockets and gears that drive the tape are not. Even if the tape is burned away, it is usually possible to establish the reading by examining the sprockets and duplicating that on an undamaged instrument.

F. DIGITAL INSTRUMENTS (GEARED). If the instrument display is mechanical number wheels, somewhat like the odometer on your automobile, readout is usually fairly easy. The wheels are rarely destroyed and if you can find any readable number, you can tell what number on that wheel should have appeared in the window. If none of the numbers are readable, remember that the racheting device connecting the wheels can index them for you.

G. DIGITAL INSTRUMENTS (ELECTRONIC). If the instrument display is electronic; either LED (Light Emitting Diode) or liquid crystal, forget it. Unless there is a memory chip in the system, there is no known way to read out the display itself.

H. CRT DISPLAYS. These are basically hopeless along with the electronic digital displays. Consider your computer. There is nothing you can do to the monitor which will help you figure out what it was displaying when you dropped it. If power was lost, anything in RAM vanishes. If the hard disk survived, maybe you can get something out of it, but not necessarily the last monitor display. So it is with CRTs in an airplane. These are essentially computer monitors. If there is any information available, it will be in the permanent (non-volatile) memory circuits.

3. EXAMINATION TECHNIQUES.

The first and simplest thing to do with an instrument is to examine the face of it. If you can still read it, and you can verify the reading with other evidence, there is no point in taking it apart. The reading isn't going to get any better. If you can't read it, the needle is probably missing. Don't give up. Here are a few techniques you might try.

A. SHADOWING. The reason the needle is missing is usually because the glass face of the instrument shattered during impact. When the shards of glass hit the face of the instrument, they left little nicks and scratches everyplace except where the needle was. This is called "shadowing." Cut a card or piece of paper to resemble a needle and see if you can place it on the face of the instrument in such as manner that it does not cover any scratches or nicks. A magnifying glass will be helpful.

B. NEEDLE POST. Examine the post where the needle used to be attached. If it is "keyed" meaning that one side of it is flat so the needle will only fit on one way, you should be able to read the instrument by lining up a new needle with the flat spot. Unfortunately, most needle posts are not keyed. If the needle post is bent, that may give you some idea of where the needle was pointing.

C. COUNTERWEIGHT IMPACT. The pointed end of the needle usually does not leave a mark on the face of the instrument when it is crushed against it. The other end of the needle may have a small counterweight which may have enough mass to leave a mark. This will appear as a small semicircular scratch the correct distance from the needle post and it represents the tail of the needle at impact.

D. LUBBER LINE INSTRUMENTS. Instruments where a circular card moves past a fixed point, usually at the top of the instrument, are much easier to read. Examples of this type of instrument are the Radio Magnetic Indicator (RMI) and the heading card for the Horizontal Situation Indicator (HSI). After impact, the card has no particular tendency to change readings and if you can find any readable number on it, you can tell what number was at the top of the instrument.

4. ULTRA-VIOLET LIGHT EXAMINATION.

Almost any text on aircraft accident investigation will recommend examining instrument faces under ultra-violet light. The logic behind this relates to how instruments used to be manufactured. In the old days, the primary cockpit instrument lighting system was ultra-violet and the instrument numerals and needles were coated with radium so they would fluoresce. The needles were dipped, which meant that they had fluorescent material on the backside, too. During impact, the needle would be crushed against the instrument face and would leave little speckles of fluorescent material where it hit. This could be detected by examining the instrument face under UV light. Unfortunately, instruments haven't been manufactured that way since the 1950s. You might find some of those old instruments still in service in equally old airplanes. Even though modern instruments do not fluoresce, their examination under different types of lighting from different angles can sometimes show things that are difficult to see in normal light. In addition to UV light, polarized light, infra-red light or sometimes red or green light can be helpful.

5. SPECIFIC INSTRUMENTS.

A. ALTIMETERS. The altimeter can usually be read fairly accurately, but we would expect it to yield field elevation of the crash site. If it doesn't, it may have been mis-set or it may have failed in flight. The Kollsman window reading or altimeter setting dial usually survives the impact and can answer the question of whether the altimeter set to the correct barometric pressure or not. It can usually be determined. The radar altimeter is slightly different. This is a syncho/gear train instrument and can be read. Remember also that the radar altimeter has a warning light that reflects actual altitude as related to the warning altitude set by the pilot. This warning light bulb can be checked for illumination using the procedures in Chapter 15.

B. ANGLE OF ATTACK INDICATOR. The AOA instrument is connected to the AOA vane or probe through a synchro mechanism. Since either the vane or probe can change position during impact, any AOA reading should be viewed with suspicion.

C. AIRSPEED INDICATORS AND MACHMETERS. These geared instruments can usually be read. If the instrument has a rotating drum, the process is somewhat easier. Also, look for pointers around the periphery of the instrument indicating limiting mach and the "speed bug," a needle set by the pilot to the airspeed to be maintained. Some aircraft may have mechanical digital ground speed or true airspeed indicators which can also be read.

D. VERTICAL VELOCITY INDICATORS. This is a geared instrument and can be read. Remember that there is a built-in delay in VVI indication and the reading may not be entirely accurate.

E. TACHOMETERS. Some of these are geared instruments and are spring-loaded to zero when power is lost. Any reading obtained should be treated as a minimum. If the tachometer is servo-motor driven, it should retain its last reading when power is lost. The source of information for tachometers is a tach generator mounted on the engine. There is no useful information available from the tach generator.

F. GYROSCOPIC INSTRUMENTS. These include the turn and bank indicator, the attitude director indication (ADI), the plain old attitude gyro, and the directional gyro. All incorporate spinning gyros which are gimbaled to move freely in one or more planes. If the impact is severe enough to lock the gimbals against the gyro, then the reading may be accurate. Keep in mind that force against a gyro causes it to precess in a direction 90 degrees to the applied force. If this occurs before capture, it can lead to some totally erroneous indications. In some instruments, the small aircraft or lubber line on the face of the instrument may be smashed against the gyro element and leave an impact mark. Look for this.

G. FUEL GAGES. Older aircraft use a float in the fuel tank which actuates a resistance circuit in the indicator. These instruments return to zero when power is lost and any reading should be considered as a minimum. Modern aircraft have capacitance circuits which measure the fuel density and incorporate a feedback potentiometer. These instruments retain their reading with loss of power. If the instrument cannot be read, it may be possible to determine the wiper position on the feedback pot. Some aircraft have a fuel totalizer which either electrically totals the individual gage (not the tank) readings or is mechanically set to the initial total fuel on board by the pilot and will mechanically count down fuel remaining as a function of fuel flow. It may be possible to obtain useful information from these.

H. POSITION INDICATORS (landing gear, flap, etc.) Some of these indicators are selsyn indicators (Up, Intermediate, Down) which are essentially meter movements. Do not trust them. After an impact, they can read anything. Other indicators (flap position indicators) are geared or synchro instruments. Readout is possible.

I. COURSE INDICATORS (localizer and glide slope.) These are meter movements and will not yield any useful information following a crash.

J. ENGINE PRESSURE RATIO (EPR). These are synchro repeaters and can be read.

K. FUEL FLOW. These are also synchro repeaters. Some information may be available from the fuel flow transmitter itself if the impact was severe enough to capture the fuel flow vane.

L. TEMPERATURE INDICATORS. These are usually galvanic meter movements using thermocouples as a source. They seldom provide any useful information.

M. PRESSURE INDICATORS. These are usually synchro repeaters and can provide useful information.

N. ELECTRICAL INDICATORS. These are almost always galvanic meter movements and seldom provide useful information.

6. SWITCHES AND LEVERS.

A. SWITCHES. Guarded switches can be trusted. Unguarded switches may be accurate, but are suspect. Although the mass of the switch knob itself is seldom enough to overcome internal resistance to movement, loose items flying around the cockpit during impact can change a lot of switch positions.

B. LEVERS. Levers which move fore and aft and have no locking device or detent are, depending on their mass, commonly found in the full forward position following impact. These include throttles, power levers, propeller controls and carb heat controls. If the lever is found full forward, it is wise to verify that indication with other evidence. Other levers which are detented or locked in place are more likely to be accurate. These include mixture controls and landing gear, flap and spoiler/airbrake levers.

C. CIRCUIT BREAKERS. These may be considered a type of switch. Impact forces seldom pop circuit breakers, but heat from the subsequent fire might, depending on the type of circuit breaker.

7. SUMMARY.

Cockpit examination is usually an essential part of any aircraft accident investigation. These steps may help.

1. Find out what was done in the cockpit area during rescue or removal of remains.

2. Keep everyone out of the cockpit area until it has been photographed and the instrument and switch positions documented.

3. Before removing any instrument from the panel, know what you are going to do with it.

4. If the instrument deserves laboratory examination by a specialist, do not compromise that examination by attempting to take the instrument apart in the field.

In addition to the instruments, look at the cockpit layout, configuration, visibility and warning devices.

15

——————————————

Light Bulb Analysis

1. INTRODUCTION.

Following a crash, it is possible to examine light bulbs involved in the crash and determine if they were on or off at the time of impact. This technique was developed back in the 1950s (or maybe earlier) and it was developed by vehicle accident investigators.

In a typical rear end vehicle collision, the driver of the rear vehicle frequently claimed that the tail or brake lights of the front vehicle were not working. Vehicle accident investigators developed techniques for examining the brake lights to determine whether they were on or off at the time of impact. Early research on this subject was done by J. Stannard Baker of the Traffic Institute of Northwestern University. His monograph on lamp examination is a classic. See bibliography.

Aircraft accident investigators quickly recognized something that had direct application to aircraft accidents; especially since airplanes had a lot more light bulbs than automobiles. The prospect of being able to determine whether various cockpit warning lights were on or off was irresistible. Definitive research in this area was done by Terry Heaslip and Max Vermij of the Canadian Aviation Safety Board (now the Canadian Transportation Safety Board.) They later continued their research as Accident Investigation & Research, Inc.

The basic principles are fairly simple. Light bulb filaments are made of tungsten. Typically, the filament is a tightly coiled tungsten wire. A current is passed through the wire and it heats up to produce both heat and light. We could do without the heat, but we have to accept it in order to get light. All incandescent bulbs work this way. Manufacturing the filament is fairly tricky, as tungsten is brittle when it is cold and it does not form well. It is this characteristic that allows us to determine if the bulb was on or off at impact.

If the filament was off or cold, then it will tend to shatter or break. It will not stretch. This is typical brittle behavior. When the filament is on or hot, though, it behaves in a ductile manner. It stretches, particularly if it was coiled in the first place. If it fails, the failed ends will show the characteristic necking-down common to ductile failures.

Aircraft bulbs are not as easy to examine as automotive bulbs. Most cockpit instrument and warning light bulbs tend to be very small. The one most commonly seen is the GE 327 bulb, or peanut bulb, as it is called. This bulb is slightly smaller than the eraser on a pencil and it is not easy to examine. It usually survives the most severe of impacts with its glass case intact and it is a good candidate for filament examination. A magnifying glass of about 10 power is very helpful.

If you intend to photograph the filament, it is quite difficult to get a decent picture of the filament without annoying reflections from the glass case. One method is to illuminate the bulb from behind by using a translucent shield that defuses the light and eliminates reflection.

In the field, you might try supporting the bulb

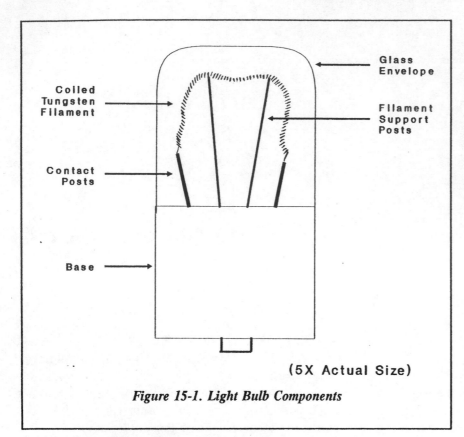

Coiled Tungsten Filament

Contact Posts

Base

Glass Envelope

Filament Support Posts

(5X Actual Size)

Figure 15-1. Light Bulb Components

inside a styrofoam coffee cup and aiming the flash through the outside of the cup. The problem with this sort of trickery is that you are never certain of success until you develop the film.

Another method is to break the glass envelope to expose the filament. This can be done with either a glass cutter or gently with a pair of pliers. When doing this, mount the bulb upside down so the pieces of glass will fall away from the filament. Suggestion: practice this a few times on spare bulbs before trying it on the evidence bulbs.

2. LIGHT BULB TERMINOLOGY.

We need to understand the basic parts of a light bulb. Consulting Figure 15-1, you will see that all bulbs have a base, a glass case or envelope, a filament, and one or more filament support posts.

3. EXAMINATION PROCEDURES.

Remove the bulb from its socket after first noting which system and which socket the bulb came from. Since you may be dealing with dozens of bulbs, this is fairly important. One method is to put bulbs (or bulb pairs) in separate plastic envelopes marked with the bulb location. A good rule to follow is to never open

more than one envelope at a time.

Using the magnifying glass, examine the bulb filament. It is helpful to have a strong light, but usually you can do it by just holding the bulb up to the sky. You are looking for evidence that the filament is stretched or bent more than it should be. If you are not sure what it should look like, go to any aircraft maintenance department and examine a few identical bulbs that are still in the box. If you want to compare a bulb to one of the same type that was exposed to the same impact forces, locate the spare bulb holder (most aircraft have one) and examine those bulbs. Alternatively, examine bulbs from systems that you know were off and use them for comparison. There are some problems to expect.

A. FILAMENT BENDING. The filament of a "cold" bulb may show some minor bending due to the mass of the filament and the impact forces. This bending will usually occur either at a filament support post or a portion of the filament that is not coiled. This is why it is helpful to use "cold" bulbs from the same accident for comparison. As a rule, the smaller the bulb, the less likely that filament bending will occur.

B. FILAMENT STRETCH. The manner in which a "hot" filament stretches depends to some extent on how the filament is oriented to the impact forces. The filament of a "hot" bulb could stretch up, down or sideways depending on the orientation. Most aircraft warning light systems have two bulbs per warning light. Always check both of them.

C. FILAMENT AGE. The age of the filament makes a difference. If the bulb has spent a lot of hours lit, it's filament will be more brittle when it is cold and more ductile when it is hot. This is not much of a problem with warning lights as they are normally off. System light bulbs which are normally on might be "aged." A bulb filament which, by visual inspection, could be either on or off, usually requires laboratory examination.

D. IMPACT FORCES. The magnitude of the impact forces make a difference. It usually takes a very high G loading to stretch the filament. Also, it is conceivable that a particular warning system could fail or be damaged early in the accident sequence and turn the bulb off before enough forces are experienced to stretch the filament. When an incandescent bulb is turned off, it cools down very rapidly in something on the order of 50 milliseconds.

E. FILAMENT BEHAVIOR. This is not exactly predictable. The filament may, for example, stretch enough to contact the glass envelope. If so, it may leave a scorched mark on the envelope (a good clue!) or it may even melt and retain a globule of glass on the filament (another good clue!) In addition to the behavior of the filament, the behavior of the support posts may add a complicating factor. The support posts may vibrate or bend during impact. In a cold bulb, this may result in a broken filament.

Figures 15-2 and 15-3 are pictures of tests conducted by the Aviation Safety Engineering Branch of the Canadian Aviation Safety Board (now the Canadian Transportation Safety Board.) They show the extremes; one bulb was obviously on and the other obviously off. These could be identified in the field. Between these, however, are a lot of other possibilities that will require laboratory examination.

4. BROKEN GLASS ENVELOPE.

If the glass envelope of the bulb breaks during impact, this can either help or hinder your investigation. You can still examine the filament for ductile or brittle behavior providing you can find it. If the force of the impact was high enough to break the glass envelope, it may also have been high enough to destroy the filament or perhaps shake it loose from the bulb so you can't find it. In this case, you may be able to examine the remaining ends of the filament for ductile or brittle behavior, but this will probably require laboratory SEM analysis.

On the other hand, the breaking of the envelope has introduced oxygen into the bulb. Consider that the inside of a light bulb is a vacuum. The only reason the filament glows but does not burn up is because there is no oxygen present. If you introduce oxygen, the

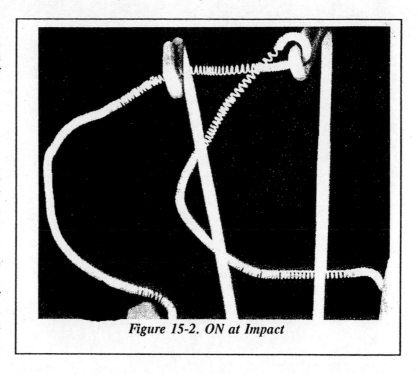

Figure 15-2. ON at Impact

filament glows brightly for a few milliseconds and is either burned through or burned up. You see this occasionally with a household light bulb. You turn on the light and it flares for an instant and then goes out. Somehow, oxygen has leaked into the glass envelope.

When a tungsten filament burns, the byproduct is tungsten oxide which is a grayish powder. If the remaining parts of the bulb are coated with a grayish powder when you examine it, the bulb was probably on at the time the glass envelope broke. It is possible that no grayish powder was produced, but the filament

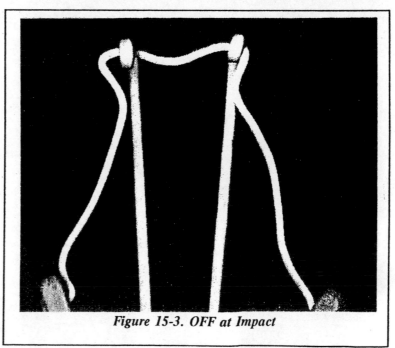

Figure 15-3. OFF at Impact

is discolored due to reaction with oxygen. The progression of color change goes from yellow to red to purple to blue. If the glass envelope is broken and there is any evidence of filament discoloration, the bulb was also probably on. If there is no evidence of discoloration, it was probably off. If the glass envelope is broken, but the filament is intact and unbroken, the bulb was definitely off.

5. SUMMARY.

All this means that examination of light bulbs demands a lot of care and patience. The act of carelessly removing the bulbs or the device containing the bulbs from the wreckage can destroy the evidence you need.

If you are successful, though, there are enormous benefits in knowing which lights in the cockpit were on or off at impact.

16

Tires and Runway Accidents

1. INTRODUCTION.

It is difficult to know what to call this chapter. Tires are only used twice per flight; takeoff and landing. Their performance (or failure) is frequently a factor in takeoff or landing accidents. It is difficult to talk about just the tires themselves without considering wheels, brakes and antiskid systems. A complete discussion must also include the runway and its condition along with aircraft performance. Since takeoff and landing performance is also influenced by atmospheric conditions, we must consider those, too. Likewise, takeoff and landing performance is influenced by items such as lift devices, lift-cancelling devices, drag devices and thrust or power reversing systems. Some of these can be covered, logically, in other chapters, but many of them need to be addressed right here. Thus this chapter is going to cover a lot of subjects in addition to tires.

2. BASIC CALCULATIONS.

A. WHEEL LOADING. Calculation of the load on a particular landing gear (or tire) is essentially a problem in moment arms. The center of gravity (CG) of the aircraft is somewhere between the main gear and the nose gear and all weight on the gear may be assumed to operate on a lever arm measured from the CG to the nose gear and to a line connecting the trunions of the main gear. The sum of the moments about the CG must be zero which means that the product of the weight on the main gear and its distance from the CG (moment arm) must equal the product of the weight on the nose gear and its moment arm. The sum of the main gear and nose gear weights must

equal the gross weight of the aircraft.

The load carried by each main gear is 1/2 that carried by both main gear. If the main gear is dual wheeled, the load on each tire is 1/2 of that on the individual gear. If the gear is a four-wheeled bogie, the load on each axle is 1/2 of that on the gear and the load on each tire is 1/2 of that figure.

B. TIRE LOADING. When a tire fails, the load normally carried by that tire must be instantly transferred to the other tire on that axle, if there is one. In the case of a four-wheeled bogie, each axle must carry 1/2 the load on the gear. Thus the load of a failed tire is transferred to the adjacent tire only and not distributed among the remaining three tires.

By knowing a tire's maximum load capabilities (from the specifications) and how it was loaded (calculations), the tire's operating condition can be determined and it may be possible to explain the sequence of tire failure. If, for example, the tire loading is near the tire's capability, it is not unusual (or unexpected) for the failure of one tire to be followed rapidly by failure of the adjacent tire on the same axle. In this case, you would only be concerned with the initial failure.

When we are dealing with an aircraft with more than one tire per landing gear, all of our calculations assume that the loads are equally distributed among tires on the same axle. If the tires are badly mismatched by manufacturer, tread design, wear or inflation pressure, they are not going to share the load equally. Tires from different manufacturers may have

DIAMETER (Inches)	MISMATCH TOLERANCE (Inches)
< 24	1/4
25-32	5/16
33-40	3/8
41-48	7/16
49-55	1/2
56-65	9/16
> 66	5/8

Figure 16-1. Tire Diameter Mismatch Tolerance

slightly different diameters or different sidewall stiffness characteristics. For this reason, it is desirable to have matched tires on the same axle. If they are not matched, this may influence the load calculations and could explain the sequence of failure.

Within the industry, the generally accepted tolerance in dual tire diameters is shown in Figure 16-1.

Measuring the outside diameter of a tire is difficult. Try measuring the circumference and dividing by π (3.1416).

Another problem with dual tires is inflation. If the tires are otherwise matched, but one is seriously underinflated; the other tire may be carrying a larger than normal portion of the load. We'll talk about inflation as a separate subject.

3. AIRCRAFT WHEEL CONSTRUCTION.

Aircraft wheels are manufactured from either magnesium, aluminum alloy or steel alloy. They operate in a high stress environment and are subject to corrosion and mechanical damage during tire mounting and normal service. Almost all modern wheels are of the split half type wherein the two halves of the wheel are bolted together across the tire beads. Due to manufacturing tolerances, it is essential that the two wheel halves be matched by both manufacturer and part number.

The wheels contain the inflation valve for the tire and the fusible plugs designed to melt and release tire pressure in a high temperature situation. Some may

also contain a pressure relief valve designed to vent excessive inflation pressure.

CAUTION. Never attempt to disassemble a wheel without first insuring that the tire is deflated. An inflated aircraft tire is a potentially lethal explosive device. At the scene of any aircraft accident, it is standard safety practice to insure that all tires are deflated before the investigation begins.

4. AIRCRAFT TIRE CONSTRUCTION.

Aircraft tires are designed to withstand very high speeds and very heavy static and dynamic loads-- intermittently. By contrast, an automobile tire is designed for much lower speeds and loads; but it is designed to run continuously at a stabilized temperature. An aircraft tire will fail under continuous operation. Exceptionally long taxi distances can fail an aircraft tire.

The tread of a tire is rubber followed by one or more layers of tread reinforcing nylon fabric. The strength of the tire is in its plies which are layers of rubber-coated nylon fabric laid diagonally and at 90 degrees to each other. Almost all aircraft tires in service today are tubeless.

5. TIRE SPECIFICATIONS.

Most tire specification information can be read off the tire sidewall. This includes the manufacturer, tire size, part number, serial number, load rating, ply rating, speed rating and number of times the carcass has been retreaded if at all. Additional information available from the manufacturer's data sheets include inflation pressure, maximum braking load, bottoming load and tire dimension and weight information. Some of these need amplification.

A. LOAD RATING. This is the maximum load (in pounds) for the ply rating of the tire.

B. PLY RATING. This is an index of the strength of the tire and is not necessarily the number of actual plies in the tire carcass.

C. SPEED RATING. This is the maximum speed (usually in MPH) to which the tire is qualified.

D. MAXIMUM BRAKING LOAD. The maximum steady braking load (in pounds) which may be applied to a tire during landing.

E. BOTTOMING LOAD. This is the approximate load (in pounds) required to bottom a correctly inflated tire against the wheel rim.

F. RETREAD HISTORY. Tire carcasses last longer than treads. If the tire has been retreaded, it will be marked with an R followed by the number of times it has been retreaded and the latest date.

G. INFLATION PRESSURE. This is the pressure required for the tire to support the rated load. If the tire is under load when inflated, then the actual inflation pressure must be adjusted for both load and temperature.

6. TIRE INFLATION.

Improper tire inflation is one of the primary causes of premature tire failure. Overinflation accelerates wear in the center of the tread. Underinflation contributes to breakdown of the carcass itself due to excessive flex heating. A tire that has been run chronically underinflated may develop a wear pattern where the tread is wearing more at the edges than in the center. In most cases, though, underinflation cannot be seen and the resulting damage is hidden until the tire either fails or is demounted. If the tire fails, it is likely that the damage which resulted in the failure occurred some time prior to that flight.

Within the industry, underinflation is considered to be anything less than 95% of the required inflation pressure. Since required inflation pressure is affected by both temperature and static load and must be measured by a fairly precise instrument, this poses some problems. Considering that modern aircraft tires operate in the 200-300 psi inflation range, knowing the correct pressure and measuring it--all within 10-15 psi is not easy.

7. INFLATION GAS.

The gas of choice for a high pressure tire is dry nitrogen. If air is used and the tire is overheated, a chemical reaction can occur between the oxygen in the air and a gas formed and released by the tire liner. This can lead to a violent explosion. This has occurred on at least two occasions in the wheel well of an aircraft following gear retraction. It could just as easily occur on the ground. Tests have indicated that the oxygen content of the gas used for inflation should be kept at less than 5%.

8. TIRE FAILURE INVESTIGATION.

The failure of an aircraft tire at high speed can be a violent event. Chunks of the tire can be thrown against the aircraft wing or fuselage structure with enough force to penetrate thin paneling. Due to the forces and speeds involved, the tire can shred itself into many pieces. Initially, this looks pretty hopeless.

First, locate the point of initial failure. You may find this clearly marked on the runway with an abrupt change from a rolling tread mark to a wobbling deflated tread mark with indications of wheel rim contact. If the aircraft is FDR/CVR equipped, the point of failure may have been captured by either recorder. If none of these work, the point of failure will be somewhere upstream of the point where the first piece of tire is found.

Next, collect all parts of the tire. Partial reassembly may be possible. While final analysis may require a tire expert, there are certain common tire failure patterns that can be seen by the field investigator.

9. TIRE FAILURE PATTERNS.

A. PHYSICAL DAMAGE. Examine the tire for excessive wear (beyond the wear cords), unusually deep cuts, areas where tread has been physically gouged from the tire, and obvious impact damage. The failure pattern resulting from impact is frequently an "X" or "Y" shaped cut on either the tread or the sidewall. Operation from unimproved runways leaves gouges or pock marks in the center tread area.

B. THROWN TREAD. It is possible for the tread to come off either a new tire or a retreaded tire. If so, the tread usually survives in two or three big chunks. You have all seen tread thrown from truck tires on the sides of the highway. If the tread was not thrown, it stays with the carcass and separates into smaller pieces. The reason why the tread was thrown will require technical examination by tire experts.

C. OVERHEAT DAMAGE. If the tire was severely overheated, due to a locked brake, perhaps, the evidence will be in the bead area where the tire contacts the wheel. Blistering and melting in this area is evidence of overheat. Severe overheat should, of course, have melted the fusible plugs in the wheel.

D. UNDERINFLATION DAMAGE. If the tire has been chronically underinflated, evidence usually appears as deterioration, distortion and wrinkling of the inner liner of the tire.

E. SKIDDING DAMAGE. A normal skid leaves an oval area or "flat spot" on the tire. The surface area is clearly abraded. A hydroplaning skid also leaves an oval area, but the surface has a blistered or bubbly appearance.

10. AIRCRAFT BRAKES.

See Chapters 13 and 30 for a description of aircraft brake systems. Keep in mind that the best set of aircraft brakes made can only stop the wheels from rotating. The actual stopping force applied to the aircraft comes from the friction between the tires and the runway.

11. ANTISKID SYSTEMS.

See Chapters 13 and 30 for a discussion of antiskid systems.

12. RUNWAY EVIDENCE.

When investigating a takeoff or landing accident, the history or sequence of events may be captured by marks on the runway. The touchdown areas of runways in heavy use are usually contaminated with rubber and the determination of which set of tracks belongs with which aircraft is sometimes difficult if not impossible. Knowledge of the width of the aircraft track and the tire tread design may help.

Assuming that the aircraft tracks can be identified, the landing or touchdown tracks can sometimes tell a lot about the landing. Initially, the tires should skid and leave marks as they are not yet rolling. In a smooth landing, these skid marks should start from a point and expand to the width of the tread over a considerable distance. If this distance is foreshortened significantly (in comparison with touchdown tracks generated by other aircraft), this would suggest a high rate of sink at touchdown. In fact, if forward velocity is known, some rough calculations of rate of descent are possible. During the time it took for the aircraft to go from initial touchdown to full tread spread (so to speak) the aircraft must have descended a distance equal to gear strut extension plus tire compression. Strut extension should be easy to determine. Tire compression can be estimated by measuring the difference between the axle to ground and axle to top of tire distances on another aircraft.

If a tire failed at touchdown, this will usually leave a distinctive mark on the runway, particularly if the tire was still skidding. This mark will be followed shortly by marks made by the wheel rims as they contact the runway.

13. HYDROPLANING.

Hydroplaning occurs on a wet or icy runway. The three recognized types of hydroplaning are dynamic, viscous and reverted rubber. Dynamic and viscous seldom leave any evidence on the runway tracks, although the high pressure water being squirted out from under the tire sometimes has a cleaning effect on the skid mark. A portion of the skid significantly cleaner than the rest might be a clue as to the initiation of hydroplaning.

Reverted rubber hydroplaning occurs in a locked wheel skid where the heat generated by friction produces steam and begins to revert the rubber on a portion of the tire back to its uncured state. As long as the steam is trapped under the tire, it literally steam cleans the runway leaving a distinctive clean spot. The mark may have a faint trace of rubber along the edges where the rubber was not reverted. If evidence of reverted rubber hydroplaning is found on the runway, there should be corresponding evidence on one or more of the tires.

The threshold velocity in knots at which hydroplaning may occur on a rolling tire is shown in Formula (1). If the tire is not rotating, use Formula (2).

$$V_{kts} = 9\sqrt{Tire\ Pressure} \qquad \textbf{(1)}$$

$$V_{kts} = 7.7\sqrt{Tire\ Pressure} \qquad \textbf{(2)}$$

If evidence of hydroplaning is found on the runway, this indicates that the tires involved were inflated at that time. Deflated tires do not hydroplane.

14. SKID MARK ANALYSIS.

As the aircraft skids down the runway during a landing accident (or even some takeoff accidents), the aircraft may yaw to the point where it is skidding sideways or even backwards. Sometimes, the aircraft may bank to the outside of the skidding turn and lift a main gear from the runway. This results in a discontinuous skid mark. As the aircraft turns on the runway, the skid marks become closer together and may even cross each other. This can be confusing to the investigator as it is sometimes difficult to visualize what the plane was doing when those marks were produced.

In this case, it may be helpful to accurately diagram the marks on a sheet of paper and construct a simple device to represent the airplane. A piece of cardboard with holes punched in the geometrically correct positions of the landing gear will work. By moving the holes over the skid mark diagram so that the marks are always visible in the correct holes, the piece of cardboard must be rotated or "slid" in the same manner as the actual aircraft.

15. COEFFICIENT OF FRICTION.

As already mentioned, the actual braking force that brings the aircraft to a stop is a function of the friction between the tires and the runway. The value used in all deceleration calculations is the coefficient of friction. This is the ratio between the force needed to slide two surfaces against each other and the force holding them together, which in this case, is the weight of the aircraft. Unless hydroplaning occurs, the tread design of the tire or its state of wear is not particularly significant in the friction calculation. The surface of the runway is very significant, however, and depends on the type of surface, cleanliness (rubber deposits), and contaminants such as water, snow or ice. Typical coefficients of friction vary from a low of .05 for ice at 32 degrees F. to a high of .75 for dry concrete. The coefficient of friction is normally designated by the Greek letter mu (μ).

The coefficient of friction for a runway is best measured by a vehicle-towed (or mounted) mu-meter or a Diagonally-Braked Vehicle (DBV). If these are not available, an approximation of μ can be obtained with an aircraft tire (or even a tire section) and a spring scale. The force needed to drag (skid) the tire along the runway is read from the spring scale while it is moving and held parallel to the runway. This is divided by the weight of the wheel or tire or tire section. The result will be the coefficient of friction. The wheel must slide, not roll, and the heavier the wheel, the more accurate the calculation is likely to be. Since the coefficient of friction will probably vary at different places on the runway, several readings should be taken and averaged. This method is not as accurate as a mu-meter or DBV, but it is likely to be

more accurate than coefficients of friction extracted from tables of average values.

16. STOPPING DISTANCE CALCULATIONS.

Assuming uniform acceleration (or deceleration), the distance required to stop the aircraft can be determined by using the formula for linear acceleration:

$$S = \frac{V^2}{2a} \qquad (3)$$

Where S = distance in feet.
V = initial velocity in feet per second (Knots X 1.689).
a = rate of acceleration (fps^2).

Acceleration is determined by:

$$a = \frac{F}{m} \qquad (4)$$

Where F = the accelerating or decelerating force.
m = mass.

Since mass equals weight (W) divided by the force of gravity (g), Equation 4 becomes:

$$a = \frac{Fg}{W} \qquad (5)$$

Decelerating force is the product of the normal force (or weight on the landing gear) and the coefficient of friction (μ).

$$F = \mu \times N; \qquad F = \mu \times W \qquad (6)$$

Substituting this into Equation 5, it becomes:

$$a = \mu g \qquad (7)$$

Notice that the weight of the aircraft has canceled out of the equation. While the weight increases the mass that must be stopped, it also increases the braking force to stop it by the same amount. Actually, weight it still there, because it influences velocity. Substituting Equation 7 into Equation 3:

$$S = \frac{V^2}{2\,\mu\,g} \qquad (8)$$

Since "g" in this case equals 32 fps^2:

$$D = \frac{(1.689\ V)^2}{64\ \mu} \qquad (9)$$

There are some inherent errors in this calculation. First, it assumes uniform deceleration, which may or may not be true. second, it assumes velocity at start of braking is known. Unless there was a Flight Data Recorder (FDR) on the aircraft, this usually has to be estimated and the equation solved for a range of possible velocities. It also assumes that all the wheels are braked and all of the weight is supported by the wheels. The calculations are pretty accurate for automobiles; less so for aircraft.

The equation is useful for situations where there is a measurable skid mark on the runway. Then the actual stopping distance can be compared with predicted stopping distance. Alternatively, velocity at the start of braking can be calculated by solving the equation for V:

$$V = \frac{\sqrt{64\ S\ \mu}}{1.689} \qquad (10)$$

The foregoing is a simplified discussion of how linear motion formulae can be applied to takeoff and landing accidents. These concepts apply to any object in motion on the ground. It doesn't make any difference whether it is an airplane, an automobile, a bus or a rock. The mathematics are the same. In aircraft accident investigation, these concepts are only applicable to a small percentage of the total accidents. In automobile accident investigation, by contrast, they apply to almost every accident. By far, the best reference sources for this type of investigation are found in manuals on vehicle accident investigation and traffic accident reconstruction. That's where the majority of research has been done. The references listed in the bibliography are highly recommended to the investigator who needs to know more on this subject.

17. SUMMARY.

Investigation of tires and takeoff and landing performance follows the same logic as other phases of the investigation. Start with everything that is known about the accident. Decide what needs to be determined. Collect all the pieces. Construct a good diagram. A good aerial photograph of the scene can be very helpful. Locate the first significant event (or failure) and determine a sequence of events to show how the aircraft got from there to where it ended up.

17

Mid-Air Collisions

1. INTRODUCTION.

As a rule, mid-air collisions are fairly easy to investigate. We already know what happened, the only question is why it happened. Because we know what happened, our work is simplified. We do not, for example, need to spend much time on the engines or the flight control systems or the maintenance history of the aircraft. We can concentrate strictly on the problem of why they collided.

2. TYPES OF MID-AIR COLLISIONS.

We generally categorize mid-airs as one of two types; associated or non-associated.

A. ASSOCIATED MID-AIR COLLISIONS. In this type of mid-air, the two aircraft were flying in each other's vicinity and knew it. These typically happen during formation flight or during military combat maneuvering. In civil aviation, mid-air collisions have occurred when one aircraft was attempting to inspect the landing gear of another aircraft.

Associated mid-airs occur because of pilot technique or the operational procedures (or lack of them) in use at the time. The thrust of the investigation is in that direction.

B. NON-ASSOCIATED MID-AIR COLLISIONS. These occur between aircraft who are not intentionally flying in each other's vicinity and neither knows the other is there. The investigation, in these cases, is toward the management of the airspace.

Where was each plane supposed to be? Who had the right of way? Who could have seen who?

In this type of investigation, the first priority is usually the Air Traffic Control records and the radar data. Second is probably the Flight Data Recorders and Cockpit Voice Recorders if either plane was so equipped. Third is usually witnesses, if any. The problem with witnesses is that most of them see the aftermath of the collision, which is the dramatic part. Few see what the planes were doing immediately before the collision, which is what the investigator would like to know.

3. INVESTIGATIVE REQUIREMENTS.

The investigative requirements for an associated mid-air don't need a lot of discussion. It's largely a matter of understanding the tactics and maneuvers involved in the collision. Two planes deliberately flying close to each other do interact aerodynamically. This interaction is well documented in the literature.

In a non-associated mid-air, we need to understand the actual geometry of the collision. What were the attitudes of the two aircraft as they collided? Which was the overtaking aircraft? What were the relative speeds? In what way was each aircraft damaged by the other? Regardless of who had the right of way, could the pilots of either of the aircraft have seen the other in time to avoid the collision? What was the role of ATC in the collision? Where, in space, did the collision take place?

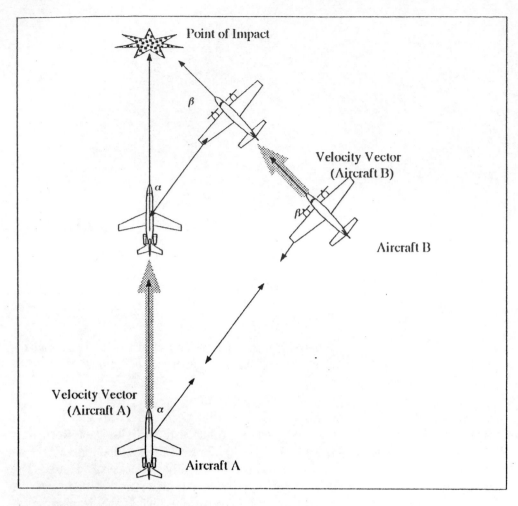

Figure 17-1. Velocity Vectors and Constant Relative Bearing

we diagram a mid-air collision by drawing velocity vectors for each aircraft. The direction of the vector depicts the aircraft's heading prior to impact and the length of the vector is scaled to the aircraft's speed. All Figures in this chapter use that method.

B. CONSTANT RELATIVE BEARING. When two aircraft are on a collision course and the two velocity vectors are constant with regard to speed and heading, then each aircraft has a constant relative bearing to the other aircraft. By establishing the velocity vectors for each aircraft, we can predict what the relative collision angles are going to be.

Constant relative bearing has another significance. If either plane is visible to the other, it will appear fixed on the windshield or transparency. It won't move. Thus the airplane that you are going to hit is the one that doesn't move around on the window. By not moving around, it doesn't attract your attention either. If it moves, your attention might be attracted to it, but you weren't going to hit it anyway. The relative bearing was not constant. This is illustrated in Figure 17-1. Aircraft A is the faster aircraft as shown by its velocity vector. The relative bearing angles, α and β, do not change.

This should come as no surprise to military pilots who learned formation flying. To join on your leader, you were taught to keep his plane steady on your windshield and you would fly right up to him. Same thing.

4. COLLISION GEOMETRY.

Generally, we can determine collision geometry from the wreckage. If we have ATC radar data, it will be helpful to know from which directions the planes approached each other as that will tell us what part of the wreckage to examine. If we have at least one FDR among the two aircraft, we know what that plane's attitude was. We should be able to determine the other plane's relative attitude from the wreckage.

When two planes collide, there will always be transfer of paint and parts. If either plane has a propeller, it will likely leave strike marks on the other plane. In some cases, investigators have found tire tread marks on the structure of the other plane. These are the clues that will allow us to determine collision geometry. Before getting deeply into this, there are a couple of factors that need to be understood.

A. VELOCITY VECTORS. For convenience,

C. PLANE OF COLLISION. This relationship of relative bearing, relative closure speed and the lack of any apparent relative motion is important to the investigator. Another important concept is the plane of collision. There are only three possible planes in which two aircraft can operate as they approach on collision courses.

HORIZONTAL. Both aircraft are in level flight or have vertical speeds which are equal. This is essentially the type of collision depicted in Figure 17-1. These collisions can be further categorized based on collision angle; which can be less than, equal to or greater than 90°.

VERTICAL. This occurs when both aircraft are flying the same course (or reciprocal course) and have different vertical speeds. If they are on the same course, these collisions can be further categorized by speed and overtake.

Reciprocal course collisions in the vertical plane can be categorized by whether the planes are nose up or nose down relative to each other.

COMBINATION (NEITHER HORIZONTAL NOR VERTICAL). This is probably the most common mid-air situation. Airspeed, vertical speed and heading are all different. To analyze this, we just graduated from plane geometry to trigonometry. This is depicted in Figure 17-2.

5. COLLISION GEOMETRY FROM FLIGHT PATH ANALYSIS.

We will use the numbers in Figure 17-2 to show how this can be analyzed.

Aircraft A is in a 20° climb at 400K. He has a vertical velocity vector of 137K and a horizontal vector of 375K on a heading of 135°. Aircraft B is on a collision course of 195° at 250K while in a 10° degree descent. B's velocity has a vertical component of 43K. Its horizontal component is separated vectorally into one paralleling A's flight path and one perpendicular to it. With the flight path vectors of A and B now separated into components which are perpendicular to each other, the relative motion and closure rates can be calculated as shown in Figure 17-2.

That's a lot of work and probably something that is best left to a computer. For simplicity, if the actual plane of collision was within 15° of either horizontal or vertical, you can assume that it was horizontal (or vertical) and the errors introduced will be less than 5 percent.

6. COLLISION GEOMETRY FROM WRECKAGE ANALYSIS.

Colliding aircraft always leave marks on each other that reflect the relative motion of the two aircraft. If aircraft A is flying North and collides with aircraft B, which is flying West, A would have damage which reflected a force from a position 45° right of the nose toward a position 45° aft of the left wing. Aircraft B would have complementary damage, i.e. from left of the nose to aft of the right wing. This is illustrated in Figure 17-3.

If the collision occurred primarily in the horizontal or vertical plane, we can establish some general rules about the damage. For simplicity, we will refer to all damage as "scratches" even though it could be paint smears, parts transfer or crumpling of metal structure.

A. HORIZONTAL PLANE.

1. If the scratches on each aircraft slope in opposite directions with respect to their longitudinal axis, then each scratch mark was made in a direction from front to rear. The smallest angle between the longitudinal axis and the scratch mark indicates the relative bearing at impact. (Figure 17-4.)

2. If the scratch marks on each aircraft slope in the same direction with respect to their longitudinal axis, then one aircraft overtook the other aircraft. The aircraft with scratches going from rear to front was the slower of the two. The faster aircraft will have scratches from front to rear. The largest angle between the longitudinal axis and the scratch mark indicates the relative bearing at impact for the slower aircraft. The smallest angle between the longitudinal axis and the scratches on the faster aircraft is its relative bearing at impact. (Figures 17-5 and 17-6.)

3. If the angles on both aircraft were equal, their speeds were equal. (Figure 17-7.)

4. If the sum of the relative bearing angles is less than 90°, the collision angle was greater than 90°. If the sum is greater than 90°, the collision angle was less than 90°. (Figures 17-8, 17-9 and 17-10.)

B. VERTICAL PLANE.

1. If the scratch marks on each aircraft slope in the same direction, the collision was head on, more or less. All scratches go from front to rear and the smallest angle between the scratches and the longitudinal axis indicates the relative bearing for each aircraft.

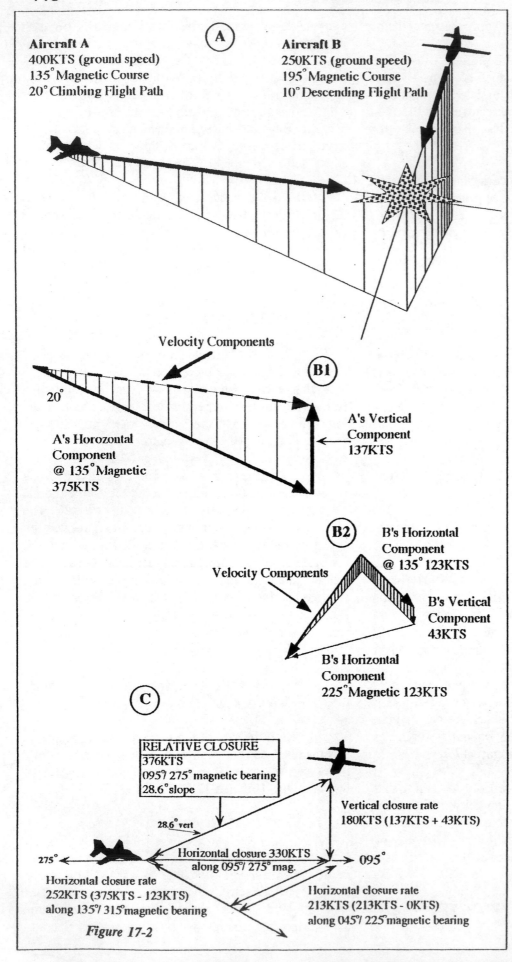

Aircraft A
400KTS (ground speed)
135° Magnetic Course
20° Climbing Flight Path

Aircraft B
250KTS (ground speed)
195° Magnetic Course
10° Descending Flight Path

Velocity Components

B1

20°

A's Horozontal
Component
@ 135° Magnetic
375KTS

A's Vertical
Component
137KTS

B2

B's Horizontal
Component
@ 135° 123KTS

Velocity Components

B's Vertical
Component
43KTS

B's Horizontal
Component
225° Magnetic 123KTS

C

RELATIVE CLOSURE
376KTS
095°/ 275° magnetic bearing
28.6° slope

28.6° vert

Vertical closure rate
180KTS (137KTS + 43KTS)

275°

Horizontal closure 330KTS
along 095°/ 275° mag.

095°

Horizontal closure rate
252KTS (375KTS - 123KTS)
along 135°/ 315° magnetic bearing

Horizontal closure rate
213KTS (213KTS - 0KTS)
along 045°/ 225° magnetic bearing

Figure 17-2

If the scratches also go from bottom to top, the aircraft were nose up; top to bottom, they were nose down. (Figures 17-11 and 17-12.)

2. If the scratch marks on each aircraft slope in opposite directions, one aircraft overtook the other. Determination of the faster or slower aircraft and the relative bearings is the same as for a collision in the horizontal plane. The aircraft scratched top to bottom was on the bottom. The top aircraft will be scratched bottom to top. (Figures 17-13 and 17-14.)

C. GENERAL RULES APPLICABLE TO ALL MID-AIRS.

1. Use the outermost or initial scratches for measurements. After the initial collision, both aircraft will be displaced and the angles will change.

2. If one aircraft left propeller slash marks on the other, measure them. Assuming the correct operating RPM for the aircraft with the propeller will allow calculation of overtake speed. See Chapter 11.

3. A series of parallel scratches most likely came from a rivet line on the other aircraft. The part on the other aircraft that made the scratches will have a rivet line with rivet spacing equal or greater than the scratches. Examination of an identical undamaged aircraft may be helpful here. If the rivet spacing of the suspect part is greater

than the distance between the scratches, just rotate the part until the rivet spacing is foreshortened and matches the scratches. This can establish impact attitude of one aircraft with respect to the other.

4. If all this sounds a little difficult, give this scheme a try. Locate a pair of plastic aircraft models, preferably of the same type and relative size of the two that collided. On each model, draw the scratches and their direction as found in the wreckage. Now try fitting the two models together in the only way that the scratches could have been produced. (Why didn't you tell me that in the first place before I read all this geometry stuff!)

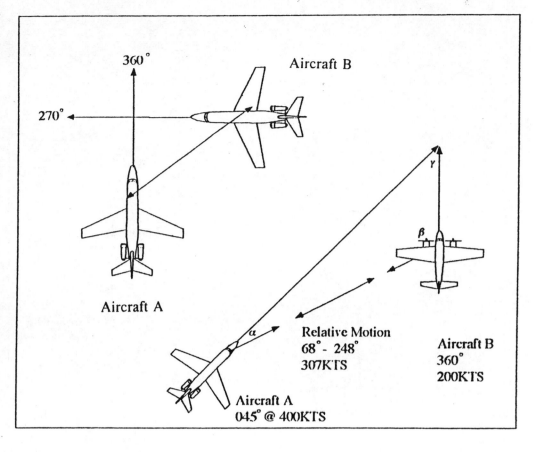

Figure 17-3. Expected Direction of Impact Damage

7. AIRCRAFT CONSPICU-ITY.

Most mid-air collisions occur in daylight in VMC conditions. The reason is that our ATC system does a pretty good job of separating IMC traffic and during night VMC conditions, the aircraft lights are highly visible and people don't run into each other so much.

This brings up the question of how visible or conspicuous each aircraft was to the other. There are a lot of conspicuity studies in the literature. Here are some general conclusions.

1. In daylight, aircraft lights including strobes and rotating beacons are of almost no value. At dusk or against the dark background of a black cloud, for example, they might be of some help.

2. Size and shape are important. All things being equal, big aircraft are easier to see than small ones. (There's a real break-through in scientific knowledge!)

3. Background is important. Any airplane can almost disappear if its paint job matches the background. That's one of the basic principles of camou-flage.

4. Color is not as important as you might think. Beyond a certain distance, colors are not distinguishable unless viewed against a contrasting background.

5. Movement is important. A maneuvering plane is easier to spot than a plane that isn't doing anything. We know that we can significantly improve the conspicuity of a helicopter by painting the top surface of one main rotor blade white. This creates a blinking stroboscopic effect that makes the helicopter easy to see from above.

6. A trail of smoke from the plane's engines makes it easier to see. In combat in Vietnam, some forward air control aircraft were rigged to produce engine smoke on demand. Aerial demonstration teams learned long ago that a trail of smoke makes their act highly visible. Unfortunately, a smoking aircraft is not socially acceptable.

7. The position of the sun makes a differ-

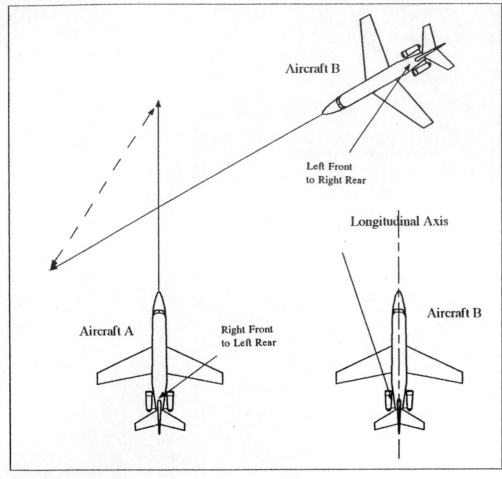

Aircraft B

Left Front
to Right Rear

Longitudinal Axis

Aircraft B

Aircraft A   Right Front
                         to Left Rear

Figure 17-4. Scratch Marks Slope in Opposite Directions

Poor visibility from the cockpit is coupled with some other problems. First, a lot of flying activity requires the pilot to have his head in the cockpit. Obviously, it doesn't make any difference what the cockpit visibility is if the pilot isn't looking. One aspect of the investigation is to determine what the pilots were most likely doing immediately before the collision. Second, aircraft speeds have gotten way ahead of our visual acuity which hasn't changed much in the last thousand years or so. General Chuck Yeager (who first broke the sound barrier) was blessed with remarkable visual acuity; at least 50 percent better than the rest of us. He credits that vision with his success as a fighter pilot in World War II. He could see enemy aircraft long before they could see him.

The rest of us can't do that. If we are flying at high speed and in danger of colliding head on with a small aircraft also flying at high speed, there is considerable doubt whether either of us could physically see the other in time to avoid the collision. To make matters worse, we are all subject to something called empty field myopia. If there is nothing out there to focus on, our eyes tend to focus at a point about 15 feet in front of our nose. We may think we are looking at a distance, but we really aren't.

Without something to focus on, it is difficult for most people to force their eyes to focus at infinity. That's really the secret to the three dimensional pictures that have become popular in the last few years. In order to see the three dimensional effect, you must deliberately focus your eyes at infinity and many people can't do that.

Another problem is what we might call the cockpit environment. Dirty windshields, the use of visors or sunglasses, cockpit lighting and the position

ence. It may illuminate the other plane better or it may blind the pilot who is trying to see it. It may also reflect off the skin of the other aircraft or produce shadows on the structure which make it more visible.

8. Finally, meteorological visibility makes a difference. Over some smoggy cities, the conditions may officially be VMC, but the visibility is still poor.

8. COCKPIT VISIBILITY.

Few aircraft outside of the military are deliberately built to provide the pilot with good visibility. If you think about it, you probably have better visibility from your automobile than you do from the average airplane and your car has rear view mirrors besides! In the old days, we never climbed or descended straight ahead. It was always done as a series of turns so we could see what was ahead of us, above us or under us. That's not done much anymore.

9. TARGET AIRCRAFT SIZE AND CLOSURE.

This brings us to an investigative technique that allows us to calculate the size of the other aircraft as it would appear in the windscreen at any given time prior to collision. This is a function of the size of the aircraft (as it appears to us) and the rate of closure. Figure 17-15 shows how this would look for the simple case of a direct head on collision.

You can construct a similar diagram for any collision. Start with the apparent aircraft size as viewed from the other aircraft. The use of a model might be helpful here. Convert the calculated closure speed to a chart of distances in feet at various times in seconds prior to impact. Calculate the half-angle view of the target at each distance. The trigonometric explanation for this is shown in Figure 17-16. Double this for the total angle subtended by the target aircraft on the windshield.

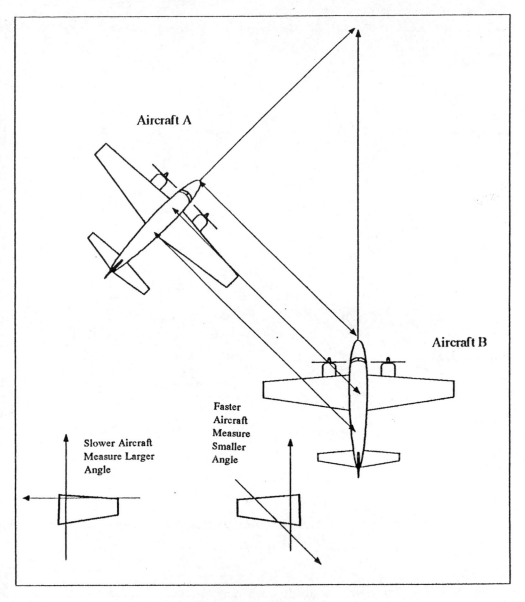

Figure 17-5. Scratch Marks Slope in Same Direction

Finally, construct a drawing similar to Figure 17-15. You could probably train a computer to do all this for you.

10. COCKPIT VISIBILITY STUDIES.

As part of the mid-air investigation, the question of which of the pilots could have seen the other aircraft and avoided the collision needs to be answered. If your collision geometry studies show that neither could have seen the other, you can skip this section.

This is normally done by using the aircraft manufacturer's "design eye reference point". This is a chart showing what the world looks like to a pilot sitting with his eyes where the manufacturer says they ought to be. An example of this is shown in Figure 17-17.

As you can see, the window frames are shaded to show where the pilot has monocular (one-eyed) vision and filled in to show where he has no vision. Based on the collision geometry and relative bearing (and any other available information) the position of the other aircraft at various times before impact is plotted on the diagram.

There are a few problems with this method. It assumes that the pilot has his eyes at the correct point

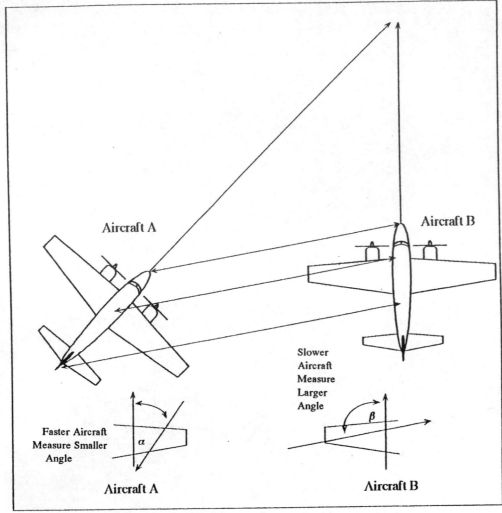

Aircraft A

Aircraft B

Slower
Aircraft
Measure
Larger
Angle

Faster Aircraft
Measure Smaller
Angle

α

β

Aircraft A

Aircraft B

Figure 17-6. Scratch Marks Slope in Same Direction

and is looking straight ahead. It does not take into account odd-sized pilots, different seat adjustment habits and the fact that he tends to move his head a lot.

In the old days, these studies were conducted by mounting a binocular camera in the pilot's seat and taking a wide angle picture. Now, we use a computer, which is helpful, because we can easily adjust the design eye location and simulate head movement.

For all its defects, this is still a useful investigative method for showing what the pilots in each aircraft could have seen (or could not have seen.) See also Chapter 20 for a discussion on the use of computers.

11. SPATIAL LOCATION OF THE COLLISION.

If you have good ATC radar plots on one or both

aircraft, this is easy to determine. If not, you might try trajectory analysis as explained in Chapter 9.

12. AIR TRAFFIC CONTROL INVOLVEMENT.

If either or both aircraft were under air traffic control, then ATC has some degree of involvement in the collision. They, after all, manage the airspace. They are also involved in runway incursion accidents and even taxi accidents. Although we seldom refer to these as "mid-airs" we really should. They fit the definition of an aircraft accident and they were definitely a collision. The fact that one or both aircraft had their wheels on the ground is really irrelevant. The same investigation techniques apply.

Investigation of ATC procedures is somewhat specialized. Important sources of information are the voice communications tapes and the radar data plots if available. If the collision occurred on or near an airport, weather records may be important. The visibility from a control tower can be examined using the same general techniques you would use on an aircraft. See also Chapter 22.

13. USE OF COMPUTERS.

This has been mentioned a couple of times already. If there is enough digital information available about the collision, it can be recreated in graphic form on a computer. One classic use of this technique was an investigation by the Canadian Transportation Safety Board into the near-collision of two air transports. Although there was no ground radar data available, both aircraft had digital flight data recorders. Both aircraft tracks, attitudes and speeds could be duplicated graphically on a computer. The aircraft passed close enough for one aircraft to be affected by the downwash from the other. The point where this occurred was

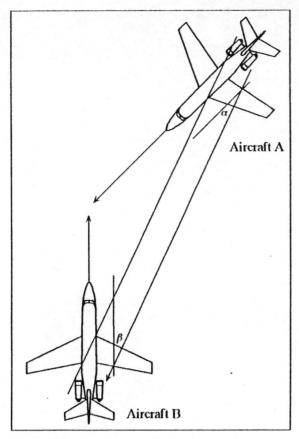

Figure 17-7. Equal Angles - Equal Velocity

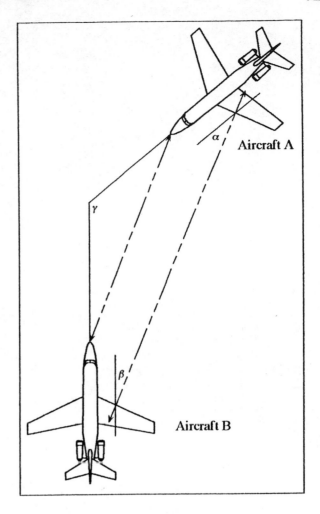

Figure 17-8.

clearly visible on the FDR data as it registered as a change in pressure momentarily affecting both airspeed and altitude. This provided the investigators with a point at which the two tracks must have crossed and allowed them to show both the dynamics of the near-collision and the view each of the pilots could have had from their respective cockpit.

14. COLLISION AVOIDANCE EQUIPMENT.

As more and more aircraft become equipped with TCAS equipment, several questions are bound to arise.

    1. Was either aircraft TCAS-equipped?

    2. If so, was the equipment on and functioning?

    3. Did the equipment provide the pilots with any warning of the impending collision?

At this writing, the techniques of investigating TCAS equipment have not yet evolved. As with almost all other investigation techniques, they probably won't until we have some accidents where TCAS was involved. Since the system uses warning lights, the

techniques described in Chapters 14 and 15 may be of some help.

15. SUMMARY.

Think about this for a moment. In automobile accidents, the dominant type of accident is the collision. Single car accidents are comparatively rare compared to all automobile accidents. In aviation, the opposite is true. Most aircraft accidents are single plane accidents and the investigation centers around the aircraft itself and how it was being flown. Mid-air collisions are comparatively rare. Thus if you really want to know more about investigating collisions, go buy a textbook on vehicle accident investigation. See the bibliography.

At the beginning of this chapter, we mentioned that investigating mid-airs was generally easier than investigating other types of aircraft accidents. The geometry can get a little messy, but that's because this is the only type of accident in which we need it.

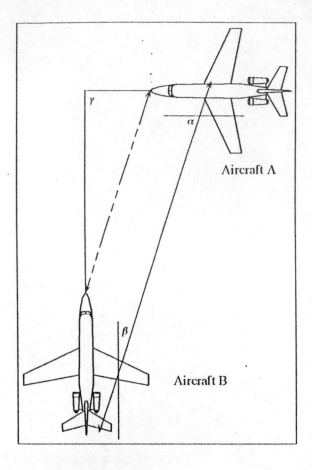

Figure 17-9.

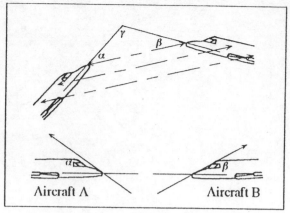

Figure 17-11. Scratches in Same Direction. Fore to Aft, Bottom to top. Aircraft Nose High Relative to Each Other.

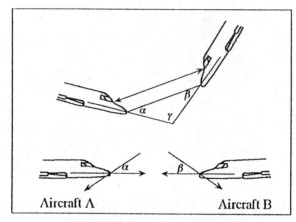

Figure 17-12. Scratches in Same Direction. For to Aft, Top to Bottom. Aircraft Nose Low Relative to Each Other.

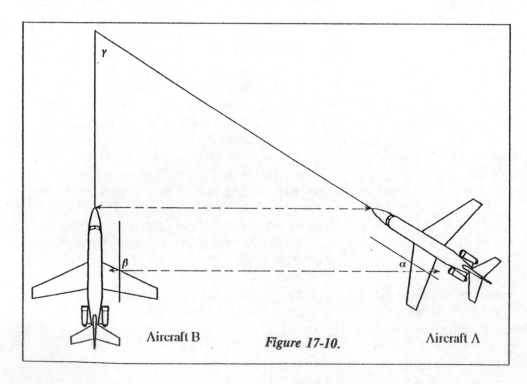

Figure 17-10.

NOTE: Figures 17-8, 17-9 and 17-10 show that the individual aircraft collision angles are complementary to the impact angle. Thus the impact angle = 180° minus the sum of the individual collision angles.

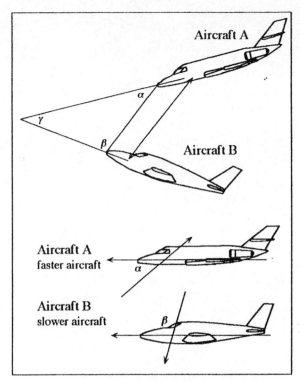

Figure 17-13. Scratches in Opposite Direction. Slower Aircraft on Bottom.

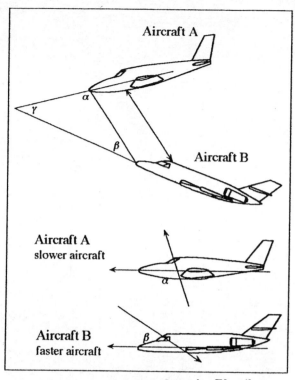

Figure 17-14. Scratches in Opposite Direction. Slower Aircraft on Top.

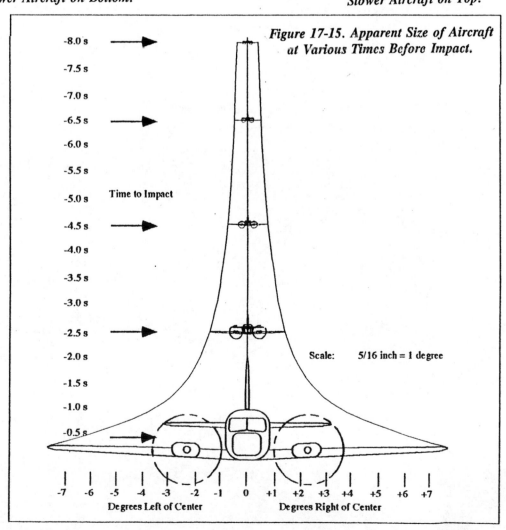

Figure 17-15. Apparent Size of Aircraft at Various Times Before Impact.

Scale: 5/16 inch = 1 degree

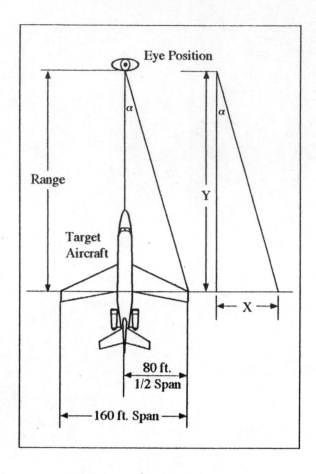

*Figure 17-16. Calculation of
Apparent Size of Aircraft*

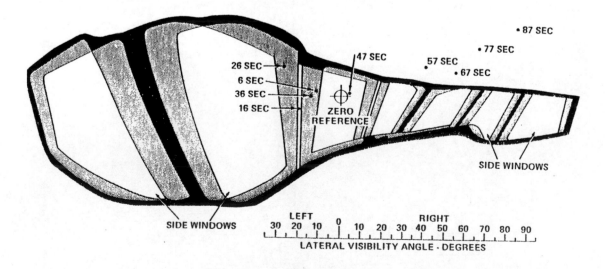

Figure 17-17. Left Seat Cockpit Visibility Depiction.

18

Aircraft Recorders

1. INTRODUCTION.

The history of aircraft recording devices is somewhat rocky. Interestingly enough, the first flight data recorder was installed by the Wright Brothers on their original airplane in 1903. They installed a device that recorded engine RPM, time and distance traveled. From then on, not much was heard about recorders until we entered the jet era.

As aircraft systems and cockpits became more sophisticated, aircraft accident investigation became harder. There was less evidence remaining after the crash. Meanwhile, aircraft accidents were becoming more expensive and less tolerable. A solution designed solely to help the investigator was the Flight Data Recorder (FDR) and the Cockpit Voice Recorder (CVR).

Neither were initially popular. They cost money and they contributed nothing to the performance of the aircraft. From the flight crews' point of view, they felt they were being spied upon and that the FDR/CVR information could be misused.

They were right. It could be misused and, in some cases, it was. That explains much of the legislation and some of the limitations on use of FDR/CVR data.

Early FDRs were fairly crude by today's stan-dards. The Digital FDR (usually called a DFDR, but we'll just call them all FDRs) has emerged as a very useful tool for monitoring both aircraft and crew performance. Many of the world's airlines are doing this very successfully. In the United States, because of built in resistance against use of the data for any purpose other than accident investigation, only a few U. S. airlines are routinely using this data. As of January, 1995, this resistance may change. As far as can be determined, CVR data is almost never used for any purpose other than aircraft accident investigation. In some countries, such use is prohibited by law.

FDRs and CVRs were first required on scheduled air transports and, over the years, they have been required on smaller and smaller aircraft. If the investigation community has its way, all aircraft will have recorders. Realistically, that is not likely to happen. In the United States after October 1991, CVRs are required on all multi-engine turbine powered fixed or rotary wing aircraft if two pilots are required for flight and they can carry six or more passengers. FDRs are required on all aircraft operating commercially and carrying ten or more passengers. It is conceivable that recorders will eventually be required on any turbine powered aircraft, fixed or rotary wing, carrying six or more passengers, but that's as far as it is likely to go. The majority of aircraft accident investigations will be conducted without benefit of aircraft recorders--at least for the foreseeable future.

2. FLIGHT DATA RECORDERS.

Early FDRs recorded five basic parameters on steel or aluminum tape. These were analog recorders and they worked by having a moving stylus scratch a trace on the tape for four of the parameters. The fifth parameter, time, was recorded by an imprint on the tape when a cockpit microphone was keyed. Since the tape did not always move at a uniform speed, this allowed correlation with some other time line or event. These FDRs were permitted on aircraft certificated prior to October, 1969. This doesn't mean manufactured prior to 1969; many of those aircraft were in production well in to the 80s and could still legally have the old foil recorders. In the United States, the existing foil recorders must be upgraded to 11 parameters by May, 1994. For all practical purposes, this means switching to a digital recorder. Aircraft manufactured (regardless of date of certification) after October 1991 which require an FDR, must have a digital FDR with a 17 parameter capability.

In spite of all this, there are still foil FDRs in use throughout the world and they are still encountered during investigations.

3. THE ANALOG (FOIL) RECORDER.

This recorded pressure altitude (+/- 100 feet), indicated airspeed (+/- 10K), magnetic heading (+/- 2 degrees), vertical acceleration (+/- 0.2G) and time. As mentioned, it recorded by scratching a trace on a roll of stainless steel or aluminum tape housed in a crash-survivable container. Readout required microscopic analysis of the traces and transfer of the data points to a graph. This was not completely accurate as the trace itself had width and two people examining the same trace might pick different points to plot and produce slightly different curves. The use of an optical reader to transfer the traces to an X-Y plotter improved accuracy significantly. The ability to read the traces was limited to the manufacturer of the recorder and major government aircraft accident investigation agencies. There wasn't much the field investigator could do with it.

4. THE DIGITAL FLIGHT DATA RECORDER.

The development of digital FDRs improved both data collection and readout accuracy. The recording medium became mylar tape and the recording parameters suddenly became anything on the airplane that could be measured and reduced to digital form.

Nominally, digital FDRs have the capability to record at least 62 different channels or parameters. Actually, the number of parameters is almost infinite as one channel can be used for several parameters. In a four-engine aircraft, for example, all four engine EPRs can be recorded on the same channel by using a sampling technique where the recorder samples data from each engine in turn. Since this sampling occurs each second (or less) any variation in the final readout due to sampling is hardly noticeable.

The parameters required to be recorded vary somewhat with the country and type, size, age and use of the aircraft. In the United States, the largest number of parameters is required for transport aircraft operated under FAR Part 121 and certificated after September, 1969 or manufactured after May, 1989. In addition, if certain other parameters are available digitally on the aircraft, they must also be recorded. The parameters listed below are current as of January, 1994 and are changing as this is written. The most current requirements are likely to be the EUROCAE specifications published in the European Community Joint Aviation Regulations (JARs).

A. AIRPLANE FLIGHT RECORDER PARAMETERS. Source: App. B, 14 CFR 121. Information on range, accuracy, sampling and resolution has been omitted.

Time
Altitude
Airspeed
Heading
Acceleration (Vertical)
Pitch Attitude
Roll Attitude
Radio Transmission Keying
Thrust/Power on Each Engine
Trailing Edge Flap or Cockpit Control Selection
Leading Edge Flap or Cockpit Control Selection
Thrust Reverser Position
Ground Spoiler Position/Speed Brake Selection
Marker Beacon Passage
Autopilot Engagements
Acceleration (Longitudinal)
Pilot Input and/or Surface Position - Pitch, Roll & Yaw
Acceleration (Lateral)
Pitch Trim Position
Glideslope Deviation

Localizer Deviation
AFCS Mode and Engagement Status
Radio Altitude
Master Warning
Main Gear Squat Switch Status
Angle of Attack (if recorded directly)
Outside Air Temperature
Hydraulic Low Pressure - Each System
Groundspeed

In addition, if there is excess recording capacity available, the FAA recommends that the following parameters be recorded. These are listed in descending order of significance.

Drift Angle
Wind Speed and Direction
Latitude and Longitude
Brake Pressure/Brake Pedal Position
Additional Engine Parameters
    EPR
    N1 RPM
    N2 RPM
    EGT
Throttle Lever Position
Fuel Flow
TCAS (Collision Avoidance System)
    TA (Target Alert)
    RA (Resolution Alert)
    Sensitivity Level (as set by flight crew)
GPWS (Ground Proximity Warning System) Alert

Landing Gear or Gear Selection Position
DME 1 and 2 Distance
Nav 1 and 2 Frequency Selection

B. HELICOPTER-UNIQUE RECORDER PARAMETERS. Helicopters have their own list of parameters which include many of those listed for airplanes. Parameters unique to helicopters are listed below.

Altitude Rate (Vertical Velocity)
Main Rotor Speed
Free or Power Turbine RPM
Engine Torque
Flight Control Hydraulic Pressure (Primary)
Flight Control Hydraulic Pressure (Secondary)

SAS Engagement Status
SAS Fault Status
Flight Control Positions
    Collective
    Pedal Positions
    Cyclic (Lateral)

Cyclic (Longitudinal)
Controllable Stabilator Position

C. ADDITIONAL PARAMETERS. As mentioned, the operator can record any parameters he can measure over and above the required parameters. The use of this capability varies among airlines. Some record only the minimum. Some record as many as 250 different parameters per aircraft. It depends on the aircraft, the recorder and, of course, the desires of the airline. In the airline business, aircraft are constantly being bought, sold, traded, leased or transferred. One byproduct of this activity is that not all aircraft in a particular fleet will have the same recording capability regardless of what the operator might want. There is no simple way to re-wire an aircraft to collect additional parameters and the operator is pretty well stuck with the recorder system installed at the time of acquisition.

D. DATA STORAGE. In most cases, the data is recorded on mylar tape and stored on the aircraft in the FDR. Access usually requires removal of the FDR and computer analysis of the tape. Some installations have a Quick Access Readout (QAR) capability where the FDR data can be electronically dumped to a portable recorder plugged into the aircraft.

Data storage and access concepts are changing. In today's world of electronic data transfer and the availability of satellite communications systems, there is no technical reason why the data cannot be transmitted, collected and stored at some location on the ground. One such system that already exists is called ACARS (Arinc Communications Addressing and Reporting System.) Airlines using this system have access to real time information about how their aircraft are performing and can plan their maintenance requirements accordingly. Obviously, this can also be a big help to the aircraft accident investigator.

E. READOUT. Since the data comes off the recorder tape in digital form, some sort of computer analysis is necessary to make sense of it. Initially, only the FDR manufacturer and some of the larger aircraft accident investigating agencies had the ability to do this. The initial computer products plotted each parameter on the X axis of a graph against a scale shown on the Y axis. With a lot of parameters, this was confusing. Since it resembled the graph obtained from the old foil recorders, though, most investigators learned to live with it.

As computers became smaller and more powerful and software became more sophisticated, readout was no longer limited to the manufacturer or the investigating agency. Airlines could readout their own FDRs which contributed to their present usefulness. Investigators with a modern PC and the right software could do their own FDR analysis. Furthermore, the readout product did not have to be a graph. Flight instruments could be depicted as graphics on the screen and the FDR data could be used to operate the instruments. The British Aircraft Accidents Investigation Branch pioneered this technique. Now, it is routinely used in many investigations. In addition, the computer could portray a graphic image of the airplane from any vantage point and, using FDR data, it could be made to fly like the actual airplane. This could even be combined on the same screen with a depiction of the flight instruments. The technical staff of the Canadian Aviation Safety Board (now the Canadian Transportation Safety Board) were pioneers in this area. The United States Air Force and the National Transportation Safety Board also developed computer analysis techniques based on FDR data.

Another use of FDR data is in simulation. This is discussed in Chapter 20.

F. INVESTIGATIVE VALUE. Obviously, the benefits of having FDR data are enormous. If there are no survivors, no witnesses, and wreckage destroyed to the point where technical investigation is impossible, the investigation is going to depend primarily on recorded data. There are four general sources for this data; air-to-ground communications tapes, ATC radar data, CVRs and FDRs. Of these, the one that provides the greatest amount of information about the airplane, its performance and its configuration is the FDR.

G. A WORD OF CAUTION. It is safe to say the modern computers can handle more data than we can collect. When available digital data is translated into something that looks nice on the computer, some smoothing or averaging techniques were probably used to remove the occasional hiccup or glitch in the data. It is possible to unintentionally (or intentionally) bias the resulting analysis and show something that looks good but does not accurately portray the facts. This problem is similar to the one described in Chapter 6 on the pitfalls of replacing photographic film with digital imaging systems. Someone has to crosscheck the computer-developed analysis with the raw data.

4. COCKPIT VOICE RECORDERS.

The CVR records on mylar tape and is much easier to install and maintain than the FDR. Thus more aircraft are likely to have them. Most CVRs have a cockpit area microphone (CAM) usually mounted on the overhead panel between the pilots. This is meant to record cockpit conversation not otherwise recorded through the radio or interphone circuits. The CVR usually has a separate channel for each flight deck crewmember and records everything that goes through those audio circuits. It may also have a channel for the cabin public address (PA) system. The recording tape is a continuous loop 30 minute tape which automatically erases and records over itself. At no time is there more than 30 minutes of recording available which means that events occurring more than 30 minutes before landing (or crash) are not recorded. There have been attempts to recover data which had been overwritten by the recording head, but these have not been very successful.

There is no technical reason why the recording is only 30 minutes in length. We could make it an hour, or two hours or the entire flight if we wanted to. Among investigators, there is strong feeling that the recording duration should be expanded and this may be changed in the future.

A. COCKPIT AREA MICROPHONE. The CAM is necessary because few flightcrews use interphone if they don't have to. They talk to each other using their normal voices. Unfortunately, the cockpit of a modern aircraft is a fairly noisy place. The level of background noise is high and fidelity of the CAM recording is always poor. The British attacked this problem in the late 70s by initiating a requirement for all transport flightcrews to use headsets and interphone below 20,000 feet. Ostensibly this was done to eliminate any need for a pilot to use a hand-held microphone during critical phases of flight. Actually, it was done to improve the quality of the CVR recording during those same critical phases. This requirement was adopted by the United States in the late 80s and it has significantly improved the CVR recordings

If the sole record of intracockpit conversation is the CAM, transcribing that conversation tends to be tedious work. It is usually done by a group that includes people who knew the flightcrew members personally and can recognize their voices and accents. The recording is transcribed word by word based on the best judgement of the group. This is after using electronic filters to mask the background noise and improve the quality of the recording.

While the ability of the CAM to record conversation is not too wonderful, it turns out that the masking background noise can contain a lot of information not available anywhere else. For this reason, the CVR would be more properly termed a CSR--cockpit sound recorder. It records everything including engine noise, cockpit switches, motors, warning chimes, stick shakers, runway noises and so on. Determining the source of the sound is a matter of isolating it and matching its signature on an oscilloscope with a known sound. From this technology, investigators have been able to determine the RPM of each engine. They have heard tires fail and wings crack on the CVR. A plane rolling over a grooved runway created a rumbling sound which (knowing the distance of the grooves) allowed calculation of acceleration. Even in aircraft where interphone is always used and there is no shouting across the console, a CAM would still be a valuable investigative tool.

B. RESTRICTIONS ON USE. In the United States, CVR information is not used by the FAA in any civil penalty or certificate action. When an accident occurs, the CVR is to be held for the NTSB for at least 60 days. If the CVR is reviewed and transcribed, the transcription cannot be released until 90 days following the accident or the public hearing, whichever comes first. Finally, the only portions of the transcript released are those pertinent to the accident.

In Canada, the CVR data is considered privileged and not releasable except under court order. The information cannot be used in any disciplinary proceeding arising from the accident.

In England, the CVR can only be reviewed under the supervision of the Aircraft Accidents Investigation Branch of the Ministry of Transport. Its contents are rarely released.

The problem is that the recording always contains information not directly pertinent to the accident which sometimes gives an erroneous impression of the cockpit atmosphere and environment. On the other hand, the CVR contains a wealth of information regarding crew resource management, cockpit discipline and communications problems; not to mention the other sounds picked up as background noise. All of these restrictions notwithstanding, some airlines in some countries will permit review of the CVR by a disinterested third party in the presence of the flight crew. This has contributed to the resolution of inci-

dents and, where this is done, flightcrews are satisfied with the integrity of the program.

5. OTHER RECORDING DEVICES.

This chapter deals mainly with FDRs and CVRs, but those are not necessarily the only recording devices that may be found on aircraft.

A. MAINTENANCE RECORDERS. These can take several different forms and may be installed on all aircraft of a fleet or selectively on specific aircraft to monitor specific problems. The simplest example is probably the Hobbs meter found on most general aviation aircraft. This merely records operating hours. Some aircraft may have event counters or "Fatigue Meters" as they are called. These measure vertical acceleration excursions at various levels and are used to predict structural service life. It is not uncommon to install recorders on turbine engines which count engine cycles and various time-temperature levels. Some maintenance recorders are fairly sophisticated and collect data on a large number of parameters. None of these are crash-hardened, but they may survive the crash and provide useful information to the investigator.

B. COMPUTERS. Modern aircraft have computer systems and computers have memories. Most of the memory chips are "volatile" which is analogous to the random access memory (RAM) in your computer. If power is removed, all data is lost. Some of the memory chips, though, are "non-volatile." They retain their memory and they can be read out. There have been some classic investigations where, for example, the non-volatile memory chips of an inertial navigation system (INS) have been read out to determine where the INS thought the plane was at the time of impact.

The trick is to be able to tell the difference between a volatile memory chip and a non-volatile one by looking at it. In an accident involving a modern aircraft, the wreckage is usually full of loose computer circuit boards which are unidentifiable to the average investigator. Help may be needed from the manufacturer who understands his board numbering system, memory chip location and chip identification. It is not unusual for the investigator representing a manufacturer to bring a notebook of full size photographs of circuit boards to assist him.

Currently, some modern aircraft have non-volatile memory chips in the Central Air Data Computer

(CADC), Flight Management System (FMS), Electronic Flight Instrument System (EFIS) and the Ground Proximity Warning Computer (GPWC). That represents a significant amount of data not otherwise available.

C. CAMERAS. The military frequently uses cameras (either film or video) to record radar scope presentations, heads up display (HUD) information and gunsight data. They have been very helpful in accident investigation and the military has developed some useful techniques for recovering data from crash-damaged film or video tape. In the civil world, installed cameras are not common, but there is always the possibility that a crewmember or a passenger will be using a personal camera. If cameras are found at the scene, always process the film or tape. If the camera is badly damaged, don't give up hope. An amazing amount of information has been recovered from severely damaged film or tape. If you find undeveloped film in the wreckage, which has obviously been exposed to light, you can probably give up on that. Loose video tape, on the other hand, may be recoverable. The best way to handle the tape is to load it gently and loosely into a paper bag without folding or creasing it any more than it is already. Do not attempt to straighten it out or wind it on anything. The tape has to be cleaned anyway and it is better to have it unwound and loose. Include all the bits and pieces of the tape. It is possible that they can be identified and matched with the rest of the tape.

6. THE FUTURE.

In the early 70s, the British produced a classic film called, "Air Crash Detective," featuring the senior Air Accident Inspectors of the British AAIB. In it, one of the inspectors (investigators) commented that what he wanted was "a cockpit video camera that would tell me not only what was said, but what was done!" He was expressing the frustration every investigator feels in not knowing what actually happened in the cockpit.

The technology to put a video camera in the cockpit has existed for a number of years. It is used extensively in NASA space shuttle flights, experimental test flights and in selected military flights. The fact that it is not used in civil transport flights is a social problem; not a technical one. Among professional safety and investigation organizations, the occasional paper or article appears advocating their development and use. At this writing, there is nothing to suggest that this investigative tool will be available anytime in the foreseeable future. Pity.

7. SUMMARY.

Aircraft recording devices, where they exist, are an essential source of information for the investigator. As aircraft cockpit instruments and systems become more sophisticated, the investigator's task of recreating the accident situation becomes more difficult--maybe even hopeless. Those electrons disappear rapidly and leave no trace. If we do not concurrently develop methods of recording and storing data, the number of aircraft accidents whose cause is "undetermined" is certain to grow.

19

Air Traffic Control Recorders

1. INTRODUCTION.

Early air traffic control radar systems were analog systems. What the controller saw on his scope was the actual radar return. There was no memory of what the scope looked like (unless someone took a picture of it) and the system had little to offer the aircraft accident investigator.

Modern ATC radar systems are digital systems and they are found in the United States and many other countries. The actual radar return is converted by a computer to electronic symbology and this is what is displayed on the controller's scope. Since this is done by computer, the data is stored in memory and the actual situation as observed by the controller can be replayed on another scope.

There are some limitations to this, as the system can only replay the raw data it has stored. It does not account for how the individual controller had his (her) scope tuned and the replay may not resemble precisely what the controller saw.

A more common use of the data is to obtain it in digital form and either use it to plot the aircraft flight path or use it to recreate the situation in a graphics computer or simulator. The data is somewhat perishable as the FAA (United States) generally erases and reuses the computer tapes after 15 days. If there has been an aircraft accident, the FAA will normally save the tapes as they know that someone will want them.

2. DATA AVAILABLE.

In the United States and many other countries, radar antennas sweep once every 12 seconds, so the available radar data consists of a series of data points at 12 second intervals. Terminal radars may sweep more often than that. The data itself consists of aircraft identification, time (to the nearest second), altitude as reported by the aircraft's encoding altimeter (to the nearest 100 feet), and location in latitude and longitude to the nearest second. Location is also reported in an X-Y coordinate system with the radar antenna as the center of the X-Y axes.

An example of the raw data as received is shown in Figure 19-1. In this example, the code column identifies the aircraft and the plot symbol column displays a code related to the accuracy of each data point.

3. ANALYSIS.

From the data available, the investigator can obtain the following directly:

1. Position every 12 seconds.

2. Ground track (average point-to-point heading).

3. Reported altitude MSL.

4. Ground Speed (distance x time).

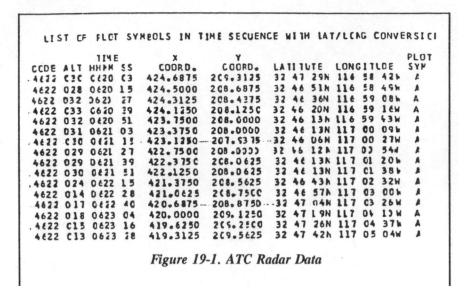

Figure 19-1. ATC Radar Data

have ever watched an ATC radar scope, you would appreciate that the symbols don't move smoothly; they jump. Furthermore, we are plotting the data on a large scale map; much larger than any aerial chart or the scale of the operator's scope. This amplifies any variations in the data points.

The accuracy of the data points is directly proportional to the distance of the target from the radar site. At the fringes of radar coverage, this can induce some substantial errors which must be taken into account in any analysis.

5. Acceleration (rate of change of ground speed).

6. Rate of climb or descent (rate of change of altitude).

By adding known wind data, these can be refined into true airspeed and aircraft heading.

By adding the basic aircraft lift and drag equations (software program available from NASA), aircraft attitude, angle of attack and "G" loading can be obtained.

4. PLOTTING.

The data is best plotted on a fairly large scale map using the latitude and longitude information. Each data point can be annotated with time and altitude. The map commonly used in the United States is the 7.5 minute topoquad with a scale of 1:14,500. (See Chapter 7.) This scale is large enough to make plotting fairly accurate and relate the aircraft track to cultural features.

The resulting plot will not be a smooth line, but will jump around a little. Remember that the radar system only takes a snapshot of the aircraft's radar return every 12 seconds. The return has size to it and the computer's decision on what latitude and longitude to assign to it leads to some small variations. Also, an occasional data point will plot out as an impossible position--completely out of the ball park, as we say. This is due to some whim of the radar system and the computer and these points must be ignored. If you

There are some methods of minimizing these errors. Plotting the data on a large scale map is a good start. Although this magnifies the variations in the individual data points, it improves plotting accuracy. Another method is to use a "moving average" technique for calculations. Groundspeed, for example, should not be calculated between two data points. It is best to calculate average airspeed over five or even ten data points and then move this calculation forward by adding the next point and dropping the first one. Headings are best measured by laying a plotter along a series of points and selecting the "best fit" heading.

Accident Investigation & Research, Inc., (see bibliography) has developed an interactive computer program for radar-based flight path reconstruction and analysis. Basically, this program uses mathematical smoothing techniques on the data, adds weather data and calculates a best fit flight path reconstruction. A functional block diagram of this method is shown in Figure 19-2.

5. INVESTIGATIVE USES.

This radar data can be used for a number of purposes.

A. FLIGHT PATH. First, obviously, we can use it to determine the flight path of the aircraft. Most commonly we are interested in what the airplane was doing prior to impact. Unfortunately, as the plane descends, it drops off the radar system and the last data point is usually some distance before the impact point. Also, the last data point tends to be a poor one and sometimes must be disregarded. The actual flight

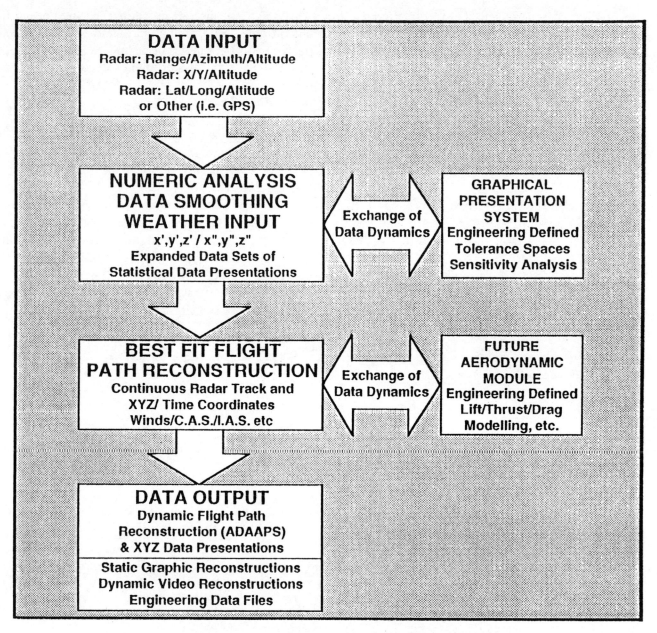

*Figure 19-2. Radar Analysis Program Functional Block Diagram
(Courtesy of <u>Accident Investigation & Research, Inc.</u>)*

path from the last (accurate) data point to the impact point is usually a matter of extrapolation, which is a fancy word for estimation. Sometimes, we get lucky. An aircraft hits power lines and we can get an exact time of impact from the power company. Since we know the exact time of the last accurate radar return and the airspeed and altitude at that point, we can calculate how the plane got from there to the power lines with reasonable accuracy.

B. AIRCRAFT PERFORMANCE. We can also use the data to tell us a lot about the aircraft's inflight performance. This can be useful if we are dealing with

an inflight structural failure.

C. MISSING AIRCRAFT. If we can't find the aircraft, plotting the last known radar location is a good place to start. Logically, the line developed by the radar plot should point to the aircraft location.

D. STRUCTURAL FAILURE. If we are dealing with an inflight structural failure, a plot of the aircraft track may give us some clues as to when things happened and in what order. If we can correlate the changes in heading, airspeed and altitude with changes in aircraft performance and aerodynamics, we may be able to predict when certain parts left the aircraft and where they might have landed. When a large component, such as an engine, falls from a plane, analysis of the radar data has assisted in locating the part.

E. ACCIDENT SEQUENCE. The radar data plot is a useful first step in determining the accident sequence. In addition to the radar data, we can add CVR and air-to-ground transmission data; witness statements, and weather data to the plot.

F. MID-AIR COLLISIONS. If we are dealing with a mid-air collision, the radar data is invaluable. Even if we only have data from one aircraft, (other aircraft VFR; not under radar control) that allows us to see what one aircraft was doing prior to the collision. (Note: It is possible that radar data on the other aircraft exists as skin paint.) More important, It permits us to locate the collision in space and know what one of the aircraft was doing in terms of heading and attitude. An examination of the wreckage should permit us to determine the actual collision geometry, since we have half the problem solved already.

6. SUMMARY.

As mentioned earlier, the radar data is in digital form and it can be used in any computer or simulator program that accepts digital data. More on this in Chapter 20.

On the whole, ATC radar data has become a useful and common investigative tool. In the United States, it is standard procedure to routinely request this data on all accidents where the aircraft was under ATC control. Useful as it is, it is important to understand its inherent errors and limitations.

20

Computers and Simulators

1. INTRODUCTION.

If there is a growth area in the field of aircraft accident investigation, it is probably in the use of computers and simulators as investigative aids. This growth started, probably, in the late 70s and must be attributed to the transition from analog devices (and data) to digital. The growth continues and it is hard to envision what uses we will see in the future.

2. COMPUTERS.

Computers were initially used to run iterative aerodynamic calculations--sort of a "what if" type of problem. Something had happened to an aircraft that defied rational explanation. If the result could be expressed in mathematical terms, then the computer could be instructed to vary all combinations of the aerodynamic equations of the aircraft until a match was found. This technique was used successfully in several investigations and it demonstrated that while the computer might not be smarter than the human investigator, it could certainly test more calculations in a short time than a room full of investigators could ever hope to test.

A. DATA CORRELATION. This is still a useful technique, but, thanks to digital equipment, we have moved beyond that. These days, an accident involving an air transport will provide digital data from at least two sources. The FDR will provide digital data on at least 17 and perhaps as many as 300 parameters (See Chapter 18.) The Air Traffic Control Radar system will provide digital data on time, location, altitude and heading. (See Chapter 19.) The aircraft manufacturer can provide all of the aerodynamic equations for the aircraft plus software containing aerodynamic models or stick figures of the aircraft.

With these data and today's computers, we can not only run the iterative calculations to examine the possibilities, we can write software to display the data we already have in any form we want. We can have graphical images of the cockpit instruments on the computer screen and we can make them respond based on the known FDR data. We can show a three-dimensional view of the aircraft and show its attitude based on FDR data. We can add terrain imagery and show its location based on ATC data. We can overlay (or impose) this with weather depictions, CVR conversations or Air-to-Ground transmissions. We can show all of this on the same screen with the instrument panel in one corner, a view of the aircraft in another and the transmissions scrolled along the bottom.

All of this spectacular imagery is not without drawbacks. It is easy to forget that the resulting depiction is not a live "picture" of what happened, but computer-created images based on the available data. While the data is fairly good in some cases, it is seldom as precise as we need to support today's computer graphics capability. Some artificial (and sometimes arbitrary) smoothing and interpolation of the data is usually necessary. Correlating data from several sources to a single precisely accurate time line is more difficult than it might seem. The common methods of defining a flight path solely from FDR data have some inherent errors, but the resulting computer depiction looks so good that no one thinks about the possible errors. It is (unfortunately) possible to use the available data accurately, but subtly bias the depiction in favor of a certain point of view. The obvious need to verify and validate the accuracy of the computer images is growing right along with the development of the images themselves.

B. MID-AIR COLLISION ANALYSIS. In spite of some misgivings about the accuracy of computer-generated images, the use of computers in mid-air collision analysis is a significant improvement over previous methods. In a mid-air collision, if we have digital data from only one of the aircraft, then we can enter that into the computer and let the computer show us where the other aircraft must have been (based on its probable airspeed) in order to produce the collision evidence found in the wreckage. If there is digital data available from both aircraft, then the geometry of the collision can be studied from any point of view.

C. COCKPIT VISIBILITY STUDIES. In some situations, particularly mid-air collisions, we want to know what the pilot could have seen if he was looking. Normally, this is done by taking the manufacturer's design eye reference point (the points in space where the pilot's eyeballs are supposed to be if the pilot is facing straight forward with the seat correctly adjusted) and plotting whatever the pilot should have seen in relation to those two points. Areas where the pilot's vision is either totally obscured or partially obscured (monocular vision--one eye only) by windshield frames are plotted on the diagram. The obvious error in this technique is that the pilot does not sit still and does not always stare at the same point. The computer allows us to simulate natural head movements and shifts of the head and body to show how the view changes along with the obscurations.

3. SIMULATORS.

The history of aircraft simulators is an interesting one. Early simulators, starting probably with the famous (or infamous) link trainer of WWII, were analog devices which didn't resemble any particular aircraft and certainly didn't fly like one. It had no value to the aircraft accident investigator as it could not be made to behave like the real airplane.

In the late 60s and early 70s, we developed full motion simulators which were digitally controlled. NASA had one of the early ones at Ames Research Center. It could be programmed to fly like any aircraft

for which NASA had the basic aerodynamic equations. Manufacturer's began developing simulators for their aircraft along with the aircraft itself as they recognized the benefit of being able to simulate various flight conditions before trying them in the real aircraft.

Today, all modern simulators are digitally controlled and we seldom hear complaints that the simulator doesn't accurately replicate the actual aircraft. It does, as long as it stays within its normal operating parameters. This means that if we have digital data from the accident aircraft (FDR), then we can program the simulator to do what it did. If we have route data (ATC Radar Input), we can program the simulator to fly the route the accident aircraft flew. Thus we can refly both the route and the accident sequence and sit there and watch what happens.

A. PILOT RESPONSE. Sometimes, we have an accident which involves a system malfunction and the questions is, "should the pilot have been able to cope with this?" By programming that malfunction into the simulator, we can test the reactions and responses of any number of pilots. This was done (for example) on the accident involving the DC10 which lost an engine on takeoff at Chicago. The engine actually fell from the aircraft and, in doing so, created an asymmetrical leading edge flap condition. Question: under the circumstances, could the flight crew have successfully recovered the aircraft? Answer: only if they had ignored flight manual procedures and maintained the airspeed they had at the time the engine failed. This was demonstrated conclusively in the simulator.

B. SIMULATED SYSTEM FAILURES. In some accidents you may suspect one or a combination of system failures, but you can't prove it. If you have a simulator that accurately replicates the aircraft, you can experiment with various combinations of system failures to see what effect they would have on the aircraft. In most cases, you could never do this type of experimentation on the actual airplane. We just don't do that with high performance airplanes.

4. SUMMARY.

Computers and simulators are emerging as very useful tools for the aircraft accident investigator. Even if you don't have any digital data for input, don't overlook their usefulness.

21

Operations and Maintenance

1. INTRODUCTION.

Investigation of the operations and maintenance aspects of an aircraft accident tends to be a little easier than the rest of the investigation. Most people involved in aviation already understand operations and maintenance and no unique technical skills are needed. Knowledge of the regulatory requirements and their intents and purposes is needed as these requirements provide the benchmarks for assessing operations and maintenance.

We also need an organized approach to the investigation and knowledge of where to obtain specific information. That's what this chapter will cover.

2. INVESTIGATION ORGANIZATION.

We could divide this topic into three basic subjects, which is sometimes called the MAN-MACHINE-MEDIUM approach. MAN refers to the crewmembers and their qualifications. MACHINE refers to the aircraft (pre-accident) and how it was maintained. MEDIUM refers to the environment in which the plane was operated. Under that topic, we should consider weather, the air traffic control environment and the airport. Although weather is frequently investigated as part of flight operations, it is a little easier to discuss along with air traffic control and airport investigations. Those will be covered in a separate chapter. For now, let's divide this topic into OPERATIONS, meaning an examination of how the plane was being flown and who was flying it; and MAINTE-NANCE, meaning an investigation into the condition of the aircraft prior to the accident.

3. OPERATIONS.

A. INITIAL ACTIONS. As a starting point, it is wise to assume that important documents were in the aircraft at the time of the accident. Pilots are supposed to have their licenses and medical certificates in their possession when they fly. Also, it is not unusual for the pilot to carry his log book with him. We should make an effort to recover these items as they are somewhat perishable. Other documents that might be in the cockpit include charts, maps, checklists, flight manuals, performance calculations, weather briefings, flight plans and miscellaneous notes.

B. PILOT QUALIFICATIONS AND EXPERIENCE. What can we learn about the pilot(s) of an aircraft?

Start with licenses and certifications. If these were destroyed in the accident, the FAA (United States) will have a record of them. There are usually two key questions. Was the pilot qualified to operate that aircraft on that flight? Did he have a current medical certificate and was he in compliance with any medical waivers?

Flying experience. If the pilots' log books cannot be located, flying experience can be recreated, somewhat. In the United States, each time the pilot renews his medical certificate, he must enter his current flying experience on the application. Flight experience may be part of his training records and some aviation operators keep separate records of their pilots' experience. Reviewing the pilot's employment application may be helpful. In evaluating experience, it is necessary to differentiate between currency and proficiency. A pilot may be technically current in an aircraft, but not proficient in it. Likewise, he may be proficient in flying the aircraft, but not proficient in the particular phase of flight or activity in which the accident occurred.

In the United States, the NTSB will routinely ask the FAA for the records on the pilot(s). This will include certification, violation records and medical information not normally available to other investigators.

C. TRAINING. Look for records of pilot training. If the pilot is flying as a commercial or air transport pilot, the training records should reflect compliance with both FAA and company training requirements. Examine the grade sheets for any formal training such as recurrent simulator training.

D. COMPETENCY. The competency of a pilot is sometimes hard to judge because there is a tendency in the industry to not keep records of incompetency. Frequently, interviews with other pilots or instructors can provide insight into competence.

E. OPERATIONAL PROCEDURES. If the aircraft was being operated as a commercial or air transport aircraft, the company should have an FAA-approved operations manual which describes company operational procedures. The key questions are: Were the procedures appropriate for the situation encountered? Were they followed?

F. AIRCRAFT LOADING. In a commercial operation, it is fairly easy to determine how the plane was supposed to be loaded. Determining how it actually was loaded is not quite so easy. It is usually done by comparing cargo and passenger manifests with what was actually on the plane and by interviewing the people who did the loading. If possible, it may be appropriate to actually weigh the cargo and passengers. Accidents wherein gross weight or center of gravity are significant usually occur during or shortly after takeoff. If the plane had flown satisfactorily for some time prior to the accident, the GW and CG were probably within limits. The actual GW and CG at the time of the accident can be calculated by assuming that the takeoff weight and balance was correct and adjusting it for fuel burned during the elapsed time until the accident.

G. FLIGHT PLAN. A copy of the filed flight plan can be obtained from the FAA. Enroute deviations can be determined by obtaining FAA records of air-to-ground communications. In these days of computers, inertial and GPS navigation systems, it may be important to know what system was in use and how it was programmed. It can also be important to compare planned track with actual track. In these cases, the ATC radar data can be helpful. See Chapter 19.

H. COMMUNICATIONS. The FAA will normally provide transcripts of air-to-ground communications they consider relevant to the accident. It is possible that communications occurred on other frequencies that contain information relevant to the accident even though the accident aircraft was not on that frequency. This may require a search of several frequencies. Also, there can be relevant information that was transmitted before the FAA transcript started or after it ended. Another point. When obtaining a communications transcript, it is essential to also obtain a copy of the tape itself. Normally, the FAA is willing to copy their tape onto an ordinary cassette tape. The reason for this is that voice tapes are transcribed by people who have transcription skills, but may not have an extensive background in aviation. What they term an "unintelligible word" may be perfectly intelligible to someone who understands the airborne situation or knows the pilot personally. If you want tapes or transcriptions other than what the FAA thinks is relevant, you have to make your request fairly quickly. The FAA recycles their tapes in as little as 15 days.

One chronic problem in aircraft accident investigation is correlating all information to the same timeline. The ATC radar data, the A/G communications tapes, the CVR and the FDR are all based on time, but it is not unusual for there to be a spread of several seconds among them. The time signals on the A/G tape is likely to be the most accurate, but, to be precise, you must deal with a copy of the tape itself; not the transcript. The CVR time line is probably the least accurate, but the keying of a cockpit microphone will be recorded on both the CVR and the A/G tape and can thus be correlated. In most FDR systems, the timeline is established by microphone keying which also allows correlation. If the radar data timeline is different from this, it is probably best to advance or retard the radar data to eliminate the error.

I. AIRCRAFT PERFORMANCE. Knowing how the aircraft should have performed involves knowing its configuration, weight and balance, and the weather. In most cases, this involves coordination with other investigators (and reference to other chapters in this book) and access to aircraft performance data. With the aid of computers, it is possible to run iterative performance calculations in an effort to match the actual flight profile.

J. OPERATIONAL TECHNIQUE. Here, the investigation gets a little murky if there is no one left to describe what was actually done. Sometimes the pilot's actions can be determined fairly accurately by analysis of the FDR. If there is no FDR, we must rely on eye witnesses and aerodynamic calculations based on weight, balance and configuration.

K. HISTORY OF THE FLIGHT. Part of the operations investigation is usually the development of a history of the flight for inclusion in the final report. This starts with the earliest event of significance to the flight and ends with the accident. It is a statement and listing of the facts and it doesn't provide any analysis or conclusions. Usually, it is necessary to confer with the other investigators to establish an accurate history of the flight.

4. MAINTENANCE.

A. GENERAL. Establishing the condition of the aircraft prior to the accident involves research into both the aircraft itself and the requirements for that aircraft.

B. REQUIREMENTS. Start with the FAA and manufacturer's maintenance requirements for that aircraft. A listing of all Airworthiness Directives applicable to the aircraft can be obtained from the FAA. Service Bulletins can be obtained from the manufacturer.

C. AIRCRAFT. If ownership of the aircraft has changed since manufacture, it might be wise to conduct a title search. This can also be done through the FAA. Keep in mind that registration number means nothing. Registration numbers can be easily (and legally) changed. Records are filed by manufacturer's model and serial number. In addition to a title search, the FAA can also provide the record of compliance with ADs as far as has been reported to them.

The airframe and engine log books reflect re-

quired inspections, major modifications, maintenance or repair and airworthiness directive compliance. They do not always reflect day-to-day maintenance activities. For these, it may be possible to determine who was actually maintaining the aircraft. Shop records in the form of work cards or invoices may provide a better picture.

D. MODIFICATIONS AND MAJOR REPAIRS. The record of any aircraft modifications or major repairs should be substantiated on appropriate forms (FAA Form 337 in the United States.) Approved modifications may require a supplemental type certificate (STC). These should be maintained by the operator.

E. WEIGHT AND BALANCE. The current aircraft weight and balance should be maintained by the operator.

F. MALFUNCTION AND DEFECT REPORTS. In the United States, serious malfunctions or defects should be reported to the FAA. The operator should maintain a copy of these reports and the report should be retrievable in the FAA Service Difficulty Report (SDR) files by aircraft serial number. See also the discussion in the next paragraph.

G. RESEARCH. In the United States, the FAA maintains two computer files that are accessible and potentially valuable to the aircraft accident investigator. These are the Accident/Incident files and the Service Difficulty Report (SDR) files.

ACCIDENT/INCIDENT FILES. These are slightly different from the files maintained by the National Transportation Safety Board because the FAA defines the term "incident" differently and they do not record the same data elements. This file can be used in two ways. First, has this particular aircraft ever been involved in any reported accidents or incidents? Second, has this particular situation ever occurred before? This second method requires some forethought. Are we interested in the situation regardless of aircraft type, or only this specific model, or perhaps all single-engine aircraft, or all aircraft using this particular engine? There are many different ways to organize the data and good research requires a basic understanding of how the data is organized in the first place.

SDR FILES. Although these reports are mandatory, the requirement is not realistically enforceable. Compliance is estimated to be somewhere around 15%. Nevertheless, the volume of reports received over time makes this a valuable file. As in the accident/incident files, it can be used in two ways. First, we can query the file for any reports logged against this specific aircraft. Next, we can ask for data on how often a particular part or component has failed in a particular manner. Again, this involves knowing how the data is entered in the first place. It also involves knowledge of how many different aircraft use the same component. Engines and propellers, for example, are almost always used on several different aircraft. If we want to study a particular engine, we have to study it in all of its applications.

ACCESS. One way to get at this data is to ask the FAA for information on the data bases, copies of the data coding sheets and samples of the report format. With these, it is possible to construct a request for data that will satisfy the investigator's needs.

Another way is to do business with an organization that can access the data bases directly and routinely provides specialized reports based on the needs of the requestor. This method tends to be somewhat faster and it starts (or should start) with a discussion with someone who understands the data base and can translate the investigator's needs into a query. One organization offering this service is listed in the bibliography.

5. SUMMARY.

The operations and maintenance investigation is fairly straight forward. Was the pilot qualified to fly this plane on this flight? Was the aircraft operated as it was supposed to be operated? Was it being operated within its performance limits? Was the aircraft properly maintained? Although some of the desired records may have been destroyed in the accident, there is a lot that can be done to reconstruct both operations and maintenance aspects of the flight.

22

Airfield, Air Traffic Control and Weather

1. INTRODUCTION.

These subjects represent the environment or the "medium" portion of the man-machine-medium investigation concept. Some specialized assistance is usually needed in these subjects, but ultimately the results have to integrated with the rest of the investigation.

2. AIRFIELD.

As a rule, it is always wise to collect all available published information about the airfield current as of the date of the accident. This includes diagrams, facilities, approach and departure charts, noise abatement procedures and published instructions to pilots. The reason is that the investigation may drag on for some time and may be reopened for litigation years after the accident. The published information about an airport changes and it is not easy to recover obsolete or superseded documents. The only information that counts, of course, is the information current at the time of the accident. While you are collecting this information, look for aerial photographs of the airport and engineering drawings of the runway/taxiway system. Both exist and it is merely a matter of locating them. In one case, the best available diagram was securely attached to the wall of the operations office. The investigator just set up his camera and photographed it.

A. AIRFIELD STATUS. The amount of information available about an airport depends somewhat on its status. In the United States, an airfield serving air carriers is referred to as an FAR Part 139 airport. This Federal Aviation Regulation requires that such airports be certified and regularly inspected. Records of this certification and inspection should be available in FAA regional offices.

In addition, a Part 139 airport must have an operations manual which describes the airport and its operation and contains an emergency plan. The operations manual should address the following subjects:

1. Operational responsibility.
2. Approved exemptions to FAR Part 139.
3. Limitations imposed by the FAA.
4. Airport grid map.
5. Runway and taxiway identification.
6. Obstructions.
7. Movement areas and safety areas.
8. Operations during construction.
9. Maintenance of paved areas.
10. Maintenance of unpaved areas.
11. Maintenance of safety areas.
12. Marking and lighting systems.
13. Snow and ice control plan.
14. Rescue and firefighting facilities and equipment.
15. Hazardous materials.
16. Traffic and wind direction indicators.
17. Emergency plan to include:
    a. Aircraft incidents and accidents.
    b. Bomb incidents.
    c. Structural fires.
    d. Natural disaster.
    e. Radiological incidents.

f. Sabotage and Hijack incidents.
g. Power failure.
h. Water rescue.
18. Airport self inspection program.
19. Control of ground vehicles.
20. Obstruction removal, marking or lighting.
21. Protection of navaids.
22. Protection of public.
23. Wildlife hazard management plan.
24. Airport condition reporting.
25. Identification and marking of unserviceable areas.
26. Other items required by the FAA.

If the airport is not subject to FAR Part 139 (United States), then it is not likely to have all that information, although it may if it is big enough. Ask.

B. POST ACCIDENT ACTIVITY. Following an aircraft accident, it is common practice for the FAA to collect the records on all facilities and navaids at the airport and recertify the navaids in use at the time of the accident. If, for example, the accident occurred during an ILS approach, it would be normal practice to take the ILS out of service until its accuracy can be tested and verified. Records of these recertifications should be available.

C. AIRPORT STANDARDS. Internationally, members of ICAO have agreed on airport and heliport standards as specified in Annex 14 (Aerodromes) to the ICAO Convention and supplemented by numerous ICAO Airport Service Manuals. While these are not binding on any one country, they have generally been adopted throughout the world. The actual standards applicable in a country are found somewhere in its laws and regulations. In the United States, FAR Part 139 applies to how air carrier airports are to be operated and FAR Part 77 applies specifically to obstructions to aerial navigation. All other information about design, construction and marking are found in FAA Advisory Circulars in the 150 series. Technically, these advisory circulars are not mandatory, but they are, nonetheless, listed as standards.

In the United States, other applicable standards are published by the National Fire Protection Association (NFPA). While not mandatory in themselves, many have been incorporated into municipal codes which makes them mandatory.

D. DISASTER RESPONSE AND RESCUE. In almost every accident on or near an airport, the question of emergency response comes up. A logical approach to this subject might go like this.

1. Were the equipment, facilities and personnel adequate for the type of flying being conducted? Minimum equipment for a Part 139 airport (United States) is specified in the advisory circulars based on maximum aircraft size and capacity. If the airport is not a Part 139 airport, there are recommended levels of crash and fire protection, but these may or may not be met.

2. If the equipment was adequate, was it all serviceable at the time of the accident? Obviously, a piece of fire equipment is of no value if it is out of service at the time it is needed.

3. Was there an emergency or disaster response plan available? Was it used? Was it adequate? Part 139 airports are required to have a plan and exercise it regularly.

4. Was the notification and response timely? In an airport environment, the control tower personnel are usually the first to know of an accident. They will keep a record of who they notified and when. Response criteria requires that fire equipment be able to reach any part of the airport within so many minutes. This response can be delayed by weather or terrain conditions.

5. Were the rescue procedures at the scene adequate? This is a technical problem and the adequacy of the effort usually varies with the training and experience of the personnel involved.

6. Was the amount of extinguishent adequate? It has happened on several occasions that things were going pretty well until the fire vehicles ran out of extinguishent--and had no system for replenishment. If you think about it, the problem of moving large amounts of fire extinguishent to remote parts of the airfield is not an easy one.

3. AIR TRAFFIC CONTROL.

Finding out what was done by air traffic controllers and tower operators is not difficult. Much of it is recorded on audio tape and much of the rest of it is logged. In the United States, all available information will routinely be collected and packaged by the FAA. This is normally available from the FAA Regional

Office after it has been delivered to the NTSB.

Typically, the FAA "Package" will contain the following information.

    1. Transcripts of Recorded Voice Transmissions

    2. Daily Record of Facility Operation
    3. Report of Aircraft Accident
    4. Pertinent Flight Progress Strips
    5. Controller Statements
    6. Pertinent Letters of Agreement
    7. Flight Plan
    8. ATIS Information
    9. ATC Radar Data
    10. Facility Accident Notification Record
    11. Position Logs
    12. "Sign-on" Log
    13. Navaid Flight Check Records
    14. Pertinent NOTAMs

This is not necessarily all the information that is available or pertinent from the investigator's point of view. Requests for additional data must be discussed with the FAA, preferably the facility manager. This needs to be done fairly early as some data is retained for only 15 days and then recycled or destroyed.

Evaluating the adequacy of the air traffic control services provided is a little more difficult. This is a specialized activity and it involves an understanding of what and how services should have been provided. In the United States, two essential reference documents are the Air Traffic Control Handbook and the Facility Operation and Administration Handbook. Both are available through the Government Printing Office (GPO).

Another approach is to collect statements from other pilots using those air traffic control services at the time of the accident. These pilots can sometimes be located by identifying their aircraft on the air-to-ground transcriptions and locating the aircraft owner through the FAA's registration files.

4. WEATHER.

This is another specialized area. On any given day, a considerable amount of information is available about the weather at any location on earth. Much of this data goes into files for future analysis. It is possible to resurrect this data and recreate the weather that existed at some earlier time and location. It takes an experienced meteorologist to know where to find this data and what to do with it.

A. SOURCES OF DATA (United States.)
    1. National Weather Service
    2. Federal Aviation Administration
    3. Private Weather Companies
    4. Airline Dispatch Offices
    5. Airline Meteorological Offices
    6. TV and Radio Stations
    7. Utility Companies
    8. Universities
    9. Air Quality Monitoring Networks
    10. Witnesses

B. TYPES OF DATA.
    1. Surface Weather Observations
    2. Weather Radar Data
    3. Upper Air Data
    4. Satellite Data
    5. Wind Gust Recorder Records
    6. Barograph Records
    7. Rotating Beam Ceilometer Records
    8. Lightning Data
    9. Transmissometer Data
    10. Rainfall Records
    11. Surface and Upper Air Charts
    12. Severe Weather Reports
    13. Low Level Windshear Alert System Data
    14. Pilot Reports
    15. ATC Radar Data (Weather Echoes)
    16. Wind Data
    17. AIRMET Data
    18. SIGMET Data
    19. Area Forecasts
    20. Terminal Forecasts
    21. Weather Briefing Records

In the United States, some of the available weather information current at the time of the accident will be furnished with the FAA "Package." Much of it must be sought. While collecting information, it is well to remember that weather data is widely distributed and all weather stations have pretty much the same charts and data. While the station on the airport where the accident occurred may be reluctant to give up their charts, the same charts are available at any other station and are usually thrown out after they are no longer current.

These days, most of the world's weather is photographed by satellite and a surprising amount of horizontal and vertical weather analysis is possible. These pictures are seldom included in the FAA package.

Once you have collected all of the available data and established the weather at the time of the accident, your next problem is to determine its affect on the airplane. You are concerned about severe weather phenomena including turbulence, hail, icing and windshear. If severe weather is not a factor (or not present), then accurate knowledge of ceiling, visibility and surface winds may be helpful in analyzing the accident sequence.

5. SUMMARY.

Because we are so good at keeping records, it is usually possible to accurately establish the environment in which the accident occurred. Analyzing the effect of those environmental factors on the accident frequently requires a specialist in that area.

23

Witnesses

1. INTRODUCTION.

The importance of witnesses varies with the accident. In some cases, they are absolutely vital. There is no recoverable wreckage, no survivors and no recorded information. In other cases, there is plenty of factual information available and the witnesses are merely corroborative. In these cases, it is interesting to note the differences between what the witnesses say and what the facts support. Are the witnesses lying? Are they deliberately trying to deceive us? In almost all cases, the answer is, "No." Rarely is the witness intentionally deceptive. The problem is in our inability to recover accurate information. This chapter will discuss some of the reasons for this and suggest an interviewing methodology.

2. DIFFICULTIES.

Psychologists estimate that we are able to recover only about 30% of what a witness actually knows. Sometimes this is the witness' fault. He has a poor ability to estimate time and distances; he has a lousy memory, poor eyesight and a limited technical vocabulary and there is not much we can do about it.

Sometimes it is our fault as interviewers. Our laws or regulations impose restrictions on how interviews will be conducted. These seldom aid the interviewer, but we must comply with them. Many of us lack good interviewing skills. We tend to interrogate instead of interview. We interrupt. We don't listen to the witness' answers. We embarrass the witness. We

inhibit him. We do not easily establish rapport with people we don't know well. People in some professions (clergy, medicine, psychiatry, etc.) tend to develop good interviewing skills because they use them so often. Also, those professions may attract people who are naturally empathetic to other people. Aircraft accident investigators tend to be technicians who suddenly find themselves in the interviewing business without any particular background or training in interviewing techniques.

Sometimes the problem is neither the fault of the witness nor the investigator. The witness had only a fleeting glimpse of the accident or he really wasn't in the right place to see what we wish he had seen. Frequently, he really didn't see the accident; he saw the results of it (the crash) which is not what we need.

Finally, the witness must translate what he saw into words which, we hope, have the same meaning to him that they do to us. To the average eye witness, for example, the airplane's rudder is the thing that sticks up on the tail. When he uses the word, "rudder," does he mean the movable part or the entire vertical stabilizer?

Whatever the reason, we should do what we can to improve our own techniques and increase the amount of information recovered.

3. TYPES OF WITNESSES.

Most investigation manuals lump all witnesses

together and treat them alike as far as interviewing them goes. This doesn't make much sense. The pilot who participated in the accident is a different type of witness from the casual observer who merely saw it happen. Here are the characteristics and methods of approach of the common witness types.

A. PARTICIPANTS. These are the people actually involved in the accident either as crew members, passengers or controllers. They may be in shock following the accident and may be suffering trauma from their injuries. They may be on medication and under the care of a physician. They know they are going to be interviewed and are likely to be defensive about their role in the accident. Recalling the accident may be a very emotional experience for them.

On the other hand, participants are frequently anxious to talk about the accident to anyone who will listen. Providing them with an opportunity to talk about it may even be helpful to their psychological recovery. The question of interviewing them while they are on medication frequently comes up. Although some associations refuse to permit their members to be interviewed while on medication following an accident, there is nothing wrong with it providing it is done with the approval of the attending physician. The fact that the witness is on medication should be noted in the record of the interview. In cases like this, the best person to conduct the interview might be a physician.

One problem with interviewing participants is dealing with their feelings of guilt. Participants may block traumatic memories of fellow crewmembers' injuries or death. Some may feel that there was more that they could have done to prevent the accident, even when there wasn't. They may even confess or accept blame for an accident even when there was nothing they could have done to prevent it. In these cases, the assistance of medical and human factors specialists is helpful.

Participants almost always have to be interviewed at least twice. It is important to get their story as early as possible as their recollection of what happened, what didn't happen, what was working and what was not working will focus your technical investigation and save you hours of fruitless effort. At this time, though, you don't really know enough about the accident to ask intelligent questions and you won't until you get into the technical investigation. Then you can go back to the participant witness with some really good questions.

B. EYEWITNESSES. These witnesses are randomly distributed regarding age, sex and background. You have no control over who they are. In general, they are perfectly willing to talk about what they saw and they are rarely intentionally deceptive. They have no reason to be. They may be transient and they may disappear rapidly. Their story will change based on what they read or hear about the accident before they talk to you. They are usually willing to help the investigator, but they don't plan on being inconvenienced. They may be sensitive about their lack of technical aviation knowledge and do not wish to be embarrassed by it. Some may resent having their privacy violated. If eye witnesses are obviously going to be critical to your investigation, getting their statements may become your number one priority. You want to talk to them before they disappear and before their recollection changes.

C. BACKGROUND WITNESSES. These could include anyone from friends of the pilot to dispatchers, weather briefers, mechanics, and so on. Although they were not directly involved in the accident and didn't see it, they are certainly aware of it and may be under some emotional stress because of their personal knowledge of the people, aircraft or situation. It is usually not necessary to interview them immediately as their story is less likely to change as time goes by. Generally, the best approach to these witnesses is to ask them for help. Make them feel that they can make a contribution to the investigation. You must expect them to be defensive about their role, however small, in the accident.

4. CHARACTERISTICS OF EYE WITNESSES.

A. SEX. The sex of an eye witness doesn't seem to make much difference. Males are better at observing things that typically interest them. Females are equally good at observing things that are of interest to them. The classic example used is that a male witness might be better at telling you how many people were in a room while a female witness might be better at telling you what they were wearing. That sounds a little chauvinistic. As a practical matter, you should remember that sex doesn't affect credibility.

B. AGE. Children make remarkably good eye witnesses. Their eyesight is sharp and their memory is excellent. Since they don't have much experience to draw on, they tend to tell you what they saw; not their analysis of what they saw. Unfortunately, children are very impressionable and it has been demonstrated that it is possible to convince them that they saw something that never happened. Also, they are likely to be inarticulate and unable to accurately describe what they saw. They don't have the vocabulary.

Teenagers can make excellent witnesses providing they are articulate enough to accurately describe what they saw. At their age, they are not yet influenced by the adult compulsion to explain things.

Adults should be good eye witnesses, but they are frequently not. As adults, they have amassed a certain amount of experience with things in general and they have a certain expectation of how an airplane is supposed to act even if they don't know much about it. They also have an almost compulsive need to explain what they saw in some rational way that fits into their experience. An adult who sees a fireball before he hears an impact will invariably tell you that the plane was on fire before it hit the ground. What he is giving you is an analysis of what he saw in terms that fit his experience. The teenager or the child is more likely to tell you that he saw the fireball; then heard the impact. They don't have any need to explain how that could occur. In dealing with an adult witness, even one experienced in aviation, the trick is to get him to tell you exactly what he saw; not what he concluded from what he saw.

Elderly witnesses tend to be unreliable for purely physical reasons. Their vision and hearing may be poor and they are losing their ability to focus on more than one thing at a time. If several things are happening, they may miss some of them. Nevertheless, they tend to be the most articulate of all witnesses and the best at explaining what they saw. Their statements should not be discounted.

C. WITNESS TRAINING. It has been argued that people with a background in law enforcement or legal work make better eye witnesses. It has been demonstrated that this isn't true. They are no more observant than the rest of us. Because of their training, though, they are better at organizing their thoughts and reporting their recollections.

5. LOCATING EYE WITNESSES.

You are investigating an accident that you know was witnessed, probably by several people. How do you find them?

Start with the people who responded to the accident first, particularly law enforcement and news media. Police officers are very good at copying down the names and addresses of witnesses. The news media have a knack for locating people who saw the accident.

If you find one witness, always ask him who else saw the accident. For some strange reason, eye witnesses tend to locate each other and exchange impressions. They literally fill in the blanks in each other's memory of the accident. It would be nice if you could locate them and isolate them from each other, but this is seldom possible.

Take advantage of communications transcripts. The air-to-ground communications tape may contain the call signs of other aircraft in the vicinity. Those pilots or passengers may have witnessed the accident.

Advertise. Don't buy an add in the classified section. Talk to the news media and give them some information about the accident or your investigation. Mention that you are interested in locating eye witnesses and anyone who saw the accident should call this number. If that is published as part of the story, you can bet that witnesses will read it. If they saw the accident, they will read everything written about it.

6. PROBLEMS TO EXPECT WITH WITNESSES.

A. FEAR OF REPRISAL. You cannot expect a witness to tell you things that are likely to result in some adverse action. The U.S. military enjoys an enormous advantage in this area as they are able to withhold witness statements and promise the witness that the statement will not be used for any purpose other than accident prevention. Even this promise of confidentiality does not guarantee that the witness will be entirely truthful, but it certainly helps.

In the civil world, promising a witness confidentiality is almost impossible. Statements collected during accident investigations become public documents and will be available to anyone who asks for them. There is no limitation on the use of the statements.

The National Transportation Safety Board has the authority to compel witnesses to make statements under oath, although it rarely finds it necessary to do this. They also have the authority to take statements in confidence, but this is almost never done. A witness being interviewed by the NTSB is entitled to be represented by counsel and the interviewer(s) must tolerate the presence of counsel if the witness so chooses. In the past, the NTSB has attempted to use a tape recorder to tape "informal statements." This practice has met with fairly stiff resistance as there is nothing "informal" about a recorded statement. Regardless of how or under what conditions NTSB statements are obtained, they all become part of the official record of the investigation and are readily available.

The net result of this situation is that the investigator has a tough time getting participant witnesses to tell him everything they know.

B. INTIMIDATION. For some people, giving an official statement is an unpleasant experience. They feel intimidated by the surroundings or the person(s) conducting the interview. If you want to get the best statement, you should remove as many of these intimidating influences as possible. A witness is likely to be more at ease and comfortable in his own surroundings; not yours. If he is likely to be awed by uniforms or rank or three-piece suits, remove them (figuratively speaking.) Dress like you expect the witness to be dressed. One problem is what we might call the gang interview. Theoretically, the ideal interview situation is one-on-one. One witness; one interviewer. In the real world, though, this rarely happens. All the investigators want to participate in the key interviews and the witness sometimes wants counsel present to represent his interests. At some point, this crowd of questioners becomes intimidating and the witness loses interest in being helpful.

The cartoon in Figure 23-1 suggests how this might appear to the witness. Although drawn to reflect a military accident board of inquiry, There are civil investigation situations which are not much different.

C. EMBARRASSMENT. No one likes to be embarrassed, but interviewers sometimes embarrass witnesses without realizing it. Suppose you are dealing with an eye witness who is illiterate or maybe just can't write or spell very well. He knows it, but he has managed to muddle through life without anyone else discovering it. His job does not require him to write. Suddenly he witnesses an accident and the next thing he knows, some investigator is handing him a witness statement form and a pen and asking him to fill it out. If he does that, he's going to be embarrassed. It's much easier to just tell the investigator that he didn't see the accident after all.

Another case. I once watched an investigator interview a witness who consistently mis-pronounced the word "aileron." Every time she mis-pronounced it, he corrected her. I could tell from the set of her jaw that she didn't appreciate being corrected. After about the third time, she just stopped talking and wouldn't answer questions with anything but "yes" or "no." For all practical purposes, the interview was over.

Another easy way to embarrass a witness is to imply that he is not telling the truth. In almost all cases, the witness is telling the truth as best he can recall it. What he saw may not fit with what you know, but that's hardly his problem.

D. WITNESS EXPECTATIONS. When we see Event A, we apply our experiences and we expect it to result in Event B, because that is normally what happens after Event A or at least that's what we think happens. In the case of an eye witness to an aircraft accident, this can lead to some strange statements. Lay people have misconceptions about the term, "stall." They may use it interchangeably to describe either an aerodynamic stall or an engine stall. Most can't tell the difference between a spin and a spiral. A large part of the population is convinced that if an engine quits, the plane immediately plunges to the ground. Thus a witness may initially see or hear an airplane and immediately develop an expectation of what is going to happen next. This is so powerful that it will sometimes show up in the statement as an observed fact, when it didn't really happen at all.

E. UNCERTAINTY. Asking an eye witness for a height, speed or distance is almost useless. Nobody including pilots can estimate those with any particular accuracy. It is better to take the witness back to where he was when he saw the accident and ask him to point to where he saw the plane. Since you probably know the flight path by now, you can measure the angle of the witness' arm and calculate the height. Likewise, if you ask the witness to indicate with his fingers how big the plane was when he saw it, you can calculate distance more accurately than he can estimate it.

F. FLEETING GLIMPSE. Aircraft accidents happen very fast and a witness is never fully prepared to watch one. Something attracts his attention to the aircraft and he may not realize he is watching an accident until it's over. Trying to go back into his memory and recall what he saw before he realized it was important is difficult.

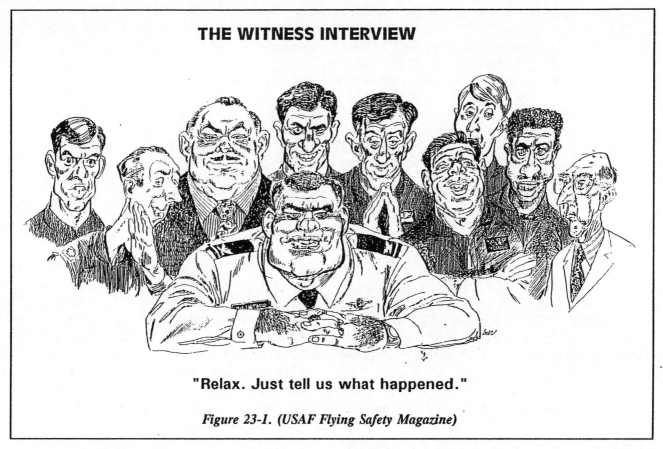

THE WITNESS INTERVIEW

"Relax. Just tell us what happened."

Figure 23-1. (USAF Flying Safety Magazine)

G. PERCEPTIONS. Like pilots, witnesses can experience temporal time distortion where time seems to stand still and the accident seems to happen in slow motion. Because of this, witnesses will consistently tend to overestimate time.

In addition, the witness' attention just naturally follows the most dramatic part of the accident; the biggest piece or the one that is burning. He may not even notice that a wing or a portion of the tail came off.

Another perception problem involves order of occurrence. The human mind is programmed to relate order of occurrence with order of perception. If we saw or heard it first, it must have happened first. Because of the difference in the speed of light and the speed of sound, this is not necessarily true. Perhaps you have had the experience of watching something that makes a lot of noise at a distance; a pile driver, perhaps. You may have noticed the illusion that the downstroke of the hammer does not coincide with the noise. Even though you know what's happening, your brain will still tell you that the events are occurring in the wrong order.

H. MEMORY. After we observe an event,

forgetting sets in almost immediately and we tend to drop out the fine details of what we saw. Furthermore, we are influenced by everything we hear and read about the event after it happened. There is another phenomena called "Retroactive Amnesia." When we witness a very dramatic event, an aircraft crash, for example, there is a tendency for the details immediately preceding the event to be blotted out of our memory. Question. Is hypnosis useful in recovering lost memory? The answer to that is well beyond the scope of this book. There are cases where it has worked, but it is controversial and the decision to try it is strictly a medical decision. The United States Air Force specifically forbids the use of hypnosis on witnesses.

7. TYPES OF QUESTIONS.

The idea behind questioning is to get the witness to tell you everything he knows without being influenced by either the question or by what he thinks you want to hear. Train yourself to follow a regular sequence or hierarchy of questions. The following types of questions are listed from the ones to be used first to the ones to be used last.

A. NON-QUESTIONS. A non-question may be just a head nod or an expectant pause. You want the witness to keep talking. If you need to say something, try, "uh-huh," or "really," or something like that. Another non-question technique is to mirror or echo what the witness says. Repeat back to him what he has just said without either agreeing or disagreeing with it. "You say you saw smoke coming from the plane?" This may encourage him to say more.

B. GENERAL QUESTIONS. These are open-ended broad questions that are useful in getting the witness to talk.

"What did you see?"

"Tell me everything you can recall."

"Tell me more about that."

C. DIRECTED QUESTIONS. We want the witness to focus on a specific subject, but we don't want to let him know what he is supposed to have seen.

"Did you notice any lights on the aircraft?"

D. SPECIFIC QUESTIONS. Here, we want information about a specific light.

"Did you notice any flashing lights?"

"What color was the light?"

E. SUMMARY QUESTIONS. Here you restate what the witness told you in your words and ask if that's correct. Frequently, the witness will add more information.

F. LEADING QUESTIONS. A leading question is one that contains or implies the desired answer. Once you ask a leading question, you have forever frozen an idea in the witness' mind about what he is supposed to have seen. Leading questions are to be avoided. If you have tried everything else, though, you might as well go ahead with leading questions.

"Was a red light flashing?"

"Was the landing gear down?"

8. THE IDEAL INTERVIEW.

There is probably no such thing as an ideal interview. We are constrained by laws and regulations;

the situation at hand and the investigative procedures in use. Nevertheless, this is how an ideal interview might be conducted.

A. TIME. As soon as possible.

B. LOCATION. The first choice is to return the witness to where he was when he witnessed the event. If he was in an airplane, put him back in an airplane or perhaps a simulator. Second choice is his home or office where he feels most comfortable. Third choice is a neutral place such as a coffee shop or a hotel lobby. Last choice is your office. That can be intimidating.

C. PREPARATION. Equipment. How are you going to record this interview? First choice is video tape. Second is audio tape. Third is stenographer or court reporter. Fourth is to reconstruct the interview from your notes. Fifth and last is a statement handwritten by the witness. Aside from the problems already discussed about written statements, psychologists tell us that once you write about an experience, you tend to forget everything that you didn't include in the written version.

In addition to recording equipment, bring whatever documents, maps or photographs you need and bring a small model airplane. A small plastic or balsa wood airplane that comes in three pieces and doesn't resemble any particular airplane works well. In a pinch, you can create something that looks like a plane by folding a piece of paper to represent wings and slipping it under the clip of a pen, which represents the fuselage. If you are dealing with a helicopter accident, you can usually buy a cheap plastic helicopter in a toy store.

Environment. You need privacy, although sometimes a booth in a coffee shop or a corner of a hotel lobby will work. If possible, arrange to sit at 90 degrees to the witness. If you sit across from him, the table becomes a barrier and everything he does with his hands is backwards to you. If you sit beside him, you have trouble maintaining good eye contact.

Dress. Skip the formal uniforms and the three piece suits. Dress like you expect the witness to be dressed.

D. ESTABLISH RAPPORT. Introduce yourself. Shake the witness' hand. Offer him something; perhaps a cup of coffee. Imply that you have all the time in the world and there is no hurry. Get on a first name basis if possible. Explain how you intend to record the interview and ask if he minds if you use a tape recorder. He probably does mind, but he will rarely say so, particularly if he is an eye witness. If he objects, point out that you are a lousy note-taker and you need the recorder for accuracy. Tell him that you will replay the tape for him and give him a copy of it if he wants one.

Question: Isn't a tape recorder or video recorder inhibiting? Answer: yes, for about three minutes, then everyone forgets that it is there. This is even more true of a video camera. Modern video cameras don't need any additional lighting and can actually can be positioned much further away from the interview than the audio tape recorder. Just put the tape recorder on the table in front of both of you, start it, ask a few casual questions and both of you will soon ignore it.

E. NARRATIVE STATEMENT. Get right into the meat of the thing. The witness expects and probably wants to tell you what he saw. If you distract him with a bunch of questions about his address and flying experience, you may make him forget what he wanted to tell you. Let him tell it his way. Don't interrupt; don't ask questions; and don't take notes (assuming you are using a tape recorder.) Just sit there and look interested and feed him an occasional non-question (head nod) to keep him going.

F. REWARD. When he's done, thank him and tell him that his statement is helpful. Remember, he wants to be helpful.

G. REPLAY TAPE. Rewind your tape and replay it for him. This serves three purposes. It relieves his anxiety about what is on the tape. It gives you an opportunity to take notes without distracting him. It gives the witness an opportunity to interrupt the tape and amplify his statement. In an interview of a key participant in the accident, this may be as far as you go at this time. You don't know enough to ask any good questions and you don't want to sow any unnecessary seeds in the witness' mind with bad questions. Thank him and tell him that you will certainly want to talk with him again after you have done some investigating. If there is going to be a second interview, bring the tape of the first one and play it for him again to remind him of what he said. Since forgetting is starting to occur, the tape is a good memory jogger.

H. QUESTIONS. This is either the second interview or a continuation of the first one. Ask your questions in the order already listed; i.e. general, directed, specific, summary and leading. Ideally, you should have prepared a script of the subjects you want to ask about and the approach you intend to take with each subject.

I. CLOSING. Now you can ask about all the identification stuff you wanted to get at the front end of the interview. In general, get a permanent address and phone number for the witness, some idea of his work schedule and his technical background with respect to airplanes. Also, always ask about other witnesses he may know of and ask if he has any opinions about the accident. His opinions may be of no value to you, but occasionally you'll get a real gem of an idea. Also, most witnesses like to be asked for their opinion. Finally, lay the groundwork for a follow-up interview if you need one. Find out the best time to contact him.

J. REWARD. You want to leave the witness with a good feeling about the interview and a willingness to talk with you again. Give him something, perhaps your business card. Ask him if he has any questions he'd like to ask you. If he does, try to answer them. Thank him for his help.

9. SOME PRACTICAL PROBLEMS.

A. "GANG" INTERVIEWS. Although the ideal interview is one-on-one, that is not the normal situation. Most of the time, there will be other investigators participating in the interview. This is not hopeless providing you discuss how the interview is to be handled beforehand. The best way is to select one investigator to do the interviewing and questioning. He sits at the table with the witness. The rest sit in the background, listen and take notes. Like the tape recorder, you eventually forget that they are there.

B. GROUP QUESTIONING. Is it possible to question more than one witness at a time? Of course it is. Does it work? Absolutely not. One witness' statement will influence the others and you will seldom get anyone to disagree with another witness' statement. You end up with a single statement which may or may not be accurate.

C. PRESENCE OF THIRD PARTY. You can exclude non-investigators, but you can rarely exclude any representative a witness chooses to bring. This representative may be a lawyer, a supervisor or a union representative. Occasionally, all three of them will show up. Now what do you do?

It doesn't do any good to object to their presence. The best approach is to talk with these people first and explain how the interview will be conducted. You mainly want to let the witness do the talking and you are not there to accuse him of anything. Resolve the issue of how the interview will be recorded before you start. Remember that as long as the witness is talking, they are not likely to object. Therefore, do as much listening as possible and keep the questions as general as possible.

10. SUMMARY.

Interviewing the witnesses could be the most important part of your investigation. It is not easy and it demands both forethought and planning. If you are part of an investigative group, pick your best interviewer and concentrate on helping him develop interview outlines for each witness. Locate the eye witnesses as soon as possible and get their statements as early as possible. Always assume that you may have additional questions for that witness at some time in the future. Leave the door open for another interview.

24

Human Factors

1.INTRODUCTION.

The search for human factors aspects in an aircraft accident investigation is usually conducted by medical personnel or individuals with specialized training in human performance areas. Since this is a general text on accident investigation, we will not attempt to thoroughly cover this highly technical subject. It is important, though, for all investigators to appreciate the contribution of human factors in aircraft accidents and understand the general investigation methodology.

2. THE ROLE OF THE AERO-MEDICAL INVESTIGATOR.

Initially, the aero-medical investigator is a gatherer of information just like any other investigator. The normal sources of information are:
    A. Wreckage Scene.
    B. Human Remains.
    C. Witness Statements.
    D. Survivor Statements.
    E. Examination of Survivors.
    F. Family Member Statements.
    G. Personal Histories.
    H. Medical Histories.
    I. Post Mortem Results.

There are usually three key questions that need to be answered in almost any aircraft accident.

    A. What human factors were involved in this accident?

    B. What were the causes of the injuries or fatalities?

    C. Was this accident survivable?

If there were fatalities, the aero-medical investigator must first deal with the human remains. This includes both their identification and the determination of the cause of death.

If there were survivors, the investigator is concerned with their medical or psychological treatment and the cause of their injuries.

Regarding the flight crew, the investigator will want to know their physical condition both before and at the time of the accident and their actions during the course of the accident.

Finally, the investigator will need to prepare a report summarizing the medical investigation and the human factors aspects. The rest of this chapter examines this process in more detail.

3. INITIAL MEDICAL INVESTIGATION.

The initial medical investigation starts with the first response of medical personnel to the scene of the accident. These are usually from local hospitals or fire departments and sometimes the coroner's office. Except for the coroner, their interest is primarily in saving lives; not investigating accidents. Nevertheless, their actions will influence the subsequent investigation.

The nature of this initial response is usually a function of accident location. If the accident occurred on an airport serving air carriers, the airport is required to have an emergency response plan and exercise it regularly. If the accident occurred near a large city, the city probably has a disaster response plan that can be put into effect. If the accident occurred away from an airport or a large city, the initial medical response may be minimal.

4. THE CORONER OR MEDICAL EXAMINER.

It is important to appreciate the role of the coroner. In the United States and many other countries, the coroner has full jurisdiction over the remains of any fatalities except for those occurring under medical care or supervision. This is true whether the coroner initially responds to the scene or not. Since the coroner must ultimately prepare the death certificates for any fatalities in his jurisdiction, it is well to acknowledge his authority.

In the United States, the National Transportation Safety Board can require autopsies of any person who dies as a result of an aircraft accident. As a rule, the NTSB will insist on autopsies of all crew members fatally injured in major accidents. This authority is unique as in most countries, the coroner makes the autopsy decisions. Following an aircraft accident, the cause of death may appear intuitively obvious to some coroners and there have been many accidents where autopsies of the crew members were not performed. The investigators were denied knowledge of the physical condition of the crewmembers and any evidence of their environment that might have been present in the remains. If this situation exists, it may become important for the aero-medical investigator to explain to the coroner why autopsies are so essential to the investigation.

An aircraft accident with many fatalities may completely overwhelm the facilities and capabilities of a small coroner function. In these cases, the common practice is to rent refrigerator truck trailers for use as temporary morgues and bring in outside help.

5. REMAINS IDENTIFICATION.

In addition to investigation requirements, there are moral and religious reasons for accurate identification of human remains. If there are a large number of fatalities, this is a time-consuming process and there are no shortcuts.

Identification starts at the wreckage scene by plotting the remains accurately on a diagram and tagging the remains with information on where they were found and what they were related to. This involves a tagging and numbering system that will allow remains to be correlated with the diagram. The accident investigators frequently assist in the construction of this diagram.

The principal means of identification are listed below in descending order of use. The first three are not positive means of identification, but they assist in the efficient use of the next four.

A. Location found.

B. Personal effects.

C. Known seating arrangement.

D. Fingerprints and footprints.

E. Dental records.

F. Physical attributes.

G. Tatoos, marks and scars.

H. Information from relatives regarding clothing, etc.

I. Other forensic techniques.

6. PROTECTION OF INVESTIGATORS.

Medical personnel understand this already, but other investigators may not. Exposure to "Bloodborne Pathogens" is a serious risk. In any accident involving multiple fatalities, there is a probability that about five percent of the remains will test HIV positive or contain some other infectious disease. Although the AIDS virus is the most feared, it is likely that Hepatitis B virus poses the greater threat because it remains active in the air longer. Some viruses are not destroyed by heat, cold or time. They are still active and any investigator at the scene is potentially exposed to them. Aside from the remains themselves, the cockpit, cabin, cabin materials and carpeting may be sources of contamination for the duration of the investigation.

In the United States, Occupational Safety and Health (OSHA) Standard 1910.1030 applies to investigators and the investigation scene. Basically, the site must be surveyed to determine the degree of hazard and entry/exit points must be established. All personnel at the scene must use appropriate personal protective equipment (PPE) and follow established rules regarding eating, drinking and smoking. When exiting the site, PPE must be controlled and disposed of properly. All equipment brought to the scene, such as cameras, etc. must be protected. Full compliance with the standard requires considerable preparation and training, which is beyond the scope of this text.

7. AUTOPSY.

Realistically, what can be learned from the autopsy of a crew member? That depends, of course, on the nature of the accident and the condition of the remains. These are some of the possibilities.

A. WHAT WAS THE PHYSICAL CONDITION OF THE INDIVIDUAL? This is determined from both the physical examination of the remains and toxicological tests. If, for example, the pilot was taking a non-prescription cold remedy, this should show up in the tox tests and could be a factor in the accident. Likewise, the presence of alcohol or drugs in the blood or body fluids can be identified. High concentrations of lactose in the brain, nervous system or muscles could indicate fatigue. Pre-existing conditions such as coronary artery disease can be identified.

B. WHO WAS FLYING THE AIRPLANE? In a multi-crewed cockpit, it is sometimes important to know who was actually flying. Frequently, the pilot who was operating the controls at impact receives distinctive wrist, finger, forearm and foot injuries. These can be correlated with the condition of the control column and rudder pedals as determined by other investigators.

C. WHAT IMPACT FORCES WERE PRESENT? A great deal of research has been done on how the human body responds to various forces or various levels of force. This is in the medical literature and the specialist knows that it takes a certain amount of force in a certain direction to produce certain injuries. He gathers his evidence by looking at intra-abdominal injures to the liver, spleen, large intestine and aorta. He also examines injuries to the skull, brain, spinal cord and neck. He is likewise interested in vertebral compression and pelvic injuries. The results of his examination can be correlated with the condition of the seats and restraint systems and the forces calculated by using the impact angle and velocity calculations listed in Chapter 36.

D. WAS THERE A TOXIC ENVIRONMENT PRESENT PRIOR TO IMPACT? If there was smoke (for example) present in the cockpit prior to impact, there should be evidence of that in the lungs. If carbon monoxide was present, it could show up as carboxyhemoglobin in the blood.

E. WHAT WAS THE CAUSE OF DEATH? Was it the impact? Drowning? Asphyxiation? Fire? Many aircraft accidents look bad, but they were actually survivable. Sometimes, it can be shown that the individual survived the impact, but died due to some post-impact problem.

F. WAS THERE AN INFLIGHT EXPLOSION? Some accidents involving bombs have been solved, or at least confirmed, because the remains retained evidence of high speed explosion fragments.

Thus autopsies are important to the investigation and the aero-medical specialist can contribute a lot of information that may be significant to other parts of the investigation.

8. SURVIVING CREW MEMBERS.

Investigating survivors is obviously easier than investigating fatalities. It is likely that the aero-medical investigator is not the only investigator who wants to talk to the surviving crew members. This part of the investigation needs to be coordinated and organized to insure that all information is collected with minimum duplication. See Chapter 23.

The investigation starts, probably, with a physical examination of the survivor(s). This would include not only present state of health and assessment of any injuries, but toxicological tests. Be careful. The rules on whether a living person has to submit to tox tests vary with country and jurisdiction. Do not automatically assume that they are mandatory.

Next, the investigator is always interested in the individual's general medical history. This would include any prescriptions, glasses or contact lenses and medical waivers.

The investigator is specifically interested in what's called the 72 hour history of the individual. This would cover all activities during the past 72 hours including food, rest, sleep, duty time, and physical conditioning. This might include inflight meals, crew facilities, lodging and transportation. Additionally, the investigator would collect information related to stress, stability, recreation, family and financial status.

Finally, the investigator would want to know something about the individual's flying history. This could include prior accidents and incidents, peer evaluations, proficiency, currency, career progression and relationships with other crew members.

9. PSYCHOLOGICAL FACTORS.

This is a difficult area to investigate. The problem is that there is seldom any positive evidence of exactly why a crewmember did or did not act in a certain way. We might suspect fatigue or complacency (or something else) and we might be able to make a strong inferential case for it, but we can never prove it. Complacency is not a measurable condition.

This shouldn't stop us from trying. If we can consistently identify these factors as being present in individual accidents, their contribution to certain types of accidents can be identified through analysis.

Within the broad subject of aviation psychology there are a number of conditions or situations that could apply to a particular accident. Here are a few of them with their definitions as developed jointly by the Life Sciences Division of the USAF Inspection and Safety Center and the USAF School of Aviation Medicine. The purpose of this list is to provide the investigator with the definitions of terms likely to be encountered when talking with human performance specialists.

AFFECTIVE STATES. These are subjective feelings that a person has about his (her) environment, other people or himself. These are either EMOTIONS, which are brief, but strong in intensity; or MOODS, which are low in intensity, but long in duration.

ATTENTION ANOMALIES. These can be CHANNELIZED ATTENTION, which is the focusing upon a limited number of environmental cues to the exclusion of others; or COGNITIVE SATURATION in which the amount of information to be processed exceeds an individual's span of attention.

DISTRACTION is the interruption and redirection of attention by environmental cues or mental processes.

FASCINATION is an attention anomaly in which a person observes environmental cues, but fails to respond to them.

HABIT PATTERN INTERFERENCE. This is reverting to previously learned response patterns which are inappropriate to the task at hand.

INATTENTION. Usually due to a sense of security, self-confidence or perceived absence of threat. BOREDOM is a form of inattention due to an uninteresting and undemanding environment. COMPLACENCY is another form due to an attitude of overconfidence, laxitude or undermotivation.

COMMUNICATIONS. Information transfer ineffectiveness such as:

MESSAGE GENERATION. Not sent.

MESSAGE TIMING. Too late or too early.

MESSAGE CONTENT. Inaccurate or ambiguous.

MESSAGE RECEPTION. Sent, but not received.

DECISION. The selection of a response to information. Decisions may be correct, incorrect, delayed or merely poor.

ERROR. An inappropriate physical or mental response such as:

ADJUSTMENT ERROR. Operating a control too slowly or too rapidly. Operating controls in the wrong sequence.

FORGETTING ERROR. Failure to check, set or use a control at the proper time.

REVERSAL ERROR. Moving a control in the opposite direction required.

SUBSTITUTION ERROR. Confusing one control with another.

UNINTENTIONAL ACTIVATION. Accidently operating a control.

FATIGUE. The progressive decrement in performance due to prolonged or extreme mental or physical activity, sleep deprivation, disrupted diurnal cycles, or life event stress.

ILLUSION. An erroneous perception of reality due to limitations of sensory receptors and/or the manner in which the information is presented or interpreted. There are perhaps a dozen different types of illusions that can affect a pilot.

JUDGEMENT. Assessing the significance and priority of information in a timely manner. The basis for DECISION.

MOTIVATION. A person's prioritized value system which influences his or her behavior.

PEER PRESSURE. A motivating factor stemming from a person's perceived need to meet peer expectations.

PERCEPTION. The detection and interpretation of environment cues by one of more of the senses.

PERCEPTUAL SET. A cognitive or attitudinal framework in which a person expects to perceive certain cues and tends to search for those cues to the exclusion of others.

SITUATIONAL AWARENESS. The ability to keep track of the prioritized significant events and conditions in one's environment.

SPATIAL DISORIENTATION. Unrecognized incorrect orientation in space. This can result from an ILLUSION, or an anomaly of ATTENTION, or an anomaly of MOTIVATION.

STRESS. Mental or physical demand requiring some action or adjustment.

10. PHYSIOLOGICAL FACTORS.

These are a little easier to investigate as there is sometimes evidence remaining. The evidence of many high altitude dysbarisms may be detectable in the individual. Also, physiological problems tend to be related to specific flight regimes. We need high altitude and loss of pressurization for hypoxia to occur. We need high G maneuvering for G-LOC (G-induced Loss of Consciousness) to occur.

Pathophysiological factors such as drugs, alcohol, nutrition, food poisoning, carbon monoxide poisoning and sudden incapacitation are likewise investigatable.

There is an area of overlap with psychological factors when dealing with problems of visual illusions and perceptions. While the result may be psychological, the sensory and perceptual mechanisms are physical.

There is an entire area of human factors called ergonomics, which is the engineering side of human factors. This relates to the interface between the man and the machine. It includes anthropometric data such as size, reach, strength and dexterity along with aircraft design factors including seats, visibility, instrumentation, and controls. If these areas are not completely investigatable as factors in the accident, they are at least measurable. Their existence can be demonstrated. It can be shown, for example, that the pilot's sleeve could have snagged on the spoiler actuator as he reached forward to reset the course indicator. Sometimes there are problems related to display interpretation, symbology, controls, lighting, glare and so on. These can likewise be demonstrated.

At one time, aircraft cockpits were designed around the size and strength attributes of the American/Western European male. Obviously, with expansion of aviation to both women and to countries where the average person does not have the size and reach of the American male, some revision in our ergonomic thinking is required. Here, we might introduce a new term used commonly in industrial safety analysis--Less Than Adequate, or LTA. It doesn't mean that the design was inadequate; it means that the design was not the best possible design for the situation and individuals involved. In an aircraft accident, it may be that the design of the cockpit instruments and controls was "LTA" with respect to the person actually flying the plane.

11. REPORT WRITING.

The reports of human factors investigations are sometimes written separately from main accident report and correlated with it later. Actually, this isn't the best way to write an accident report as the human factors portion should be an inter-disciplinary correlated with other investigative evidence. In some cases, though, the aero-medical investigation is done separately and apart from the main investigation. If so, the reports tend to follow a fairly standard protocol.

A. Scope of the investigation.

Condition at the scene.

Investigative protocols and procedures.

B. Psycho and physiological data collection.

72 hour histories.

Medical and personal histories

C. Analysis of results.

Autopsies

Medical examination of survivors.

Examination of wreckage.

Psycho and physiological data

D. Correlation with other investigative data.

Crash dynamics.

System failures.

Witness statements.

Accident sequence.

FDR/CVR data.

E. Conclusions

Causes of death or injury.

Crash survivability.

Presence of other human factors.

F. Recommendations

12. SUMMARY.

Like the subject of aircraft accident investigation itself, the subject of human factors is almost endless. This chapter did no more than scratch the surface of a very complex subject. As said earlier, the aircraft accident investigator does not have to be a specialist in human factors, but he does have to understand the scope of the subject and how it can be applied to the overall investigation. One statement can be made with certainty. Human factors have been involved somehow in every single aircraft accident that has ever occurred. They will continue to be involved in all future accidents and their investigation will become increasingly important.

25

Helicopter Accident Investigation

1. INTRODUCTION.

Perhaps 90 percent of the investigation techniques discussed in this text will work equally well for fixed or rotary wing aircraft. Helicopters, obviously, have some differences and different techniques apply in some areas.

This chapter will discuss only the differences. The subjects covered will be organized to generally match the order of the chapters in the rest of the text.

2. WRECKAGE DISTRIBUTION.

All helicopter impacts are low speed impacts (comparatively) so we do not find the total breakup and destruction we see in a high speed fixed wing accident. The helicopter impact can be either high angle or low angle and the same general rules on dispersion will apply with one exception.

If a fixed wing aircraft was intact at the time of impact, you can count on all the parts being downstream of the impact point. In a helicopter impact, this is not necessarily true for the main rotor blades which can be flung a large distance in any direction. An accurate wreckage diagram is usually helpful in determining the sequence of blade failure and verifying that they did not fail prior to impact.

3. FIRE INVESTIGATION.

Most fixed wing aircraft carry fuel in the wings.

Carrying it in the belly is not considered a great idea from a safety point of view.

Most helicopters are top heavy because of the weight of the engine(s) and transmission which are frequently mounted on top of the fuselage. To balance this, fuel is sometimes carried in tanks under the floorboards. This changes the post impact fire characteristics somewhat. The tanks are vulnerable to impact damage and, if a fire starts, it is going to spread rapidly throughout the entire fuselage.

It is no wonder that research on crashworthy fuel tank and fuel system design began with helicopters. As is pointed out in Chapter 36, the designers have done a remarkably good job. These crashworthiness features are rarely found on fixed wing aircraft. One aspect of any helicopter investigation involving fire is an examination of the crash worthy tank fittings and fuel plumbing to determine what worked and what didn't work.

4. STRUCTURAL INVESTIGATION.

Structural investigation is about the same for either type of aircraft. It may even be a little easier on a helicopter, because the fuselage wreckage is usually not widely scattered. Impact forces can be determined and analyzed by examining the damage to the skids (or wheels) and direction in which the seats collapsed. Since the transmission is usually on top and is a heavy piece of machinery, the manner and direction in which it failed is indicative of the forces present at impact.

The rule about dispersed wreckage still holds true. A rotor blade found far from the main wreckage can be a clue to inflight failure.

When a blade does come loose in flight, it may contact other blades or the fuselage or tail boom in the process. The remaining blades are not going to last long and the flight is essentially over. There is too much imbalance in the rotor. The remaining blades may interact with each other and the fuselage. Figuring out which blade failed first is important as it can explain the dynamics of the remaining blades.

5. RECIPROCATING ENGINES.

There are two principal differences between a fixed wing recip and a nearly identical engine mounted on a helicopter. First, the helicopter, since it hovers a lot, cannot depend on ram air flow to cool the engine. This is usually done with a belt-driven fan. The fan may pick up rotational damage during the impact and can aid in determining whether the engine was turning or not.

The other difference is that helicopter reciprocating engines are sometimes mounted vertically. This requires a modified oil system to keep the oil flowing to the top cylinders.

6. ROTOR AND TRANSMISSION SYSTEMS.

A. ROTOR SYSTEMS. The two types of rotor systems in general use are semi-rigid (teetering) and fully articulated. The teetering system is found on most two-bladed helicopters. In it, the entire rotor system tilts in the direction of desired flight. In the fully articulated systems, the rotor head does not tilt. The individual blades are connected to the rotor head through hinges and direction of flight is controlled by adjusting the plane of each blade to the desired plane of rotation.

MAST BUMPING. If the entire rotor head tilts, (semi-rigid rotors) it is possible for it to tilt too far. This is usually the result of cyclic control inputs by the pilot, particularly in low or zero G flight or during over reaction to some other displacement. Under normal conditions, the rotor head never comes close to its limits.

If the limits are exceeded, one or both limiting stops on the rotor head contact the mast. At best, this will leave dents on the mast. At worst, it can fail the mast within one revolution. The clue that this has happened is a pinched or oval-shaped fracture on both sides of the mast at the location where the rotor head limit stops would have hit it.

In a severe case of mast bumping, the main rotor blades are likely to continue flexing down far enough to contact the helicopter structure; usually the tail boom. Certain model helicopters have characteristic "signature" patterns where the main rotor slices through the tail.

ROTOR BLADE SLASH MARKS. As with propellers, rotor blade slash marks at the scene can be useful in determining velocity. This is not wonderfully accurate because the plane of the rotor is not perpendicular to the ground and the direction of flight. Nevertheless, it is possible, knowing rotor RPM, to calculate the time between each blade strike. With this knowledge, the standard V = D/t formula can be used to calculate velocity. The way to measure the distance between the slash marks is to measure them on a line that is parallel to the impact heading and comes closest to representing a radius of the arcs of the slashes. You may have to arbitrarily extend the slash marks a little to establish this line. Measure them this way and also measure the distances at the end of the real slash marks on a line parallel to the impact heading. See what difference it makes, if any. Here are the equations.

$$t = \frac{60}{RPM \times N} \qquad (1)$$

Where:

t = time in seconds between each rotor blade at that RPM
RPM = Rotor RPM
N = Number of blades on the rotor

$$V = \frac{D}{t} \qquad (2)$$

Where:

V = Velocity in feet per second
D = Distance in feet between blade slash marks
t = time in seconds from Equation 1.

DIRECTION OF ROTOR ROTATION. If you are dealing with a helicopter manufactured in the United States, the main rotor rotates counter clockwise from the point of view of someone looking down from above the helicopter. (From the pilot's point of view, the blades to his right are moving forward and the blades to his left are moving to the rear. This means that tail boom strikes should always come from the left side if the blade was still attached to the rotor head at the time. If the strike was on the right side, something else happened. The blade was probably no longer attached at the time it hit the tail.

Many French-built helicopters have rotor systems that rotate in the opposite direction. It really doesn't make any difference except that it changes the direction of torque. Thus the pilot has to reverse his normal anti-torque (pedal) habits as he switches from one type of helicopter to the other.

The tail rotor can be mounted on either side of the tail (tractor or pusher) and made to rotate in either direction depending on how the blades are pitched. The tail rotor drive shaft can likewise turn in either direction depending on how it is geared. It isn't hard to figure out which way everything is turning just by looking at which side of the blades has the leading edges and how the tail rotor drive shaft is geared at either end.

The tail rotor drive shaft is normally a thin walled tube. If either end comes to an abrupt stop while the other end tries to keep turning, this load is generally soaked up by the shaft itself in the form of torsional damage. If you see twisting or diamond-shaped 45° buckles on the drive shaft, you can tell which end stopped first by using the handkerchief trick.

This involves using a handkerchief, a rag or (if you are really desperate) a cigarette. Fold the rag a number of times so it resembles a long tube. The cigarette has already been folded in this manner for you. Hold it along the failed shaft and hold one end still while rotating the other end in the same direction the shaft turns. If the resulting wrinkles in the rag resemble the torsional damage to the shaft, that's what happened! If they don't, hold the other end of the rag still and rotate the first end. Those are the only two ways to get that kind of torsional damage and one of them must match.

B. TRANSMISSIONS. The helicopter transmission is a fairly complex piece of machinery and definitely a key element in the rotor system. It must take power from the engine and send it to both the main and tail rotor in the correct amounts. In addition, it must be able to disengage from the engine and allow the main rotor to free wheel so the helicopter can autorotate. If the main rotor comes to a sudden stop due to impact, this force is fed back through the transmission and may displace the transmission in a direction opposite the direction of impact.

Some people think that an autorotating helicopter rotor is analogous to a windmilling propeller. Not so. The helicopter engine can drive the rotor, but the rotor cannot drive the engine. If it had to pick up that load, the helicopter would never be able to autorotate. This brings up two points of interest to the investigator.

First, when a helicopter reciprocating engine quits, it really quits! There is no automatic or instant restart because there is nothing to maintain the engine RPM. It doesn't windmill. (A turbine engine might windmill a little depending on the configuration.)

Second, this ability of the rotor system to disconnect itself from the engine and become free wheeling is managed by a clutch, usually called a sprag clutch. The helicopter specialist, by examining the sprag clutch, can tell whether it was being driven by the engine or was free wheeling. If it was free wheeling, start checking the engine.

Transmissions are sensitive to both lack of lubrication and high temperature. Most transmissions have chip detector plugs and a warning light in the cockpit. If there is obvious evidence of high temperature operation on the transmission case, it undoubtedly occurred well before impact.

7. TURBINE ENGINES.

As discussed in Chapter 7, turbine engines don't much care whether they are being used as reaction engines or used to drive a propeller or a rotor system. They still have the same basic components and operate in the same way.

In the case of a helicopter, the power turbine, the one that powers the transmission, is almost always a free turbine in that it rotates independently of the engine itself.

Another difference can be in the engine mounting. Although many helicopter engines are aligned with the longitudinal axis of flight, they don't have to be. The engines in Hughes/McDonnell Douglas helicopters are in the rear of the fuselage and canted upward at an angle. Obviously, this would change the pattern of expected impact damage.

Due to its location and the fact that helicopter impacts are low velocity events, the turbine engine frequently survives the impact and has been known to continue to run for some time after the accident. This is a pretty good clue that the problem is not with the engine.

8. HELICOPTER SYSTEMS.

A. FUEL SYSTEMS. This was addressed briefly under the heading of Fire Investigation. Consider that if the fuel is located beneath the engines, there must be a reliable method of getting the fuel up there. Check this.

Another point. If you are checking for the presence of fuel, start up high with the engine and work down. If you crack a line down by the tank, you'll drain it of fuel and you won't know whether the engine was getting any or not.

B. FLIGHT CONTROL SYSTEMS. The cyclic control is the one between the pilot's knees. Through a swash plate, this controls the direction the helicopter goes by either tilting the entire rotor head (semi-rigid rotor system) or by changing the pitch of the individual blades as they turn past opposite points.

The collective is the one in the pilot's left hand and it moves up and down to control the pitch of all the blades simultaneously by moving the swash plate up and down.

The pedals control the anti-torque system by adjusting the pitch of the tail rotor blades.

C. ANTI-TORQUE SYSTEMS. If the helicopter has two main rotor systems which rotate in opposite directions, torque is cancelled out and not a problem. A single rotor system produces significant torque which is counteracted by the tail rotor or, more recently, by engine exhaust thrust delivered out the tail perpendicular to the line of flight. This is the NOTAR (NO TAil Rotor) helicopter developed by McDonnell-Douglas.

9. FLIGHT RECORDERS.

In the United States as of October, 1991, Cockpit Voice Recorders (CVRs) are required on all turbine powered helicopters if two pilots are required for flight and they can carry six or more passengers.

As of the same date, Digital Flight Data Recorders (DFDRs) are required on all commercial helicopters that can carry ten or more passengers. The required parameters to be recorded are listed below.
- Time
- Indicated Airspeed
- Altitude
- Magnetic Heading
- Vertical Acceleration
- Longitudinal Acceleration
- Pitch Attitude
- Roll Attitude
- Altitude Rate (Vertical Velocity)
- Main Rotor RPM
- Power Turbine RPM (each engine)
- Engine Torque (each engine)
- Flight Control Hydraulic Pressure
- Radio Transmitter Keying
- Autopilot Engaged
- SAS Status
- SAS Fault Status
- Pedal Position
- Collective Position
- Cyclic Position (Lateral)
- Cyclic Position (Longitudinal)
- Controllable Stabilator Position

See Chapter 18 for recorder analysis techniques.

10. HELICOPTER OPERATIONS AND MAINTENANCE.

A. WEIGHT AND BALANCE. The Center of Gravity of the helicopter is under the main rotor mast and within a few inches of it. Considering that the engines and transmission may be on top, the CG could be up near the ceiling of the cabin.

The natural layout of the helicopter tends to keep the CG fairly well under control. A serious CG problem is immediately obvious to the pilot as he tries to lift the helicopter off the ground.

Weight tends to be more of a problem. Because of the way the helicopter is used, it is not common practice to calculate a new weight and balance for every flight segment even though one might be technically required. There is a pervasive feeling in the industry that if it can be lifted off the ground it is probably OK.

It may be OK if the helicopter is flown gently and not taken up to a higher density altitude where the extra weight may not be OK at all. The investigator should always check the actual weight on board the helicopter at the time of the accident. This is the weight that will determine helicopter performance at various altitudes and determine whether the helicopter is being operated above its weight limits.

B. HEIGHT-VELOCITY CURVES. A successful autorotation depends on having a certain amount of forward speed in the first place. Each model of helicopter has its own height-velocity curve (sometimes called a "Deadman's Curve") which represents combinations of height and velocity wherein a successful autorotation is highly unlikely. The helicopter either doesn't have enough forward speed to autorotate and is too high to fall safely to the ground, or it doesn't have enough forward speed and is not high enough to obtain it before it hits the ground. An examination of any of these curves will show that there are certain combinations of airspeed and height which are best avoided.

Unfortunately, they can't be completely avoided and still get the job done. The single engine rescue helicopter hovering at 100 feet and zero velocity over the rescue scene has no chance of successfully autorotating if something bad happens. Sometimes knowledge of the height-velocity curve and knowledge of what the helicopter was doing is all you need to explain what happened. Now, you need to explain why it happened, i.e. why was autorotation necessary.

11. HELIPORTS.

There are plenty of sources of standards for helipads and heliports; both from ICAO, the FAA and probably most other countries. The problem is that the helicopter is a fly anywhere-land anywhere machine and it frequently operates from locations that don't meet anyone's standards. The most common problems are physical size of the helipad, lighting, ground obstructions and approach/departure obstructions; usually in the form of wires. These problems account for a fairly large percentage of helicopter accidents.

12. HELICOPTER GROUND HAZARDS.

A. DYNAMIC ROLLOVER. The helicopter knows a cute trick where it suddenly rolls over sideways on the ground and thrashes itself to death. This is called dynamic rollover. If it just tipped over the

way you might tip over a wine glass by getting the CG out of line with the base of it, that would be a static rollover. Here, we're talking about something that requires some dynamics to understand.

It is rumored that dynamic rollover only happens on sloped landing areas. Not so. The sloped landing area will certainly make it more likely, but it can happen on a perfectly level helipad.

The situation is that the helicopter is light on the skids. It is just ready to fly (or not fly) and one skid or wheel is in contact with the ground. If there is a lateral cyclic input for any reason, the helicopter will normally roll around the CG which is somewhere up in the cockpit. With one skid on the ground, though, the helicopter now rolls about that point of contact and the extra leverage provided by that new fulcrum just overwhelms the lateral control system. Very shortly, the pilot becomes a passenger because the helicopter is going to roll over regardless of what he does.

The reason this works better on an upslope is because the helicopter always comes off the ground one skid at a time; the perfect situation for dynamic rollover. The same thing can happen, though, if one skid is caught in a rut, a tie-down rope or just anchored hard enough to the ground to provide a pivot point.

B. DOWNWASH. Everyone knows that the helicopter kicks up a fierce downwash during takeoff or landing. Besides blowing up a lot of dust and ruining the pilots' visibility, it can create a couple of other problems.

If there is anything fairly lightweight on the ground, a tarpaulin or large piece of plastic for example, the downwash can literally pick it up and blow it up to the level of the rotors. A plastic sheet caught in the tail rotor will bring it to an abrupt halt. This type of accident is not difficult to investigate.

Second, and perhaps more important, is the phenomenon called power settling or "vortex ring state." Aerodynamically, this results from air that is flowing upward at the roots and tips of the blades and downward at the center of the blades. This results in a near-zero rotor thrust condition and has been described as, "settling in your own downwash." This is most commonly associated with steep landing approaches at low forward speeds. At shallow descent angles, the downwash is behind the helicopter. At steep angles, it can also be behind the helicopter if the forward speed is high enough. That's why, except on TV, you don't see too many helicopters hovering straight down to a landing from a couple of hundred feet up.

Sometimes power settling is incorrectly called settling with power. This latter term refers to a situation where power required is greater than power available. The helicopter is being operated beyond its capabilities.

13. SUMMARY.

As said at the beginning of this chapter, most investigation techniques are common for both fixed wing aircraft and helicopters. Understanding the helicopter isn't that difficult and time spent studying the pilots handbook and system schematics is usually time well spent.

On the other hand, the dynamics of helicopter flight can be fairly difficult to understand. Consult the bibliography (Appendix A) for some references in this area.

PART III

Technology

The previous parts of this book dealt with investigation procedures and techniques. This part covers some technical material that is basic to an understanding of aircraft performance and some of the hazards that are frequently factors in accidents.

We will also get into some concepts on aircraft loads and stresses and some common failure patterns.

These chapters are written from the point of view of the investigator; not the engineer or the pilot. Not all investigators need all of this knowledge, so this part can be read selectively.

These subjects have never been particularly popular with investigators; or even pilots, for that matter. "Why do I have to understand aerodynamics or metallurgy? We have specialists in those areas and there is no point in both of us learning it."

A common complaint; usually voiced by someone who didn't like math and physics as a child and hasn't changed his mind as an adult.

In a way, the complaint is valid. If you are not already an aeronautical engineer or a metallurgist, reading this book won't make you one. If you are, reading this book won't teach you anything you don't already know. So why bother?

The problem is that the specialist only looks at the specialty and is not always capable of relating that to the whole accident. We could define the "whole accident" as being any or all of the subjects covered in Part II. It's a bit much to expect the metallurgist (for example) to see how his conclusions fit into everything else known about the accident.

That's the job of the investigator and it is a job he cannot do if he doesn't understand what the aeronautical engineer or the metallurgist is trying to tell him. In Part III, we are trying to provide that understanding.

All of the aerodynamic and operational factors discussed in this part have been involved in aircraft accidents. An understanding of these factors is an important tool in an aircraft accident investigators toolkit. Some of the subjects covered are often directly involved in the accident sequence. Other subjects are included because they allow the investigator to better understand the phenomena involved in the sequence. Still others are included because they allow the investigator to discover facts which contribute to the understanding of the accident sequence and identification of the root causes of the accident.

Although the material presented assumes an understanding of basic physics and aerodynamics, brief reviews of some fundamentals are provided where necessary. For those readers who wish to review these concepts at a pilot's level (as opposed to a engineering level) the authors recommended either of two texts;

Aerodynamics for Naval Aviators by H. H. Hurt or Flight Theory for Pilots by C. E. Dole. See the bibliography for the full citation.

Aerodynamics for Naval Aviators is extremely complete and slightly more technical. Although written some time ago, the principles covered in the text remain as applicable today as they were when written. Flight Theory for Pilots is not as deep technically, but is better illustrated and more current. Both are excellent references. They complement each other and should be included in every aircraft accident investigator's reference library.

The chapter on crash survivability introduces basic crash worthiness concepts and deals with methods for calculating G loads encountered during an impact and correlating these with injuries to occupants.

26

Determination of Aircraft Performance

1. INTRODUCTION.

This chapter includes a brief discussion on the relationship between an airplane's true airspeed, bank angle, G load, turn radius and turn rate.

2. REVIEW OF AIRCRAFT MANEUVERING PERFORMANCE.

Occasionally an aircraft accident investigator needs to reconstruct the final maneuvering performance of an airplane which has been involved in a crash. Perhaps the investigator wants to know the minimum turn radius for an airplane flying up a dead end canyon. Perhaps the aircraft failed structurally inflight. Based on the circumstances, can we determine the G load or the airspeed or both? Can we tell if the aircraft was in or out of its envelope? Perhaps the investigator would like to know the G load on the aircraft so that the total drag can be estimated.

In order to make these and similar calculations, it is necessary to have an understanding of the forces acting on a maneuvering aircraft and how these forces affect the turn radius and rate of turn of the airplane. This discussion will be limited to airplanes in steady coordinated turns and will not go into the more complicated subject of airplanes which are both turning and making significant changes in altitude. This is a situation in which the plane of the turn is significantly inclined to the horizon and there is an interplay of both the airplane's kinetic and potential energy. If the aircraft's maneuvering plane is within eight degrees of

horizontal, the errors induced in turn geometry will be less than one percent and are not significant.

3. DETERMINING TURN RADIUS FROM VELOCITY AND G LOAD/BANK ANGLE.

When an airplane is in a steady coordinated, constant altitude, constant true airspeed turn, the relationship between the radial acceleration (the acceleration towards the center of the turn which gives us the sensation of G force) is referred to as centripetal force. The speed of the airplane, and its turn radius is given by the equation:

$$a_r = \frac{V^2}{r} \qquad (1)$$

Where:
  a_r = radial acceleration in feet/second^2
  V = speed of the airplane in feet per second
  r = turn radius of the airplane in feet

The force necessary to produce this acceleration can be determined by remembering Newton's second law, which is commonly referred to as "F = ma" (Force equals mass times acceleration). If we are interested in radial acceleration, the force necessary to cause the radial acceleration (centripetal force) is given by the equation:

$$CF = ma_r = \left(\frac{W}{g}\right)\left(\frac{V^2}{r}\right) \qquad (2)$$

where:

CF = centripetal force which is causing the airplane to turn (lbs.)

m = mass of the airplane (slugs)
a_r = radial acceleration of the airplane (fps)
W = weight of the airplane (lbs.)
g = acceleration due to gravity (32.2 fps^2)
V = speed of the airplane (fps)
r = airplane's horizontal turn radius (ft)

Let's put this equation on hold for a moment and look at the forces acting on our turning airplane. The horizontal acceleration the airplane is experiencing (which is at a right angle to the airplane's velocity vector and causes only the airplane's heading to change; not its speed) is generated by the lift vector's horizontal component. A force acting at a right angle to an airplane's velocity vector and causing its heading change while maintaining a constant speed is referred to as a centripetal force. The relationship between an airplane in a steady turn, its bank angle, weight and centripetal force is depicted in the Figure 26-1.

From the geometry shown in Figure 26-1, it should be obvious that the centripetal force (CF) is equal to the lift (L) multiplied by the sine of the airplane's bank angle (ϕ). In other words:

$$CF = L \sin \phi \qquad (3)$$

where:

CF = centripetal force causing the horizontal acceleration
L = total lift generated by the wing
ϕ = the airplane's bank angle

Substituting into equation (2) produces:

$$L \sin \phi = \left(\frac{W}{g}\right)\left(\frac{V^2}{r}\right) \qquad (4)$$

Looking at Figure 26-1, we can see that in order for the airplane to maintain a constant altitude, the effective lift will have to equal the airplane's weight. This relationship can be expressed by the equation:

$$L \cos \phi = W \qquad (5)$$

and:

$$\frac{L}{W} = \frac{1}{\cos \phi} = G \qquad (6)$$

where:

W = weight of the airplane (lbs.)
L = total lift generated by the wing (lbs.)
ϕ = airplane's bank angle (degrees)
G = G load on the aircraft (no units)

If we divide equation (4) by equation (6):

$$\frac{L \sin \phi}{L \cos \phi} = \left(\frac{W}{g}\right)\left(\frac{V^2}{r}\right)\left(\frac{1}{W}\right) \qquad (7)$$

or:

$$\tan \phi = \frac{V^2}{gr} \qquad (8)$$

or:

$$r = \frac{V^2}{g \tan \phi} \qquad (9)$$

where:

r = airplane's turn radius (ft)
V = airplane's speed (fps)
g = acceleration due to gravity (32.2 fps^2)
ϕ = bank angle of the airplane (degrees)

Since we normally deal with airplane speeds in terms of knots, and the equations shown above deal with speeds in terms of feet per second, a simple conversion factor will allow us to use the more convenient knots. It should be remembered that we are dealing with knots of true airspeed and not knots of indicated airspeed. At the same time we can get rid of the "g" in the equation by substituting 32.2 fps^2. The end result is shown in the equation shown below:

$$r = \frac{V_k^2}{11.26 \tan \phi} \qquad (10)$$

where:

r = turn radius of the airplane (ft)
V_k = speed of the airplane (kts)
ϕ = bank angle of the airplane (degrees)

We can apply these relationships to an airplane in a level coordinated turn in the following ways.

A. If we know the airplane's true airspeed and bank angle, we can calculate its turn radius, which can be useful in determining the horizontal space it needs to turn.

B. If we know the airplane's true airspeed and turn radius, we can calculate its bank angle, which can be useful in determining its G load and stall speed.

C. If we know the airplane's bank angle and turn radius we can determine its speed.

4. RATE OF TURN FROM VELOCITY AND G LOAD/BANK ANGLE.

Sometimes an airplane's rate of turn is of interest to the accident investigator. This relationship is useful in the following situations.

A. Estimating how long it took to complete a turn when the average true airspeed and bank angle are known.

B. Estimating the average bank angle (and therefore the G load) when
the average true airspeed and rate of turn are known.

C. Estimating the average true airspeed given the average rate of turn and bank angle.

Using the relationships discussed in the previous paragraphs and some basic physics concerning rotational motion you can come up with the following relationship:

$$ROT = \frac{1091 \tan \phi}{V_k} \qquad \textbf{(11)}$$

where:
ROT = the airplane's rate of turn (rate of heading change).

ϕ = the airplane's bank angle in degrees.

V_k = the airplane's speed in knots.

5. DERIVATION OF AIRPLANE PERFORMANCE DATA FROM RADAR DATA.

(See also Chapter 19.)

Most ground based ATC radar equipment today (both civil and military) provides digital data and incorporates some means of recording the aircraft position data it receives. This radar information typically includes aircraft latitude, longitude and (when encoding altimeters are used) altitude to the nearest 100 feet. This information can be useful to accident investigators attempting to determine an aircraft's performance in terms of heading, ground speed and turn radius. With this, other data discussed in the preceding paragraphs can be estimated.

For example, radar information may be used to determine the airspeed and maneuvers of an aircraft which experienced an inflight break-up. It would be nice to know if the aircraft's velocity could have contributed to the break-up.

It should also be noted that while information from only one radar site (normally the nearest) is being used to provide information to the center controller, several different radar sites (the number depends on the center, terrain features, and aircraft altitude) may be tracking an aircraft. This information is temporarily saved on discs and can be used to improve the accuracy of estimated ground track. There are some limitations associated with the use of radar.

Obviously, coverage is limited and radar information on many aircraft flights just does not exist. Sometimes, though, the data exists as "skin paint" and

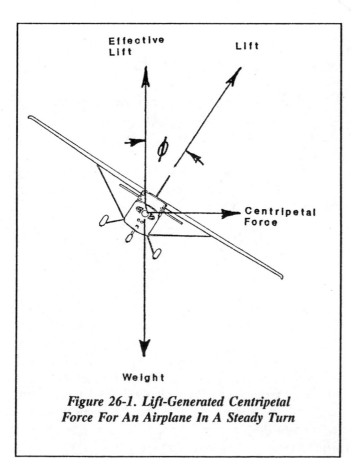

Figure 26-1. Lift-Generated Centripetal Force For An Airplane In A Steady Turn

is recorded in the computer memory even though it was not seen by the controller. Also, there may have been other radars tracking the aircraft. It is not unusual for military radar systems to also have information on the aircraft.

The information is perishable, and if not preserved, it can be lost. The medium on which the data is recorded is normally reused by erasing and re-recording. If investigators fail to request the information promptly, it will be destroyed.

The time between radar sweeps (typically 10-12 seconds for center radar and 4-6 seconds for terminal radar) makes it impossible to accurately reconstruct the flight path of aircraft engaged in extreme maneuvering.

Even with moderate maneuvering, averaging techniques must be used. It is also possible that two or more radars are "painting" the same aircraft. The more information about the aircraft's track, the better the aircraft's maneuvers can be estimated.

Because of errors associated with the radar's beam width, there are degrees of uncertainty associated with the exact position of the aircraft.

Despite these limitations, radar information can be a valuable source of information for the aircraft accident investigator. Although there now are computer programs available which automatically convert raw radar data into the ground tracks, speeds, G loads, rates of turn, and so on, don't bet that this will automatically be done for you. It is best to assume that you will have to deal with the raw data and a basic understanding of what can be done with it is a nice thing to have. These basic concepts are described next.

6. DETERMINING GROUND SPEED FROM LATITUDE/LONGITUDE POSITIONS.

Probably the simplest way to turn the raw radar data into useful information is to simply plot the positions and associated times on the maps, much the same as you would plot your position during a cross-country flight. The typical choices concerning the map to use are either a sectional published by NOAA or a topographical published by U. S. Geological Survey. The USGS maps, commonly called "topos", come in 15 minute and 7.5 minute varieties. The 15 minute has a scale of 1:12,000 and the 7.5 has a scale of 1:24,0-00. Although the 15 minute topos cover more of the

country, the 7.5s are more readily available. The decision on which map to use will hinge on the exact purpose of the investigation.

For example, the higher fidelity of the topographical map and its detail concerning terrain features will probably be more useful when investigating a collision with terrain accident. However, the aeronautical navigation information published on the sectional may be more useful if there are questions concerning the use of navigational aids. In addition, topos are updated only infrequently and may contain decades old information concerning man made features like roads and buildings.

Once the positions are plotted, an estimated course line can be drawn, connecting the various positions. This can be a series of straight lines (which makes measuring distance between the points easier), or a curved line (which probably makes the distance between the points more accurate). Once the course line is drawn, distances can be measured and ground speeds calculated. If winds aloft are available, the ground speeds can be adjusted to true airspeeds. One thing you may want to do to correlate the radar data with the topo is to locate the impact point and other wreckage (if it is significantly remote from the main wreckage) with one of those hand-held GPSs.

7. ESTIMATING BANK ANGLE AND G LOAD FROM LATITUDE/LONGITUDE POSITIONS.

There are two ways to estimate the bank angle and G load on the aircraft using the course line you have just plotted.

The first method uses estimations of the aircraft's turn radius and true airspeed to calculate the bank angle and associated G load.

The second method uses estimations of the aircraft's turn rate to calculate the bank angle and associated G load.

A thorough investigation includes estimations using both methods. When using turn radius to estimate other turning performance parameters, the point about which the aircraft is turning can be established by drawing a series of lines which are perpendicular to the turn radius. One method used to determine the center of the turn is to draw several tangent lines (line tangent to the curving course line) and then (using a right triangle) draw a line perpendicular to each tangent line from the point of tangency toward the inside of the curve. See Figure 26-2.

If you do this for three or four points which appear to have the same radius of curvature, the perpendicular lines should cross at approximately the same point. This point is the center of the turn and the distance from the point to the course line is the radius of the turn. The aircraft's bank angle during this portion of the course can be calculated using the equation:

$$\phi = \tan^{-1}\left(\frac{V_k^2}{11.26\ r}\right) \qquad (12)$$

where:

ϕ = the aircraft's bank angle in degrees
V_k = the aircraft's estimated speed in knots
r = the aircraft's estimated turn radius in feet
11.26 is a conversion factor which allows us to mix knots, feet and degrees.

The average G load on the aircraft during the turn can be estimated by using the equation:

$$G = \frac{1}{\cos\ \phi} \qquad (13)$$

where:

G = average G load the aircraft experienced during the turn.
$\cos\ \phi$ = the aircraft's average bank during the turn in degrees.

The other method which can be used to estimate these same parameters involves measuring the heading change during any known period of time. This could be the course change estimated between two radar plots. If the change in course is divided by the time it took to change the course, the average rate of the aircraft heading change during this time interval can be estimated. Knowing the aircraft's average rate of turn we can estimate the bank angle using the equation:
where:

$$\phi = \tan^{-1}\left(\frac{V_k\ ROT}{1091}\right) \qquad (14)$$

ϕ = the aircraft's average bank angle in degrees.
V_k = the aircraft's estimated true airspeed in knots.

ROT = the aircraft's estimated rate of turn in degrees per second.

The aircraft's average G load can be estimated using the equations given in the previous paragraph.

8. SUMMARY.

When considering an aircraft's manuevering performance, the three key elements are true airspeed, bank angle and turn radius.

If we know (or can reasonably estimate) any two of those, we can calculate the third and also establish the G load on the aircraft.

This type of calculation has been used a number of times to establish performance conditions at the instant of inflight structural failure.

If radar data is available, it is possible, with a minimal amount of information, to make some fairly accurate assessments of the aircraft's attitude, turn radius, G load and ground speed.

This is a "pencil and paper" technique which does

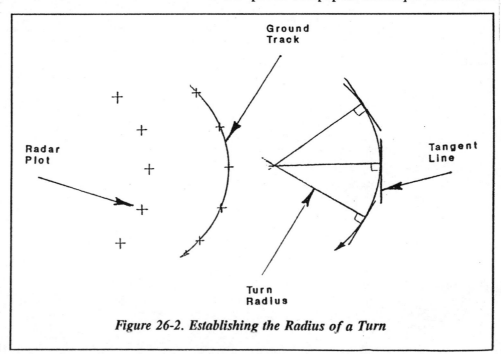

Figure 26-2. Establishing the Radius of a Turn

not require any exotic computer software.

There are, of course, limitations to this technique. The aircraft has to have been "painted" by ground radar for the number of "sweeps" necessary to establish its course. With current radar technology, this is possible only for aircraft in steady state flight. This won't work for aircraft whose speed and heading are rapidly changing. In addition, analysis of the radar data must always consider the errors associated with beam width, range and other system tolerances. See Chapter 19.

27

Stalls and Spins

1. INTRODUCTION.

Probably the first bug-a-boo to strike fear in the heart of the potential pilot is the stall and its progeny the spin. Although stall/spin accidents may have claimed many of the early pioneers of aviation, the frequency of stall/spin accidents has decreased over the history of powered flight. In 1947 the Civil Aeronautics Board (the FAA's predecessor) eliminated the requirement for candidate private and commercial pilots to demonstrate their proficiency in spin recov-

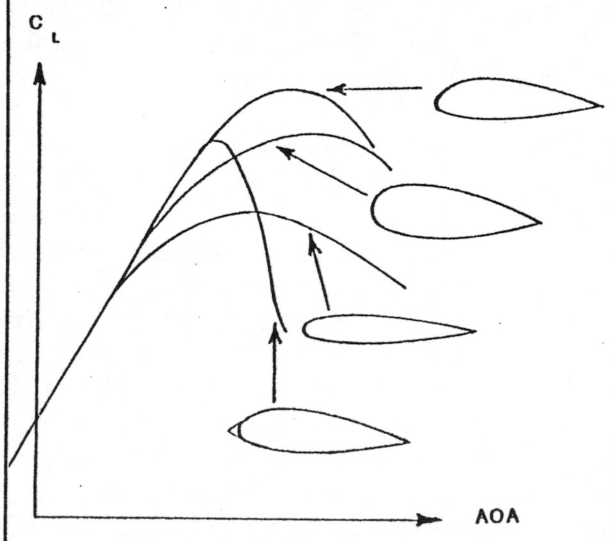

Figure 27-1. Effect of Airfoil Shape on C_L - AOA Curves

eries. In 1976, despite being prompted by an NTSB report recommending spin training for all pilot applicants, an FAA study found that accident statistics did not justify mandatory spin demonstrations for all pilot applicants. However, stall/spin accidents still occur. They remain a source of concern to safety professionals. As an aircraft accident investigator, you need to look not only at the physical evidence in the wreckage following a stall/spin accident, you should also fully understand the aerodynamic factors which contributed to the stall, post stall gyration, spin sequence, and why the pilot failed to recognize and correct them.

2. STALL HAZARDS.

Not all stall accidents are the same. They don't all have the same causes and they don't all have the same consequences. For instance, some stall accidents occur when the wing exceeds its critical angle of attack and the C_L falls while C_D increases. The lift decreases to the point where the airplane can't maintain level flight and drag increases to the point where the aircraft can't accelerate. The aircraft maintains a roughly wings level attitude while it "mushes" into the ground. This type of accident is often survivable. On the other hand, during another accident, one wing might stall before the other with the resulting asymmetric lift causing the airplane to roll rapidly toward the stalled wing. If the airplane is too close to the ground to recover from this "out-of-control" condition, it will crash into the ground at an extreme bank and perhaps pitch attitude. This is normally a non-survivable crash. If the aircraft is at a higher altitude it might transition

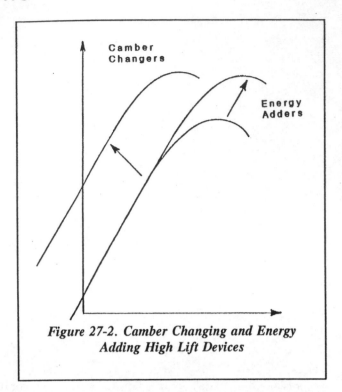

Figure 27-2. Camber Changing and Energy Adding High Lift Devices

stalled. Three of the stall characteristics which are of interest to the accident investigator are the maximum coefficient of lift produced by the airfoil (C_{Lmax}), the abruptness of the stall (the rate at which C_L falls once the AOA for C_{Lmax} has been exceeded) and the point along the airfoil at which the airflow first starts to separate. Since the minimum speed at which a given airplane can fly or the number of Gs its can pull is established by its C_{Lmax}, airfoils with a high C_{Lmax} are good when you are interested in flying slow or pulling lots of Gs. Unfortunately, high C_{Lmax}s come with the penalty of high drag.

Abrupt stalls are an undesirable feature for airfoils. Lift falls off suddenly with a small increase in angle-of-attack. In addition to placing the airplane in a stall with little warning, one wing is more likely to stall before the other with the resulting asymmetric lift condition producing an out-of-control roll. If the airflow around an airfoil separates first near the leading edge, it tends to have abrupt stall characteristics.

The shape of a two dimensional airfoil determines its C_{Lmax} and influences the abruptness of its stall. Four general examples are sketched in the coefficient-of-lift versus angle-of-attack curves shown below. It should be obvious that the slope of all four lift curves is the same. In fact the slope of the C_L versus AOA curve for all two dimensional airfoils is approximately 0.1 increase in C_L for each 1.0 degree increase in AOA. It should be noted that wing planform parameters such as aspect ratio, taper, and sweep do influence the shape of C_L versus AOA curves for three dimensional airfoils. The safety impact of the effects will be discussed later in this section.

The first airfoil shape shown is a medium thickness airfoil with a matching leading edge radius. This airfoil is typical of most general aviation airplanes and has a reasonable C_{Lmax} and a moderately abrupt stall. Airflow separation occurs first at the trailing edge and moves forward with increasing AOA. The point at which the airflow separation first occurs shows up on the lift Vs AOA curve and can be identified by the point at which the C_L versus AOA first starts to deviate from the constant slope line. As the separation point moves forward on the wing, the slope of the C_L versus AOA curve decreases even further. The wing is considered stalled when the slope of the C_L versus AOA curve decreases to zero. The AOA at which the slope of the C_L versus AOA curve is zero is the stall AOA. Also note that the separation point moves even

into a spin, rolling and yawing into the most deeply stalled wing. Here, inertia and aerodynamic forces are balanced, maintaining the airplane in a downward spiraling helix. If recovery is not completed at sufficient altitude, the airplane will impact the ground in a nose low attitude with very little or no forward speed.

Yet another type of stall accident involves a more complex interaction between all of the airplanes surfaces which produce aerodynamic forces and pitching moments. In this "deep stall" condition, nose-up pitching moments developed at extreme angles of attack can't be overcome by full nose-down flight control commands, preventing the pilot from reducing the angle-of-attack and breaking the stall. The resulting high sink-rate can result in an non-survivable crash. All of these examples involve stalls, but all involve different causes, sequences and consequences.

3. INFLUENCE OF AIRFOIL SHAPE ON THE LOCATION OF AIRFLOW SEPARATION.

An airfoil (wing, horizontal tail, vertical tail or any shape capable of producing an aerodynamic force) is considered to be stalled when an increase in angle-of-attack (AOA) results in a decrease in lift as the airfoil becomes increasing less efficient in performing its intended purpose. Another way of defining a stall is to identify the angle-of-attack at which the airfoil produces its maximum coefficient of lift (C_L) and simply state that at any AOA above this, the wing is

further forward as AOA moves even higher above the stall AOA.

The next airfoil shape depicted is extremely thick (the maximum thickness exceeding 16 percent of its chord) with a generous leading edge radius. It is not very efficient, and produces a lower C_{Lmax} due to the separated flow which occurs forward of the trailing edge, even at low angles-of-attack. Despite its inefficiency in turning dynamic pressure into lift, this airfoil shape has one redeeming safety related characteristic; its stall is gradual. C_L falls off relatively slowly as the critical AOA is exceeded.

The third airfoil shape has a relatively low maximum thickness and is more typical of those used in high speed airplanes with swept wings. As its AOA is increased toward its critical angle of attack, a vortex should form shortly aft of its leading edge. Although this vortex will normally reattach the flow and delay the stall, it can also "burst" and allow the airflow over virtually the entire wing to separate. The sudden flow separation will cause an abrupt stall.

The last airfoil shape to be covered takes the typical general aviation wing airfoil and gives it a relatively small or sharp leading edge radius. This smaller leading edge radius will precipitate premature separation, and it will do it in a rather abrupt manner. This not only decreases C_{Lmax}, it also increases the abruptness of the stall, causing the C_L versus AOA to drop sharply as the AOA moves past the critical AOA. When properly designed into the airfoil, a sharp leading edge over a small portion of the wing can provide the pilot with a passive (it has no moving parts and requires no power) stall warning device. Look at the stall strip on the leading edge of many general aviation aircraft and even some military airplanes. However, when mother nature modifies the leading edge of an airfoil (wing or otherwise) the consequences can be tragic.

4. HIGH LIFT DEVICES.

Before we leave the subject of airfoil shape, C_{Lmax}, and stall AOAs, the effects of using high lift devices need to be addressed. Airplane designers are faced with the conflicting objectives of achieving efficient high speed flight for cruise (which mandates a relatively thin airfoil with relatively poor low speed characteristics) and safe low speed for take-off and landing (which mandates a thicker airfoil with more camber and inefficient high speed characteristics). High lift devices are used for the purposes listed below.

A. INCREASE AN AIRFOIL'S CAMBER. This increases the speed of the airflow over the lift producing surface. This change causes the C_L Versus AOA curve to shift up and to the left, providing a higher C_{Lmax} and, in general, decreasing the AOA necessary to produce any given C_L. Although high lift devices increase C_{Lmax}, the AOA at which C_{Lmax} is achieved is normally decreased. In other words, the stall AOA is decreased. These features allow an airplane to not only take-off and land at lower speeds, they allow the airplane to take-off and land at lower AOAs, affording the pilot better over-the-nose visibility and relaxing the requirement for the engineer to provide a design which avoids tail strikes during high pitch attitude take-offs and landings. The most typical examples of "camber changer" high lift devices are wing flaps.

B. ENERGIZE THE BOUNDARY LAYER OF THE AIRFLOW. This delays airflow separation and stall. Delayed separation extends the C_L versus AOA curve up and to the right increasing C_{Lmax} by allowing the airfoil to operate, without airflow separation, at higher angles of attack. It should be noted that "energy adder" high lift devices require increased AOAs to approach C_{Lmax}. Typical energy adder high lift devices

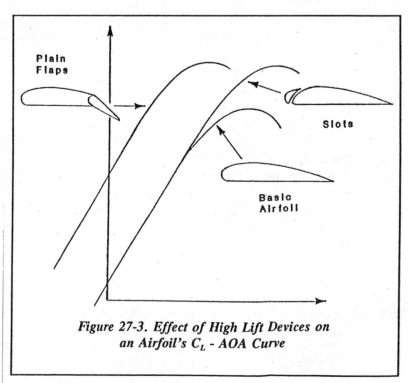

Figure 27-3. Effect of High Lift Devices on an Airfoil's C_L - AOA Curve

are:

BOUNDARY LAYER CONTROL.

There are two general methods of boundary layer control. First, the low kinetic energy air in the boundary layer can be removed, by suction, though small holes in the wing. With the low speed air which was close to the surface removed, the higher speed, higher energy air which has not been slowed by skin friction is free to move closer to the airfoil's surface, delaying separation. The second method involves blowing high speed air across the lift producing surface in regions where separation will first occur. This type of high lift device not only delays separation, it increases the velocity and quantity of air moving over the lift producing surface, significantly increasing lift.

VORTEX GENERATORS.

Vortex generators are relatively small airfoils which are placed on and extend perpendicular to the primary airfoil's surface. They extend from the primary airfoil's surface (where the boundary layer air is relatively slow and has low kinetic energy) up into the airflow and into or through the upper ranges of the boundary layer (where the airflow has relatively high speed and high kinetic energy). The vortex generators, as their name implies, generate a vortex which both sweeps low energy air away from the primary airfoil's surface and replaces it with high energy air which has not been slowed by skin friction. In addition to being used for high C_{Lmax}, vortex generator's are often used in front of control surfaces to delay separation of the airflow over the controls when they are deflected while the primary airfoil is at high angles of attack.

SLOTS.

Slots are fixed leading edge devices which allow relatively high pressure air below and behind the wing's leading edge to flow to the upper surface during high AOA conditions. When the airfoil is at high AOA its forward stagnation point moves under the leading edge of the wing and provides a source high pressure air which can be used to re-energize the boundary layer on the lift producing surface. At low angles of attack there is relatively little airflow through the slot. Although, today, wing slots are pretty much reserved for STOL aircraft, you can see slots on the horizontal stabilizers of some modern aircraft. When used on the horizontal tail, which normally has its lift producing surface on the bottom, the slot vectors air from the upper surface of the airfoil to its lower surface.

C. DEVICES WHICH COMBINE THE FEA-

TURES OF BOTH CAMBER CHANGERS AND ENERGY ADDERS. Slotted flaps and slats are examples of such devices. They both change the camber of the airfoil and use the energy associated with relatively high pressure on the non-lifting surface to re-energize the boundary layer on the lift producing side of the airfoil.

For the accident investigator, the concern with high lift devices is mostly "Did they fail to produce lift when they were supposed to?" or "Did the high lift devices on one wing function properly while those on the other side failed?" or "Did they function to produce lift when they were not supposed to?" In order to know when to ask these questions and determine their answers, the investigator will have to understand how and why they work. The purpose of this section it to provide some understanding of why they work. How they work (and can fail) will have to be discovered in the individual airplanes flight and technical manuals.

5. ACCELERATED STALLS.

When an airplane stalls at a speed greater than its one G stall speed, it is referred to as an accelerated stall. Some people link all stalls with a specific airspeed and wrongfully assume that an airplane stalls whenever its speed is less than that speed and it will not stall above that speed. Wrong! Wing stall is normally considered to be strictly a function of the airfoil's angle of attack and is only indirectly connected to the airplane's speed through the air. A wing on an airplane moving down a runway at well below its "stall speed" will not be stalled if its AOA is less than its critical or stall AOA. Likewise, an airplane flying at twice its "stall speed" can stall if it is forced to an AOA which exceeds its critical or stall AOA. The key to the issue is to think of angle-of-attack when you think of stalls. The "stall speeds" you find in an airplane's Flight Manual or Pilot Operating Handbook only relate to stalls in one G flight while at a specific weight. The stall angle of attack is valid at all weights and G loads. Only changes in the airfoil's shape will cause its stall AOA to change. The aircraft accident investigator should be able to, given an airplane's one G stall speed at a specific weight, calculate:

1. the airplane's stall speed at any other G load and/or weight.

2. the G load at which the airplane, at any gross weight, will stall.

A couple of simple equations can solve these problems. Both equations start with the basic lift equation. Shown below is this lift equation which has been algebraically manipulated to solve for, in one case V_s (stall speed) and in the other case G load required to stall.

$$V_s = 17.2 \sqrt{\frac{W \times G}{C_{Lmax}\, \sigma S}} \qquad (1)$$

where:

V_s is the stall speed of the airplane in KTAS at a given weight and G load

W is the weight of the airplane in pounds.

C_{Lmax} is the airfoils maximum or stall C_L.

σ is the ambient air's density ratio.

S is the airplane's wing area.

$$G = \frac{C_{Lmax}\, \sigma V^2 S}{295 W} \qquad (2)$$

where:

G is the G load required to stall the airplane.

C_{Lmax} is the airfoils maximum or stall C_L.

σ is the ambient air's density ratio.

V is the airplane's true airspeed in knots.

S is the airplane's wing area.

W is the airplane's weight.

Don't let the above equations confuse you. We can live without most of the information listed above. What we need we can get out of the airplane's Flight Manual or Pilot Operating Handbook. What we need is its one G stall speed at one or more gross weights (for given wing configurations).

To find the stall speed at other G loads and/or gross weights we can use the equation:

$$V_{S2} = V_{S1} \sqrt{\frac{W_2\, G}{W_1}} \qquad (3)$$

where:

V_{S2} is the stall speed at the new G load and/or gross weight.

V_{S1} is the stall speed we obtained from the Flight Manual/Pilot Operator Handbook.

W_2 is the new gross weight.

G is the G load in which we are interested.

W_1 is the gross weight which generates the one G stall speeds in the handbook.

To find the G load required to stall an airplane, use the equation:

$$G = \left[\frac{W_1}{W_2}\right]\left[\frac{V}{V_S}\right]^2 \qquad (4)$$

where:

G is the number Gs the airplane can pull before it stalls.

V is the speed of the airplane (this can be KIAS, KCAS, KEAS, or KTAS).

V_s is the stall speed of the airplane in the same units as the actual speed.

It is important for an accident investigator to realize that the stall characteristics of an airplane experiencing an accelerated stall may be

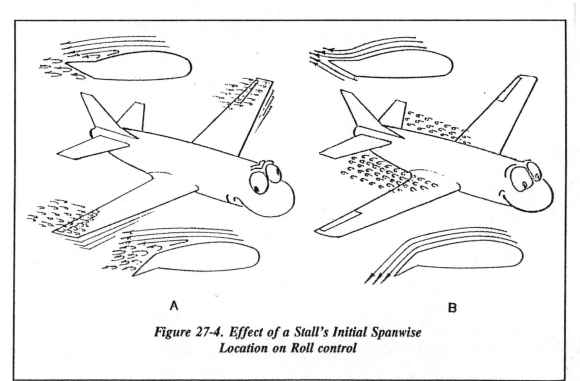

Figure 27-4. Effect of a Stall's Initial Spanwise Location on Roll control

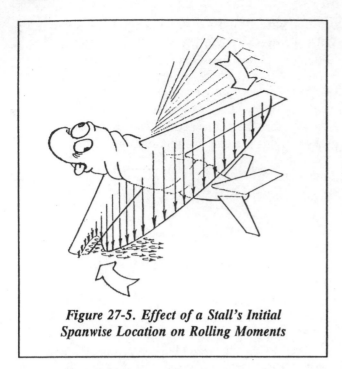

Figure 27-5. Effect of a Stall's Initial Spanwise Location on Rolling Moments

very different from its stall characteristics at one G, power off flight. While in high speed flight at high power settings, the asymmetric aerodynamic forces created when one wing stalls before the other can couple with the gyroscopic forces generated by the rotating components in the power plant. This can result in an exciting ride for some with sufficient altitude to recover before ground impact. Although

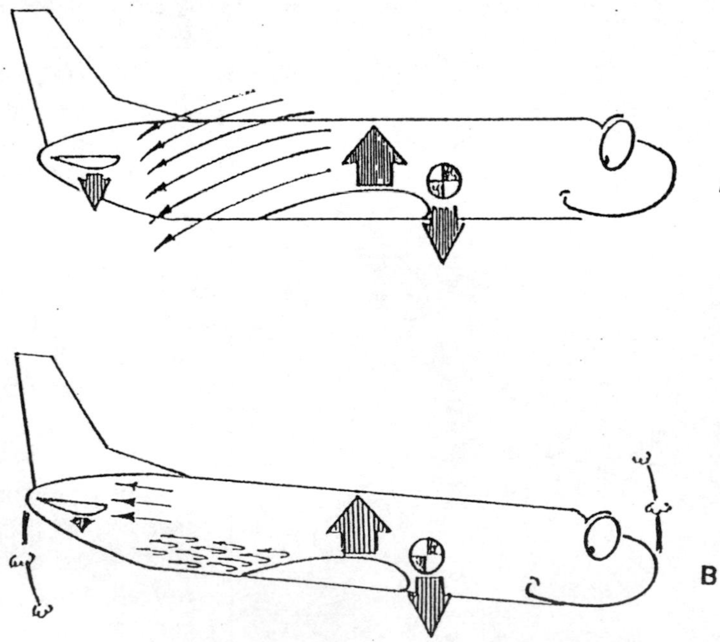

most competent pilots are familiar with the behavior of the aircraft during typical take-off and landing situations, the stall characteristics at cruise speeds and maximum power settings may be well known to only company test pilots.

6. HIGH ALTITUDE AND DYNAMIC STALLS.

Many novice pilots have it drummed into their heads by their instructors that the stall angle of attack of a given airfoil section is constant and will not change. Although this statement is an excellent rule of thumb, like all rules there are some exceptions. In this case, two; high altitude flight and dynamic stalls.

First, high altitude stalls. It turns out that the stall AOA of an airfoil is sensitive to the Reynold's number (RN) of the airflow over the airfoil. You remember Reynold's number, its that parameter that's used to predict whether boundary layer air is turbulent, laminar or transitioning between the two.

Well, it turns out that airfoils with low RNs (laminar boundary layers) stall at slightly lower angles of attack than airfoils with high RNs (turbulent boundary layers). It also turns out that higher kinematic viscosity associated with high altitudes lowers the boundary layer's Reynold's number which in turn decreases the airfoil stall angle of attack, and its C_{Lmax} and changes the abruptness of the stall. Like accelerated stalls, high altitude stalls may not be exactly the same as the low altitude, low speed, low power cousins. More on this in the "coffin corner" section of this chapter.

A dynamic stall is a phenomena that occurs when an airfoil is very rapidly rotated from an angle below it stall AOA to an angle above its stall AOA. If the rotation rate is fast enough (greater than 50°/sec), separation can be momentarily delayed and the stall angle-of-attack momentarily increased. Naturally C_{Lmax} will also increase, even if only for a fraction of a second. NASA and university tests have shown increases in C_{Lmax} of up to 30 percent. Although the rotational rates necessary to create this phenomena are above those achievable in civil (and probably military) aircraft, flight into extremely strong vertical wind gusts (such as suddenly penetrating wing tip vortices) could cause and the angle-of-attack to increase at rates consistent with the rates necessary to cause a dynamic stall. This could

Figure 27-6. Effect of a Stall's Initial Spanwise Location on Pitching Moments

cause the wing to develop a higher C_L than its normal C_{Lmax} before it stalls.

7. INFLUENCE OF WING PLANFORM ON STALL CHARACTERISTICS.

Some of the variations in the stall characteristics of different airplanes can be explained by examining the differences in the planforms of their respective wings. A wing's planform (aspect ratio, taper and sweep) affect the spanwise location of the point where airflow over the wing first starts to separate, the slope of the C_L versus AOA curve, and the abruptness of the airfoil's stall. All of these factors have some influence on the hazards associated with a stall.

Generally speaking, wings which stall first at their root (as opposed to their tip) are considered to be safer for four reasons.

First, stalls which start first near the wing's tip can cause loss of effectiveness of the ailerons and associated reduction in lateral or roll control. Stalls which start first at the root are not associated with this loss of control.

Second, this hazard can be become more critical

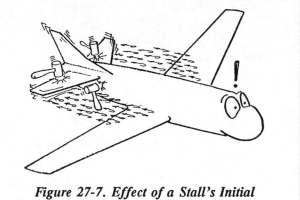

Figure 27-7. Effect of a Stall's Initial Location on Natural Warnings to the Pilot

if only one wing stalls or stalls first (due to any number of factors). Because of the large rolling moment created when one wing tip stalls and the other does not, rapid roll rates coupled with degraded roll control could result in extreme bank angles in a very short period of time. Once again, stalls which occur at a wing root do not cause these large rolling moments and uncontrollable roll rates.

Third, root-first wing stalls tend to naturally give the airplane a nose down pitching moment and tend to break the stall by lowering the airplane's nose. On the other hand, tip-first wing stalls may, in some cases, cause a nose-up pitching moment with the potential to aggravate the stall by increasing the airplane's AOA. The nose down pitching moment for root-first stalls occurs when the decease in down-wash angle associated with the stall causes a decrease in the angle-of-attack of the horizontal tail and an associated reduction in its downward lift. Stalls which occur first at the wing-tip do not normally affect the downwash angle at the horizontal tail.

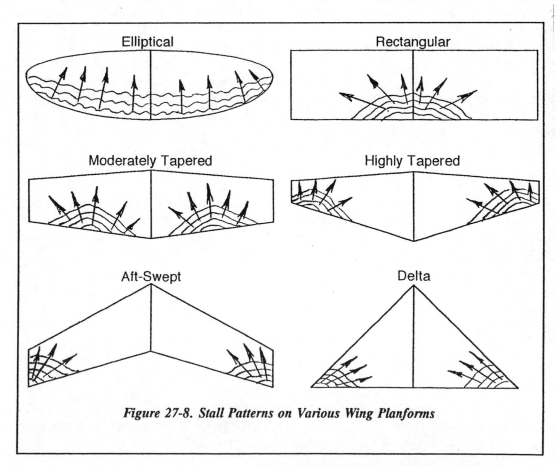

Figure 27-8. Stall Patterns on Various Wing Planforms

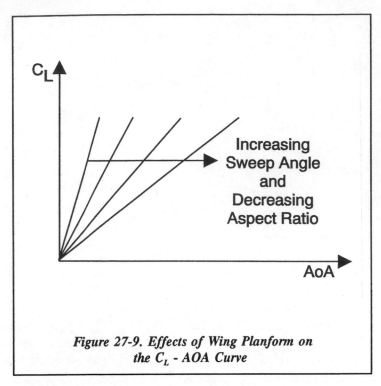

Figure 27-9. Effects of Wing Planform on the C_L - AOA Curve

swept wings cause airflow over the top of the wing to be deflected toward the wing-tip, in turn causing a thick, low energy boundary layer toward the wing-tip. This further increases the propensity of tapered and pointed tips, which are also swept, to stall first near the wing-tip. Wing-tip stalls on aft-swept wings cause the center of lift to move forward, inducing a nose-up pitching moment which can increase the angle-of-attack, further aggravating the stall. Where tapered, pointed and aft-swept wing planforms are used, other means must be used to move the location of the initial separation inboard. Sometimes triangular "stall strips" are placed on the inboard leading edge of the wing, causing a premature, abrupt stall on a small section of the inboard wing. Some other techniques include varying the thickness and/or camber of the airfoil sections along the span and twisting the outer portion of the wing leading edge down (washing out). The objective is to create a design which causes the aircraft to stall in a manner which allows the pilot to safely recover the airplane to level, controlled flight.

Finally, stalls which start first at the root tend to provide the pilot with a natural warning of the impending full stall as soon as the airflow starts to separate. This is due to the turbulent flow associated with the initial airflow separation striking the aircraft's empennage and even the sides of the fuselage. If flow separation first occurs on the wing tip, there is no feedback through the tail to the pilot.

Let's take a look at how the wing's planform affects the location where the stall starts. Rectangular wing planforms have strong downwash near their tips. This results in large induced angles of attack and low airfoil section angles of attack near the wing tip. As a result, the airfoil sections near the wing-root are at a higher angle-of-attack than the airfoil sections near the wing-tip. The root of rectangular airfoils, therefore, tends to stall first. On the other hand, wings with pointed tips or high degrees of taper tend to stall first at or near their tip. This is due to the downwash angle being greater at the wing root, putting the wing sections near the root at a lower angle of attack than those near the wing tip. With elliptical or moderately tapered wings, the downwash angle is pretty constant along the entire wingspan. As a result the stall starts along the entire trailing edge at pretty much the same time. Although aft-swept wings are often tapered and can even be pointed, having some of the stall pattern characteristics these planforms, aft-swept wings also induce another factor. Spanwise pressure patterns on

Although variations in a wing's aspect ratio do not effect the stall patterns on a wing, they do have some other safety-related factors. Earlier in this section, it was mentioned that the slope of an airfoil section is approximately 0.1 increase in C_L for each 1.0 degree increase in angle-of-attack. This is true only for a theoretical wing of infinite aspect ratio. Real wings with finite aspect ratios, have lower slopes than this. The lower the aspect ratio, the lower the slope of the C_L versus AOA curve.

The safety implications of a lower slope for the C_L versus AOA curve are that:

1. it takes higher AOAs to produce the high C_Ls necessary for take-off and landing. This reduces the pilot's over the nose visibility of the runway and its environment (which is not good).

2. it reduces the sensitivity of the aircraft to vertical wind gusts (which is good).

Decreasing the aspect ratio of a wing also tends to make the stall angle of attack less defined, making the decrease in lift as the AOA increases through the stall AOA more gradual. This reduces the potential for asymmetric stalls and their accompanying roll-off into the stalled wing. Design decisions concerning high versus low aspect ratios will probable be driven more by mission and structural consideration.

8. INFLUENCE OF THE LEADING EDGE VORTEX ON LIFT/STALL OF SWEPT WINGS.

As a wing of a given airfoil section is swept aftward, the relative thickness of the wing decreases because, as the air flows aftward over the wing the longitudinal distance it has to travel increases while the vertical distance it has to travel remains the same.

As mentioned earlier, thin wings have a tendency to have the airflow over their surfaces

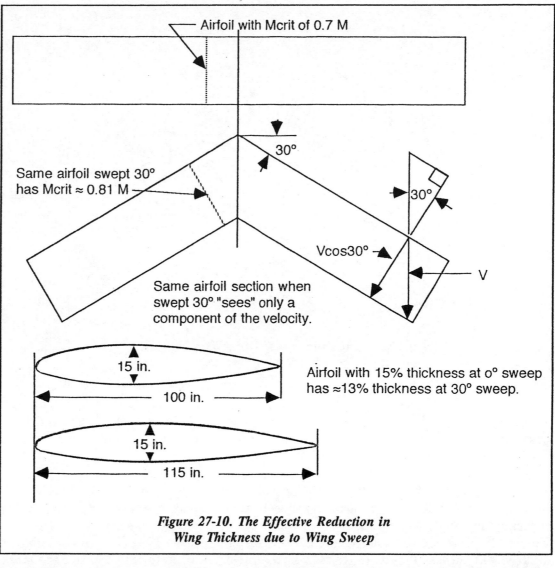

Figure 27-10. The Effective Reduction in Wing Thickness due to Wing Sweep

separate first near their leading edge. If the angle of attack is not increased much above the point where separation first occurs, the boundary layer will re-attach itself to the wing. The separated flow just aft of the leading wing edge will remain as a "bubble" in the flow without significantly affecting the wing's coefficient of lift. If a significantly large wing sweep is combined with a relatively thin wing this "bubble" will develop into a spanwise vortex, flowing spanwise, just behind the leading edge, away from the wing root and toward the wing tip. This vortex flow has relatively high energy and its existence can delay separation of the airflow behind it and the stall of the wing. When the vortex loses energy it will "burst", increasing rapidly in size and decreasing rapidly in intensity. If a wing's leading edge vortex bursts before it departs the wing, it can cause the section of the wing behind it to stall. A leading edge vortex is very sensitive to irregularities along the leading edge. Some airplanes use saw tooth and saw-cut leading edges to maintain strong

vortex flow along the entire leading edge. On the other hand, poor designs, modifications or even damage may cause premature vortex burst or departure from the leading edge. Irregularities along the leading edge therefore may result in vortex flow being shed by the wing well inboard of the wing tips. If this vortex passes over the tail of the airplane it can have undesirable stability and control consequences. Downward flow of the horizontal tail can cause a "pitch-up" or a deep stall (see the next paragraph). In a yawed attitude, unequal vortex flow across the vertical tail can cause undesirable directional stability/control characteristics.

9. DEEP STALLS.

"Deep" stall is a term that is occasionally misused. While some people refer to a deep stall as a stall in which the pilot maintains the wings in a stalled or partially stalled condition by keeping the stick or yoke

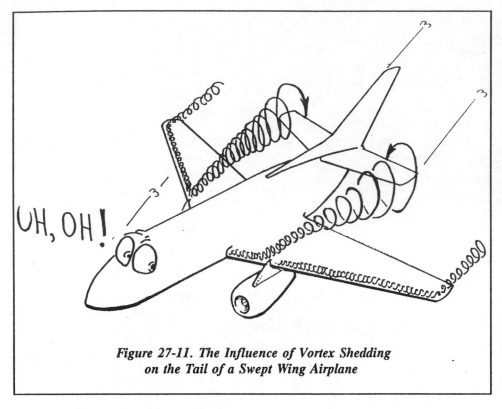

*Figure 27-11. The Influence of Vortex Shedding
on the Tail of a Swept Wing Airplane*

guessed it, even further reduced its pitch and yaw stability. Luckily, this aircraft's computerized, fly-by-wire flight control system could quickly be modified by reducing the maximum AOA achievable on these airplanes. This same airplane could be "tricked" into a deep stall by flying straight up and allowing the airspeed to drop off to virtually nothing. As the airplane pitched forward, its AOA could go well above that which the computer would have allowed and the airplane could enter what could be an unrecoverable deep stall, mushing downward until it impacted the ground. In this case too, the solution was found in the computerized fly-by-wire flight control system which could be tricked into providing control inputs which could break the stall.

at the full aft position, a true deep stall is a more complex condition. When an airplane is truly in a deep stall, the wing's angle-of-attack is well above the stall angle-of-attack and the combined aerodynamic forces and pitching moments generated by the airplane will prevent even full forward control movements from reducing the wing's angle-of-attack to break the stall. Unless some method is found to break the stall, the airplane will mush stall into the ground with a relatively low forward speed and a very high sink rate. Obviously a deep stall is a very dangerous and undesirable condition.

When a properly maintained airplane is operated in accordance with its Flight Manual or Pilot Operating Handbook, the airplane should not be able to enter a deep stall. However, it is sometimes possible to modify an airplane's pitch stability (more on this in Chapter 31) or maneuver an airplane into an attitude which allows it to be placed in a deep stall. For instance, if an airplane's whose center-of-gravity is well aft of its aft limit is fully stalled, it may settle into a deep stall. One version of a USAF fighter was found to be able to enter a deep stall following a design change which slightly reduced its pitch and yaw stability. This was combined with an external stores configuration which further reduced its pitch and yaw stability, and the airplane was operated at high altitudes and very high angles-of-attack which, you

The first diagram shown below provides an illustration of how full aft, neutral and full forward pitch control inputs could generate nose up and down pitching moments at AOAs from zero to well above the stall AOA. From this diagram you should see that the nose can be lowered to break the stall, even when the AOA is well above the stall. This airplane can't be deepstalled (at least at the AOAs shown). The second diagram represents an airplane which can be deep stalled. Once the airplane's AOA exceeds 21 degrees, even full forward control inputs cannot generate the nose down pitching moment necessary to break the stall.

One of the design features which may induce deep stall is aft mounted engines. In order to balance the weight of the engines at the aft of the airplane, the wings have to be mounted fairly well aft on the fuselage. This leaves a long length of fuselage extending in front of the leading edge of the wing. At high AOAs, the fuselage can generate vortices which are shed off the sides of the fuselage. The vortices rotate in the same direction as their respective wing-tip vortices. The two vortices cause a general down flow of air above the aft centerline of the fuselage. If the horizontal tail becomes enveloped in this down-flow, the resulting nose up moment may be enough to prevent the nose from being lowered. Although generalizations such as those just mentioned are possible, the stability issues which cause deep stalls is more complex and the assistance of specialists, and test flights is most often

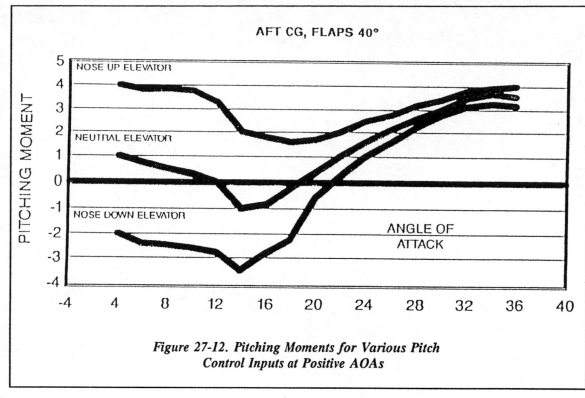

Figure 27-12. Pitching Moments for Various Pitch Control Inputs at Positive AOAs

necessary to determine the exact factors involved in the phenomena.

10. INFLUENCE OF YAW IN STALLS IN SWEPT WING AIRPLANES.

When a swept wing airplane is yawed, the maximum coefficient of lift the airplane can achieve will be reduced, therefore increasing the airplane's stall speed. This is due in part to the decreased sweep angle of leading wing (the wing on the side toward which the airplane is skidding), the increased sweep angle of the trailing wing (the wing on the same side as the direction of the yaw, and the displacement of the ailerons (or spoilers) necessary to prevent the airplane rolling in the direction of the yaw. The concept is somewhat tedious in following through, but fairly simple from an aerodynamic perspective. If a swept wing airplane is yawed to the right, the sweep angle of the left wing will decrease and the sweep angle of the right wing will increase. Refer to Figure 27-13. Sweeping

a wing has the effect of reducing the C_L of the wing at any given angle of attack. Therefore, the left wing will produce more lift than the right wing. In addition, the airflow over the right wing may be partially blocked by the fuselage, resulting in an additional loss of lift on this side. If un-corrected, this will cause the airplane to roll into the direction of the yaw (yaw right - roll right).

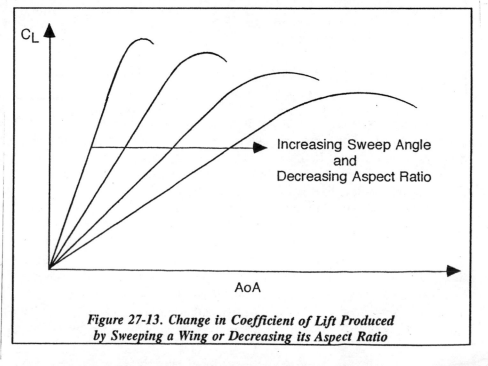

Figure 27-13. Change in Coefficient of Lift Produced by Sweeping a Wing or Decreasing its Aspect Ratio

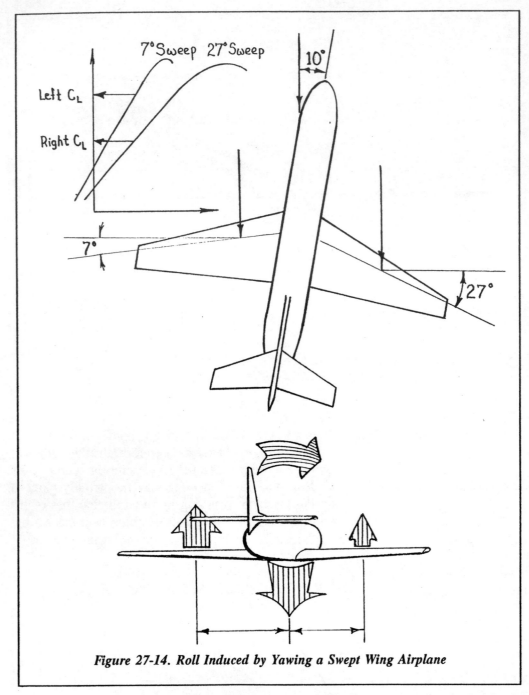

Figure 27-14. Roll Induced by Yawing a Swept Wing Airplane

the boundary layer air near the wing-tip, and increase the wing's maximum coefficient of lift. When coupled with the deflection of the ailerons, the down aileron on the trailing side effectively increases its AOA toward an already reduced stall AOA. On the other side of the airplane, the upward deflected aileron has increased the margin between the AOA and the stall AOA.

In addition, the reduced sweep of the leading wing will cause a reduction in induced drag due the effective reduction in this wing's aspect ratio. The induced drag generated by the trailing wing will increase even more. Coupled with the increased drag generated by the downward deflected aileron on the trailing wing and the reduced drag created by the upward deflection of the leading wing, the net effect is to increase the yaw angle of the airplane, further compounding the problem. The end result of this situation could be the stalling of the wing tip of the trailing wing and the resulting rapid roll of the airplane in this direction. This could be unrecoverable if it occurred at the low altitudes associated with takeoffs and landings.

Of course we have vertical tails on most airplanes, one of the purposes of which is to prevent such occurrences. Some airplanes even have autopilots and yaw dampers which can further reduce the potential for such an occurrence. However these devices have sometimes failed to do their job. In addition to the obvious, but thankfully rare departure of the tail due to a structural failure (yes, it has happened) inflight

Down aileron on the trailing side and up aileron (or spoiler) on the leading side are necessary to keep a wings level attitude. If the airplane were at a speed close to the stall speed when this all started, a potentially hazardous situation could develop. The aftward sweep of the trailing wing will cause an increase in the spanwise flow toward the wing tip. This will thicken and de-energize the boundary layer of the air flowing over the wing-tip, increase the propensity of the wing to stall, and reduce the wing's maximum coefficient of lift. On the other hand, sweeping an aft swept wing toward the un-swept position will decrease the spanwise flow toward the wing-tip, increase the energy of

icing can cause the vertical tail to stall, vastly reducing its effectiveness in providing directional stability for the airplane. Some failure modes (hopefully resulting from more than a single component failure) could also contribute to such an occurrence.

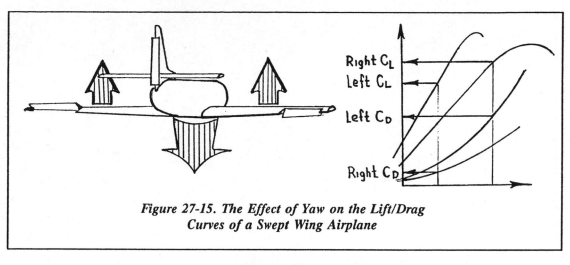

Figure 27-15. The Effect of Yaw on the Lift/Drag Curves of a Swept Wing Airplane

11. SPINS.

If there is anything a novice pilot fears more than stalls, it is probably spins. At one time all student pilots had to demonstrate proficiency in spins before they were awarded their private pilot's certificate. Aspiring commercial pilots had to demonstrate an even higher level of proficiency before they were given their ticket. Today only the candidate CFI has to have an endorsement in their log book which verifies their competency in spins. The emphasis during training is now on recognizing and preventing the conditions which can cause a spin to occur. In addition, design improvements have resulted in aircraft which are more docile and don't snap off into a spin at the slightest provocation. In fact most general aviation airplanes have to have pro-spin controls continuously applied in order for the airplane to remain in a spin. Simply releasing the controls in these cases will break the spin. Nevertheless, spin accidents still occur. Some pilots will apparently hold pro-spin controls all the way into the ground. Some airplanes will continue to spin because their pilots failed to initiate the correct recovery controls in the correct sequence and at the correct timing. Others will spin into the ground because the recovery could not be completed before the ground arrived. Yet others will spin when the pilots allowed them to be flown when the center of gravity was out of limits or modifications to the airplane invalidated the original airworthiness of the airplane.

A spinning airplane is best described as an airplane whose wings are experiencing an aggravated stall and whose resulting aerodynamic forces cause the airplane to "autorotate", a condition where the airplane is continuously rolling and yawing at a rate where the pitch and bank angle are relatively constant. In a fully developed spin, the aerodynamic forces on the airplane are balanced by the inertial forces created by the rolling and yawing motion.

It is fairly easy for an investigator to identify the wreckage of an airplane that has "spun in". There is little or no evidence of forward motion. Although the fuselage probably impacted at a steep nose down attitude (flat spins are an exception.), it is likely that there is evidence of a wing tip striking the ground before the nose. The down-going wing will normally strike the ground before the up-going wing, providing one clue as to the direction of the spin. Both the fuselage and the wings will probably have damage which reflects both a high sink rate and yaw. Tall, thin objects on the ground, like trees and fence posts, are likely to penetrate the airplane almost from bottom to top, reflecting the almost vertical trajectory of the airplane. Undamaged objects may be found immediately behind the trailing edges, again indicating the vertical path of the airplane.

The flight path of a spinning airplane will normally follow a helix whose axis is oriented vertically. Imagine a long, loosely wound coil spring hanging from the sky. Perhaps a loosely wound spring or "Slinky" dangling out of a cloud. Now imagine an airplane sliding down the spring, its fuselage yawing to the left or right in order to remain aligned with the spring. The airplane is also banked toward the center of the spring constantly rolling toward the inside wing but at a rate which keeps the bank angle from increasing. Although this text will describe a fully developed spin as having relatively constant pitch and bank attitudes and relatively constant yaw and roll rates,

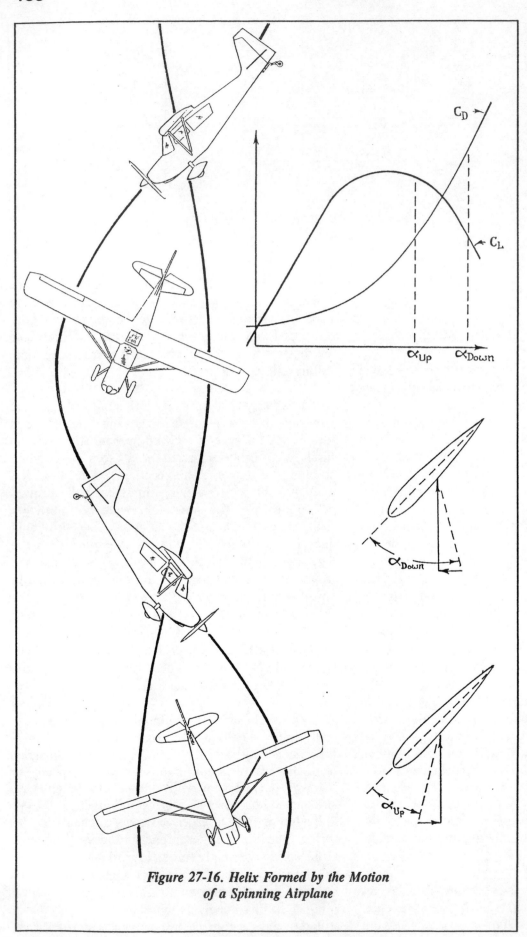

*Figure 27-16. Helix Formed by the Motion
of a Spinning Airplane*

some military jet fighters have a fully developed spin mode which is anything but constant and looks more like a series of tumbles, rolls, and extreme yaws. Although the dynamics associated with this type of motion are much more difficult to analyze, the aircraft's ultimate trajectory will still be vertical, with the wings stalled, the aircraft yawing and rolling into the direction of the spin.

12. THE NECESSARY INGREDIENTS.

In order for a spin to occur the wings must stall and the nose must yaw. Most spins result when the critical or stall AOA is exceeded while the pilot is performing an uncoordinated turn. Too much back stick or yoke and either too little or too much rudder. In a skidding turn where too much bottom rudder (or not enough top rudder if top rudder is used) the spin will be in the direction of the turn, or under the bottom. Bank increases rapidly and the pitch attitude will decrease rapidly. Altitude loss will be rapid during the first turn and if the spin was entered at traffic pattern altitudes, recovery is unlikely before ground impact. In a slipping turn where too much top rudder (or not enough bottom rudder if bottom rudder is necessary to maintain coordinated flight) is used the spin

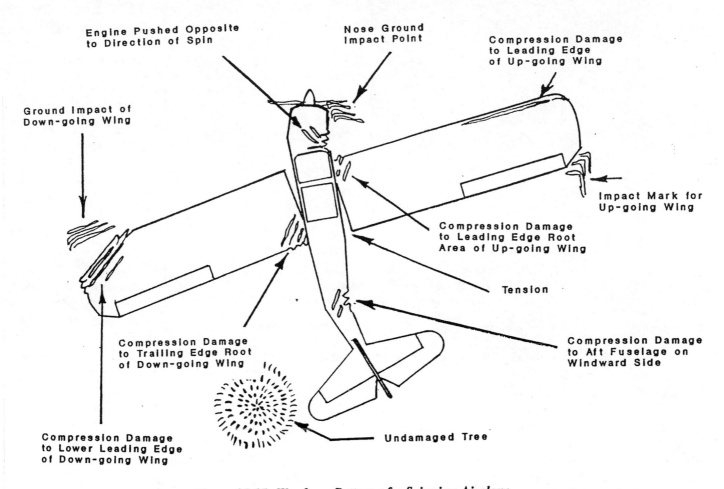

Figure 27-17. Wreckage Pattern of a Spinning Airplane

will be in the direction opposite to the initial turn, or over the top. Bank decreases rapidly toward, and then through, wings level. The nose will not drop rapidly at first and may even come up a little. If the pilot is quick enough to recognize the error, relax the back pressure to break the stall and coordinate aileron and rudder to stop the roll, the pilot may recover with little loss of altitude. However, if the pilot fails to react almost immediately and maintains the initial control positions, the aircraft will continue to roll and yaw, the nose will drop and the airplane will end up flying a similar flight path as the airplane which entered the spin from a skidding turn. The direction of this spin will be opposite to the direction of the first spin described.

13. THE INCIPIENT SPIN OR POST STALL GYRATION.

Between the time when an airplane first stalls and the time when the airplane enters a fully developed

spin, the airplane will normally go through a series of transient pitch, roll and yaw movements. These unsteady motions could last from one to over three rolling, yawing turns before motion settles down to the relatively constant yaw and rolls rates which describe a "fully developed" spin. Four to six turns may be required before the airplane settles down to its maximum steady state roll and yaw rates. The nature of these "post-stall gyrations" depends on a number of different factors; the aerodynamic and inertia properties of the airplane, airplane weight, CG, power setting and speed, and the position of the flight controls, to name some of the more important. The behavior of the airplane during the post-stall gyration can change dramatically if one of these factors is changed. For instance, if an airplane is in a skidding turn when the stall occurs, bank angle will probably increase suddenly and the post stall gyration could immediately involve extreme nose low attitude. On the other hand, if the airplane is in a slipping turn, bank will suddenly decrease and pitch attitude may actually increase mo-

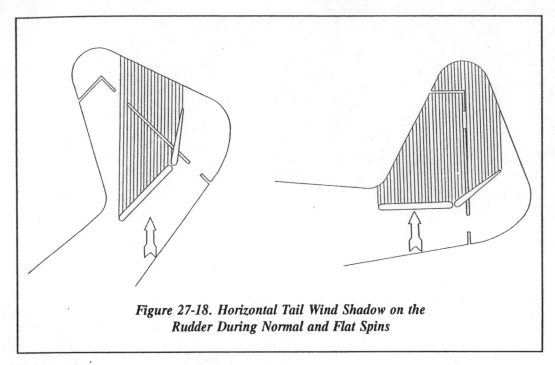

Figure 27-18. Horizontal Tail Wind Shadow on the Rudder During Normal and Flat Spins

described earlier. A "flat" spin differs from a "normal" spin in several ways. First, the AOA of the flat spin will be higher. While the average AOA in a normal spin may be 45°, the average AOA in a flat spin may be at or above 70°. This difference in AOA can cause the airflow over control surfaces during a flat spin to be quite different from those in a normal spin. Controls which could effectively recover an airplane from a normal spin may be ineffective in a flat spin.

A flat spin will occur if the aerodynamic forces trying to push the nose down are overpowered by the gyroscopic forces attempting to pull the nose toward the horizon. In a flat spin the axis of the helix about which the airplane is spinning moves closer to the airplane's center of gravity and the radius of the helix is smaller. Because the pitch attitude of the airplane is higher, the roll rate into the spin is lower and the airplane's motion will be more characterized by yaw than the rolling yaw of a normal spin. If the attitude of the airplane becomes completely flat, the yaw rate will increase and the airplane's motion can become more erect. The vertical tail will stall and the airplane's wings can act similar to the rotors of a helicopter during an autorotation, providing much of the autorotational force necessary to maintain the spin. It is interesting to note that the descent rate of a flat spin can be less than the descent rate in a normal spin. However, an airplane which can recover from a normal spin may not be able to recover from a flat spin. The wreckage pattern of an airplane which failed to recover from a flat spin would exhibit less damage associated with nose down attitudes and more damage associated with yaw rotation. In addition, the flat spin crash would have even less evidence of damage associated with forward velocity, and the normal spin didn't have much.

mentarily as bank goes through the wings level attitude. An airplane with an extremely forward CG may resist entering the spin while an airplane with a CG close to or behind the aft limit may jump at the opportunity to enter the spin. The only way to predict what the maneuver will involve is to try it under very controlled conditions (test pilot time), or locate reports which documented the conditions in which you are interested.

14. THE FULLY DEVELOPED SPIN.

A spin is commonly defined as an aggravated stall which results in autorotation. The AOA of both wings will be well above the stall AOA and the airplane will, on its own, maintain a roll and yaw into the direction of the spin. The fully developed spin can be further classified by the bank attitude (erect or inverted) and the pitch attitude (flat, normal or accelerated).

By the way, although the motion of an airplane in a spiral dive may resemble that of a spin, there are important differences. the wings are not stalled, the airspeed is higher and the aircraft will respond normally to control inputs. When the a spiraling airplane impacts the ground the impact angle will be shallower, the energy higher and the wreckage pattern spread further along the direction of the flight path vector.

15. THE FLAT SPIN.

The characteristics of normal, erect spins were

If an accident investigator can distinguish between

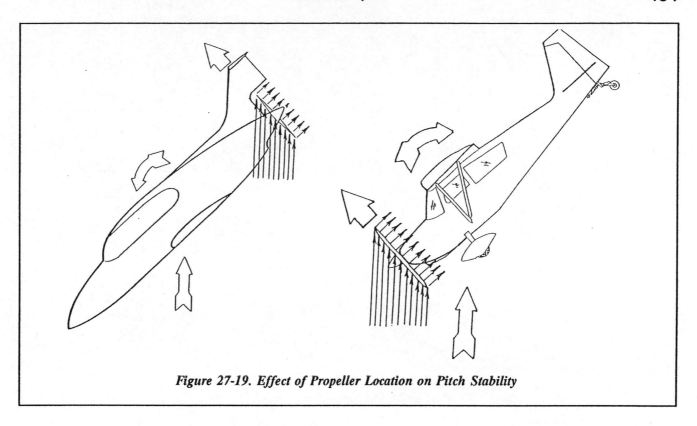

Figure 27-19. Effect of Propeller Location on Pitch Stability

a crash involving a normal and a flat spin and knows the flat spin may be more difficult or even impossible to recover from, the next question should be, "What makes an airplane enter a flat spin?" Aside from the basic design of the airplane, over which the pilot normally has little control, there are two pilot controlled factors which can encourage a spinning airplane to transition into a flat spin.

The first is flying with a excessively aft center of gravity. It should be remembered that center of gravity limits are expressed in terms of percent mean aerodynamic chord (MAC). And on small general aviation airplanes and especially some home built aircraft, the range of allowable center of gravity travel can be easily exceeded by careless loading of baggage, passengers and fuel.

The second factor that could encourage a normal spin to transition into a flat spin is power. Aircraft with propellers in the front (often called a "tractor" configuration) are destabilizing in pitch. In other words, location of a propeller forward of the aircraft's CG will cause changes in pitch to be exaggerated. At high angles of attack, the air passing through the propeller arc is deflected downward, resulting in an upward force. See Figure 27-19, "Effect of Propeller Location on Pitch Stability". Since, in a tractor design, this force is forward of the CG, the nose will be

pitched up as power (thrust) is increased and pitched down as power (thrust) is decreased. In virtually all airplanes of the tractor configuration, pulling the power to idle is normally the first step in the recommended spin recovery procedure. Failure to pull the power "off" may hold the nose in an unusually high attitude, precluding recovery from the spin.

One "hangar story" has an experienced aerobatic pilot in an unrecoverable spin when he can't reduce the power because the throttle linkage has failed. Being a quick learner and understanding the principles involved, our astute aviator pulls out on the mixture, shutting down the engine until the spin is broken. With pusher configurations, this procedure may be totally wrong! The change in momentum imparted by the propeller is stabilizing in pitch, and at high angles of attack will produce a pitching moment about the CG which will tend to reduce the angle of attack.

The gyroscopic forces created by an airplane engine at high power settings may also be a player in determining the angle of attack of a spinning airplane. If an airplane is in a left spin, the precessional forces created by an engine with clockwise (as viewed by the pilot) rotating propellers will cause the pitch to increase. Increasing power will increase the precessional force, pushing the nose to a higher pitch attitude an perhaps a flat spin. Since "P" factor generated by a

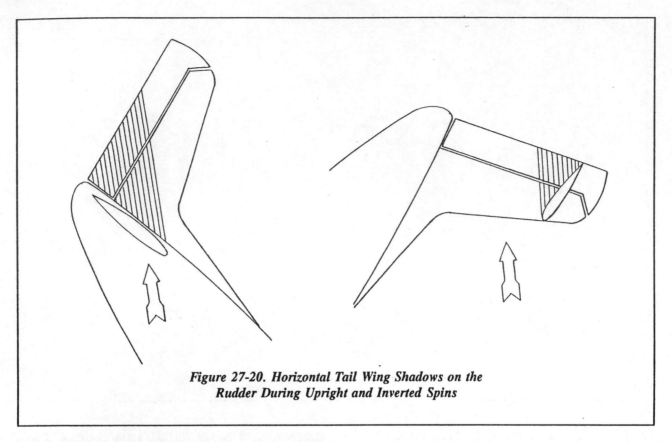

*Figure 27-20. Horizontal Tail Wing Shadows on the
Rudder During Upright and Inverted Spins*

propeller rotating in this direction also pushes the nose to the left, increasing the power while in a left spin may be more dramatic than increasing the power during a spin in the other direction. The first step in most spin recovery procedures is to retard the power. Evidence concerning center of gravity location and engine power settings are important when investigating accidents involving failure to recover from flat spins.

16. INVERTED SPINS.

If an airplane experiences an aggravated negative G stall (the wings are at a negative angle of attack) and sufficient auto-rotational forces are present, the airplane can enter an inverted spin. The most probable reason for a non-aerobatic pilot to encounter a negative G stall is the use of excessive forward stick (yoke) while recovering from an erect spin. If the rudder is already being applied (as part of the erect spin recovery attempt) as the negative G stall is encountered, conditions are set for the inverted spin. Just because an airplane can recover from an erect spin doesn't mean that it can recover from an inverted spin. During an inverted spin the airflow patterns across the rudder will be very different. The wind shadow of the horizontal tail will be totally different and could eliminate the effectiveness of the rudder in stopping the spin rotation.

The wreckage pattern of an inverted spin should be very obvious to the accident investigator. If the airplane is not equipped with an inverted fuel system, the engine should not exhibit the characteristics of an engine producing power. Unless the spin occurred at very low altitude, there will probably be little evidence of rotational damage to the propeller. The sustained outward forces generated during a multi-turn upright spin can also cause the engine to quit due to fuel starvation. In this latter case, however, it is due to the un-porting of the fuel tank pickup points located in wing roots being starved of fuel when the fuel is flung toward the wing tips.

Most erect spin recovery techniques for subsonic airplanes involve some variation of the PARE tech-

POWER	IDLE
AILERONS	NEUTRAL
RUDDER	OPPOSITE THE DIRECTION OF TURN
ELEVATOR	FORWARD OF NEUTRAL

*Figure 27-21. The PARE Procedure
for Spin Recovery*

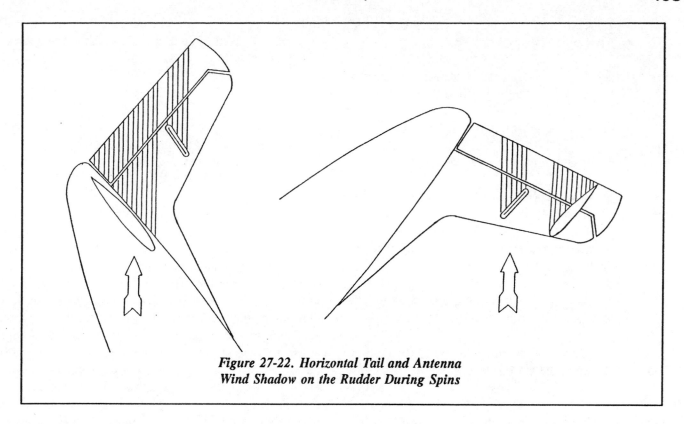

Figure 27-22. Horizontal Tail and Antenna Wind Shadow on the Rudder During Spins

nique. This involves first reducing engine power to idle. This action will normally cause an immediate reduction in the airplane's angle of attack. The reduction in the airflow over the elevators will reduce their effectiveness in maintaining a stall angle of attack. The reduction in the nose-up pitching moment created by the downward deflection of the air passing through the propeller arc is also reduced. Aileron is next neutralized, this time reducing the roll and adverse yaw contributions which could assist in establishing the spin. Next full rudder is applied opposite to the direction of the spin. Hopefully, with pro-spin inputs eliminated, anti-spin rudder will stop the autorotation. Finally, the stick (yoke) must be moved to a position forward of neutral, ensuring that the stall has been broken.

17. DETERMINING THE REASONS FOR SPIN ENTRY.

Once it has been determined that an airplane crashed while it was in a spin, the investigation should attempt to determine the factors that caused the airplane to enter the spin and, if sufficient altitude was available to recover from the spin, and why a safe recovery was not performed.

In attempting to determine the causes for entry into the spin both the likelihood of the attempted maneuver to excite a spin and the susceptibility of the airplane to enter a spin need to be examined. Some of the factors which should be addressed are listed below.

1. Did the maneuver involve high angles of attack and yaw?

2. Was the pilot's skill level adequate for the maneuver?

3. Were high power settings used during the maneuver?

4. What was the configuration of the airplane when entering the stall?

5. Where was the center of gravity of the airplane?

6. Were there any malfunctions in the flight control or autopilot systems?

When the investigator attempts to discover why the pilot failed to recover from the spin prior to ground impact, this line of questioning should extend down at least two paths; were the pilot's skills capable of recovering in the altitude available and was the airplane capable of recovering in the altitude available?

1. What was the pilot's training concerning recovery from post stall gyrations?

2. What was the pilot's training concerning recovery from fully developed spins?

3. How valid was this training?

4. How recent was this training?

5. How applicable was this training to the type of airplane being flown?

6. Are the instructions for spin recovery contained in the POH/FM?

7. Are the instructions contained in the POH/FM complete and valid?

8. How much altitude was required to recover from the post stall gyration phase?

9. How much altitude was required to recover from the fully developed spin phase?

10. Had system modifications to the airplane degraded its ability to recover from:
    a post stall gyration?
    a fully developed spin?

11. Had aerodynamic modifications to the airplane degraded its ability to recover from:
    a post stall gyration?
    a fully developed spin?

17. SUMMARY.

Investigation of stall/spin accidents starts with recognition of the stall/spin impact characteristics and determination of the type of stall or spin. With that knowledge, it is possible to use a logical approach to determine why the stall or spin occurred, why the sequence was not interrupted by the pilot, and whether recovery was possible. Accidents involving fully developed spins are rare. However, accidents involving stall and post stall gyrations still plague aviation and point to a continuing need for attention on the man-machine interface during these critical phases of flight.

28

Downwash and Wingtip Vortex Hazards

1. INTRODUCTION.

Induced drag is a necessary price for the creation of lift by any wing. Accompanying this induced drag are vortices trailing behind the wing tips and a downwash behind the wing. The effects of these vortices and downwash create hazards which have been involved in accidents and incidents. Vortices left by aircraft have caused trailing aircraft to roll out of control or to be exposed to excessive G loads.

The first part of the discussion concerning wing tip vortices will deal with the nature of the beast; what it is, when it is, and how long does it stay around to feed. It should be pointed out that there are three factors which determine the magnitude of risk to trailing aircraft associated with wing tip vortices: the intensity of the winds in the vortex, the size (diameter) of the vortex and the length of time the vortex exists. Although helicopter rotor blades generate similar hazards, this discussion will limit itself to analyzing vortices formed by airplane wings.

The second portion of this section on vortices will concentrate on the specific hazards associated with wing tip vortices; i.e. how can it eat an airplane up and spit it out. A close cousin of wing tip vortices is downwash. Althoug the effects of downwash are not nearly as dangerous as vortices, these effects have exacerbated difficulties associated with takeoffs and landings performed during emergency situations.

2. FACTORS DETERMINING THE MAGNITUDE OF VORTICES AND DOWNWASH.

Wing tip vortices form when relatively high pressure air on the bottom of the wing "leaks" toward the relatively low pressure air on the top of the wing. When coupled with the aircraft's forward velocity this "leakage" results in a swirling flow of air trailing

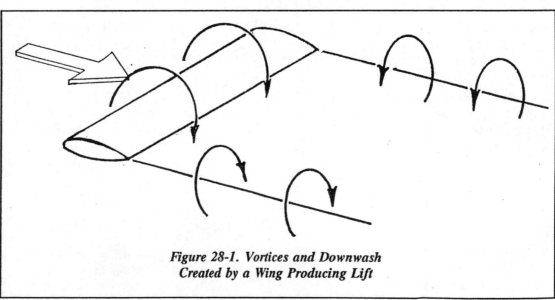

Figure 28-1. Vortices and Downwash Created by a Wing Producing Lift

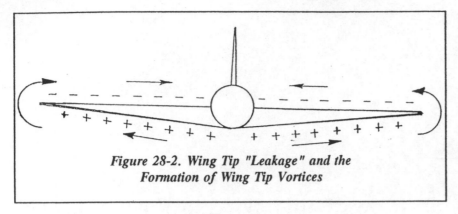

Figure 28-2. Wing Tip "Leakage" and the Formation of Wing Tip Vortices

behind each wing tip.

Basically, there are four factors which determine the intensity of these wing tip vortices; 1) the amount of lift, 2) wing span, 3) air speed and 4) air density.

3. LIFT MAGNITUDE VERSUS TIP VORTEX INTENSITY.

As the lift produced by a wing increases, the intensity of the wing tip vortices produced also increases. In other words, the kinetic energy associated with the vortex increases. Likewise, if the lift produced by a wing decreases, the intensity of the vortices produced by the wing also decrease. A wing which is producing no lift does not produce vortices. By the way, in one G flight the lift produced by a wing is approximately equal to its weight (actually a little more due to the down load created by the horizontal tail). All other things being equal, the heavier of two aircraft of the same model will produce vortices of greater intensity. However, if the lighter of the two were to be pulling enough Gs to cause the lift produced by the wings to exceed the weight of the other airplane, its (the lighter of the two) vortices would be stronger. Thus, it is not necessarily the weight of the airplane that's important, it's the lift (weight multiplied by the G load) that's important. We tend to lose sight of this factor because the most likely, and probably the most dangerous place

to encounter wing tip vortices is over the ends of the runway where the airplane is in one G flight.

4. SPAN VERSUS TIP-VORTEX INTENSITY.

The next factor to be considered is the span. The larger the span, the smaller the intensity of the vortices. That's because the low pressure area which produces lift is spread over a greater span. The total energy in the vortex is spread over a greater area and thus its intensity is decreased. The lift and span factors are often combined into a single term referred to as span loading, expressed normally in terms of pounds of airplane weight per foot of wing span. As span loading increases, the intensity of the wing tip vortices increases. As span load decreases, the intensity of the wing tip vortices decreases. We can increase the span loading of an airplane by either increasing the weight of the airplane, increasing the G load on the airplane, or decreasing the wing span of the airplane (by sweeping the wings or by flying between two trees). The importance of span loading is not always recognized. A smaller, lighter airplane with a small wing span can produce just as intense a vortex as a larger, heavier airplane with a large wing span. The vortices of the smaller airplane will be smaller in size and therefore less probable to encounter, but the intensity of the wind in the smaller, lighter airplane's vortices can be just as strong as vortices produced by the heavier, larger airplane. Concern about wing tip vortices should not be limited to those produced by "very heavy" airplanes.

5. AIRCRAFT SPEED VERSUS TIP-VORTEX

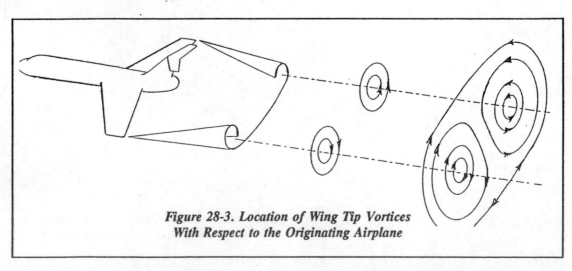

Figure 28-3. Location of Wing Tip Vortices With Respect to the Originating Airplane

INTENSITY.

The next factor to be considered is the airplane's speed. The slower an airplane is flying, the greater the intensity of its wing tip vortices. This is related to the higher angles of attack associ-

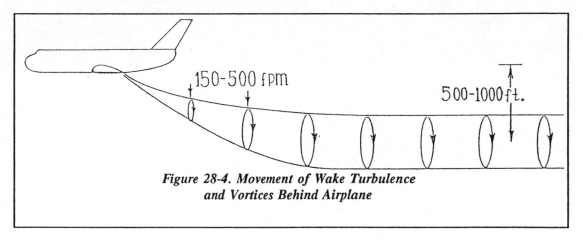

Figure 28-4. Movement of Wake Turbulence and Vortices Behind Airplane

ated with slow speeds, and the higher downwash angles these high angles of attack produce. Another way of thinking about it is to consider that the disturbed air necessary to keep the airplane in the air is spread over a large distance when the airplane is moving fast, and concentrated when the airplane is moving slow. Again, the hazard of wing tip vortex encounters shows up in its most severe condition over the ends of the runway, just where it is most likely to be encountered.

6. AIR DENSITY VERSUS TIP-VORTEX INTENSITY.

The last factor to be considered is air density. If the air is less dense, the intensity of the vortex will be less. In addition, since the dynamic pressure (or "q") of an airstream is a function of airspeed and density, air speed and air density can be combined in a single parameter, dynamic pressure.

7. FACTORS AFFECTING THE SIZE AND MOTION OF TIP-VORTICES.

As the tip-vortex forms it trails behind the aircraft, both growing in size and descending to a position below the aircraft's flight path before dissipating. A general understanding of the movement and growth in size of the vortex is important in the understanding the hazards/risks associated with wing-tip-vortices.

The vortex first forms as a downwash behind the wing and a spanwise flow around the wing tip. It then trails aft of the tip. The axis of the vortex remains parallel to the longitudinal axis of the airplane but moves inward until the vortex is about 80% of the way from the fuselage to the wing tip. The degree of rotation in left and right vortex increases until, at a distance of about two to four wing spans aft of the wing, complete rotation has occurred. At this point the whirlpool-like flow has been established. As the vortices trail further behind the airplane they grow in

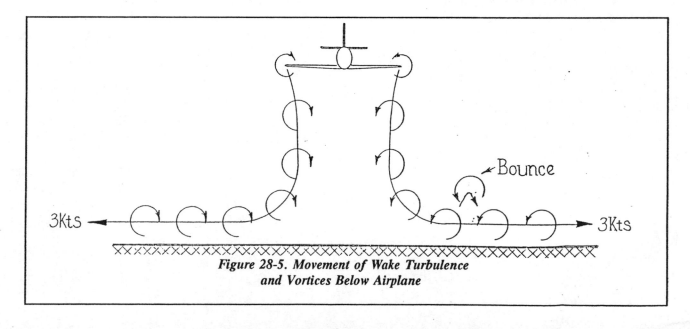

Figure 28-5. Movement of Wake Turbulence and Vortices Below Airplane

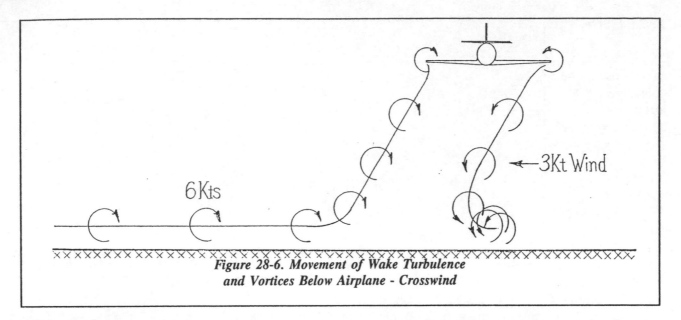

*Figure 28-6. Movement of Wake Turbulence
and Vortices Below Airplane - Crosswind*

diameter until they begin to interact with each other. As this happens they normally begin to descend below the original flight path of the airplane. The diameter of the combined vortices will grow to about two wing spans. The air at the center of the vortex will have significantly higher speeds, but be much smaller in diameter. Tangential velocities of approximately 325 feet per second have been reported for a 757 while descending for landing while in the landing configuration. The tangential speeds of the air in the surrounding portions of the vortex are significantly less. The diameter of the highest speed air in the vortex core will be from about one tenth, to only one twenty-fifth of the wingspan. For smaller airliners with spans around 100 feet (e.g. MD-80, B-727 and B-737) the vortex "core" would range from 4 to 10 feet in diameter. For the "big birds" with spans approaching 200 feet (e.g. B-747, A-330 and A-340) the vortex "core" would range from 8 to 20 feet.

In nice stable air the vortices could descend at a rate from roughly 500 fpm for a heavy jet transport aircraft, 350 fpm for a medium sized jet to a slower rate of around 150 fpm for a light transport airplane. The exact rate will be a function of the airplane's dimensions, weight, configuration and speed as well as the local atmospheric conditions and density altitude. The downward motion of the vortex will be resisted by the surrounding air and eventually the vortex will cease its vertical motion. Although 900 feet of vertical travel is a common example, airplane specifics and atmospheric conditions can cause this number to vary both upward and downward. The vortices of a 270,000 pound KC-135 at 310 KIAS are reported to descend at 1,500 feet per minute when the airplane is in one G flight at an altitude of 25,000 feet. The vortices' rate of descent will also be influenced by the airplane's proximity to the ground. As an airplane approaches the ground and enters ground effect, the vortices rate of descent will decrease. In addition, as vortices generated at higher altitudes descend toward the ground, their rate of descent will slow as they approach a height of one-half wingspan and begin to move apart. Under some atmospheric conditions the vortex has been observed to "bounce" back into the air, moving upward to a height of as much as two wing

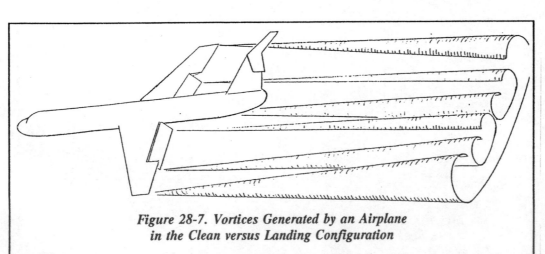

*Figure 28-7. Vortices Generated by an Airplane
in the Clean versus Landing Configuration*

spans. If the vortices descend to the ground and do not bounce, they will probably move away from each other at a total speed of 6 knots (the left vortex moving leftward at 3 knots and the right vortex moving to the right at 3 knots). If there is a cross-wind when the vortices reach the ground, the

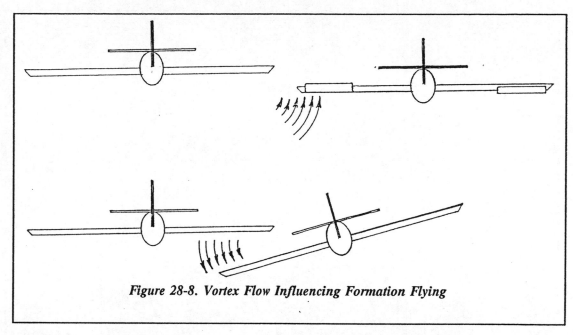

Figure 28-8. Vortex Flow Influencing Formation Flying

movement of the vortices through the air mass could cause movement across the ground to be different. For instance, with a 3 knot left cross-wind, the left vortex could remain stationary over the runway while the right vortex moves to the right at 6 knots. Any cross wind component in the region of 3-8 knots should alert the investigator to the possibility of wake vortex involvement.

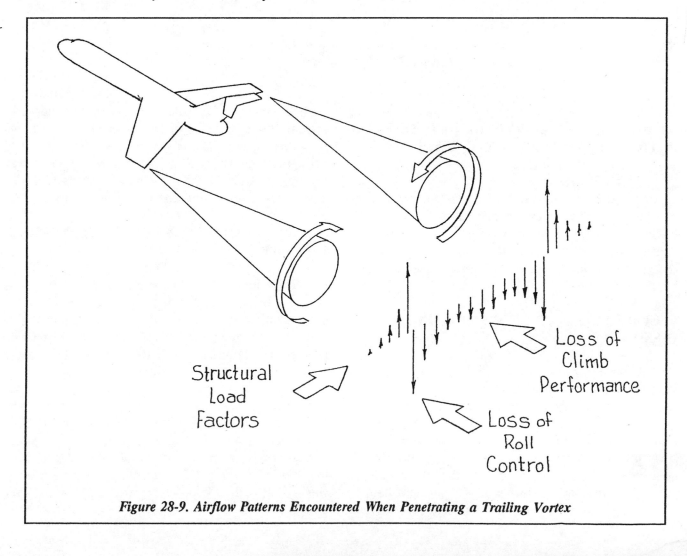

Figure 28-9. Airflow Patterns Encountered When Penetrating a Trailing Vortex

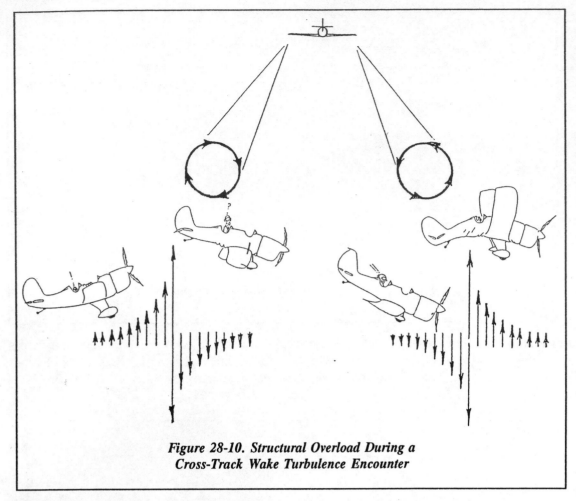

*Figure 28-10. Structural Overload During a
Cross-Track Wake Turbulence Encounter*

supporting wing mounted engines, extension of wing spoilers and wing leading edge irregularities can also send additional vortices trailing aftward. All these additional vortices add confusion to the flow trailing the airplane and tend to result in the tip vortex breaking up sooner than it otherwise would. High engine power settings can also add turbulence and heat, further reducing the life time of the vortex. Thus, landing behind an airplane making a no-flap landing and using low power settings will increase the probability of encountering a vortex. On the other hand, landings behind aircraft in dirty configurations and using a relatively high power setting may establish a false sense of security (complacency) concerning wing tip vortex encounters.

8. FACTORS AFFECTING THE RATE OF DISSIPATION OF WING TIP VORTICES.

Since wing tip vortices are only a hazard so long as they exist, the factors which control the rate of dissipation of the vortex are of interest to the investigator. Basically there are two factors (the airplane's configuration and atmospheric conditions) which determine the amount of time it takes for the vortex to break up and dissipate.

9. EFFECT OF CONFIGURATION ON RATE OF DISSIPATION.

When an airplane, whose wings are uncluttered with leading edge discontinuities and pylons, is in its cruise or clean configuration, there tends to be just one vortex trailing from each wing. There is no competition for the available energy. On the other end of the spectrum, a wing with landing flaps extended will have additional vortices extending from the lateral ends of the flaps. The landing gear will also set up a trailing wake of air extending back into the vortex. The pylons

10. EFFECT OF ATMOSPHERIC CONDITIONS ON RATE OF DISSIPATION.

The other factor influencing a vortices life is the atmosphere. A vortex will hang around longer if a calm, neutrally stable atmosphere exists. Luckily, this condition doesn't normally exist and the life will be prematurely terminated. If we increase turbulence, wind shears or create inversions or non-standard lapse rates, the vortex will break up sooner. By the way, some studies have shown that wing tip vortices will maintain most of their intensity until just before they break up. Up to 90 % of its strength can be maintained until it has used up 85% of its life. You can't assume that if a vortex has lived out half its expected life, it has reduced in intensity to half its original strength.

Although vortices have been measured to last several minutes, the length of time is very much a function of the two variables we have just discussed. An airplane with a "dirty" wing design, in a landing configuration on a windy day will have its wake dissipate much sooner than a clean winged airplane in a cruise configuration on a calm day with neutral atmospheric conditions.

11. HAZARDS ASSOCIATED WITH WAKE VORTEX ENCOUNTERS.

Airplanes encountering wing tip vortices generated by other airplanes are subject to several different hazards. These hazards can be broken into two general areas; hazards associated with formation flight and flights through trailing vortices.

A. HAZARDS ASSOCIATED WITH FORMATION FLIGHT. The first case deals with two airplanes intentionally flying in formation, normally with one airplane in an echelon position relative to the other (see Figure 28-8). Examination of the vortex flow generated by the leading airplane shows that as the airplane flying the wing (trailing) position moves toward the leading airplane, the up flow portion of leader's vortices will initially cause the airplane flying the wing position to bank away from the leader. In order to maintain position, the pilot flying the wing position has to apply aileron controls which would, with greater lateral spacing, cause the wing airplane to bank into the leader. If however, the airplane flying the wing position moves in even closer, so that wing tip separation is not maintained, the wing airplane will encounter the down flow portion of the leader's airplane. If not immediately corrected by the pilot of the airplane flying the wing position, this airplane will roll toward the lead air-

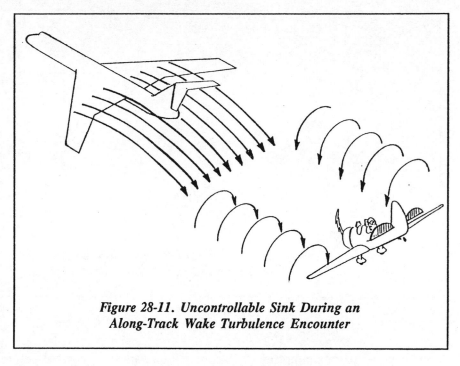

Figure 28-11. Uncontrollable Sink During an Along-Track Wake Turbulence Encounter

plane and a mid-air collision is a distinct possibility. Probably the most notorious of this type of accident was the collision of one of the two USAF XB-70s and a USAF F-104. The accident occurred during a routine mission designed to show a number of high performance USAF airplanes (one XB-70 and four fighter/trainer aircraft), all powered by engines produced by one of the major USAF jet engine manufacturers. All of the pilots involved in the flight were test pilots.

If highly experienced test pilots, the guys with the "right stuff", can lose control while flying in close

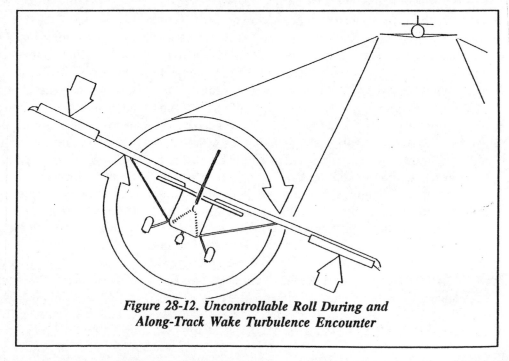

Figure 28-12. Uncontrollable Roll During and Along-Track Wake Turbulence Encounter

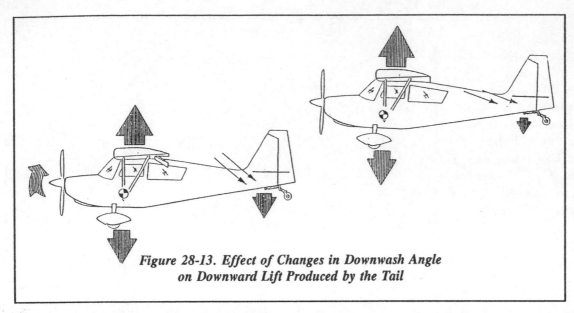

*Figure 28-13. Effect of Changes in Downwash Angle
on Downward Lift Produced by the Tail*

TRACK EN-COUNTERS WITH WAKE TURBULENCE.

The primary hazard associated with an airplane crossing at 90° and behind another airplane at or slightly below that airplane's altitude is over G. The airplane will encounter the vortices as a series on up, down, and then up again wind gusts. The airplane will first be pitched upward, then downward and finally upward before it is turned loose of the monster. The intensity of these gusts will be a function of the intensity of the original vortex, the rate of dissipation of the vortex, the distance the encounter occurs behind the originating airplane and the exact location of the penetration. It should be obvious that if the crossing airplane penetrates the area of the vortex containing the highest speed air, the encounter will be more violent than one where the crossing airplane barely penetrates the wake area. Another important factor to consider is the wing loading of the crossing airplane. The G load generated by a vertical wind gust is inversely proportional to the encountering airplane's wing loading.

High wing loads smooth out the ride, low wing loadings yield bumpy rides. Since the wing loadings of modern airliners can be ten times higher than the wing loadings of general aviation airplanes, if each were to encounter the same gust at the same speed, the general aviation airplane would experience a gust induced G load ten times higher than that of the airliner. However, the general aviation airplane will normally be flying at a lower speed. Since the gust imposed G load is also directly proportional to the airplane's speed, the G load experience by the light airplane will also be influenced by its speed.

If the light aircraft with a wing loading of only 10% of that of the airliners were flying at a speed that is one half the airliner's speed, it would experience a G induced gust load five times higher than that experienced by the airliner.

formation, it should not be surprising that novice pilots can make the same mistake. Perhaps these wannabe fighter pilots were attempting to learn to fly formation "on the cuff". Or maybe one pilot was attempting to check over a fellow airman's troubled airplane (perhaps the landing gear does not indicate all three are down and locked).

B. HAZARDS ASSOCIATED WITH TRAILING, LINGERING VORTICES. A more common, and insidious, hazard associated with wing tip vortices is an encounter further from the source. If we imagine the vortex flow field perhaps a mile behind the airplane which generated it, it would probably look similar to that shown in Figure 28-9. The two circulation flows join immediately aft of the airplane to create a relatively large area of general down flow. Two relatively small cores of air rotating with high tangential speeds separate the large down flow area from two smaller, up-flow areas near the outside of the turbulent area. What happens to an airplane entering this turbulent wake is very much a function of where the airplane enters the wake. Generally these encounters can be classified as cross-track or along track encounters. Of course there are all sorts of combinations, but let's keep it simple and look at only tracks which either intersect at about 90° or are parallel. During the investigations involving possible wake turbulence penetrations, it must be remembered that the location of the penetration is critical. Crossing the wake of an airplane at the edge of the vortices will be an entirely different experience than sticking your nose into the high speed air at the center of either vortex.

12. GUST-INDUCED OVER G DURING CROSS-

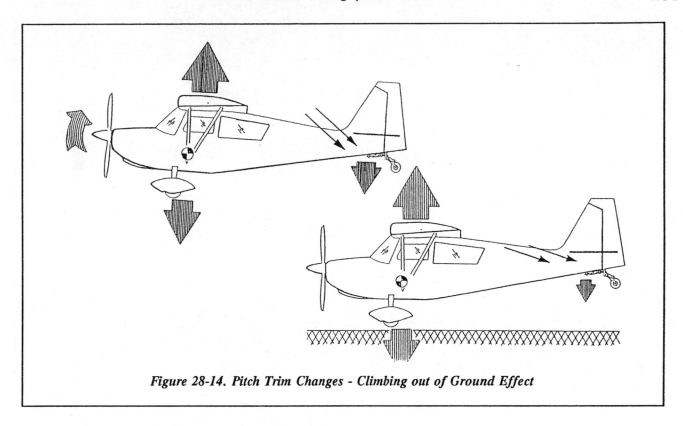

Figure 28-14. Pitch Trim Changes - Climbing out of Ground Effect

13. HAZARDS DURING CROSS-TRACK WAKE TURBULENCE ENCOUNTERS.

One other factor to consider when examining the hazard potential of this event is the control inputs of the pilot. When the airplane first pitches nose up, it is not unreasonable to assume that the pilot will attempt to lower the nose. If the pilot does this as the airplane is entering the down gust, the ensuing nose down pitch will be even greater than that generated by the down gust. If the pilot again attempts to fight the gusts and is pulling aggressively on the yoke when the airplane enters the final up gust, the combined effects of the pilot induced G load and the gust induced gust load could cause structural failure. One last complication to add to this mixture. Remember "dynamic stalls" as discussed in Chapter 27. If the angle of attack of an airfoil increases fast enough, the stall angle of attack and C_{Lmax} can increase beyond those used to calculate V_a (maneuver speed). And, although pitch control authority on even high performance fighters is not sufficient to produce these rates, flying into a high speed, sharp edged gust associated with a wing-tip-vortex could do it. Several years ago the outer wing of a military fighter was bent sufficiently to put it out-of-control when it flew through another fighter's vortex. Thus even if an airplane is flying at a speed below maneuver speed, it could, if it ran into a very sharp

edged, strong gust, be over Geed!

Fortunately Mother Nature doesn't make gusts like this very often. Cross-track penetrations of a high span loaded, low speed airliner by a low wing loaded general aviation airplane can do it. They can cause the failure of the general aviation airplane's primary structure. The probability of structural overload and failure will be directly related to: the intensity and suddenness of the vortex encounter; the wing loading and speed of the airplane penetrating the vortex; and the ability of the airplane's structure to dampen the load.

14. HAZARDS GENERATED BY ALONG-TRACK ENCOUNTERS WITH WAKE TURBULENCE.

Loss of roll control and degraded climb performance are the two different hazardous areas for "along track" encounters with wing tip vortices. Looking back to the illustration showing the flow patterns encountered by a airplane following in the wake turbulence generated by another airplane, you should be able to see three general patterns. A general up flow outboard of the leading airplane's wing tips. That's not really a problem. A general down flow immediately behind the leading airplane. This could be a problem if the down flow exceeds the trailing airplane's ability to maintain its desired flight path. Finally in the area aft and

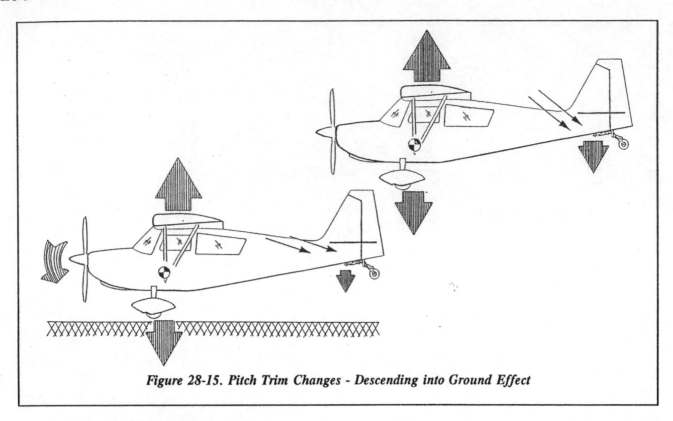

Figure 28-15. Pitch Trim Changes - Descending into Ground Effect

perhaps a little inboard of the wing tips lies the core of the vortex and its swirling mass of air similar to a mini-tornado. If an airplane flies into this tornado-like air mass while flying parallel to its axis, the rolling motion imparted by the flow can exceed the roll control of the airplane and it will roll out of control with the flow.

Lets go back to the area of high sink immediately behind the leading airplane. Tests of Boeing 727 airplanes are reported to have detected down flows of 1,400 feet per minute, two minutes after passage of the airplane. 1,200 feet per minute down flows remained three minutes after passage. These sink rates exceed the climb capability of most general aviation airplanes. Although this area of sinking air obviously can't continue all the way to the ground, it can force the airplane into obstacles which rise above the ground or lead the inexperienced pilot to stall the airplane during a hopeless attempt to level the airplane. Airplanes with higher span loads and more efficient wing shapes will generate greater sink rates in the air immediately behind it. Airplanes with larger wing spans will spread this sinking air over a larger area.

Now on to what many people consider to be the primary hazard associated with wake turbulence encounters: uncontrollable roll during an along track encounter with the core of the wing tip vortex. There

are four primary factors which determine the severity of the rolling upset of an airplane encountering an along track encounter: the rotational energy of the vortex core, the ability of the vortex core to envelop the trailing airplane's wings; the natural roll inertia of the trailing airplane; and the ability of the trailing airplane to generate roll rates which exceed those produced by the core vortex.

The core vortex which has lingered two minutes after the passage of a Boeing 727 at approach/departure speeds is of sufficient rotational energy and size to roll the average general aviation airplane at a rate of about 30 degrees per second. It requires full roll control displacement in a general aviation airplane to develop similar roll rates. In other words, the roll rate imparted by the 727 could be just about canceled by full application of the general aviation airplane's controls. Unfortunately, application of full anti-roll controls will normally lag behind the rolling motion of the airplane. It takes a finite length of time for the pilot to recognize the developing situation, decide on a course of action and then command muscles to move the controls into the desired position. By this time the airplane may be at 90° of bank. And, unless the pilot is an airshow pilot flying a special airshow airplane, holding 90° of bank and maintaining altitude is not possible. It would not be surprising if most general aviation pilots would either stall the airplane or "split S" into the ground if the roll excursion impacted by the vortex core caused the bank attitude to exceed 90 degrees. Airplanes which have high roll inertia such as wing mounted multi-engine configurations and/or wing tip fuel tanks will have a natural tendency to automatically resist the rolling input caused by a vortex core. This is good. However, once the roll has started the roll inertia works against safe recovery of the airplane, requiring more time and greater control displacements to stop the roll.

The velocities and size in the core of a wing tip vortex depend on many different things (e.g. airplane weight, wing span/geometry, G load, speed, and air density). In an investigation where the velocities and size of the wing-tip-vortex core is an issue, specific information will be hard to obtain. One NASA study reports that a light transport can produce a vortex with a greater tendency to produce roll in a trailing airplane than the vortex produced by a 727. It is also important to realize that although loss of roll control can result in a collision with the ground accident if experienced at low altitude, loss of roll control has also precipitated inflight structural failures. Some years ago, an Air Force KC-135 (similar to the Boeing 707) encounter the wake turbulence of another KC-135 it was following by something over a mile. The resulting out of control roll and pull out caused sufficient inertia loads to rip both of the engine pylons from the left wing. The airplane was successfully landed with both the number 1 and number 2 engines and their associated pylons missing.

The potential for loss of roll control by a "trailing" airplane which experiences an along-track encounter with the wake turbulence generated by another airplane can be represented by the relationship shown in the following equation. The hazard "number" generated by this equation has not, to the authors' knowledge, been used to establish high or unacceptable risk. The equation does, however, provide valuable insight into the factors involved in initiating a trailing vortex induced loss of control event.

$$Hazard = \left[\frac{G \times W}{\rho \times V_G \times b_G} \right] \times \left[\frac{1}{b_E^2 \times A_E} \right]$$

$$(1)$$

where:

G = the G load of the airplane generating the vortex.

W = the weight of the airplane generating the vortex.

r = air density

V_G = speed of the airplane generating the vortex.

b_G = wing span of the airplane generating the vortex.

b_E = wing span of the airplane encountering the vortex.

A_E = Aileron power of the airplane encountering the vortex.

One last comment about vortex core encounters. If you look again at the illustration shown above you will see that to be in a position where the airplane is subjected to the full rolling energy of the vortex core, the fuselage must be close to the core's axis. In order to get to this position the airplane would first have to maneuver either laterally through a location where only part of the airplane's wings were in the vortex core or vertically where the airplane would react to the vortex core by yawing. It is highly likely that while moving through this "partially in and partially out" position, the rolling/yawing motion imparted by the vortex core would cause the airplane to be "spit" out of the vortex before it arrived at the center of the vortex core and an extreme bank attitude was achieved. Encounters where an airplane was spit out of vortex core probably account for the vast majority of the encounters. As a result, it shouldn't be surprising that many pilots don't give these potential killers the respect they deserve. After encountering the relatively low rotational energies outside the vortex core or being spit out of the vortex core before the bank angle became excessive, they think they have mastered the beast.

15. FORECAST.

The majority of the wake turbulence encounters occur either during the final approach to landing or during the takeoff leg. This can be attributed to the fact that a lot of airplanes utilize this small chunk of sky. Eye-balls during VMC, and precision navigation aids during IMC, necessarily keep the dispersion of the three dimensional aircraft tracks small. Improved math models of vortex behavior are providing some protection against this hazard. However, decreasing aircraft spacing to increase airport utilization can be safely achieved only when new technology provides the capability for real time detection and avoidance of wake vortices.

15. TRIM AND PITCH CONTROL EFFEC-

TIVENESS CHANGES DURING TAKEOFF AND LANDING.

This last hazard associated with wing-tip-vortices is very different from those discussed earlier in that it does not threaten aircraft flying along side or following the aircraft generating the hazard. Instead, changes in wing-tip-vortices caused by flight at very low altitudes (less than one wingspan in the air) does present a hazard to the aircraft generating the vortex. The intensity of a wing's downwash and the vortices the downwash induces are directly influenced by the airplane's proximity to the ground.

When a wing is within approximately one wing span of the ground, the downwash created at higher altitudes starts to be influenced by the presence of the ground. At altitudes below one wing span the airplane is said to be in "ground effect." At approximately one wing span above the ground there is an approximately 1% decrease in the amount of induced drag. At one half span above the ground induced drag is reduced by approximately 8%; at one-tenth of a span the induced drag is only one-half of what it was before the airplane entered ground effect. Since the induced drag produced by a wing is directly proportional to the downwash angle of the wing, a reduction in induced drag also reduces the downwash angle.

A wing's downwash directly influences the pitch trim of an airplane. The horizontal tail of an airplane with natural static pitch stability produces lift in downward direction. Essentially, the horizontal tail performs like an upside down wing. Most airplanes have horizontal tails which can be in the wing's downwash. Exceptions could be T-tails on airplanes with low mounted wings, although even T-tails can be influenced by the wing's airflow when at exceptionally high angles-of-attack.

If the wing's downwash increases as an airplane climbs out of ground effect during takeoff, the following sequence of events takes place:

    1. the horizontal tail's angle of attack increases,

    2. the downward lift produced by the horizontal tail increases and

    3. the airplane experiences a nose up pitch trim change.

Likewise, if the downwash decreases, the downward lift produced by the tail decreases and a nose down pitch trim change will occur. What this means to us is that as an airplane climbs out of ground effect, all other things being equal, it will experience a nose up change in pitch trim. Similarly, as it descends into ground effect it will experience a nose down change in pitch trim.

For a normally functioning airplane this should not be a problem. However, if the airplane has already suffered performance or controllability degradation due to system or component failures, this change may increase the pilot's tasking to the point of overload. For instance, on two occasions military aircraft making emergency landings without the benefit of elevator control pitched down as they entered ground effect. The nose gear first landings, in one case, resulted in the complete destruction of the airplane and major damage to it in the other case. In another case a military airplane attempting a takeoff after suffering engine failures late in the takeoff roll, stalled and crashed shortly after takeoff. The stall was attributed to four related factors. First, the pilot appeared to be holding the recommended engine out climb speed in a very heavy airplane. Not much margin for error. Second, as the airplane climbed out of ground effect, the change in the downwash angle caused a change in the airspeed indicator's position error, in turn causing the indicated airspeed to increases without an accompanying increase in the airplane's true airspeed. Third, the nose up pitch change associated with the climb out of ground effect provided the pilot with what appeared to be a natural correction for the increase in indicated airspeed. Finally, as the airplane climbed out of ground effect, the natural increase in induced drag coupled with the extra induced drag associated with a slower than recommended true airspeed exceeded the airplane's available thrust. The airplane quickly fell onto the reverse side of the thrust curve, decelerated, stalled and crashed.

16. SUMMARY.

Downwash and wing tip vortex hazards are out there. They exist. They have been causal factors in accidents in the past and it seems likely that their contribution may increase in the future. The investigator needs to understand these phenomena.

29

Loss of Performance

1. INTRODUCTION.

When talking about "performance" we will be talking about an airplane's ability to accelerate to a higher speed, climb to a higher altitude or maintain flight without sacrificing altitude or airspeed. This will involve analyzing interactions of the aerodynamic forces acting on the airplane (in terms of lift and drag), thrust and weight. Sooner or later every airplane accident investigator will be faced with trying to

determine why an airplane couldn't out-climb the terrain or lost necessary airspeed while attempting a turn. In other words, the investigator will be trying to understand the interaction between the aerodynamic forces (lift and especially drag), thrust and weight.

2. REVIEW OF THE DRAG/THRUST REQUIRED CURVE.

While in level flight, the thrust required to

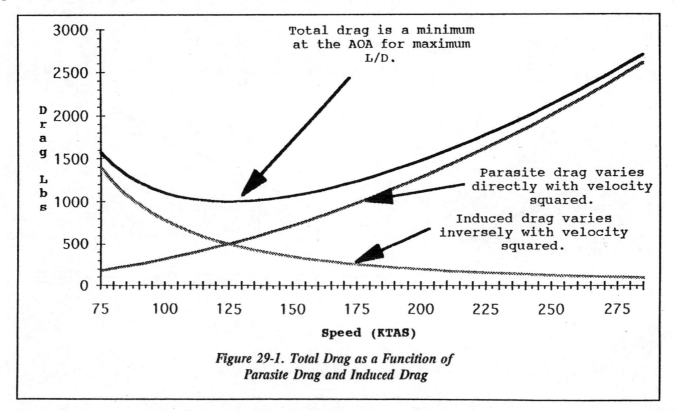

Figure 29-1. Total Drag as a Function of Parasite Drag and Induced Drag

maintain both level flight and airspeed in an airplane is equal to the total drag being produced by the airplane. At its maximum level flight airspeed, the total drag being developed by the airplane is equal to 100% of the thrust available from the engine. More important to the airplane accident investigator, the sustained climb angle of an airplane depends on the amount by which the thrust produced by the engine exceeds the total drag produced by the airframe. In order to achieve the maximum steady state (sustained) climb angle, the airplane must be at an airspeed where the thrust available exceeds the drag (or thrust required) by the maximum amount. In a pure jet airplane, where the thrust produced by the engine is relatively constant, the maximum climb angle is achieve at the airspeed where the drag is minimum. For airplane's whose engines produce thrust which varies significantly with airspeed, computations to determine the speed for maximum climb angle is complicated by the fact that the thrust available curve is not a straight line. More on how to determine the maximum sustained climb angle and the speed necessary to achieve this angle later. For now, we'll concentrate on how to determine the drag or thrust required to maintain level flight at normal operating speeds for sub-sonic airplanes.

The drag on a subsonic airplane exists in two forms: induced drag which is inversely proportional to the square of the airplane's speed; and parasite drag which is directly proportional to the square of the airplane's speed. Induced drag is the drag associated with the production of lift. No lift, no induced drag. Increase the lift (increased weight or more probably increased G load) and parasite drag increases. When in one G flight, induced drag is highest at the airplane's stall speed and constantly decreases at an ever slower rate as the airplane's speed increases. Increasing lift be it either increased G load or weight will also increase induced drag.

Parasite drag is the drag not associated with the production of lift. When lift produced by an airplane is zero, all its drag is parasite drag. Parasite drag is zero when the speed of the airflow over the airplane is zero and increases at an ever increasing rates as speed increases. Parasite drag is composed of:

　　1. form drag (drag due to low pressure areas which act on rearward facing surfaces),

　　2. friction drag (which, as the name implies is due to friction forces acting on the airplane as it moves through the air),

　　3. momentum drag (due to intentional flow through the airplane, such as engine cooling, and unintentional flow, such a leakage) and

　　4. interference drag (due to interference

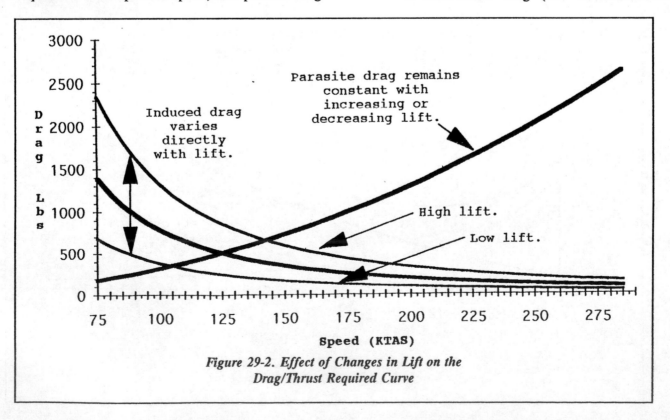

Figure 29-2. *Effect of Changes in Lift on the Drag/Thrust Required Curve*

between the boundary layer flows over the various external shapes.

Total drag is the sum of the parasite and induced drag. These relationship is shown in Figure 29-1.

It should be obvious from reviewing the illustration shown above that total drag is very high when the airspeed is very low (just above the stall) and very high at very high airspeeds (just below the airplane's maximum speed). There is also a speed in-between the these speeds where drag is minimum. It turns out that at this speed parasite and induced drag are equal. It also turns out that, given some normally available information, an airplane's minimum drag is not to difficult to estimate. And, knowing this point, the average accident investigator, using nothing more than a simple hand calculator, can estimate the entire total drag curve for an airplane in ten minutes. The method is shown below.

1. Decide on the weight for which total drag is to be calculated.

2. Use the airplane's Flight Manual (FM) to determine the speed necessary to achieve the maximum engine out glide ratio at the weight determined in step one.

3. Use the airplane's FM to determine the airplane's maximum engine out glide ratio in terms of height/distance.

4. Calculate the cosine of the maximum engine glide angle.

5. Calculate minimum drag using the following equation:

$$D_{min} = \frac{Height}{Glide} \times Weight \, [\cos(\angle \, of \, descent)] \qquad (1)$$

where:
Height/Glide was determined in step 3. above.
The cosine of the descent angle was determined in step 4. above.

NOTE: For quick estimates, the cosine of the descent angle can be estimated as 1.

6. On a chart of true airspeed (horizontal) versus drag (vertical) plot the value for minimum drag at the speed determined in step 2. above.

7. Use the following equation to determine the drag at several different speeds both above and below the speed for minimum drag. These estimates should be valid for speeds between 1.3 times the stall speed to at least Mach 0.7.

$$Drag = \frac{D_{min}}{2} \left[\left(\frac{V_1}{V_2} \right)^2 + \left(\frac{V_2}{V_1} \right)^2 \right] \qquad (2)$$

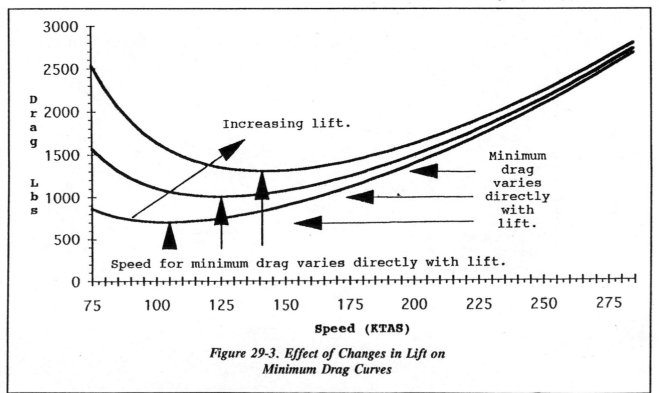

*Figure 29-3. Effect of Changes in Lift on
Minimum Drag Curves*

where:

D_{min} is the minimum drag calculated in step 5. above.

V_1 is the speed determined in step 2. above.

V_2 is the new speed for which drag is being calculated.

Drag is the total drag at the new speed.

8. Plot the values calculated in step 7. on the chart started in step 7. above.

The curve plotted in the last step of this process, is called the drag curve, and it establishes the airplane's total drag in the configuration described in the Flight Manual:

1. at airspeeds from 1.3 times the stall speed to about Mach 0.7, the speed at which the effects of drag produced by local super sonic flow start to make themselves known, and

2. in one G flight.

If the event being investigated involves an airplane in turning flight, the curve drag curve calculated above will have to be compensated for the effects of increased induced drag. Induced drag will increase because more lift is required to maintain level flight in a turn and provide the centripetal force necessary to make the airplane turn. This will cause the total drag curve to shift up and to the right when lift is increased

at higher G loads as when making a level coordinated turn (see Figure 29-2). The new curve can be calculated either by starting from scratch using a new weight where the new weight is equal to the original weight multiplied by the G load or by using the method shown below.

1. Calculate the G load. In a level coordinated turn this will be:

$$G = \frac{1}{\cos \phi} \qquad (3)$$

where:

ϕ is the airplane's bank angle.

2. Calculate the lift produced by the wings by multiplying the airplane's weight by the G load.

$$Lift = Weight \times G\ Load \qquad (4)$$

3. Calculate the speed necessary to maintain the angle of attack which establishes minimum drag.

$$V_2 = V_1 \sqrt{\frac{L_2}{L_1}} \qquad (5)$$

where:

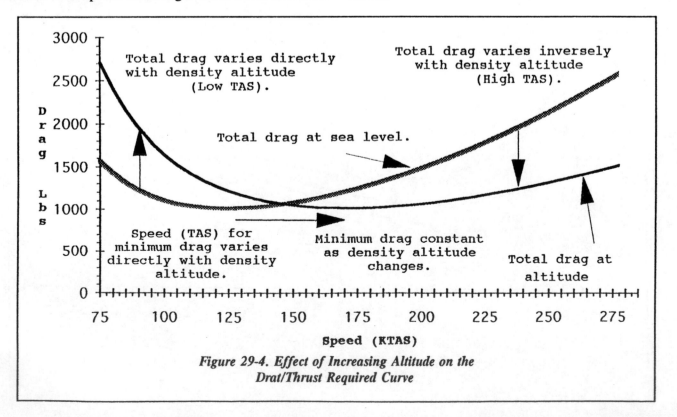

Figure 29-4. Effect of Increasing Altitude on the Drat/Thrust Required Curve

V_1 = the original speed for minimum drag.

V_2 = the new speed for minimum drag.

L_1 = the original lift produced by the airplane's wings.

L_2 = the new lift produced by the airplane's wings.

The new lift produced by the wings can be due to either increased G (typically created during a turn) or a change in weight.

4. Calculate the new minimum drag.

$$D_2 = D_1 \frac{L_2}{L_1} \qquad (6)$$

where:

D_1 = the original minimum drag.

D_2 = the new minimum drag.

L_1 = the original lift produced by the wing.

L_2 = the new lift produced by the wing.

5. You now have the new minimum drag point for the heavier airplane and the speed necessary to achieve this minimum drag. This is the lowest drag point on a new drag/thrust required versus airspeed curve. To complete the curve, use the equation below to compute the new curve.

$$Drag = \frac{D_{min}}{2}\left[\left(\frac{V_1}{V_2}\right)^2 + \left(\frac{V_2}{V_1}\right)^2\right] \qquad (7)$$

In general terms, if the lift produced by a wing changes (either due to a change in the actual weight or a change in the G load created) the drag/thrust curve will move. If lift increases, the drag/thrust required curve will shift up and to the right. If the lift is decreased, the drag/thrust required curve will shift down and to the left. The changes in the curve are due to the changes in induced drag. Greater lift creates higher values of induced drag, shifting the curve up and to the right. Reducing lift creates lower values of induced drag, shifting the curve down and to the left.

Now, given only an airplane's best engine out glide range, and the airspeed necessary to achieve this range at a given weight, you should be able to plot the airplane's drag/thrust required curve for that weight in one G flight and other weights and G loads. But this is only good for sea level. The performance related accident you are likely to investigate probably won't be at sea level. Not to worry. A simple correction can take care of this situation.

To correct the drag/thrust required curve for altitudes other than standard day - sea level, it is first necessary to determine the density ratio (annotated as "σ" and pronounced as "sigma") for the altitude in question. This can be determined from a chart of atmospheric conditions by using the ambient pressure altitude and temperature. Remembering that the best

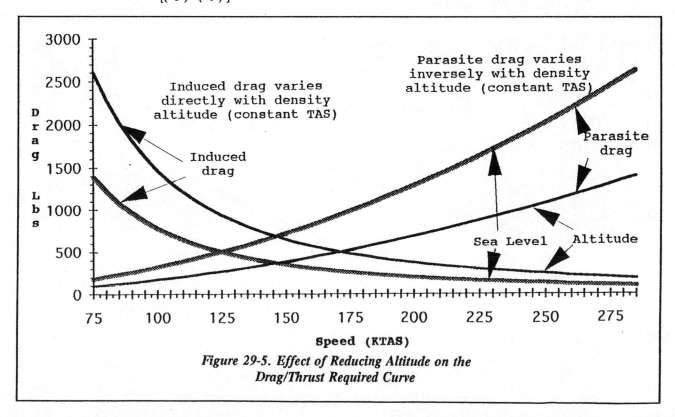

Figure 29-5. Effect of Reducing Altitude on the Drag/Thrust Required Curve

lift over drag (L/D$_{max}$) conditions are obtained at a specific angle of attack, the equation shown below can be used to determine the true airspeed necessary to achieve the angle of attack necessary to achieve L/D$_{max}$ at higher altitudes.

$$V_2 = V_1 \sqrt{\frac{1}{\sigma}} \qquad (8)$$

where:

   V_2 = the true airspeed necessary to achieve L/D$_{max}$ at altitude.

   V_1 = the true airspeed necessary to achieve L/D$_{max}$ at sea level.

   σ = the density ratio at altitude.

It turns out that the increase in induced drag due to the lower equivalent airspeed at altitude exactly cancels the decrease in parasite drag due to the lower equivalent airspeed and the total drag is exactly the same as it was at sea level. Since only the airspeed for best L/D$_{max}$ changes, plotting the drag/thrust required curve is a little easier. Essentially the drag/thrust required curve shifts to right as the airplane climbs to altitude.

The parasite drag produced by an airplane can be increased by "dirtying-up" the airplane, lowering landing gear and flaps, extending speed brakes or spoilers, or even opening the cowl flaps. Cleaning-up the airplane by raising the landing gear and flaps, retracting speed brakes or spoilers, will decrease the parasite. An increase in parasite drag shifts the drag/thrust required curve up and to the left, increasing D$_{min}$ and decreasing the speed at which D$_{min}$ and L/D$_{max}$ occur. Unless the engine out glide range and the speed for achieving this glide range for a specific gross weight are given in the airplane's Flight Manual, you will have to obtain these figures from the manufacturer. Once known, you can go through the process discussed above to plot the drag/thrust required curve.

3. REVIEW OF THE POWER REQUIRED CURVE.

The power required to overcome total drag is a key factor when estimating the maximum steady state rate of climb achievable by an airplane. Normally you will first calculate the drag/thrust required values for an airplane's flight conditions (airspeed, density altitude, G loads, and configuration) before attempting to calculate the power required. That's because the power required to maintain steady state flight depends on the thrust required to maintain steady state flight. If you remember from freshman physics, work is defined as a force acting over some distance (F x d), while power is defined as the amount of work done per unit time (F x d / t). And, since distance traveled per unit time (d

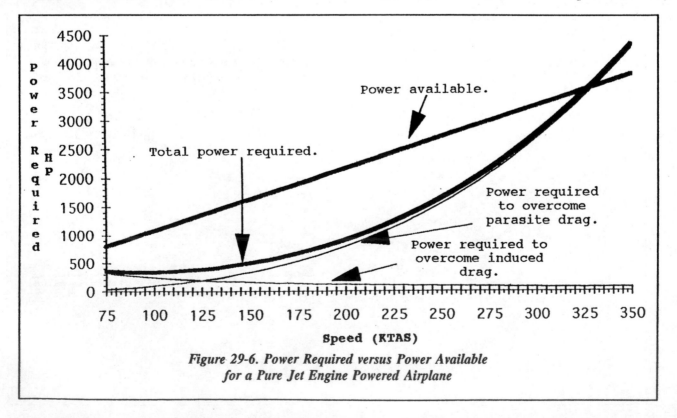

Figure 29-6. Power Required versus Power Available
for a Pure Jet Engine Powered Airplane

/ t) is also defined as velocity, power can also be considered as a force multiplied by the velocity at which it is moving (F x V). Thus, in order to determine the power required for an airplane to maintain level unaccelerated flight at the various speeds within its performance envelop, all that is necessary is to multiply the drag at any true airspeed by that speed. Of course units and conversion factors are necessary, so for knots, pounds and horsepower, the relationship is:

$$Horsepower\ Required = \frac{Drag\ (lbs) \times Speed\ (KTAS)}{325} \quad (9)$$

It may take a couple of minutes to convert the drag/thrust required curve into a power required curve when using a pocket calculator. Less time is required if the calculator is programmable and the operator is proficient in its use. Once the power required to maintain level unaccelerated flight has been plotted for all airspeeds in the airplane's performance envelope,

the next step is to establish the power available. For a pure jet engine powered aircraft this is fairly easy. Simply multiply the thrust of the engine in pounds by the speed of the airplane in knots of true airspeed and divide by 325. The answer will be in horse power.

One slightly surprising fact may surface the first time you try this. The power produced by a jet engine when it's not moving is zero! In addition, the slope of the power available curve for the pure jet will be constant, a straight line. Double the speed and the horse power output of the engine doubles. Triple the speed and the horsepower triples. The faster you go the more power the engine delivers. Of course the thrust is constant while all this is happening and the drag produced by the airplane is increasing by almost the square of the velocity and the power required to overcome the drag is increasing by almost the cube of the velocity. That means that despite the constantly increasing power available, the airplane can't keep accelerating until it exceeds its structural limits. A generalized chart of this relationship is shown below.

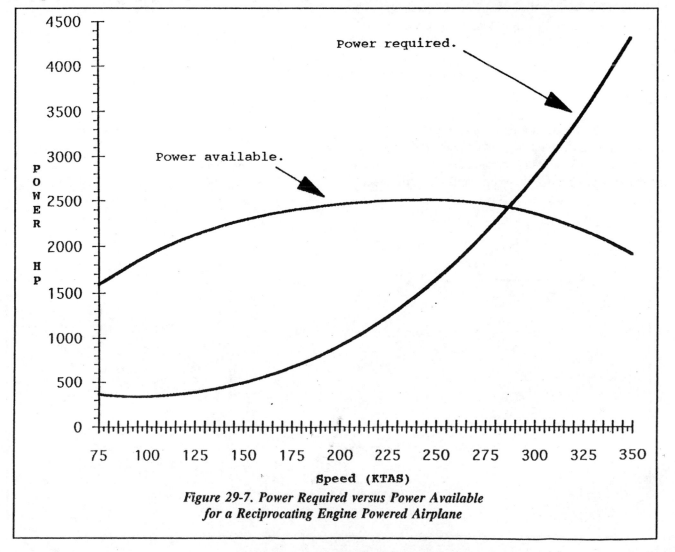

Figure 29-7. Power Required versus Power Available
for a Reciprocating Engine Powered Airplane

It should be kept in mind that the thrust and power output of aircraft engines are rated for sea level and the thrust/power of an unboosted engine will be reduced roughly in proportion to the decrease in air density at higher altitudes.

For airplanes with reciprocating engines powering propellers the situation is changed. For these airplanes the thrust developed by the engine varies quite markedly as a function of speed. The thrust available from this engine typically is high at low speed and drops as speed increases. The power available (because it is multiplied by the speed) starts low, increases fairly quickly to a plateau and drops off fairly quickly as speed increases. The exact shape of these curves will depend heavily on the propeller's efficiency at the various speeds at which the airplane is expected to fly. While the conversion from thrust to power for a jet engine is relatively simple, the conversion from thrust to power for a reciprocating/propeller engine is more complex and will normally require special assistance from the engine and/or propeller manufacturers. Very high by-pass engines fall into a middle ground. Their thrust does fall off at higher speeds, although not so fast as reciprocating engines.

4. DETERMINATION OF CLIMB PERFORMANCE.

Now that we understand how we can easily determine the thrust and power required to maintain level flight for all the airspeeds in the airplane's subsonic performance envelope and we understand the problems associated with determining the thrust and power available at various speeds, we can get to the heart of the matter. How can we use this stuff to estimate the expected performance of the airplane and determine whether or not it was a factor in the accident?

If we look at the illustration shown below, we can identify the forces acting on an airplane in a steady, constant speed, constant angle climb.

The climb angle is represented by the Greek symbol γ or "Gamma". If we look at the forces acting along the longitudinal axis of the airplane we can see that the thrust of the airplane is balanced by the drag of the airplane plus the component of weight acting along the longitudinal axis.

Σ Horizontal Forces = Zero = Thrust - Drag - Weight sin γ
or symbolically:

$$T - D = W \sin \gamma \qquad (10)$$

Rearranged:

$$\frac{T - D}{W} = \sin \gamma \qquad (11)$$

where:
    T = Thrust available (lbs)
    D = Drag or thrust required (lbs)

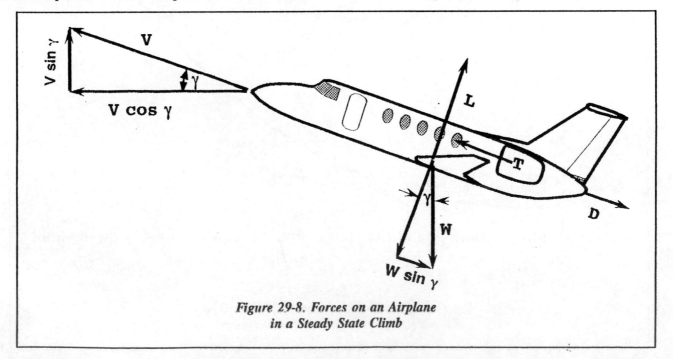

Figure 29-8. Forces on an Airplane in a Steady State Climb

W = Airplane weight (lbs)

γ = Climb angle (degrees)

Contemplating the equation shown, and considering the problem on maximizing the climb angle (γ), we have three potential options:

1. increase thrust available (can't do because we are assuming maximum thrust/full power is already being used).

2. decrease weight (can't do, jettisoning fuel/equipment is not authorized).

3. decrease drag (now this is a real option).

If we fly the airplane at a speed where drag is a minimum (that's the speed for L/D_{max}) we can maximize the value of T - D. This term (T - D) is the value by which thrust available exceeds thrust required and is referred to as excess thrust. Excess thrust can be used to make the airplane climb. If the airplane is flown at the airspeed where excess thrust is maximized (the airspeed where drag is minimized) the airplane's steady state climb angle will also be maximized. This is true for both jet and reciprocating engines. If we want to calculate the airplanes maximum climb angle at any airspeed we can use the equation:

$$\sin^{-1}\left(\frac{T-D}{W}\right) = \gamma \qquad (12)$$

where:

γ = the airplane's climb angle.(degrees)

T = the airplane's available thrust (lbs) at the speed in question

D = the airplane's drag (lbs) at the speed in question.

W = the weight of the airplane.

As mentioned earlier, g will be at a maximum when the airplane is flown at the speed for minimum drag. This is the speed for best engine out glide range which, for jet airplanes, is also the speed for maximum endurance.

Now let's take a look at rate of climb. By referring to the illustration shown below, you should be able to see that rate-of-climb will be maximized when V sin γ is maximized.

Since g is equal to (T-D)/W, the rate-of-climb will be maximized when V(T-D)/W is maximized. Since V times T is equal to power available (P_a) and V times D is equal to power required (P_r) and we have no control over weight (W), rate-of-climb will be maximized when power available (P_a) minus power required (P_r) is a maximum. In other words, maximum rate-of-climb occurs at an airspeed where there is maximum excess power. This occurs where the slope of the power available curve is parallel to the power required curve.

5. EFFECT OF WEIGHT, G LOAD, ALTITUDE AND CONFIGURATION ON CLIMB PERFORMANCE.

Aircraft weight, G load, density altitude and configuration will affect both the speed at which maximum climb performance is obtained and the actual values for the angle and rate-of-climb. If an airplane's drag/thrust required versus angle of attack has been plotted assuming maximum gross weight, one G flight, a density altitude about sea level or a configuration other than that used to compute minimum drag, the drag, thrust required curve will have to

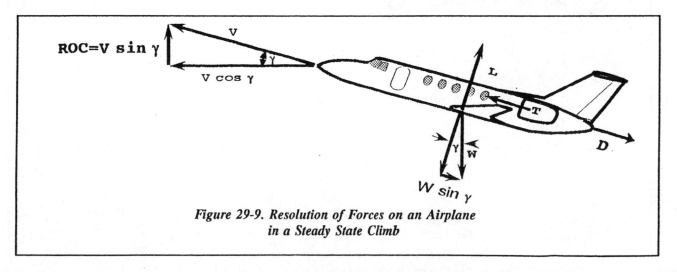

*Figure 29-9. Resolution of Forces on an Airplane
in a Steady State Climb*

be modified to account for these variances.

When an airplane's weight is changed, both the speed at which minimum drag (or thrust required) occurs and the value of the minimum drag change. Increases in weight cause an increase in both the speed at which minimum drag is experienced and an increase in the value for minimum drag. As you might expect, a decrease in weight will result in a decrease in the speed at which minimum drag is experienced and a decrease in the value of minimum drag. The new values for minimum drag and the speed for minimum drag can be estimated using the following equations.

$$V_2 = V_1 \sqrt{\frac{W_2}{W_1}} \qquad (13)$$

where:

V_2 = the speed for minimum drag at the new weight.

V_1 = the speed for minimum drag at the original weight.

W_1 = the original weight.

W_2 = the new weight.

$$D_2 = D_1 \frac{W_2}{W_1} \qquad (14)$$

where:

D_2 = the new value for minimum drag.

D_1 = the original value for minimum drag.

W_1 = the original weight.

W_2 = the new weight.

If only the aircraft's weight has changed, then the new drag/thrust required versus speed curve can be plotted using the equation:

$$D_T = \frac{D_{min}}{2} \left[\left(\frac{V_1}{V_2} \right)^2 + \left(\frac{V_2}{V_1} \right)^2 \right] \qquad (15)$$

The drag/thrust required curve will shift up and

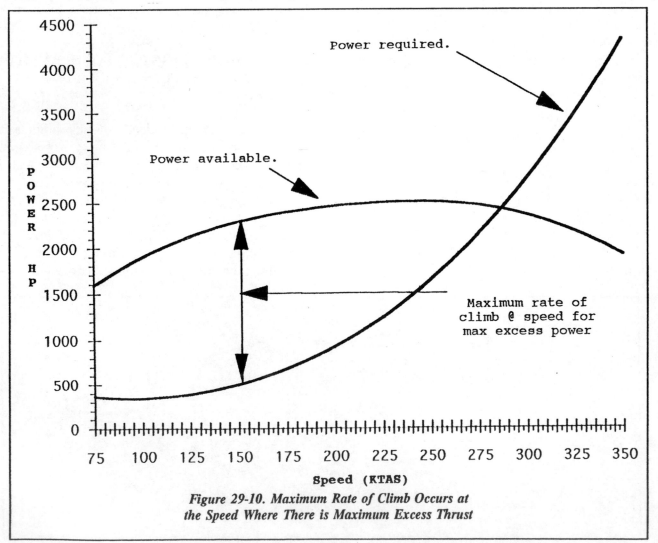

Figure 29-10. Maximum Rate of Climb Occurs at the Speed Where There is Maximum Excess Thrust

to the right when weight increases and down and to the left if weight decreases. Thus the speeds at which and airplane can achieve maximum climb angle and maximum rate of climb will increase with increased weights. Likewise, the value of the maximum climb angle and rate of climb will decrease. More likely, we will be dealing with airplane weights that are less than the maximum gross weight and the speed for best angle and rate will be lower while the values for the best climb angle and rate will be higher.

The effects of increasing or decreasing the G load on an airplane are very similar to the effects of increasing or decreasing weight. Increasing the G load shifts the drag/thrust required curve up and to the right, increasing both the speed for minimum drag and the value for minimum drag. However, while we normally see weight decreasing during a mission, steady state turns will normally result in increased G loads, higher speeds for minimum drag and higher values of drag/thrust required. In addition, there is a definite relationship between G load and bank angle when an airplane is in a steady state turn. These relationships are given in the equations shown below.

$$G = \frac{1}{\cos \theta} \qquad (16)$$

where:

G = the G load on an airplane during a steady state turn.

θ = the airplane's bank angle

$$D_2 = D_1 G = D_1 \frac{1}{\cos \phi} \qquad (17)$$

where:

D_2 = the minimum drag when G load is being applied.

D_1 = the minimum drag at one G.

ϕ = the bank angle.

Just as when examining increases in weight, increases in G load result in higher speeds for minimum drag and higher values of minimum drag/thrust required. The drag/thrust curve moves up and to the right when G load increases.

Next, let's adjust the drag/thrust required curve for air density. As an airplane climbs into the less dense air at higher altitudes, it has to fly at a higher true airspeed in order to maintain the angle of attack necessary to achieve minimum drag. However, due to the lower air density at altitude, there is no net in-

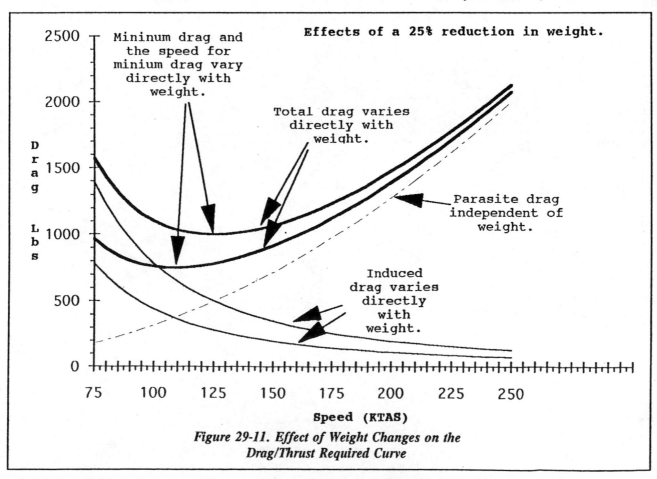

Figure 29-11. Effect of Weight Changes on the Drag/Thrust Required Curve

crease in total drag at the higher altitude. The drag/thrust required curve shifts to the right but does not move up. The equation to determine the true airspeed for minimum drag is shown below.

$$D_2 = D_1 \sqrt{\frac{1}{\sigma_2}} \qquad (18)$$

where:

D₂ = the total drag at altitude.
D₁ = the total drag at sea level (standard day)
σ₂ = the density ratio at altitude.

There is one other factor to consider when considering the effects of decreasing air density (increasing altitude) on the climb performance of an airplane. For normally aspirated (non-boosted) engines, both the thrust available and power available from the engine decreases as altitude increases. Although the exact decrement in engine performance due to decreased air density is influenced by the engine's design and must be obtained from data provided in engine performance manuals, a pessimistic estimate can be made by equating the reduction in performance to the new density altitude. In other words, since the

density ratio at 10,000 feet (standard day) is about 0.74, the thrust produced at 10,000 feet density altitude should be at approximately 74% of the thrust produced at sea level.

As mentioned earlier, the effect of changes in configuration depend on the specific type of airplane involved in the investigation. Although the effect of various configuration changes on L/D$_{max}$ is not normally available in airplane Flight Manuals, it can be estimated through aerodynamic analyses which are well beyond the scope of this text or may be based on tests performed by the airplane's manufacturer. In any case, increases in drag will result in a decrease in the speed at which minimum drag occurs and an increase in the value of the minimum drag.

6. CONSTRUCTION OF THRUST REQUIRED AND POWER REQUIRED CURVES USING AN AIRPLANE'S MAXIMUM ENGINE-OUT GLIDE RANGE, WEIGHT AND GLIDE SPEED.

Let's run through a typical problem. While we are doing this, let us also think about why we are doing this and what the limitations of this process are. First, this section of the book is designed to give you

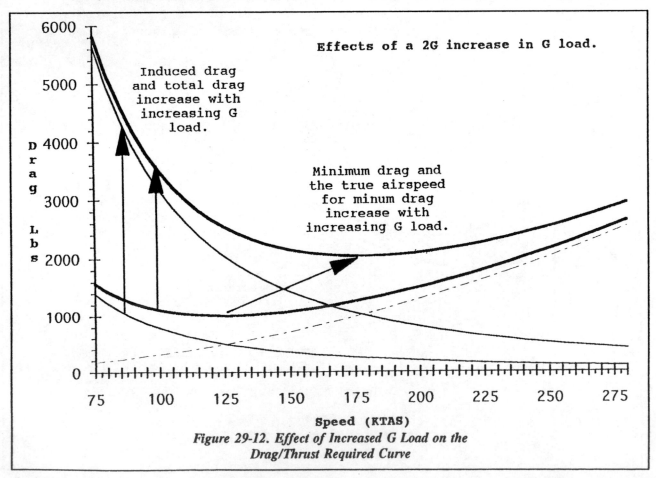

Figure 29-12. Effect of Increased G Load on the
Drag/Thrust Required Curve

a better understanding of the types of things engineers with expensive computers and lots of data can do for you. The engineers will go through a similar process to estimate climb performance. If you understand this material you will have a pretty good idea of where those engineering estimates came from. Second, and equally important, you should be able to come up with some ball-park estimates while in the field and without a great deal of effort. Armed with a simple hand-held calculator (the $10.98 variety) a piece of graph paper, a pencil and a Flight Manual, you should be able to come up with a fair estimate of an airplane's climb performance.

Let's run through a hypothetical problem. Suppose we are investigating the crash of a twin engine corporate jet. The airplane had just taken off on runway 35 from a high altitude airport (5900 feet MSL) when the pilot radioed that the right engine had failed. The pilot indicated to the tower and that they were going to immediately turn back to the field for an emergency landing. Witness statements indicated the engine was making "popping" sounds as the plane broke ground and that it was cleaned up before the pilot started what appeared to be a 30°-45° banked left turn at about 200-300 feet above the ground. The airplane did not appear to gain altitude after it rolled into the turn. The airplane crashed on a heading of 170°, approximately 3600 feet west of the field, abeam the takeoff end of the runway. Radar data indicates the airplane's turn radius was fairly constant and the ground speed of the airplane was approximately 130 knots. Winds were calm. About 26 seconds elapsed from the time the tower saw the airplane roll into its left turn until the time it was observed to strike the ground. The airplane hit the ground at a small ridge, left wing low. Trimming of small trees and brush prior to the main impact suggest that the airplane was in a 35° left banked turn just prior to impact. Crash forces appear to have trapped one airspeed indicator at 112 knots. The other indicator although damaged by the post crash fire was still free to rotate. The right engine did not show any indications of rotational energy at impact. The left engine appeared to have been operating at high RPM at impact. The field altimeter setting was 29.92 and the temperature was +40°C (104°F). What can we deduce from this information? Was the crew doomed when the engine failed? Did they have enough thrust to complete the turn? What happened?

First, let's take a look at the geometry of the

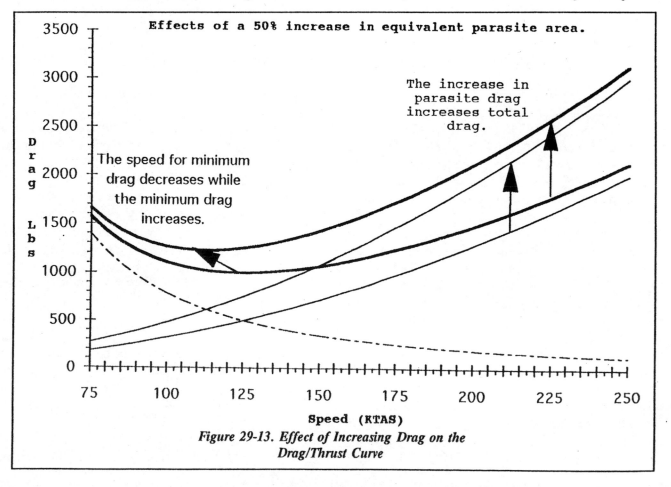

Figure 29-13. Effect of Increasing Drag on the Drag/Thrust Curve

crash. The radar plots indicate the airplane was flying at about 130 KTAS. The trapped airspeed indicator shows 112 KIAS. Let's assume position error at this speed and configuration are negligible. At the speeds and pressure altitudes involved, compressibility error is also insignificant. However, correction for pressure altitude and non-standard temperature will make a difference. At pressure altitude of 5900 feet and a temperature of +40°C, the density ratio turns out to be about 0.761 and an indicated airspeed of 112 knots yields a true airspeed of about 130 knots. Not bad for starters, but in real life not all the facts agree. The radar plots indicate a fairly constant true radius of about 1800 feet. And (remember this is practice) the location of the impact point agrees with this value. Knowing the airplane's speed and turn radius, we can estimate its bank angle and rate of turn.

Estimating Bank Angle.

$$\gamma = \tan^{-1}\left(\frac{V^2}{11.26r}\right) = \tan^{-1}\left(\frac{130^2}{11.26(1800)}\right) = \tan^{-1}(0.834) = 39.8° \quad (19)$$

Estimating Rate Of Turn.

$$ROT = \frac{1,091 \tan\phi}{V} = \frac{1,091 \tan(39.8°)}{130} = 7°/sec \quad (20)$$

Luckily, both of these figures seem to agree with other evidence gathered so far. The witness statements put the calculated bank angle in the middle of their estimates and the 7°/second would result in the airplane completing the 180° turn in close to the same time provided by the tower folks and their tapes. So far so good.

Now let's take a look at performance. Let's assume that we have found out that this airplane has a L/D_{max} of 12:1 (clean). That would yield a maximum engine out, no wind, glide range of just under two nautical miles for every one thousand feet of altitude. Therefore, the airspeed required to achieve minimum drag is 120 KIAS when the airplane is at its maximum gross weight of 60,000 pounds. Assuming no position error in the pitot static system, this would yield 120 KTAS at sea level on a standard day. Some other facts about our hypothetical airplane; it has two jet engines each producing 6,175 pounds of thrust at standard day, sea level conditions, but only 4,945 pounds of thrust

at our 10,000 foot density altitude (yes, that's what 5900 feet pressure altitude coupled with +40°C yields). Knowing what we now know we can make a pretty good guess as to what the performance capabilities of the airplane were. Of course those other folks with the main frame computers and giga-bites of data will have an even better guess, but they are miles away and will take paper and time to get you their answer. What we are going to do here can be done with that cheap pocket calculator.

First let's look at the airplane's maximum climb angle with both engines operating. To do that we need to know the airspeed for minimum drag, which will be lower than given for the airplane's maximum weight.

$$V_2 = V_1\sqrt{\frac{W_2}{W_1}} = 120\sqrt{\frac{50,000}{60,000}} = =110 \; KTAS \quad (21)$$

Thus, 110 KTAS is the speed at which our airplane, at 50,000 pounds, will produce minimum drag. But, since this is only good for sea level and we need to correct it for the existing density altitude of 10,000 feet.

$$V_2 = V_1\sqrt{\frac{1}{\sigma}} = 110\sqrt{\frac{1}{0.7385}} = 128 \; KTAS \quad (22)$$

where σ is the density ratio.

Now we know we have to be going at 128 KTAS when we are at 50,000 pounds and 10,000 feet density altitude in order for the airplane to achieve minimum drag and a maximum angle of climb. The airplane's minimum drag can be closely estimated if we know the airplane's maximum lift to drag ratio (L/D_{max}) and the airplane's weight. For airplanes whose engine out glide ratio is less than about 9°, L/D_{max} is essentially equal to the airplane's maximum glide ratio (the horizontal distance traveled for each unit of altitude).

$$D_{min} = \frac{Weight}{L/D_{min}} = \frac{50,000 lbs}{12} = 4,167 lbs \quad (23)$$

If our airplane had been flying at 128 KTAS its climb angle would have been:

$$\gamma = \sin^{-1}\left(\frac{T-D}{W}\right) = \sin^{-1}\left(\frac{9,890 lbs - 4167 lbs}{50,000 lbs}\right) = \sin^{-1}(0.114) = 6.6° \quad (24)$$

Its rate of climb at this speed would have been:

$$ROC = 101.3 V_k \left(\frac{T-D}{W} \right) = \frac{101.3(130KTAS)(5723lbs)}{50,000lbs} = 1,507 Ft/Min \quad (25)$$

However, our airplane was flying a little bit faster and its drag would have been slightly higher and its rate of climb slightly lower. How much? Let's see. (Equations 26 and 27.)

$$D_T = \frac{D_{min}}{2} \left[\left(\frac{V_1}{V_2} \right)^2 + \left(\frac{V_2}{V_1} \right)^2 \right] = \frac{4,167}{2} \left[\left(\frac{128KTAS}{130KTAS} \right)^2 + \left(\frac{130KTAS}{128KTAS} \right)^2 \right] = 4169$$

(26)

$$ROC = 101.3 V_k \left(\frac{T-D}{W} \right) = \frac{101.3(128KTAS)(572lbs)}{50,000lbs} = 1,503 Ft/Min \quad (27)$$

These values are only slightly less than those for 128 KTAS. No big deal, if the aircraft had both engines operating. Let's see what would happen if only one engine were operating. With only one engine operating, excess thrust (T-D) will be less due to the loss of the engine. The angle of climb straight ahead with only one engine operating will be reduced to:

$$\gamma = \sin^{-1} \left(\frac{T-D}{W} \right) = \sin^{-1} \left(\frac{4,945lbs - 4,169lbs}{50,000lbs} \right) = \sin^{-1}(0.016) = 0.92° \quad (28)$$

The rate of climb straight ahead with only one engine operating will be reduced to:

$$ROC = 101.3 V_k \left(\frac{T-D}{W} \right) = \frac{101.3(128KTAS)(776lbs)}{50,000lbs} = 201 \ Ft/Min$$

(29)

That's a substantial reduction from what we saw with two engines operating. But the plane is at least still going up. Let's see what happens if when the aircraft rolls into a 30 degree banked turn.

First, the increase in bank will cause the lift required to maintain level flight to increase.

$$L = W \times G = \frac{W}{\cos\phi} = \frac{50,000lbs}{0.866} = 57,735lbs \quad (30)$$

The value for minimum drag will increase:

$$D_{min} = \frac{Lift}{L/D_{max}} = \frac{57,735lbs}{12} = 4,811lbs \quad (31)$$

The speed at which minimum drag will occur will also increase to:

$$V_2 = V_1 \sqrt{\frac{W_2}{W_1}} = 128KTAS \sqrt{\frac{57,735lbs}{50,000lbs}} = 138KTAS \quad (32)$$

Since the aircraft is actually flying slower than the speed for minimum drag, the drag will be higher. At the 130 KTAS the radar plots estimated, the actual total drag in a 30° banked turn would be:

$$D_T = \frac{D_{min}}{2} \left[\left(\frac{V_1}{V_2} \right)^2 + \left(\frac{V_2}{V_1} \right)^2 \right] = \frac{4811}{2} \left[\left(\frac{138KTAS}{130KTAS} \right)^2 + \left(\frac{130KTAS}{138KTAS} \right)^2 \right] = 4,845$$

(33)

The angle of climb would be:

$$\gamma = \sin^{-1} \left(\frac{T-D}{W} \right) = \sin^{-1} \left(\frac{4,945lbs - 4,845lbs}{50,000lbs} \right) = \sin^{-1}(0.002) = 0.11° \quad (34)$$

And the rate of climb would be:

$$ROC = 101.3 V_k \left(\frac{T-D}{W} \right) = \frac{101.3(128KTAS)(100lbs)}{50,000lbs} = 26 \ Ft/Min$$

(35)

Rolling into a turn destroyed the little climb capability left with one engine out. Also, the calculations shown above assume that the airplane is perfectly coordinated, not a likely situation with asymmetric thrust. With rudder and aileron deflected to control the yaw and roll into the dead engine, and probably some side slip thrown in, the actual minimum drag would be even higher and the angle of climb and rate of climb would be negative. In a 30° banked turn the airplane would be going down. The only real option the pilot had was to climb straight ahead until sufficient altitude existed to safely make a shallow turn. We could further second guess our pilot by calculating the maximum bank angle the pilot could use without sacrificing altitude (that's a zero rate of climb), and the airspeed necessary to maintain level flight, while in the engine out situation. However, if we are simply trying to explain what happened we have gone far enough. When the right engine failed, the airplane did not have the thrust available to maintain a steady 30° banked turn without losing altitude. If the airplane had continued straight ahead, it did have the thrust available to maintain a very shallow, but positive climb.

7. SUMMARY.

All this may seem like a lot of work, but it isn't that bad. An understanding of how aircraft performance is affected by various factors can enable the investigator to run some back-of-the-envelope calculations on whether the plane could do what it was attempting to do.

Turning that around, it is also possible to predict the results of any particular performance situation. An understanding of how the various performance factors are related is probably more important than the actual mathematics.

Finally, knowledge of aircraft performance relationships can also provide a better understanding the performance data provided by aircraft manufacturers.

30

Ground Operations Hazards

1. INTRODUCTION.

Although the vast majority of today's aircraft are of the tricycle landing gear configuration, a configuration where the airplane is equipped with a nose wheel, there was a time in the authors' memories that this was not the case. In our youth, most airplanes came equipped with a tail wheel, and since this was the standard, this landing gear configuration was called "conventional". Between now and then, there were airplanes whose landing gear were arranged in tandem, in a configuration referred to as "bicycle" or "roller skate". This latter configuration was reserved pretty much for military aircraft such as bombers which required both a large bombbay and thin, high speed wings (leaving only the space in front of and behind the bombbay for the gear) or other special use aircraft with similarly unusual requirements. Today most airplanes being manufactured are of the tricycle configuration with a swiveling nose gear in front of the main load carrying gear. The reason why this configuration has gained overwhelming popularity is largely due to their positive static directional stability while moving on the ground. True, there still are some "tail draggers" being produced in factories for the sport aviation crowd, and even more being built in garages or hangars around the world, but if you look at insurance premiums, the insurance companies' pilot experience requirements, and the ground accident rates for these airplanes, the reasons for the tail dragger's fall from grace is obvious. Ground operations of conventional gear configured airplanes involves more risk than ground operations for tricycle gear config-

ured airplanes. First, its harder to see over the nose of these airplanes. "S" turns while taxiing are required in many whose over the nose visibility is severely restricted. And even the most skillful and talented can let down their guard. Several years ago, at a national air race, one of the best of the best taxied his big shiny tail dragger right into a parked airplane! Reportedly, he thought the shouts and waves of the crowd were in appreciation for another outstanding flight. Wrong! But, even more serious than the reduced over the nose visibility is the lack of the tail dragger directional stability while on the ground. This propensity for wanting to suddenly reverse headings has led to the expression that if you fly tail draggers, the question is not if you will ground loop, but when you will ground loop.

2. DIRECTIONAL STABILITY ON CONVENTIONAL GEAR (TAIL DRAGGER) AIRCRAFT.

The tail dragger's reputation for ground loops was earned scientifically. It is unstable, at least when the tail wheel is free to swivel (i.e. not locked), a condition that exists on most small tail draggers all the time. The reason for this undesirable situation can best be explained by looking at a diagram of what is happening when a tail dragger enters a turn while on the ground.

Assume, for some reason, a tail wheel equipped airplane experiences a heading change while moving on the ground. This could be due to a gust of wind, the propeller's gyroscopic force as the nose is lowered during takeoff or raised during landing, a little asymmetric brake action or even a slight pot hole on a dirt runway. It could certainly be caused by an unwise rudder input by the pilot. As the nose of the airplane moves off the original heading a side load will be applied to main wheels, resisting the skidding motion that would otherwise occur. The airplane will begin to turn, following its nose. The outward momentum of the turning airplane can be considered to act at the airplane's center of gravity and is resisted by the frictional force of the wheels, all three of them if the tail wheel is locked and unable to swivel. If, however the tail wheel IS free to swivel, it will align itself with the direction the tail wants to go and allow the tail to swing out, tightening the turn. The tighter turn increases the outward force, sending the tail out even further. Each succeeding heading change causes an increase in the outward force caused by the turn and a corresponding out swinging of the tail and an even greater heading change. At high airspeeds the increasing angle of attack of the vertical tail creates a force in the direction which will realign the airplane with its original direction of flight (i.e. stop the turn). If the airplane is at a low airspeed or in a pitch attitude which blanks the vertical tail, the restoring force of the vertical tail will not be available. High speed taxiing with a strong tail wind is especially dangerous. The

momentum forces associated with even small changes in heading can be high because of the high ground speed while the natural restoring force of the vertical tail while in a skid is low. You should be able to see that even a very small heading change will result in side loads which cause ever increasing heading changes. In other words the airplane is directionally unstable. If the pilot fails to correct the ever- increasing rate of heading change in time, the airplane could depart from the runway/taxiway, collapse the main landing gear, blow a tire, roll onto the wing-tip on the outside of the turn or two or more of the above.

Conventional landing gear (tail draggers) are naturally directionally unstable on the ground and are always looking for an excuse to perform a ground loop. Pilots have to be constantly on the alert from the time the vertical tail starts to lose its efficiency in keeping the pointy end in the front till the time when the airplane is going slow enough to accept a sudden heading change without tipping over or running into something. It's easy to blame a ground loop on the pilot, and most of the time this is where the blame belongs. However it is still important for the conscientious investigator to look for the factors which may have made the pilot's task a bit harder than it should have been or even presented a situation which most

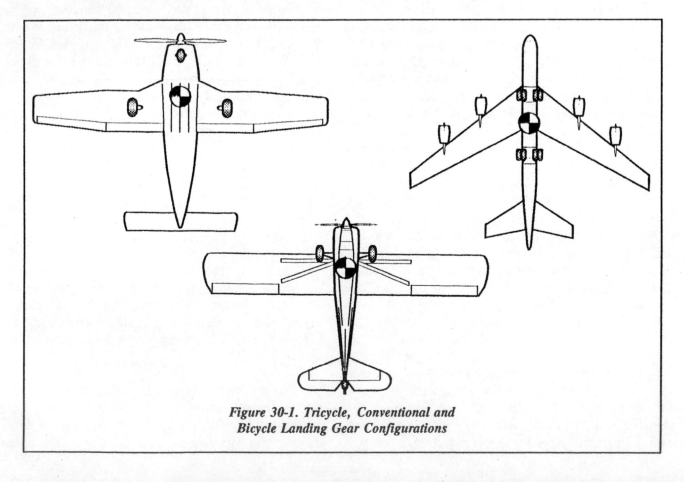

Figure 30-1. Tricycle, Conventional and Bicycle Landing Gear Configurations

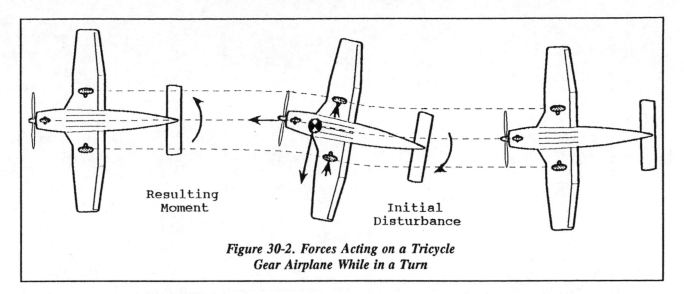

*Figure 30-2. Forces Acting on a Tricycle
Gear Airplane While in a Turn*

pilots could have handled. Were there wind gust across the field? Was a brake grabbing? Were the controls to the rudder functioning properly? These and similar questions should be answered before the blame is placed on pilot skills and control application.

3. DIRECTIONAL STABILITY OF TRICYCLE LANDING GEAR AIRCRAFT.

Unlike conventional landing gear which are naturally directionally unstable on the ground, tricycle landing gear configurations are naturally stable. The pilot doesn't have to constantly work at keeping the airplane taxiing on a straight line. It will naturally tend to do this. To understand why it is necessary to look at the dynamics of this geometry if a taxiing airplane is suddenly deflected to the left or right of its straight line path. Assume, for some reason, a nose wheel equipped airplane experiences a heading change while moving on the ground. As when we discussed the tail wheel configuration, this could be due to any number of reasons. As the nose of the airplane moves off the original heading a side load will be applied to main wheels, resisting the skidding motion that would otherwise occur. The airplane will begin to turn, to follow its nose. The outward momentum of the turning airplane acts at the airplane's center of gravity which is forward of the main wheels. This will cause the nose to want to move outward from the turn. If the nose wheel is free to swivel, it will align itself with the direction the nose wants to go and allow the nose to swing back toward its original heading. Thus any deviation from the original heading will create a force which move the nose back toward its original heading. The pilot doesn't have to constantly work to keep the airplane on a straight line while taxiing.

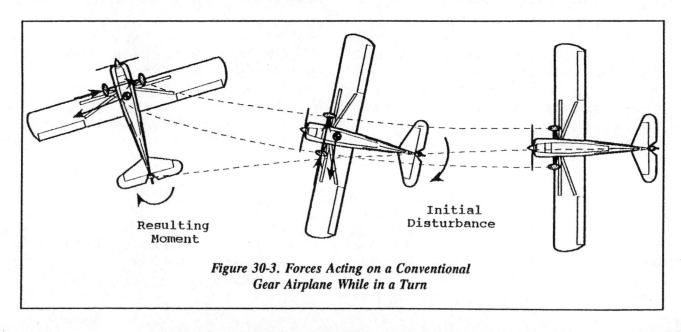

*Figure 30-3. Forces Acting on a Conventional
Gear Airplane While in a Turn*

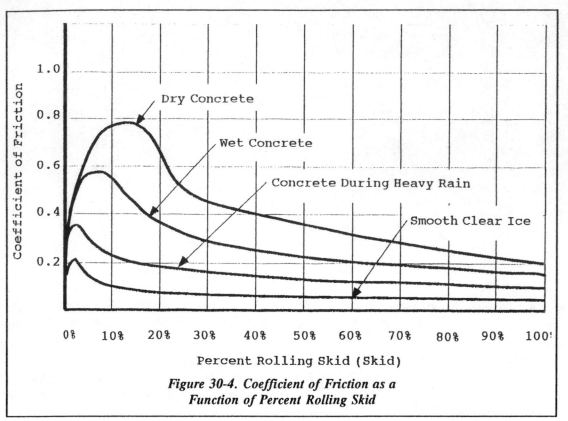

*Figure 30-4. Coefficient of Friction as a
Function of Percent Rolling Skid*

to increase the back pressure on the pitch control and start steering with the rudder pedals, it is likely that the airplane will depart the runway. Normally this will be at sufficiently low speed to limit damage to the airplane and injuries to the crew and passengers. The pilot, however, may suffer from the long term effects of chronic embarrassment.

4. BRAKING.

You can demonstrate this to yourself with your automobile. Going forward, the car is directionally stable and it tends to recover from a slight turn and go straight. Going backwards, the car is directionally unstable and it tends to increase a slight turn until it winds itself up into a tight spiral called a ground loop.

Back to the tricycle gear airplane, there can be problems if an unskilled pilot misapplies the controls. One of the most common is called "wheel barrowing". If a pilot, typically during a landing roll-out applies forward controls, an excessive amount of weight can be shifted to the nose wheel. This puts more of the rolling friction force out in front of the center of gravity, tending toward an unstable condition. If the pilot then attempts to "steer" the airplane like a car, the slightly degraded directional stability can be further degraded. If the pilot "steers" right (twists the yoke right) because the nose is headed left, the left wing will be aerodynamically raised by the lowered left aileron, the right wing will be lowered by the raised right aileron and additional weight will be shifted to the nose wheel and right main. This shifting of weight to the outside of the turn is amplified by the high center of gravity of high wing airplanes and can cause the nose wheel to turn further to the left and increase the roll toward the outside of the turn. If the pilot fails

Before discussing the problems associated with refused takeoffs and landing over shoots, it is necessary to discuss some issues concerning the braking forces created by rubber wheels rolling over paved surfaces.

A. BRAKING AND CORNERING PERFOR-MANCE. The ability of an airplane's tires to create a breaking force when stopping and lateral forces when steering is directly related to the friction force created between the tire and the runway surface. Without this friction force, the ability of the pilot to control the aircraft's speed and direction can be severely compromised. You may remember from a long ago physics class that the friction force developed when an object is pushed across a surface is proportional to the normal force between the object and the surface and the coefficient of friction between the surface and the object. This can be represented by the equation:

$$F = \mu N \qquad (1)$$

where:

F = the friction force developed.

μ = the coefficient of friction.

N = the force normal (perpendicular) to the surface.

From this simple relationship it can be concluded the we can increase the friction force by either increasing the coefficient of friction (μ) or increasing the normal force (N). We can change the coefficient of friction by changing the nature of either of the two surfaces which are in contact with each other. We can change the magnitude of the normal force by decreasing the amount of lift created by the wings and/or shifting weight from un-braked wheels to those wheels which have brakes.

A coarse surfaced concrete runway will have a higher coefficient of friction than a well worn asphalt runway with a layer of oil-covered rubber on it. Drag racers spend considerable effort to ensure that their rear tires have a maximum grip on the roadway. We should also remember from that long ago physics class that the coefficient of friction comes in two flavors, static or sliding. If we push hard enough on our object to get it sliding, the force required to keep it sliding will be less than the force originally required to get it sliding. The coefficient of sliding friction will normally be less than the coefficient of static friction.

Question! Can you name one part of a Corvette that isn't moving as the 'vette is motivating down the highway at 130 per? Hint...it's on the other side of the part that's moving at 260. Answer... the part of the tire that's in contact with the highway, and the top side of the tire is moving at 260 mph. The part of a rolling tire that is contact with road (or "contact patch") is not moving relative to the road. Remember those tire commercials on TV showing an earthworm's view of a tire moving through a layer of water? Gently apply the brakes and a retarding friction force is generated and the wheel's contact patch is still stationary relative to the road. Increase the pressure and the friction force increases. Increase it some more and the tire starts to skid, the contact patch is moving at a low speed across the road. Increase it some more and the skid becomes more pronounced. The tire is still rotating but the contact patch is sliding down the road at a good clip. Finally, force down on the brake pedal hard. Blue smoke is emerging from the tires and an ear shattering squeal splits the air. The wheel has stopped rotating and the contact patch is skidding over the road at 130 mph, or what ever the speed of the 'vette is. Look back down the road and you can see that the rubber from that flat spot on the tire has been deposited in a long black line which traces the motion of the tires.

It turns out that the coefficient of friction for the tire is a function of the degree to which it is skidding

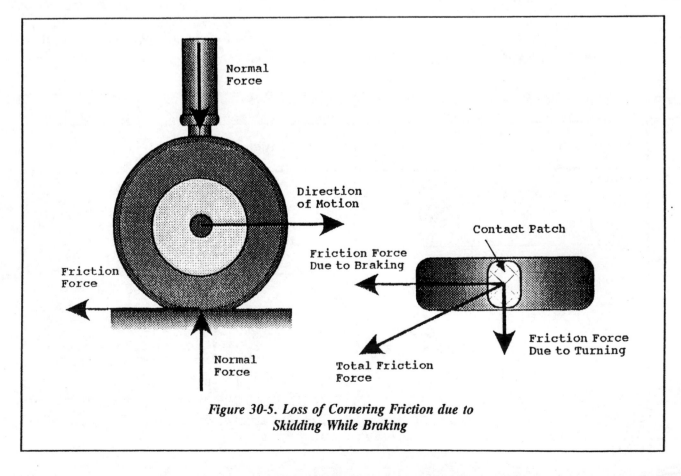

Figure 30-5. Loss of Cornering Friction due to Skidding While Braking

over the road (or runway). Although the coefficient of static friction is high, the coefficient of friction is even higher when the tire is skidding across a dry surface at 10% to 20% of the vehicles speed.

This is the famous 10% to 20% "rolling skid" that shows up in many references as the "optimum braking" condition. Although some rubber is being laid down, the tire is still turning at 80% to 90% of the speed it would be if the wheel were rolling freely and the rubber is coming from all around the tire. As the tire's skid increases above 20%, the coefficient of friction drops to the point where, at a 100% skid, the coefficient of friction is just a fraction of its static friction value. The tire's contact patch is stationary and no longer dissipating heat as it moves around the tire's circumference. The heat generated at the contact patch breaks down the rubber, melting it, and reducing its ability to adhere to the surface.

Another factor which can affect the safety of an airplane during a high speed abort, is its ability to corner or steer. As a wheel is braked and starts to skid, the ability to create or resist a side force decreases. This means that if an airplane which is undergoing strong braking experiences side forces, it will have little ability to resist side forces such as crosswinds or gravity forces resulting from the effects of crowned runways.

B. HYDROPLANING. The paragraph above discusses what happens when a tire is being braked on a dry surface. If we add a layer of water to the surface we get a whole new picture. The presence of a layer of water can reduce the coefficient of friction in three different ways; viscous hydroplaning, dynamic hydroplaning, and reverted rubber hydroplaning. All three can degrade both the braking and cornering ability of the airplane, but while viscous and reverted rubber hydroplaning can occur only during braking or cornering, dynamic hydroplaning can occur any time sufficient speed and water depth exist. On the other hand, viscous and reverted rubber hydroplaning don't require much water and can occur when a runaway is simply damp and airplane speeds are relatively low. Hydroplaning affects both the stopping distance and directional control of an airplane. The loss of cornering or side-force capability when braked wheels are operated at slip ratios greater than 25% can account for the tendency of airplanes to weathervane into the wind when braking on wet runways during crosswind operations.

Loss of tire braking and cornering ability during operations on damp (less than 0.01 inches thick) or wet (0.01 to 0.1 inches thick) runways is predominantly attributable to viscous hydroplaning. The conditions are ripe for viscous hydroplaning occur when relatively thin films of water reduce the coefficient of friction between the tire and the runway. In the most simple terms, it makes the runway slippery. This thin layer of water can reduce the braking and

	Minimum Speed Necessary	Minimum Water Depth Necessary	Evidence
Viscous	low speeds	<0.01 inch	Little to none.
Dynamic Wheel Rotating	$9\sqrt{p}$	~0.1 inch for worn tires	Witness statements of standing water and/or rainfall rate.
Wheel Stationary	$7.7\sqrt{p}$	~0.3 inch for new tires	Depth of tire tread. Lack of "Grooving" & "Crowning" of Runway
Reverted Rubber	low speeds	<0.01 inch	Each tire leaves a double black stripe with "steam" cleaned center.

Figure 30-6. Comparison of Viscous, Dynamic and
Reverted Rubber Hydroplaning

cornering ability of a tire by reducing the coefficient of friction between the tire and the runway surface. The texture of the runway, the skid resistance of the exposed aggregate in the runway and the tire's tread determine just how slippery the runway will be. On smooth runways, a layer of water only 0.01 inches thick can significantly reduce the runway's coefficient of friction. This reduction on friction can occur at any tire speed.

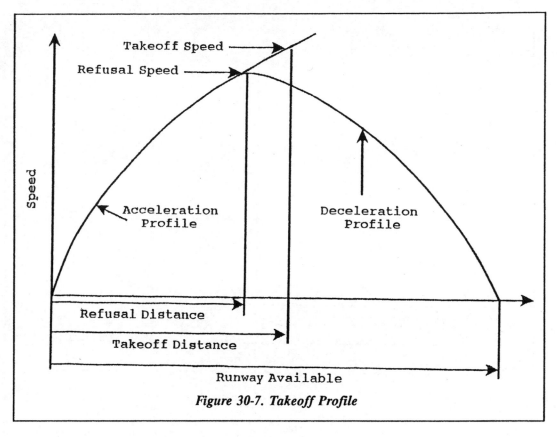

Figure 30-7. Takeoff Profile

Dynamic hydroplaning occurs when the forward speed of the tire is coupled with a relatively deep layer of water which raises part (partial hydroplaning) or all (total hydroplaning) of the tire's contact patch off the runway surface. When total hydroplaning occurs the tire will no longer be in contact with the runway and essentially all braking and cornering ability are lost. The depth of the water required to support dynamic hydroplaning varies from 0.1 inch for a well worn aircraft tire to 0.3 inches for new tires with full tread depth. The speed required to generate water pressures sufficient to raise the tire's contact patch off the runaway depends on the tire's pressure and whether or not the tire is rotating when the tire either enters the water or reaches its hydroplaning speed. If a tire is rotating at a speed which cancels motion between the contact patch and the runway as the tire moves through water sufficiently deep to support hydroplaning, the minimum hydroplaning speed (in knots) will be nine times the square root of the tire pressure. This is typical of an airplane on the takeoff roll, decelerating after a refused takeoff or decelerating after the wheels have spun up after a landing. If the tire is not rotating when it enters the water, the minimum hydroplaning speed (in knots) will only be 7.7 times the square root of the tire pressure. This is typical of an airplane touching down in water that is deep enough to cause hydroplaning. That means that the hydroplaning speed for an airplane will be lower during touch down than during either the takeoff roll or during the landing roll after the wheels have "spun up". An airplane with a tire pressure of 30 psi can hydroplane when touching down at speeds greater than 42 knots, while during the takeoff roll or during the landing roll after the wheels have spun up, the aircraft will have to be moving at least 50 knots before hydroplaning can occur. If the tire pressure is higher, hydroplaning speeds will be higher while if tire pressures are lower, hydroplaning speeds will be lower. Cold soaking tires at high altitude can reduce tire pressures, which when coupled with the naturally lower touchdown hydroplaning speeds can cause an even greater reduction in hydroplaning speeds.

It should be remembered that dynamic hydroplaning affects more than braking. It can also degrade the pilot's ability to steer the airplane or even maintain a straight track down the runway. Strong crosswinds coupled with standing water on a runway can result in an airplane literally being blown off the side of the runway. Crowned and grooved runway can assist in the rapid draining of a runway, limiting the depth of water standing on the runway. Tire tread also influences the depth of water necessary to support dynamic hydroplaning.

Reverted rubber hydroplaning occurs during

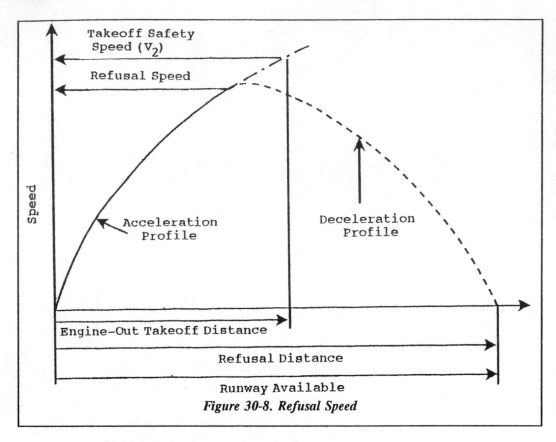

Figure 30-8. Refusal Speed

airplanes with relatively flexible landing gear, an experienced pilot may be able to feel the subtle changes in deceleration necessary to achieve maximum braking. However, for large airplanes, the assistance of an anti-skid braking system is necessary to achieve high levels of braking without locking the wheels and blowing out the tires.

When maximum braking is commanded, these systems will provide braking pressure until the wheels indicate the start of a skid and then pressure will be released. When examining the skid marks made by a tire being braked by a properly functioning anti-skid system, the path of the tires will be shown as a trail of skid marks alternating with no skid marks, reflecting the alternating high pressure and relaxed pressure being applied to the wheels.

braking when the heat of friction developed at the contact patch causes the reversion of the rubber to its un-cured state and heats into steam the moisture in a damp runway. The pressure of the steam is apparently sufficient to raise the center of the tire off the runway while the edges remain in contact. All of this naturally greatly reduces the coefficient of friction available during braking. Proof of reverted rubber shows up in the skid marks laid down by the tires. Two black tracks where the reverted rubber on the edges of the contact patch is laid down on the runway and between the black stripes a clean section of the runway where it has literally been steam cleaned. On a concrete runway this steam cleaned stripe will look almost white.

C. BRAKING TECHNIQUE. When it comes to braking, the trick is to keep the tire rotating at the speed where the contact patch is slightly skidding down the runway, developing the maximum amount of friction while maximum weight is being applied to the wheel. If too little braking torque is applied to the wheel, the wheel will turn with too little retarding force being developed and stopping distance will increase. On the other hand, if too much braking torque is applied to the wheel, the tire will either skid, or viscous or reverted rubber hydroplaning can occur and stopping distance will again increase. In small

5. REJECTED TAKEOFFS AND LANDING OVERSHOOTS.

Each year aircraft run off the end of the runway while attempting to stop. Some of these overshoots occur during landing and are in part attributable to long landings, landings with excessive speed, tailwinds and/or a downhill runway gradient. The effects of these factors will be covered in the next section. This section will concentrate on factors which can affect decisions involving refusal speed and critical field length.

A. REFUSAL SPEED. The maximum speed an airplane can accelerate to and then stop on the remaining runway is referred to as its refusal speed. The length of runway it takes to accelerate to takeoff speed and then decelerate to a stop is referred to as the accelerate-stop distance. The refusal speed and accelerate-stop distance are based on four primary factors:

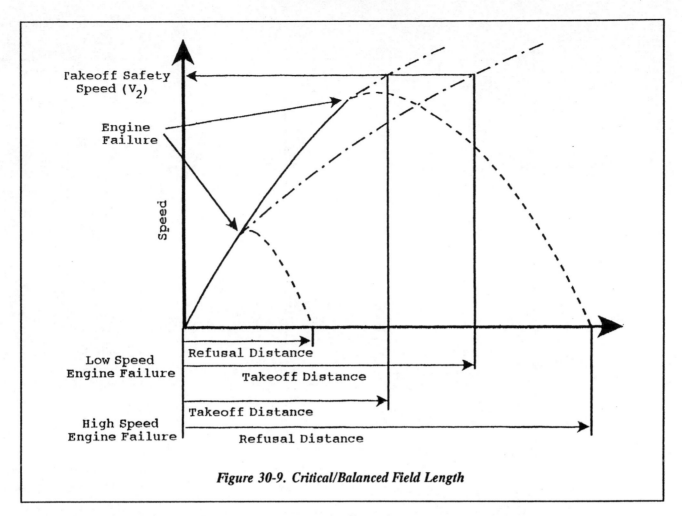

Figure 30-9. Critical/Balanced Field Length

1. The rate at which the airplane can accelerate from a stop to takeoff speed.

2. The length of time it takes the pilot to transition the airplane from accelerating down the runway at full power to decelerating down the runway at maximum braking.

3. The ability of the airplane to decelerate to a stop. .

4. The length of runway available to accelerate, transition to braking and then stop.

Generally speaking, it is nice if an airplane can accelerate to takeoff speed and then decelerate to a stop on the remaining runway. In a single engine airplane, we would like to be able to stop on the runway should the engine fail during the takeoff roll. In multi-engine airplanes however, the issue of continuing the takeoff with partial power is definitely an option that should be considered. Continuing the takeoff with partial power after experiencing an engine failure late in the takeoff roll may be less risky than

attempting a high speed, heavy weight abort. I n order to provide multi-engine pilots with better information upon which to make the continue/abort decision, the concepts of critical field length and critical engine failure speed were developed.

B. CRITICAL FIELD LENGTH. The concept of critical field length and critical engine failure speed is a little more complicated than that of refusal speed and accelerate-stop distance. Consider a multi-engine airplane which experiences an engine failure early in its takeoff roll and has the ability to takeoff and climb on the remaining engine(s). It should be obvious that it will take less runway to decelerate to a stop than it will take to continue to takeoff speed under partial power. The safest option would be to abort the takeoff. Now lets look at the other extreme, engine failure shortly prior to achieving takeoff speed. In this situation it will take less distance to continue to takeoff speed on the remaining power. Attempting to stop from just below takeoff speed will eat up a lot of runway and, in most large multi-engine airplanes, expose those on board to more risk than continuing the takeoff. Now let's look at the situation between these

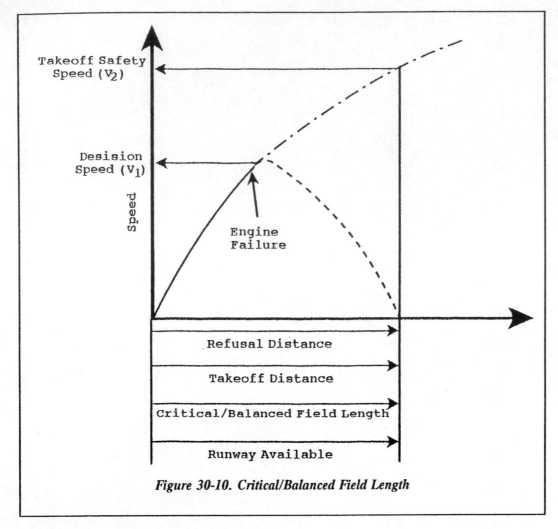

Figure 30-10. Critical/Balanced Field Length

when to continue the takeoff is not totally resolved within the aviation industry.

The FAA no longer uses the term "critical field length" and "critical engine failure speed" in the rejected takeoff decision process. Instead the term "decision speed", or V_1 is used to establish to point beyond which the pilot should, having experienced an engine failure, continue the takeoff. The difference between the old "critical engine failure speed" and "decision speed" is subtle but significant. As currently used, decision speed refers to the maximum speed at which the pilot should initiate decel-

two extremes where the decision is relatively easy. At some point during the takeoff roll the airplane is at a speed where it will take exactly the same distance, following an engine failure, to either accelerate to takeoff speed or decelerate to a stop. This speed is referred to as the "critical engine failure speed." If an engine fails prior to accelerating to this speed, it is normally safer to abort the takeoff. If an engine fails after accelerating through this speed, it is generally safer to continue the takeoff on the remaining power.

The distance required to accelerate to this speed and then either decelerate to a stop or accelerate to takeoff speed (these last two are by definition equal) is referred to as the critical field length. It sounds easy, and at this level of complexity it is simple; abort if an engine fails before reaching the critical engine failure speed and continue if an engine fails after passing critical engine failure.

But, unfortunately, things are not quite that simple and the whole issue of when to refuse a takeoff and

eration for a refused takeoff. It is not really a "decision speed" at all. It is based on the pilot having already reached a decision, determined the actions necessary to implement the decision, command muscles to move and then have their various appendages do their job.

What this means is that if the pilot hasn't already reacted to initiate the refused takeoff by the time V_1 shows up on the airspeed indicator, it is generally better to continue the takeoff. Basically that means that the engine failure has to occur some time prior to V_1. Three seconds is the time allocated by the FAA for the pilot to recognize that an engine failure has occurred, decide on the proper course of action, and then initiate actions to discontinue the takeoff and start the deceleration process.

This number is considered optimistic by many safety professionals. In addition, some pilots do not understand the full meaning of V_1. Perhaps more unfortunate, is the industry's failure to make the refused takeoff decision making process easier for the pilot.

Another issue arises from the fact that only engine failure is considered as an initiating event for the continue/refuse decision. Other reasons for the decision, such as blown tires (which can greatly increase stopping distance) and warning signals (which may or may not compromise a continued takeoff), occur at least as frequently as engine failures, but are not considered.

Before we examine some of the things that may go wrong and invalidate the calculations made to compute and validate decision speeds, it is necessary to consider the fact that most refused takeoffs are initiated for reasons other than engine failure. These reasons could range from warning signals to blown tires. Some failures may affect the ability of the airplane to fly, others are false warning systems which, if ignored, have no effect and others may affect the ability of the airplane to stop.

When the failure comes at high speeds, in two of these three situations it is more dangerous to attempt to stop than to continue the takeoff. A blown tire can definitely affect an airplane's ability to stop. When an airplane undergoes refused takeoff certification, the airplane's kinetic energy at the moment the brakes are applied is largely converted into heat energy which has to be absorbed by brakes alone.

If one or more brakes are not available for braking due to tire or brake system failure, the remaining brakes have to pick up the load. If one of the remaining brakes becomes overheated and then fails, the braking load is then transferred to the others which sets up the potential for a chain reaction of brake failures. Stopping distance increases and directional control is degraded.

Blaming the pilot for failing to determine the cause of a malfunction and initiating the proper corrective action in less than three seconds, may make an example of the pilot, but will not effect the overall outcome.

C. COMPLICATING FACTORS. History has shown that handbook- based predictions rarely match actual accident experience. The problems center on three different areas: the predictions of the airplane's acceleration performance, the reactions of the crew in the transition from takeoff to abort, and the decelera-

tion performance of the airplane.

D. ACCELERATION LESS THAN PREDICTED. It is possible for an airplane to reach V1 with more runway behind it (and therefore less remaining in front) than originally planned. This can be due to pilot actions, environmental conditions and/or airplane performance. In any case, the airplane will arrive at V_1 further down the runway.

Some of the factors that should be evaluated when investigating the acceleration portion of a refused takeoff accident are listed below.

1. Aircraft position on the runway at the start of the takeoff roll. Did the pilot position the airplane at the end of the runway prior to takeoff or was the airplane already part way down the runway when takeoff power was applied? Was an intersection takeoff attempted?

2. Rate of application of thrust/power to specified setting. How long did it take for takeoff thrust/power to be achieved? Was auto-throttle available and used?

3. Actual thrust/power achieved during the takeoff roll. Did the crew set the thrust/power at the correct takeoff values. Were the engines capable of developing advertised takeoff thrust/power?

4. Extra drag during the takeoff roll. Were brakes dragging during the takeoff roll? Was the parking brake still set? Questions such as these are especially critical since it will not only affect the acceleration portion of the accident sequence, the heat energy absorbed during the acceleration phase could cause brake failure during the deceleration phase. Was the airplane properly configured (flaps, speed brakes, etc.)? Was the runway surface contaminated? Standing water and/or snow on the runway can cause a retarding force. Was the airplane prematurely rotated to a nose high attitude?

5. Unplanned for runway gradient and/or tail wind caused V1 to be achieved further down the runway than planned.

E. ABORT PROCEDURES COMPLETED AFTER THE AIRPLANE HAS EXCEEDED V_1. A successful high speed abort on a runway whose length is at or slightly longer than the airplane's critical field length depends on the pilot's ability to reduce power and initiate maximum braking within a short period of time. The exact timing of the interval during which the pilot must perform these tasks is the subject of some debate and is not totally standardized among aircraft manufacturers and certification agencies.

Some manufactures argue that V_1 is the speed at which the pilot needs to have retarded power and is applying brakes. Others imply that the pilot has three second from V1 to retard the power and start applying the brakes. In any case, the procedure used to calculate pilot reaction times and validate their adequacy should be examined by the investigator.

1. At what speed did the pilot in control of the airplane start the abort sequence?

2. Were speed call outs as specified?

3. Did the cockpit crew communicate the nature of the emergency and intended procedures?

4. Was the abort sequence executed as specified?

5. How long did it take to complete the abort sequence?

F. AIRCRAFT RATE OF DECELERATION LESS THAN PREDICTED. There are many factors which can cause the aircraft to decelerate slower than predicted and therefore consume more runway before coming to a stop. Like the failures of the airplane to accelerate as predicted, they can be divided into three different categories: pilot actions, environmental conditions and airplane performance.

1. Aircraft configured for maximum braking. Was power adjusted to the appropriate setting? Were appropriate drag devices deployed in the proper sequence? Were aerodynamic devices which place maximum weight on the braked wheels properly positioned? Deployment of wing mounted spoilers, retraction of wing flaps and nose up pitch commands (short of raising the nose wheel off the runway) are all techniques which have been used in various airplanes to place more weight on the braked wheels thereby increasing the braking force available for wheel brakes. When aircraft configuration is established through "squat" or "WOW" (weight-on-wheels) switches in the landing gear, proper functioning of the switches and the linkages which cause them to function should be confirmed.

2. Maximum braking attempted. Did the crew apply maximum braking at the start of the abort? It is not always apparent that maximum braking will be required to stop the airplane in the remaining runway.

3. Energy available in the wheel braking system. The kinetic energy consumed by the wheel brakes is converted into heat energy stored in the brake discs. There are limits to the energy each brake can absorb before failing. Two commonly encountered limitations are:

- the thickness of the brake discs. Worn discs are thinner and can therefore absorb less energy before failing. Once a brake fails it not only increases the deceleration distance because of its failure to contribute to the retarding force, the energy it would have absorbed will have to be consumed by the other brakes, causing a chain reaction of brake overheating and failures.

- the temperature of the brake discs before starting the takeoff (long, down hill, down wind taxiing, especially in airplanes with high residual thrust at idle can consume a significant proportion of the a brake's energy storage capability).

4. Energy to be absorbed during braking. It is sometimes incorrectly assumed that the energy to be absorbed by the brakes is equal to the kinetic energy of the airplane at the start of the abort minus energy subtracted through aerodynamic braking and reverse thrust. The effect of thrust produced by the fan section of a high by-pass turbo fan engine as it spooled down from takeoff power to idle was overlooked during the refused takeoff certification of a jet transport. The extra energy contributed during the seconds it took to spool down plus test procedures which allowed the use of new, full thickness brake discs lead to refused takeoff performance figures which were optimistic and brake failures and runway overshoots during actual usage.

6. FACTORS AFFECTING TAKEOFF AND LANDING PERFORMANCE.

Although exact predictions and analysis of an aircraft's takeoff and landing performance require detailed information about the airplane, its load and the environment, it is possible to make some general statements about the effects of changes in weight, density altitude, wind and runway slope. Armed with this information an investigator can make some general conclusions in the field without the need to dig out the books. Of course these rules of thumb should sooner or later be verified the long way.

A. EFFECT OF WEIGHT CHANGE ON TAKEOFF AND LANDING DISTANCE.

It should be obvious to pilots that when an airplane's weight is increased, its minimum takeoff and landing rolls are longer. Likewise, if its weight is decreased its minimum takeoff and landing rolls are shorter. But how much? Examining some simple principles of physics and aerodynamics can provide us with rules of thumb and quick approximate answers.

The gross weight of an airplane affects length of a takeoff roll in two ways.

First, if an the airplane's weight changes, its acceleration will experience a change which is inversely proportional to the change in weight. Increase the weight, decrease the acceleration; decrease the weight, increase the acceleration. The change in acceleration will then cause the distance required to accelerate to a given speed to change. Heavily loaded airplanes will accelerate slower than lightly loaded airplanes, lightly loaded aircraft faster.

Second, the takeoff speed of an airplane is directly proportional its weight. Higher weight, higher takeoff speeds. And, all other things being equal, it takes a longer roll to achieve these higher speeds.

The net result is that the a change in weight will cause a change in takeoff distance which is inversely proportional to the square of the weight change ratio. This relationship is shown in equation 2. It should be remembered that this equation is approximate, it will get you in the ball park when a quick answer is necessary or if a "TLAR" (That Looks About Right) is necessary to verify a computer output. Aircraft performance manuals and computer programs will still provide more accurate answers, but equations like the one shown below can provide a "big picture" perspective.

$$S_2 = S_1\left(\frac{W_2}{W_1}\right)^2 \qquad \textbf{(2)}$$

where:

S_2 = the new takeoff roll.
S_1 = the original takeoff roll.

W_2 = the new takeoff weight.
W_1 = the original takeoff weight.

For landing the situation is a little different.

First, heavier weight will mean higher landing and touchdown speeds And the higher speeds will mean longer landing rolls. However, the higher landing weights do not necessarily mean slower rates of deceleration during the landing roll. By examining the friction forces developed by an object sliding across a surface, it can be shown that the friction force developed is independent of the weight and depends only on the coefficient of friction between the object and the surface.

Second, during the takeoff roll, there was only one force causing acceleration, during the landing roll there can be several forces causing deceleration (e.g. brakes, aerodynamic drag, reverse thrust, drag chutes). If we could assume that decreases in aerodynamic drag during the landing roll were balanced by increases in drag produced by the wheel brakes and the airplane slowed and that the effects of reverse thrust or drag chutes where not significant it can be shown that the changes in the landing roll will be directly proportional to changes in weight. If all of the above were true, and that is often a reasonable assumption, and the brakes all functioned properly during the landing we could show that changes in landing weights will cause proportional changes in the landing distances.

$$S_2 = S_1\left(\frac{W_2}{W_1}\right) \qquad \textbf{(3)}$$

where:

S_2 = the new landing roll.
S_1 = the original landing roll.
W_2 = the new landing weight.
W_1 = the original landing weight.

B. EFFECT OF DENSITY ALTITUDE ON TAKEOFF AND LANDING DISTANCE.

Changes in pressure altitude or temperature will cause changes in density altitude which in turn affect both the thrust/power produced by normally aspirated engines and the true airspeed for the airplane at takeoff. Higher density altitudes will reduce the thrust of these engines, reduce the airplane's acceleration and therefore increase the distance required to accelerate to a given airspeed. Since higher density altitude means lower density air, higher true airspeeds are needed to generate the dynamic pressures necessary to produce the lift required to support the full weight of the aircraft.

For normally aspirated engines, the thrust and power produced by the engine will vary inversely with the density altitude. Higher density altitude means lower thrust/power, slower acceleration and longer takeoff rolls. The equation shown below shows how changes in takeoff roll attributed to changes in density altitude will be directly proportional to the square of the change of density altitude. Once again a word of caution. This equation shows general relationships which are useful in determining the "big picture". The equation should not be used when precise answers are required.

$$S_2 = S_1 \left(\frac{\sigma_1}{\sigma_2} \right)^2 \qquad (4)$$

where:
   S_2 = the new takeoff roll.
   S_1 = the original takeoff roll.
   σ_1 = the original density altitude (the density altitude for a standard day at sea level is 1).
   σ_2 = the new density altitude.

For supercharged engines whose thrust/power output for takeoff is independent of density altitude, the effect of density altitude on takeoff roll will be less severe. The higher density altitude will still require a longer takeoff roll to achieve the higher true airspeed required for takeoff. However, the airplane's rate of acceleration will not suffer as it did for the normally aspirated engine.

$$S_2 = S_1 \left(\frac{\sigma_1}{\sigma_2} \right) \qquad (5)$$

where:
   S_2 = the new takeoff roll.
   S_1 = the original takeoff roll.
   σ_1 = the original density altitude (the density altitude for a standard day at sea level is 1).
   σ_2 = the new density altitude.

The effects of changes in density altitude on landing roll distance are pretty much the same as effects of changes in weight. Increasing density altitude will increase landing speeds and therefore

landing distances. However, if deceleration is achieved primarily through the use of wheel brakes (additional drag devices such as thrust reversers are not used), changes in density altitude will not cause significant changes in the airplane's rate of deceleration.

Therefore any changes in landing roll which are attributable to a change in density altitude will be directly proportional the change in density altitude.

$$S_2 = S_1 \frac{\sigma_1}{\sigma_2} \qquad (6)$$

where:
   S_2 = the new landing roll.
   S_1 = the original landing roll.
   σ_1 = the original density altitude (the density altitude for a standard day at sea level is 1).
   σ_2 = the new density altitude.

C. EFFECT OF WIND ON TAKEOFF AND LANDING PERFORMANCE.
When taking off into a head wind, the ground roll necessary for an airplane achieve to its takeoff airspeed will be less than the ground roll required for a no-wind condition. Likewise, the takeoff roll required for an airplane to accelerate to its takeoff speed while experiencing a tail wind will be more than the ground roll required for a no-wind takeoff.

While the mathematical relationship between the head/tail wind and the change in takeoff roll is more complex than those shown for weight and density altitude, it turns out that it is the same for both landing and takeoff. The equation used to estimate the effect of a head wind on a no-wind takeoff roll is shown below.

$$S_2 = S_1 \left(1 + \frac{V_W}{V_1} \right)^2 \qquad (7)$$

where:
   S_2 = the new takeoff or landing roll distance.
   S_1 = the original takeoff or landing roll distance.
   V_W = the head wind's velocity.
   V_1 = the takeoff speed.

The equation used to estimate the effect of a head wind on a no-wind takeoff roll is shown below.

where:
   S_2 = the new takeoff or landing roll.

$$S_2 = S_1\left(1 - \frac{V_w}{V_1}\right)^2 \qquad \textbf{(8)}$$

S_1 = the original takeoff or landing roll.
V_w = the tailwind's velocity.
V_1 = the takeoff speed.

The equation shown above can also be used to approximate the increase in the landing roll associated with touchdowns at speeds above the recommended touchdown speed. Simply use the estimated speed above recommended touchdown speed as the tailwind component and crank the numbers. Perhaps a more straight forward approach is to use the equation shown below.

$$S_2 = S_1\left(\frac{V_2}{V_1}\right)^2 \qquad \textbf{(9)}$$

where:
S_2 = is the new landing roll.
S_1 = is the original landing roll.
V_2 = the actual touchdown speed.
V_1 = the planned touchdown speed.

D. EFFECT OF RUNWAY SLOPE ON TAKE-OFF PERFORMANCE. Takeoff up hill and down hill will also affect the length of a takeoff roll. When examining the illustration shown below it can be seen that the component of the airplane's weight which acts parallel to the runway surface will cause an additional acceleration force if the runway slopes down hill and a drag force if the runway slopes up hill.

The rule of thumb for the effect of runway gradient is that each degree of up hill slope increases the takeoff distance by 5%, while each degree of down hill slope decreases the takeoff distance by 5%.

E. WIND-SHEAR EFFECTS ON TAKEOFF AND LANDING. Horizontal wind shears which are parallel to the runway can cause significant hazards during takeoff and landing. During takeoff the primary concern is a situation during which the departing airplane either climbs out of a head wind or climbs into a tail wind.

In either case the airplane can experience a sudden loss of airspeed, which, if the airplane's speed is sufficiently low, could cause either stall or loss of altitude. During the final approach for landing an airplane descending out of an air mass with a strong head wind component into a relatively calm air mass will experience a sudden loss of airspeed.

If the pilot is slow to react with additional power or if there is insufficient altitude for the pilot to react, the airplane can stall. If on the other hand, the descent of the airplane on final approach results in the transition from a relative tail wind to a relative head wind, the crew may find themselves with more airspeed than they know what to do with. It is quite likely that the airplane will end up fast and high as it approaches the runway threshold. If improperly handled, the airplane could end up landing further down the runway than originally planned and at a higher speed. If coupled with a short or wet runway, this situation could end up as a runway overshoot.

Another crew might pull all the power to idle when encountering the head wind, lower the nose and set up a high sink rate, low thrust/power attempt to set the airplane down on the original intended touchdown point. If misjudged, an excessively hard landing could occur.

8. SUMMARY.

Aircraft accidents resulting from directional instability on the ground or from runway undershoots or overshoots are not as simple as they first appear. A number of factors can influence the situation and all of them should be evaluated.

This chapter complements Chapter 16 on Tires and Runway Accidents. Chapter 16 addressed some concepts on wheel and tire loading which are not covered here.

That chapter also covered stopping distance calculations using linear acceleration formulae and pointed out some of the inaccuracies of this type of calculation.

Chapter 16 made a basic point about this type of investigation which is worth repeating here.

We put wheels on airplanes to increase their utility; not to make the airplane fly any better. As a matter of fact, the wheels make it fly worse by adding weight and drag. Airplane designers would prefer to eliminate them altogether.

We only need the wheels twice per flight; takeoff

and landing. (Well, OK, also taxiing.) Considering that a fair percentage of aircraft accidents occur during takeoff or landing, the types of calculations described in these two chapters come up fairly often.

We should remember that if the plane is on the ground and the weight is firmly on the wheels, it tends to behave more like an automobile than an airplane. The majority of the research on skids, cornering forces, coefficients of friction and so on have actually been done on automobiles and most of that research applies to aircraft.

If you want to fully understand the dynamics of a rolling or skidding object on the ground, consult the textbook on Traffic Accident Reconstruction referenced in the bibliography.

31

Stability and Control

1. INTRODUCTION.

Stability and control are not real big players in the airplane accident investigator's life. Increased knowledge and the rigorous testing required for the certification of new airplanes have minimized the number of accidents which have lack of stability or controllability as a contributing factor or cause. When inadequate stability or controllability does raise their ugly heads, it is normally the result of a failure to operate the airplane in accordance with approved procedures; a failure of equipment or structure which provide the airplane with its necessary stability or controllability; and, occasionally, an oversight failure in the certification process.

This portion of the text will not try to turn you into an aeronautical engineer nor will it be a course in stability and control. Although we will talk about what stability and controllability are, the primary thrust will be on identifying the factors which influence stability and control. This will allow the accident investigator to know when to raise the questions, "Was the designed in stability and controllability of the airplane adequate?" Or, "Was stability and control compromised?" And, "Was stability and control involved in the chain of events leading to the accident?"

2. STABILITY VERSUS CONTROL.

In addition to having the ability to climb or accelerate, an airplane must be able to maneuver safely. This chapter will review the various factors which affect the handling qualities of an airplane, its stability and its controllability. Where the performance of an airplane is directly related to its excess thrust and power, the controllability and stability of an airplane deals with the ease with which an airplane's flight vector or attitude can be disturbed and its tendency to return to its original flight vector or attitude. If you think about it, controllability and stability tend to be opposites. If a designer makes an airplane more controllable, in other words makes it easier for the pilot to change the airplane's pitch, roll and yaw attitude, the airplane's natural reluctance to depart from its original flight vector must be reduced. On the other hand, if a designer increases an airplane's stability, the ability of the pilot to change the direction of flight will be decreased. The battle between, what has been in the past, the mutually exclusive attributes of controllability and stability has been going on since the dawn of powered flight. Samuel Langley, inventor of the "Aerodrome", an airplane which had the potential to become the first powered airplane to fly, crashed on takeoff just days before the Wrights made their first flight. While the design of Langley's "Aerodrome" emphasized "stability", the Wright's "Flyer" emphasized "controllability". Where the "Aerodrome's" design attempted to make it easy for the pilot to maintain straight and level flight (and therefore more sluggish to maneuver), the "Flyer's" design relied on the pilot's skill to keep the airplane on the intended course. Although the Wright's years of experimentation with un-powered gliders provided them skills necessary to control the airplane most of the time, the airplane and wind gusts did occasionally get the best of them. A days flying was often concluded not by sunset, but by a crash which could not be repaired before sunset. Traditionally, designers of fighter type airplanes have sacrificed stability in order to improve the pilot's ability to rapidly turn and point the airplane. Designers of transport and bomber airplanes, on the other hand, have emphasized stability, sacrificing the ability to rapidly change the airplane's attitude and flight path vector Today's modern fly-by-wire airplanes have finally provided the opportunity to achieve a high level of controllability while artificially inducing stability. More on that later. Let's start, and spend the major portion of this discussion, talking about how the natural design of an airplane affects the stability and controllability of an airplane.

3. STATIC VERSUS DYNAMIC STABILITY.

When we talk about an airplane's stability, we are talking about its natural tendency to return to its original flight path after it is disturbed from that flight path. No fair for the pilot to raise a wing or pull up the nose after they fall. The airplane has to recover all by itself. Stability is divided into two different types by time. The initial tendency of the airplane to correct for the disturbance is referred to as *static* stability.

How the airplane corrects over a period of time is referred to as *dynamic* stability. For instance, if an airplane's pitch attitude and flight path vector were increased due to a wind gust, and the nose naturally tended to lower itself, we would say the airplane had positive static pitch stability. The airplane's initial tendency was to correct for a nose up disturbance by lowering its nose. If the nose eventually returns to the original position the airplane has positive dynamic stability. It would be nice if the nose position returned to the original position in one steady motion, but it will still have dynamic stability if it oscillated through the original position a couple of times. It would be nice if all airplanes always exhibited positive static and positive dynamic stability. But that isn't always the case. In fact, there are three different types of static and three different types of dynamic stability. And two of each are undesirable.

Negative Static Stability - following a disturbance from equilibrium, the initial tendency is to continue in the direction of the disturbance. Bad thing!

Neutral Static Stability - following a disturbance from equilibrium, the there is no tendency to either increase or decrease the magnitude of the disturbance. Not good!

Positive Static Stability - following a disturbance from equilibrium, the initial tendency is to return toward the direction of equilibrium. Good thing!

If an airplane exhibits either neutral or negative static stability it cannot have positive dynamic stability. Without an initial tendency to return toward equilibrium, there is no way for the airplane to return, unas-

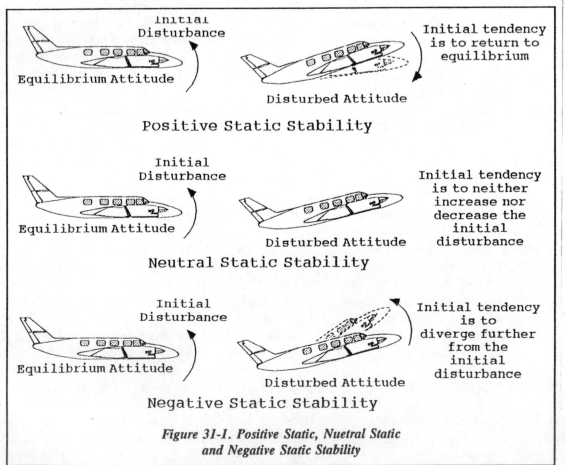

Initial Disturbance

Equilibrium Attitude

Initial tendency is to return to equilibrium

Disturbed Attitude

Positive Static Stability

Initial Disturbance

Equilibrium Attitude

Initial tendency is to neither increase nor decrease the initial disturbance

Disturbed Attitude

Neutral Static Stability

Initial Disturbance

Equilibrium Attitude

Initial tendency is to diverge further from the initial disturbance

Disturbed Attitude

Negative Static Stability

Figure 31-1. Positive Static, Nuetral Static and Negative Static Stability

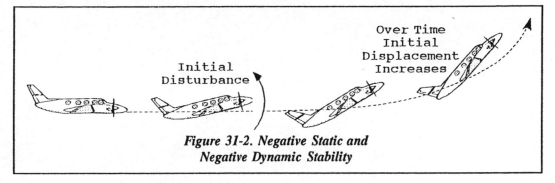

Figure 31-2. Negative Static and Negative Dynamic Stability

sisted to its equilibrium condition. On the other hand, having positive static stability does not guarantee that positive dynamic stability will exist. For, although the initial tendency to return to equilibrium may, in time, result in a return to equilibrium, the airplane could continue to oscillate on either side of equilibrium. The amplitude of these oscillations could remain constant, or even increase in size. Thus, positive static stability could result in positive, neutral or even negative dynamic stability.

A. NEGATIVE STATIC STABILITY - NEGATIVE DYNAMIC STABILITY.

If, following a disturbance from equilibrium, the initial tendency is to continue in the direction of the disturbance, the airplane exhibits negative static and will therefore always have negative dynamic stability. The tendency is for the magnitude of the disturbance to increase over time.

B. NEUTRAL STATIC STABILITY - NEUTRAL DYNAMIC STABILITY.

If, following a disturbance from equilibrium, the there is no tendency to either increase or decrease the magnitude of the disturbance, the airplane is said to exhibit neutral static stability and will therefore always exhibit neutral dynamic stability. The tendency over time is for the amplitude of the disturbance to remain constant.

C. POSITIVE STATIC STABILITY - VARYING DEGREES OF DYNAMIC STABILITY.

If, following a disturbance from equilibrium, the airplane's initial tendency is to return in the direction of the equilibrium, the airplane exhibits positive static stability. However, over time the maximum amplitude of the disturbance could increase, remain the same or decrease. Dynamic stability could be positive, neutral or negative.

If the distur-

bance's amplitude eventually returns equilibrium, the airplane's dynamic stability is said to be positive. If it returns to equilibrium in one oscillation, it is also said to be "dead beat". If the return to equilibrium takes more than one oscillation it is said to exhibit "damped oscillations".

If the disturbance's amplitude returns to equilibrium, but the oscillations do not decrease in amplitude, the airplane's dynamic stability is said to be neutral.

If the disturbance's amplitude returns to equilibrium, but the oscillations steadily increase in amplitude, the airplane's dynamic stability is said to be negative.

Stability exists around the airplanes three axes; pitch or longitudinal stability, roll or lateral stability, and yaw or directional stability.

D. POSITIVE STATIC LONGITUDINAL (PITCH) STABILITY.

Using the concepts we just discussed we would say that an airplane trimmed for level flight exhibits positive static longitudinal (pitch) stability if, following a nose up displacement (due to a vertical wind gust or a bump of the controls), the nose dropped toward its original attitude. Similarly, if the nose were momentary displaced downward, its initial tendency following the disturbance would be to rise toward its original position. Positive longitudinal stability, either natural or artificially induced by a stability augmentation system is essential for modern

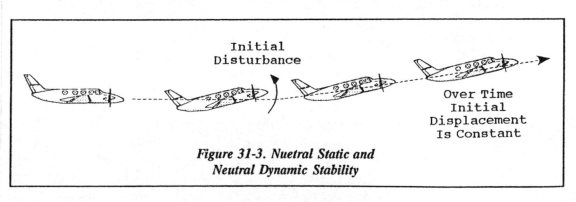

Figure 31-3. Nuetral Static and Neutral Dynamic Stability

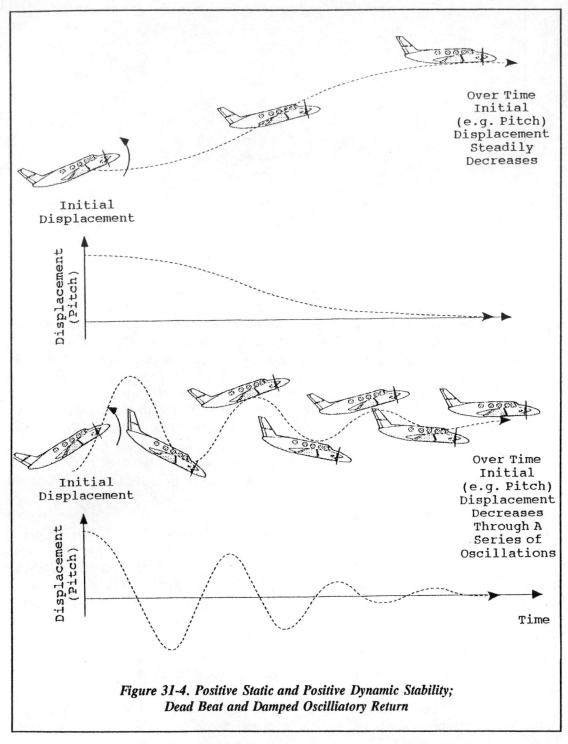

*Figure 31-4. Positive Static and Positive Dynamic Stability;
Dead Beat and Damped Oscilliatory Return*

move freely) and stick fixed (the pilot is holding the controls rigidly in the neutral position) static longitudinal stability. Since the stick free condition will allow the elevator to streamline into the wind, the horizontal tail surface available to force the return of the pitch attitude to its trimmed attitude is reduced. On the other hand, if the pilot's pitch control (stick or yoke) is held in the neutral position, more horizontal tail surface is available and positive static longitudinal stability is increased.

E. POSITIVE DYNAMIC LONGITUDINAL STABILITY. We should know by now that in order to have positive dynamic longitu-

airplanes. While small amounts of negative static longitudinal stability will simply increase the pilot's workload and degrade their ability to handle normal chores, serious degradation of static longitudinal stability will allow inadvertent nose displacements to grow so fast as to preclude corrective actions by the pilot and subsequent loss of control or over G'ing of the airplane. For airplanes with un-powered flight controls, there will be a difference between the airplane's stick free (the pilot is allowing the controls to

dinal stability we have to have positive static longitudinal stability. In other words, if the nose is inadvertently displaced upward from the trimmed pitch attitude, following the initial displacement, the nose will have to start downward. If the nose moves steadily toward the initial pitch attitude, or goes through a series of decreasing amplitude oscillations toward the initial pitch attitude, the airplane has positive dynamic longitudinal stability. Pitch displacements from the trimmed pitch position will, over time, be self correcting and the airplane will naturally return to the trimmed pitch attitude.

There are two modes of dynamic pitch stability with which the accident investigator should be familiar. The first mode of dynamic longitudinal stability is referred to as the Phugoid and is characterized by slow but noticeable changes in altitude, pitch attitude and airspeed while the angle of attack remains essentially constant. The period of the oscillation is typically long, lasting from 20 to 100 seconds with a very slow changes in pitch attitude. Although the relatively slow changes in pitch attitude can be corrected by the pilot, if this mode does exist and strict altitude control has to be maintained (the norm in today's IFR environment), pilot fatigue during long flights can be significant. Airplanes equipped with autopilots incorporating an operable altitude hold function can take care of this problem.

Dynamic longitudinal stability's second mode is relatively short period oscillation which involves relatively rapid and noticeable changes in pitch attitude and angle of attack while the airplane's velocity and altitude remain relatively constant. The typical period of this oscillation is from 0.5 to 5 seconds, with the amplitude of the oscillations decreasing by one half in about 0.5 seconds in a properly designed and maintained airplane. This latter time is referred to as the time half amplitude. This second mode does have

some critical safety aspects and is therefore of more than passing interest to the accident investigator. Since there are significant and rapid changes in AOA, there will be significant and rapid changes in the G induced load on the airplane's structure. In addition, it is possible for a pilot's attempts to speed the damping of the oscillations to progressively increase the magnitude of the oscillations. This has resulted in both loss of control accidents and catastrophic structural failure due to G induced overload. This latter complication is due to the fact that humans do not react instantaneously to external stimuli. In fact, the reaction time of humans (pilot's included) in this type of situation is roughly one second. The pilot's flight control inputs might be made during the phase of the oscillation where the input (intended to decrease the amplitude of the oscillation) is actually timed so as to increase its amplitude. In other words, with the perfectly "normal" pilot "in the loop", an airplane's natural positive dynamic longitudinal stability may become negative dynamic longitudinal stability. Loss of control and over Gs (both positive and negative) become real possibilities. Again, with a properly designed and maintained airplane, this should not be a problem. However, natural undesired aerodynamic characteristics are sometimes controlled through the use of control augmentation systems. And if these systems fail, the crew and passengers could be in for an interesting ride.

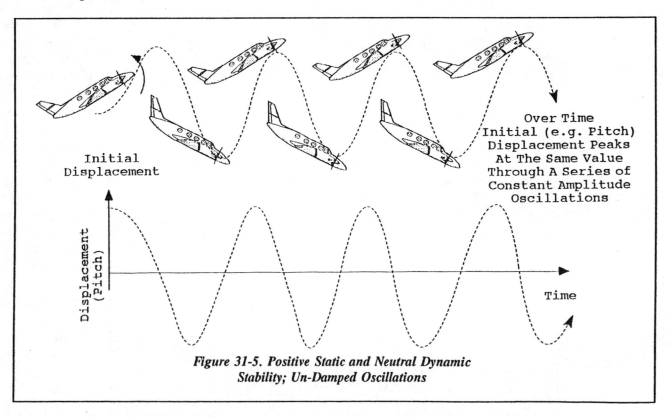

Initial Displacement

Over Time Initial (e.g. Pitch) Displacement Peaks At The Same Value Through A Series of Constant Amplitude Oscillations

Displacement (Pitch)

Time

Figure 31-5. Positive Static and Neutral Dynamic Stability; Un-Damped Oscillations

4. FACTORS INFLUENCING LONGITUDINAL STABILITY AND CONTROL.

In order for an airplane to exhibit positive static longitudinal stability, the airplane must respond to pitch disturbances by correcting back toward the trimmed pitch attitude. In other words, if the disturbance raised the airplane's nose, the natural tendency of the airplane would be to lower the nose. If the disturbance lowered the airplane's nose, the natural tendency of the airplane would be to raise the nose. These actions should take place without the assistance of the pilot. Some parts of the airplane assist this process (trying to lower the nose if it is raised and raising the nose if it is lowered). And some parts of the airplane work against it (trying to lower the nose further if it is lowered and raising the nose further if it is raised). The effect of the various parts of an airplane are summarized in the illustrations shown in Figure 31-8.

When a new airplane is being tested as part of its certification process, the sum of the pitching moments contributed by all of the airplanes parts must yield a pitching moment versus C_L curve the general characteristics shown in Figure 31-9. If the C_L increases due to an increase in AOA, the pitching moment of the airplane should become more negative. And, if the C_L should decrease (you guessed it, due to a decrease in AOA) the airplane should develop a nose up pitch attitude.

5. FACTORS INFLUENCING DIRECTIONAL AND LATERAL STABILITY AND CONTROL.

Directional (or yaw) stability involves an airplane's natural tendency to return its nose to the trimmed position when it is momentarily displaced to either the right or left. In other words, directional stability is the natural tendency of the airplane to rotate about its vertical axis and weathervane into the relative wind. Lateral or roll stability, on the other hand, involves the tendency of an airplane to roll when a yaw is induced. When an uncoordinated roll in an airplane with lateral static stability causes a side-slip, the side-slip will in turn cause a rolling moment in a direction which will cause the airplane to reverse its roll. Thus it can be seen that side slips cause both yawing and rolling moments. Most pilots learn this during their first few hours of flight instruction. If the right wing drops while both hands are busy with other duties, a gentle shove on the left rudder pedal will bring it up. "Look Ma! Roll control with no hands!" Generally, airplanes will exhibit greater positive directional stability than lateral stability. Some of the factors that determine the directional (yaw)

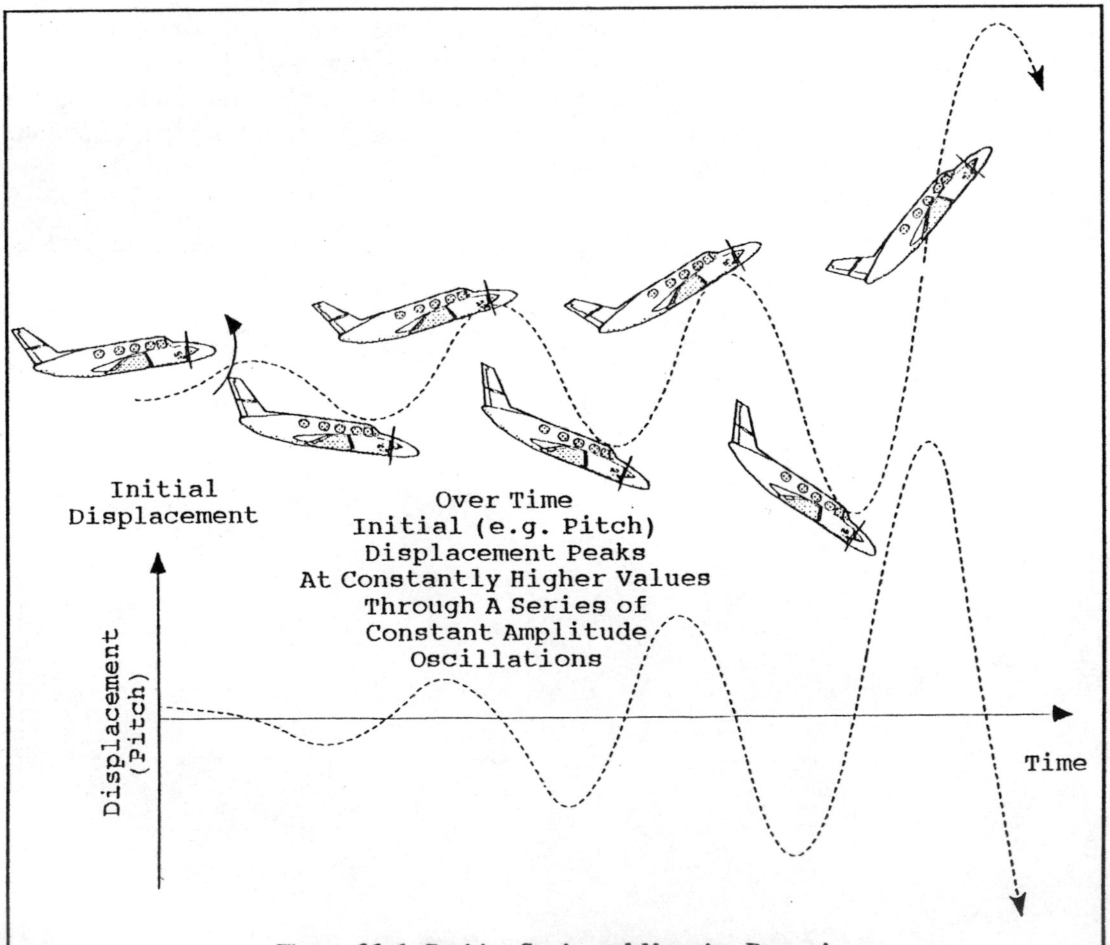

Initial Displacement

Over Time Initial (e.g. Pitch) Displacement Peaks At Constantly Higher Values Through A Series of Constant Amplitude Oscillations

Displacement (Pitch)

Time

Figure 31-6. Positive Static and Negative Dynamic Stability; Divergent Oscillations

stability and lateral (roll) stability are listed in Figure 31-10.

6. DEGRADATION OF AIRPLANE STABILITY AND CONTROL.

The stability which is designed into an airplane can, unfortunately, be offset by component or system failures or the actions of its operators. Failure of structural components which generate the aerodynamic forces necessary to produce positive static and dynamic stability can create piloting demands which exceed the abilities of even the best of pilots. Where marginal natural stability is augmented by autopilot and/or stability augmentation systems, lost of this assistance can again overload the pilot. Operation of the airplane outside of its approved operating envelope can also present the pilot with requirements which exceed the abilities of any pilot.

A. VIOLATIONS OF CENTER OF GRAVITY LIMITS. Both the controllability and stability of an airplane are affected by changes in the location of an airplane's Center of Gravity. Generally speaking an aftward movement of the C.G. decreases an airplane's longitudinal (pitch) stability. As the C.G. moves beyond its aft limits, the airplane will at first experience neutral stability and the pilot will have to manually correct for any inadvertent changes in AOA from the trimmed position. The pilot workload necessary to maintain the desired pitch attitude will increase. If the C.G. increases further aft, pitch stability will eventually become negative and any deviation from the trimmed AOA will cause an even larger deviation from the trimmed AOA. As the C.G. moves aft, the pilot's ability to control the pitch attitude can also be degraded. More and more trailing edge down elevator or stabilator is necessary to maintain a constant AOA. The airplane may run out of nose down control before it becomes statically unstable in pitch. In addition to causing problems in cruise flight, aft C.G.s can cause problems during takeoff and landing. Because of the extreme sensitivity of the airplane to pitch inputs while flying with an excessively aft C.G., rotation during

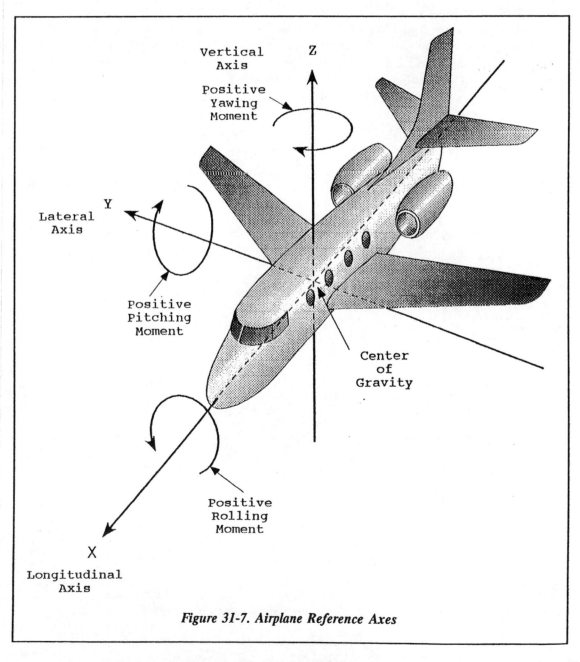

Figure 31-7. Airplane Reference Axes

takeoff and landing may be excessive, leading to premature lift-offs and stalls during takeoff and excessively rapid or high flares and stalls during landings.

Forward C.G.s bring their own set of problems. Takeoff speed and distances will increase because more down force will be required from the horizontal tail to raise the nose to the takeoff attitude. This increase in down force will mean more total lift is required, increasing takeoff speed slightly. In addition, more trailing edge up elevator and greater control forces will be necessary to raise the nose. If the pilot is not expecting these changes in control forces and travels, takeoff distances could be further increased. If the C.G. is far enough forward, the nose down moment created by the forward C.G. could exceed the tail down (or nose up if you like) moment created when nose up controls are applied and the takeoff distance could exceed that available on the runway. During landing, the pilot's control over nose up changes in pitch attitude can also be degraded. Normal control pressures and displacements may be insufficient to rotate the airplane from the pitch attitude on the final approach glide path to the touchdown attitude. This could result in nose wheel first, hard, or short landings. A dangerously forward C.G. during landing could result when an airplane takes off with a forward C.G. (but not enough to cause real problems) and then has the C.G. move further forward as the fuel burns down during the flight. During landing, the C.G.

could be well forward of its takeoff condition. If coupled with a steeper than normal final approach or a lower than normal initiation of the flare, this could set up a condition where a touchdown in a nose low attitude, at a high sink rate and at a position well short of the intended touchdown point is inevitable.

B. SYSTEM FAILURES. Any airplane system that contributes to the stability and control of an airplane also has the potential to degrade the airplane's stability or control. For instance, if a stability augmentation system relies on airspeed information to provide the correct gain between cockpit controls and movement of the flight controls, then a failure in the pitot static system could degrade the stability of the airplane. In one case a jet fighter traveling at high speed and relatively low altitude experienced a separation in the total pressure line which connected a pitot port to the airplane's central air data computer (CADC). Prior to the failure, the CADC was telling the stability augmentation system that the airplane was traveling at about 500 knots and the control surface movements were relatively small when compared to cockpit control movements. When the failure occurred, ambient static pressure instead of dynamic plus static pressure was provided to the CADC. The CADC then told the stab/aug system the airplane was flying at zero knots when it was actually flying at about 500 knots! The gain between cockpit control movements and flight control surface movements increased and the airplane quickly became dynamically unstable. Fortu-

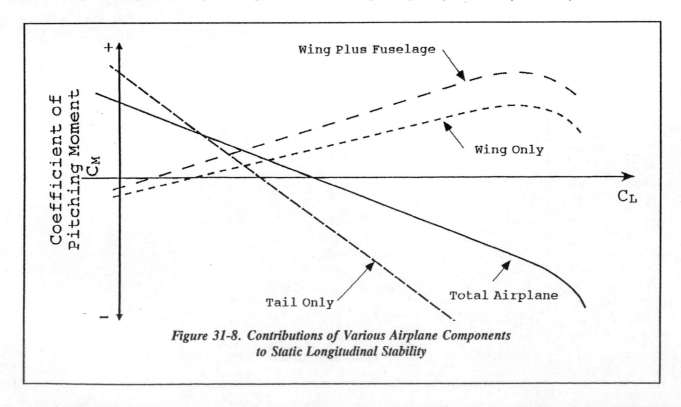

Figure 31-8. Contributions of Various Airplane Components to Static Longitudinal Stability

nately the pilot was able to disconnect the stability augmentation system and recover the airplane before catastrophically over G'ing the airplane or colliding with the terrain. The separation in the total pressure line would have been virtually impossible to discover if the airplane had crashed. The only clue would have been a reading of zero knots in the CADC while the wreckage pattern would have indicated the airplane was flying at very high speed. The message here is that if post accident evidence indicates that stability augmentation problems were a factor in the accident, the investigator should not limit the investigation to the stability augmentation system alone, but should also look at the systems which provided information and power.

C. HIGH ANGLE OF ATTACK OPERATION.
Airplanes which exhibit satisfactory stability and control characteristics at normal operating AOAs may become unstable or uncontrollable at high angles of attack. Static directional stability can be reduced when the stabilizing influence of the vertical tail is degraded by the effects of the airflow coming off the wing and fuselage. In addition, vortices generated by the forward part of the airplane, when coupled with yaw, can cause the tail to produce destabilizing side forces. Since the stall angle of attack of an airfoil is increased when a wing is swept aft, the effects of high AOA are more pronounced in swept wing airplanes. In

fact, the limiting AOA in these types of airplanes is often determined by the AOA at which directional stability becomes marginal, and not the AOA at which stall occurs. Changes in the location of the center of lift and the pitching moment created by the airplane's horizontal tail can cause the airplane to pitch-up or enter a deep stall (See Chapter 27). Any hazards associated with very high AOA operations are normally discovered during the flight testing associated with airplane certification and the airplane is either redesigned to eliminate the hazard or devices or warnings developed to minimize the risks generated by the hazard. The adequacy of these hazard controls should be addressed during the investigation.

On the other hand, subtle changes in an airplane's design which typically occur over time, may lead to degradation of an airplane's stability. A modern, front line jet fighter was discovered to have a deep stall characteristic which had been induced by the an increase in engine thrust which necessitated a larger inlet duct. A larger inlet duct, located forward of the C.G., was naturally destabilizing in pitch. When coupled with a stability degrading external fuel tank, the degraded stability associated with high Mach number, altitude, and AOA operations during typical fighter air-to-air operations, the deep stall mode was discovered not on the test range, but in the hands of a junior pilot. The synergistic effects of multiple small

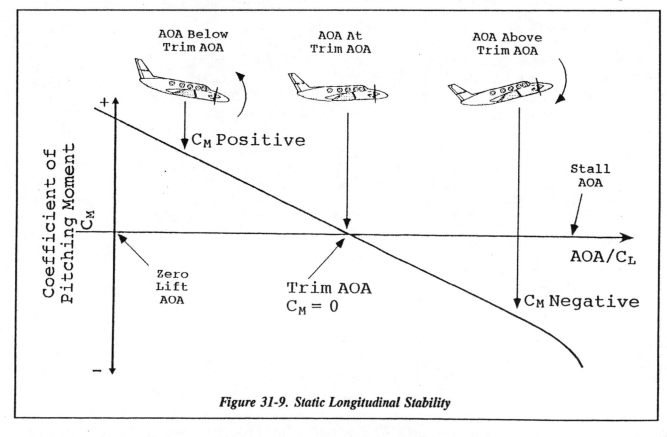

Figure 31-9. Static Longitudinal Stability

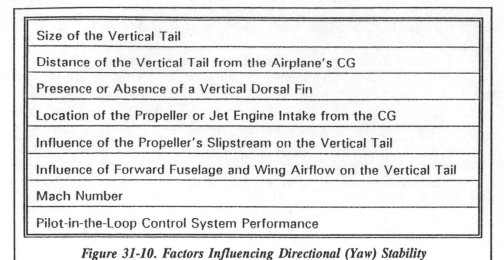

Size of the Vertical Tail
Distance of the Vertical Tail from the Airplane's CG
Presence or Absence of a Vertical Dorsal Fin
Location of the Propeller or Jet Engine Intake from the CG
Influence of the Propeller's Slipstream on the Vertical Tail
Influence of Forward Fuselage and Wing Airflow on the Vertical Tail
Mach Number
Pilot-in-the-Loop Control System Performance

Figure 31-10. Factors Influencing Directional (Yaw) Stability

changes had been overlooked by the development and discovered only after the system had been cleared for operational use.

D. DUTCH ROLLS. Dutch Roll can occur when an airplane's lateral (roll) stability is excessive with respect to its directional (yaw) stability, causing the yaw correction following sideslip to lag behind the roll correction which also follows a sideslip. This can lead to excessively high workloads on the pilot and the inability to maintain coordinated flight. If a tendency toward Dutch Roll is discovered during the certification testing of an airplane, it can be corrected by either increasing the directional stability of the airplane naturally (e.g. increasing the size of the vertical tail) or artificially (e.g. use of the airplane's autopilot or its stability augmentation systems). The FAA does not provide precise definitions of certification criteria concerning pilot perception/decision/reaction times. Instead, test pilots provide subjective opinions concerning the adequacy of he dynamic man/machine interface. These decisions on rare occasions may be optimistic and can place the three sigma low pilot having a three sigma low day outside their personal performance limits.

Although small amounts of Dutch Roll may be present in some airplane designs while at low (approach) speeds, loss of control due to Dutch Roll is normally associated with structural or system failures. Loss of a significant portion of an airplane's vertical tail has lead to severe Dutch Roll and the final out-of-control condition prior to collision with the terrain. Severe Dutch roll leading to complete loss of control has also occurred when a jet military cargo airplane flying at high altitude (and Mach number) experienced failures in both its autopilot and stability augmentation systems and the pilot could not control the resulting yaw/roll excursions. Luckily, during the resulting high speed dive, the increase in directional stability due to the lower Mach numbers associated with lower altitudes allowed the pilot to recover before collision with high terrain in the general area. Operation of some airplanes at very high altitudes can degrade natural aerodynamic damping of Dutch roll to the point where yaw damping is necessary.

7. ENGINE OUT OPERATION IN MULTI-ENGINE AIRPLANES.

When a multi-engine airplane with non-center line thrust experiences an engine failure, there is a speed below which the aerodynamic forces produced by the flight controls are inadequate to control the yawing moment created by the asymmetric thrust. If the airplane is flown below this speed, the pilot will not be able to control the yaw and the airplane will yaw and then roll uncontrollably into the dead engine. The FAA defines an airplane's Minimum Control Airspeed (V_{MC}) as "the calibrated airspeed, at which, when the critical engine is suddenly made inoperative, it is possible to recover control of the airplane with that engine still inoperative and maintaining straight flight either with zero yaw or, at the option of the applicant, with an angle of bank not more than five

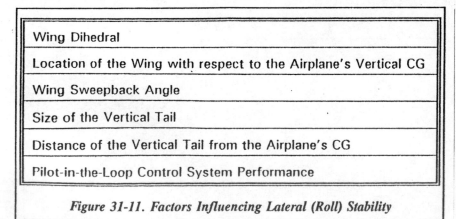

Wing Dihedral
Location of the Wing with respect to the Airplane's Vertical CG
Wing Sweepback Angle
Size of the Vertical Tail
Distance of the Vertical Tail from the Airplane's CG
Pilot-in-the-Loop Control System Performance

Figure 31-11. Factors Influencing Lateral (Roll) Stability

degrees. The method used to simulate critical engine failure must represent the most critical mode of power plant failure with respect to controllability expected in service."

For reciprocating engine-powered airplanes, V_{MC} must not exceed 1.2 V_{S1}. V_{S1} is determined at the maximum takeoff weight with:

- takeoff or maximum available power on the engines.

- the most unfavorable center of gravity.

- the airplane trimmed for takeoff.

- the maximum sea level takeoff weight (or any lesser weight necessary to show V_{MC}).

- flaps in the takeoff position.

- landing gear retracted.

- cowl flaps in the normal takeoff position.

- the propeller of the inoperative engine either:

- windmilling.

- in the most probable position for the specific design of the propeller control.

- feathered, if the airplane has an automatic feathering device.

- the airplane is airborne and the ground effect is negligible.

For turbine engine-powered airplanes, V_{MC} must not exceed 1.2 V_{S1}. V_{S1} is determined at the maximum takeoff weight with:

- takeoff or maximum available power or thrust on the engines.

- the most unfavorable center of gravity.

- the airplane trimmed for takeoff.

- the maximum sea level takeoff weight (or any lesser weight necessary to show V_{MC}).

- the airplane in the most critical takeoff configuration except with the landing gear retracted.

- the airplane is airborne and the ground effect is negligible.

At V_{MC} the rudder pedal force required to maintain control may not exceed 150 pounds, and it may not be necessary to reduce power or thrust of the operative engines. During recovery the airplane may not assume any dangerous attitude and it must be possible to prevent a heading change of more than 20 degrees.

Since multi-engine airplanes must be able to maneuver following an engine failure, the FAA has established the following additional requirements.

It must be possible to make turns with 15° of bank both towards and away from an inoperative engine, from a steady climb at 1.4 V_{S1} or V_y, with:

- one engine inoperative and its propeller in the minimum drag position.

- the remaining engines at not more than maximum continuous power.

- the rearmost allowable center of gravity.

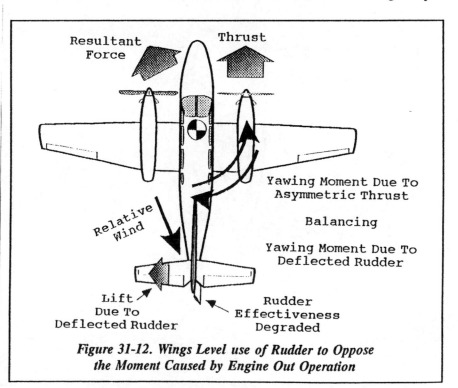

Figure 31-12. Wings Level use of Rudder to Oppose the Moment Caused by Engine Out Operation

- the landing gear extended.

- the landing gear retracted.

- the flaps in the most favorable climb position.

- maximum weight.

It must be possible, while holding the wings level to within 5°, to make sudden changes in heading safely in both directions. This must be shown at 1.4 V_{s1} or V_y with heading changes up to 15° with the:

- critical engine inoperative and its propeller in the minimum drag position.

- remaining engine(s) at maximum continuous power.

- landing gear retracted.

- landing gear extended.

- flaps in the most favorable climb position.

- center of gravity at its rearmost allowable position.

Figure 31-12 shows how the yawing moment (around the airplane's C.G.) produced by the operating engine can be countered by the moment produced by the lateral force produced by a rudder displaced toward the operating engine. Deflection of the rudder causes an increase in the camber of the rudder, and the resulting lateral force produces the opposing moment around the airplane's C.G. Thus, a multi-engine airplane with one engine inoperative can counter the yaw developed by the operating engine while in wings level flight. It should be obvious that a reduction in dynamic pressure (airspeed) will cause a reduction in the maximum horizontal force available.

It is also possible for the wings to produce an additional lateral force by placing the airplane in a side slip. By banking the airplane slightly into the operating engine, the wings will develop a small lateral force and cause the airplane to slip toward the operating engine. The positive angle of attack of the airflow over the vertical tail will create a lateral force, even without displacement of the rudder toward the operating engine. The resulting moment around the C.G. which will counter the moment produced by operation with one engine inoperative and the other engine producing thrust. This situation is depicted in the Figure 31-13. Again, as speed is slowed, the aerodynamic force

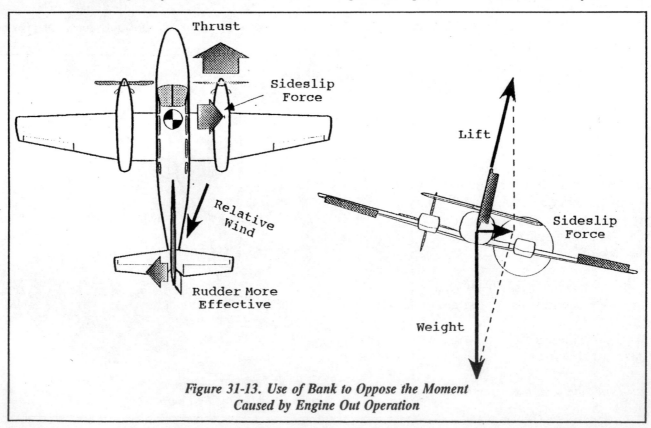

*Figure 31-13. Use of Bank to Oppose the Moment
Caused by Engine Out Operation*

produced by the rudder will decrease while the force produced by engine thrust will remain relatively constant.

A combination of rudder and bank into the operating engine will be necessary to maintain control of the airplane as speed is decreased toward V_{MC}.

By reviewing Figures 31-12 and 31-13, it should be apparent that any additional yawing moment that enforces the yawing produced by the operating engine will increase V_{MC}. Any additional yawing moment that opposes the operating engine's yaw will decrease it.

For twin engine airplanes whose engines rotate in the same direction, the engine whose propeller is moving downward while it is closest to the airplane's C.G. will be the critical engine since its failure will produce the highest V_{MC}. The reason for this phenomena is related to the "P" factor associated with operation at high angle of attack in single engine airplanes. At the higher angles of attack normally associated with engine out operation, the downgoing propeller will produce slightly more thrust, the upgoing propeller will produce slightly less thrust and the engine's center of thrust will shift toward the downgoing propeller. Thus, failure of the "critical" engine (the one whose inboard propeller is moving down) will result in operation on the engine whose downgoing propeller is furthest from the airplane's longitudinal axis, thereby creating the maximum yawing moment. On the other hand, the yawing moment created by the vertical tail will not change as a function of the engine which has failed. It should be obvious that loss of a "critical" engine will present the pilot with a more challenging (and dangerous) situation. The concept of a "critical" engine is shown in

Figure 31-14.

For twin engine airplanes whose propellers rotate in opposite directions, both engines are equally critical with respect to the magnitude of asymmetric thrust they produce. It should be noted that, all other things being equal, airplanes whose counter-rotating propellers move downward when they are furthest from the longitudinal axis will have higher V_{MC}s than airplanes whose propellers are moving downward when they are closest to the longitudinal axis. Since this second propeller configuration results in a higher and therefore more dangerous V_{MC}, there should be strong aerodynamic reasons for the selection of this propeller rotation configuration.

The critical engine in a four engine airplane will always be one of the outboard engines. If all propellers turn in the same direction, the "most critical" of the outboard engines will be the one whose propeller is moving downward when it is furthest from the airplane's longitudinal axis.

Operators of jet powered airplanes don't have to worry about the influence of "P" factor or the creation of "critical" engines. And, because jets don't have to worry about propeller clearances, they can be mounted

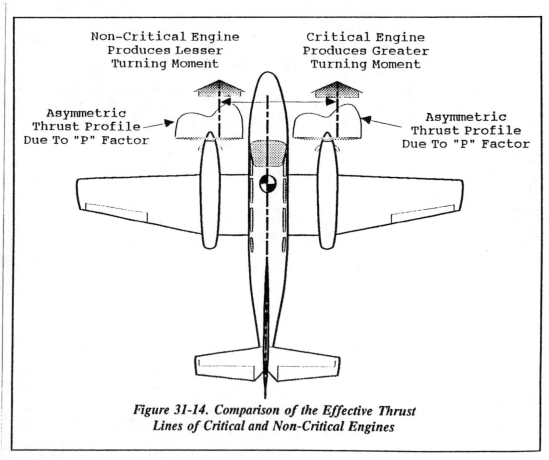

Figure 31-14. Comparison of the Effective Thrust Lines of Critical and Non-Critical Engines

much closer to the airplane's longitudinal axis, minimizing the problems associated with asymmetric thrust. The minimum control speeds associated with jets are normally well below the airplane's stall speed and are therefore not a significant hazard. However, system failures and the maneuvers performed can cause an increase in V_{MC} to the point where it is an issue even in jet airplane. Some of the factors which can affect V_{MC} are listed below:

- the density altitude at which an airplane is operating.

- an engine failure which degrades the ability of the airplane's flight control system to control yaw and roll.

- an engine failure which degrades the ability to change the airplanes configuration to one which creates less yawing moment or rolling moment toward the inoperative engine.

- pilot induced rolling and/or yawing maneuvers into the inoperative engine.

A. DENSITY ALTITUDE EFFECTS ON V_{MC}.
As density altitude increases, the thrust produced by a normally aspirated engine decreases, thereby reducing the maximum yawing moment into the inoperative engine. Since the yawing moments available from the aerodynamic surfaces are a function of the dynamic pressure, the calibrated airspeed necessary to maintain airplane control is decreased. This is shown for an airplane in level flight in Figure 31-15. V_{MC} also decreases at high density altitude when a shallow bank into the operating engine is used to maintain airplane control.

It is important to remember that as density altitude increases the calibrated airspeed at stall will remain constant. V_{MC}, on the other hand will, for the normally aspirated engine, be constantly decreasing. A potentially catastrophic situation can occur when a pilot, intending to practice V_{MC} maneuvers at high density altitudes, unexpectedly encounters a stall. With the critical engine at maximum power and the other engine at idle (feathered?), the yawing moment during the stall will be quite significant and the potential for entry into a spin is maximized. Once into the spin, failure to reduce the high power setting on the operative engine will place the airplane's attitude in a flatter, and normally less recoverable attitude while the yawing moment produced by the operative engine will increase yaw rates. This situation can occur when an airplane which has a V_{MC} which is higher than stall speed at low altitudes, is used to provide V_{MC} training at high enough density altitudes to cause the airplane to stall just as or before the pilot reaches V_{MC}. A non-recoverable spin may then be encountered. The critical relationship between V_{MC} is depicted in Figure 31-16. It should be remembered that the "book" V_{MC} is determined by using the engine's maximum available power. If less than full power is used deliberately, the V_{MC} for that lower power setting will be lower than the "book" value.

B. EFFECT OF CENTER OF GRAVITY ON V_{MC}. As can be seen from Figure 31-17, moving the C.G. aft reduces the moment arm between the C.G. and the vertical tail's center of lift. But, since VMC is originally determined using the C.G. "at its rearmost allowable

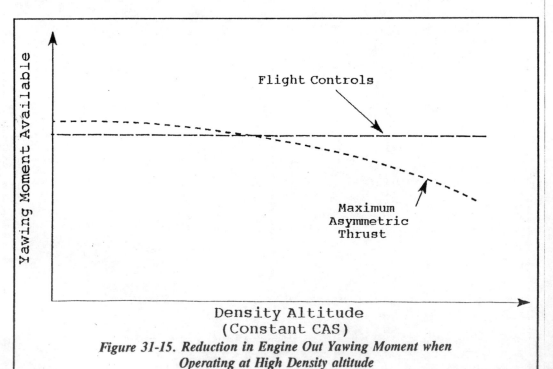

Figure 31-15. Reduction in Engine Out Yawing Moment when Operating at High Density altitude

position", C.G.s forward of this location should result in an even lower V_{MC}. However, airplanes have been known to fly when loaded with a C.G. aft of their rearmost position, and so it is possible from C.G. location to cause an airplane's V_{MC} to be higher than that advertised in the information manual. Although the allowable C.G. range for some military cargo airplanes can be significant, the allowable C.G. range for light general aviation twin engine airplanes is relatively small and will therefore not cause a significant variation in V_{MC}.

C. EFFECT OF BANK ANGLE ON V_{MC}.

V_{MC} for a given airplane/configuration will be achieved when the airplane is in a shallow bank toward the operating engine and cross-controlled rudder ailerons are used to maintain a constant heading. The shallow bank (a maximum of 5° is allowed during the certification) creates a lateral component of lift and a result velocity (sideslip) toward the operating engine. This sideslip creates a favorable angle of attack for the vertical tail, generating lift and a moment to counter the asymmetric thrust moment, even without the application of rudder. Of course application of rudder toward the operating engine will create an even greater amount of lateral force and moment, allowing controllable flight at an even lower speed. A general rule of thumb for light airplanes is that V_{MC} will increase approximately 3 KCAS for each degree of bank the bank angle is off the bank angle which was used to demonstrate the V_{MC} published in the Pilot's Operating Handbook. By the way, minimum drag, and therefore best climb performance will occur with zero side slip while the airplane is slightly banked into the operating engine. Ideally, the lateral component of lift will just balance the horizontal force created by the vertical stabilizer/rudder. A wings level or banked into the dead engine attitude will increase V_{MC}.

D. EFFECT OF AIRPLANE CONFIGURATION ON V_{MC}.

V_{MC} is demonstrated with the flaps in the takeoff position and landing gear up. Generally this configuration gives a somewhat pessimistic estimate of V_{MC}. The increased velocity of the airflow behind the operating engine while the flaps are in the takeoff flaps position will cause that wing to produce significantly more lift than the wing behind the inoperative engine. This will cause the airplane to want to roll into the inoperative engine. The use of aileron to hold the

bank into the operative engine will cause adverse yaw in a direction toward the inoperative engine. The reaction force of the propeller slipstream will be impacting the flap behind the operating engine, reducing the thrust and associated yawing moment it is producing. Finally, the lower AOA associated with takeoff flaps (as opposed to no-flaps) will reduce the asymmetry of the trust produced by operating engines. This reduction in AOA will make the critical engine less critical since lowering the AOA will move its thrust line toward the airplane's longitudinal axis. However, the reduction in AOA will make non-critical engines more critical since reducing the airplane's AOA will move the engine's thrust line outboard away from the longitudinal axis. It this sounds confusing, its because it is. The bottom line is that the flap position which gives the lowest V_{MC} (that's good) will depend on the specific airplane's design.

Extension of the landing gear will normally increase the parasite drag behind the airplane's C.G., increasing the static directional stability and reducing V_{MC}. Dropping the gear will however, increase the requirement for thrust, and perhaps eliminating the potential for reducing V_{MC} by reducing power on the operating engine.

E. EFFECT OF SYSTEM FAILURES ON V_{MC}.

Failures of the systems which are used to minimize the yawing moment into the inoperative engine or maximize the pilot's ability to bank or yaw the airplane into

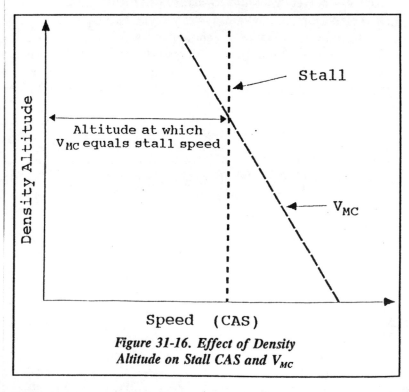

Figure 31-16. Effect of Density Altitude on Stall CAS and V_{MC}

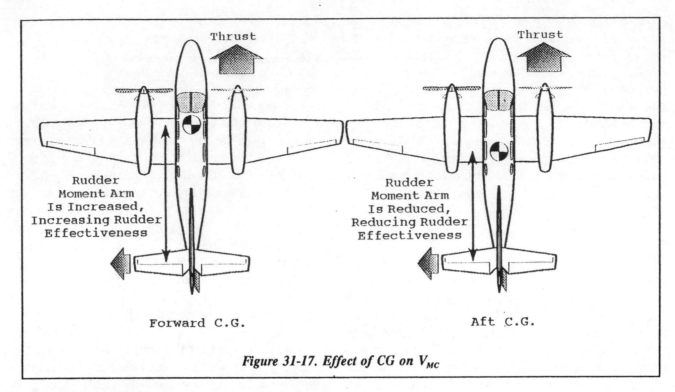

Rudder
Moment Arm
Is Increased,
Increasing Rudder
Effectiveness

Rudder
Moment Arm
Is Reduced,
Reducing Rudder
Effectiveness

Forward C.G.   Aft C.G.

Figure 31-17. Effect of CG on V_{MC}

the operating engine obviously influence an airplane's actual V_{MC}. For instance, if a twin engine airplane has an automatic feathering device, it is perfectly acceptable for the manufacturer to determine the published V_{MC} with the propeller on the most critical engine feathered. If, however, the auto-feather function fails, the drag produced by the windmilling engine can cause a significant yawing moment into the inoperative engine. If loss of an engine also limits the hydraulic or electrical power available to achieve normal control movement, the pilot's ability to restrict banking or yawing into the inoperative engine could be limited. Catastrophic engine failure could also damage control systems normally used when demonstrating V_{MC} and perhaps even increase the drag on the "wrong" side of the airplane.

When investigating V_{MC} accidents. there are a number of questions the investigator should address. Why did the pilot allow airspeed to decay below the V_{MC} published in the Pilot's Operating Handbook?

What was the actual V_{MC} for the accident airplane? Did the pilot really understand how various performance and configuration factors affected V_{MC}? Could the pilot have been reasonably expected to keep the airspeed above V_{MC}? What was the criteria used to establish the time required for the pilot perceive the problem, decide what to do and then react?

8. SUMMARY.

Any airplane with an airworthiness certificate should exhibit satisfactory stability and control within its operating envelope. As we have seen, there are many factors which could change the handling qualities of the airplane and could be factors in an accident.

Control of a multi-engine airplane in an engine-out condition is a unique type of control problem. Failure to understand and cope with minimum control speeds has been a factor in many accidents. The investigator needs to appreciate the various influencing factors.

32

Ice, Rain and Microbursts

1. INTRODUCTION.

Ice, both on the ground and inflight, heavy rain and microbursts have produced accidents in the past and probably will in the future. To some degree, we have designed ourselves into these problems. Modern high performance jet airplanes are aerodynamically less tolerant of airfoil contamination than planes of an earlier era. During takeoff and landing, they operate fairly close to a stall and turbine engines don't accelerate fast enough to provide that instant burst of recovery power that you get with a reciprocating engine. Such is the price of progress.

From an investigation point of view, investigators need to understand the phenomena associated with these hazards and the investigation techniques that go with them.

We'll start with the hard one--icing.

2. GROUND ICING.

Federal Aviation Regulations (FARs) prohibit airplanes from taking off when snow, ice, or frost is adhering to the airplane's wings, control surfaces, propellers, engine inlets and other critical surfaces. In addition, the FARs state that it is the pilot-in-command's (PIC's) responsibility to ascertain that all of the airplane's critical components are free of frozen

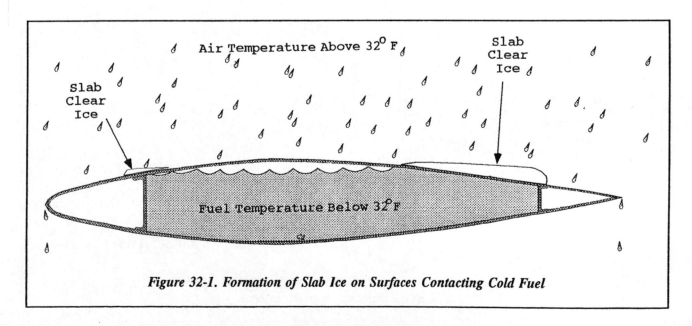

Figure 32-1. Formation of Slab Ice on Surfaces Contacting Cold Fuel

contaminants. In order to ensure that this requirements is complied with, a visual inspection of the airplane's critical surfaces and components needs to be accomplished shortly prior to takeoff. If frozen contaminants are found, de-icing and perhaps anti-icing needs to be performed to ensure that the contaminants are removed and additional accumulations do not collect prior to takeoff. Since the conditions that lead to ground icing, the locations where the ice collects, the hazardous consequences of the ice, and potential hazard controls are different from those associated with in-flight icing, investigations of accidents in which ground icing is suspected will involve different avenues of inquiry and will therefore be reviewed in this separate section.

A. CONDITIONS CONDUCIVE TO GROUND ICING. Ground icing comes in three different forms. Frost, precipitation which freezes to the upper surfaces of the airplane and ground water, slush and snow which is picked up and collected on airplane components and structure.

Frost can accumulate on airplane surfaces when the surface is below the freezing temperature and there is enough moisture in the air to cause the water vapor to sublimate directly out of the air, forming small crystals of ice. These crystals are similar in appearance to the frost which may form in your refrigerator and forms without first showing up as visible precipitation. Frost may form selectively on the airplane, accumulating on some surfaces while ignoring others. Most pilots know that if an airplane is left on the ramp during a subfreezing night, when there is sufficient moisture in the air, frost will appear in the early morning on the upper surfaces of the airplane. The upper surfaces radiate heat into the black night sky while the lower surfaces have radiant heat re-radiated back to the airplane from the tarmac. It shouldn't be surprising that frost is less likely to form when an airplane is stored under a simple shade hanger.

Another less obvious situation occurs when an airplane is flying at high altitude. Much of the airplane's structure, and fuel in its tank is cooled to the

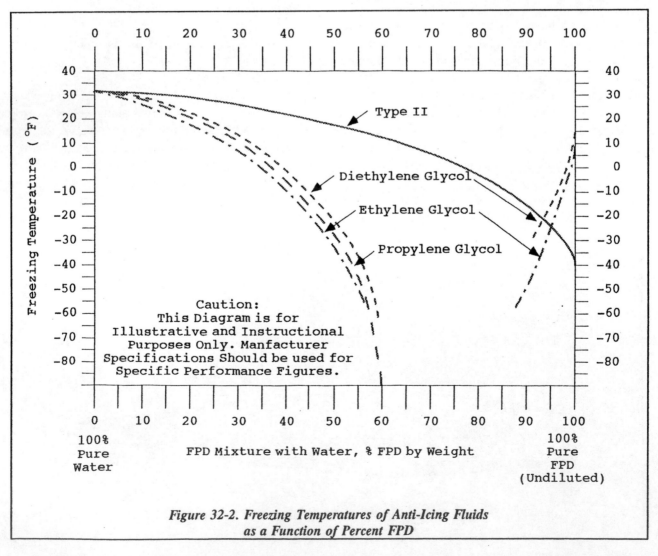

*Figure 32-2. Freezing Temperatures of Anti-Icing Fluids
as a Function of Percent FPD*

9ambient air's subfreezing temperature. When the airplane lands, the cold fuel and the walls of the fuel tank will heat slower than the rest of the airplane structure. If warm fuel is not pumped into the tanks, it is possible for frost to form on airfoil surfaces which are also fuel tank walls. This can occur on calm days even if the ambient ground temperature is above freezing. In this situation, partially filled tanks, may cause frost to form on the bottom of the wing.

Another fairly unusual situation can occur if fuel is pumped out of an underground storage tank where the fuel temperature is below freezing. The subfreezing temperature of the fuel can cool the integral wing tanks to subfreezing temperatures, allowing frost to grow on adjacent wing surfaces.

B. PRECIPITATION WHICH FREEZES TO THE UPPER SURFACES OF THE AIRPLANE.

Freezing rain is super cooled water which freezes as soon as it makes contact with a surface which is at or below water's freezing temperature. Although it provides a relatively smooth coating on the surface, variations in the surface can seriously degrade the aerodynamic performance of airfoils (that's wings, tail feathers and propellers), decreasing its lift/thrust producing capabilities while increasing drag. Freezing rain is a hazard both on the ground and in the air. While in the air it strikes first on leading edges, and normally freezes while it flows back with the airstream. On the ground, the flow is downward.

In addition to modifying airfoil aerodynamic characteristics, freezing rain can increase aircraft weight, jam flight controls and hamper the pilot's visibility. Although dry snow should blow off an airplane as soon as if gains sufficient airspeed, snow which has melted and then refrozen is a much more serious issue. Its irregular shape can seriously disrupt the airflow over airfoils, decreasing lift and increasing drag. This frozen snow can be easily hidden below the fluffy white stuff. The temperature of fuel in an airplane's wings can also be a player in the formation of ground ice. Fuel temperatures from underground storage tanks in extremely cold regions can be well below freezing even when the ambient air temperature is above freezing. When the fuel is pumped into airplane wing tanks (the upper and lower walls of

OAT		Type II Fluid Concentration Neat-Fluid/Water [% by Volume]	Approximate Holdover Times Anticipated Under Various Weather Conditions (Hr:Min)				
°C	°F		Frost	Freezing Fog	Snow	Freezing Rain	Rain on Cold Soaked Wing
0 and above	32 and above	100/0	12:00	1:15-3:00	0:25-1:00	0:08-0:20	0:24-1:00
		75/25	6:00	0:50-2:00	0:20-0:45	0:04-0:10	0:18-0:45
		50/50	4:00	0:35-1:30	0:15-0:30	0:02-0:05	0:12-0:30
below 0 to -7	below 32 to 19	100/0	8:00	0:35-1:30	0:20-0:45	0:08-0:20	Caution! Clear ice may require touch for confirmation
		75/25	5:00	0:25-1:00	0:15-0:30	0:04-0:10	
		50/50	3:00	0:20-0:45	0:05-0:15	0:02-0:05	
below -7 to -14	below 19 to 7	100/0	8:00	0:35-1:30	0:20-0:45		
		75/25	5:00	0:25-1:00	0:15-0:30		
below -14 to -25	below 7 to -13	100/0	8:00	0:35-1:30	0:20-0:45		
below -25	below -13	100/0 if 7°C (13°F) Buffer Maintained	A buffer of at least 7°C (13°F) must be maintained for Type II used for anti-icing at OAT below -25°C (-13°F). Consider use of Type I fluids where SAE or ISO Type II cannot be used.				

These figures reflect 1992 data for only SAE or ISO Type II FPD fluids.

Figure 32-3. Anticipated Holdover Times for Type II Fluid Mixtures as a Function of Weather

which are normally the upper and lower wing skins) the wing's skin can be cooled to sub-freezing temperatures. If liquid water comes in contact with this sub-freezing structure a flat thin slab of ice can form.

If this slab forms on the inner most upper wing surface of an airplane with aft mounted engines, the situation is ripe for a post takeoff, low altitude engine failure. Some years ago a jet transport aircraft with two aft mounted engines lost power on both engines shortly after takeoff. Slabs of ice had formed on the upper wing surfaces located just above the aft, inboard corner of the fuel tanks. Ice was shed from the wings shortly after takeoff, perhaps due to the flexing of the wings.

C. GROUND WATER, SLUSH AND SNOW WHICH COLLECTS ON COMPONENTS AND STRUCTURE. On January 13, 1982, Air Florida Flight 90, a Boeing 737, crashed while taking of from Washington National Airport. Although the crash, like most others, was due to a number of factors, snow and ice played a major role.

First, failure of the crew to use engine anti-ice resulted in the crew using significantly less power than required for takeoff. This was caused when the probes which measured the dynamic pressure of the air entering the engines became blocked with ice. As a result, when the airplane was accelerating down the runway, the engine pressure ratio (EPR) gauge was showing normal takeoff readings while the actual EPR was only a fraction of that shown on the gauges. In addition, ice adhering to the upper surface of the wing changed the stall characteristics of the airplane. Frozen contamination on the upper surface of a wing can reduce its C_l at high angles of attack, decrease C_{Lmax}

and increase stall speed. On the 737's wing it can also induce a nose up pitching moment that can drive the airplane deeper into the stall.

The airplane had accumulated a quarter to a half inch of snow on the wings since deicing and departure from its gate, legally precluding the crew taking off. However, there is a chance this snow might have blown off during the takeoff roll. If, additionally, the crew had used full takeoff power (instead of using the faulty EPR readings) it is possible Flight 90 might have successfully completed its takeoff. However, the actions of the crew, which resulted in some of the snow melting and then refreezing at critical locations, aggravated the situation.

First, the crew used reverse thrust to back the airplane out of its gate. This not only blew ground water and snow forward onto the wing, it would have caused some melting of the snow and later refreezing to the wing. Further melting and freezing probably occurred when the airplane was taxied into the exhaust of the airplane ahead of it in the lineup waiting for takeoff.

This could have been when the freezing water was blown onto the pressure ports in the engine inlets.

D. AERODYNAMIC CONSEQUENCES OF FROST ACCUMULATION ON AIRPLANE AIRFOILS. Although the effects of frost accumulation on the lift producing surfaces is not as significant as the effects of the formation of ice, even small amounts of frost can have a pronounced affect on their ability to produce lift and can also create drag. And, although frost does not have the ability to significantly change a wing's aerodynamic shape, its rough surface can greatly affect the nature of the boundary layer, slowing it and increasing its thickness. Airflow separation will occur at lower than normal angles of attack and coefficients of lift will be reduced at high AOAs. The formation of a hard layer of

OAT		Approximate Holdover Times Anticipated Under Various Weather Conditions (Hr:Min)				
°C	°F	Frost	Freezing Fog	Snow	Freezing Rain	Rain on Cold Soaked Wing
0 and above	32 and above	0:18-0:45	0:12-0:30	0:06-0:15	0:02-0:05	0:06-0:15
below 0 to -7	below 32 to 19	0:18-0:45	0:06-0:15	0:06-0:15	0:01-0:03	Caution! May require touch
below -7	below 19	0:12-0:30	0:06-0:15	0:06-0:15		for confirmation

These figures reflect 1992 data for only SAE or ISO Type I FPD fluids.

Figure 32-4. Anticipated Holdover Times for Type I Fluids as a Function of Weather Conditions

thick frost on the leading edges and upper surfaces of a wing have been reported to reduce C_{Lmax} by as much as 50% A quick math exercise will show that a 50% reduction in C_{Lmax} will result in over a 40% increase in stall speed. Although these numbers may reflect worst case conditions, they cannot be ignored. Stalls induced by frost will also occur at lower than normal angles of attack. Thus not only will stall speeds increase by, the accuracy of stall warning devices, which depend on either airspeed or AOA, will be degraded.

It is possible for an accumulation of frost to cause the wings to stall when a pilot attempts to takeoff at normal takeoff speeds. Even worse, if the accumulation is asymmetric, only one wing may stall, causing the airplane to roll rapidly and impact the ground at an attitude which decreases the chances of crew or passenger survival.

E. AERODYNAMIC EFFECTS OF FREEZING RAIN OR SNOW ACCUMULATION ON AIRPLANE

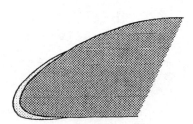

32 to 30 F.
Large water droplets freeze slowly, flowing aft before freezing. Smooth surface has smallest effect on airfoil's aerodynamic performance.

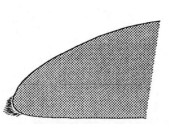

30 to 25 F.
Large water droplets freeze more rapidly. Double horn ice forms as two ragged ridges on the leading edge. Most severe effects on airfoil's performance.

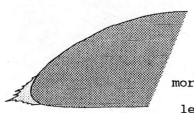

25 to 14 F.
Smaller water droplets freezes almost immediately. Single ragged ridge of ice forms at stagnation point. Severe effects on airfoil's performance.

Below 14 F.
Very small water droplets freeze on contact. Frost like ridges form along stagnation point. Severe effect on airfoil's performance.

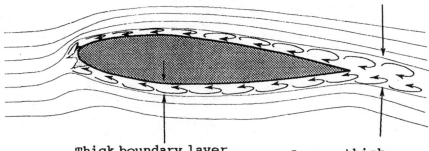

Thick boundary layer.

Large, thick, high drag wake.

Figure 32-5. Leading Edge Ice Shapes as a Function of Temperature and Drop Size

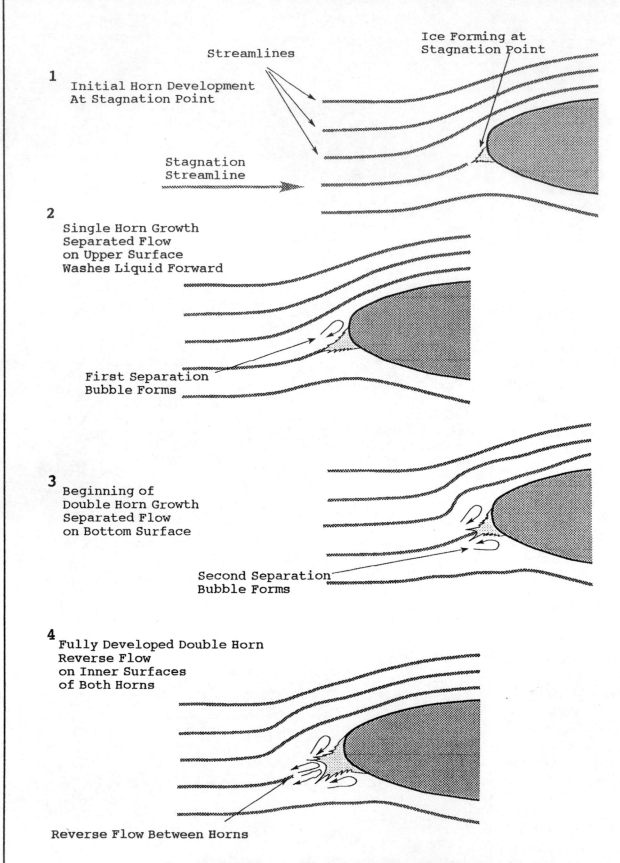

1 Initial Horn Development At Stagnation Point

Streamlines

Ice Forming at Stagnation Point

Stagnation Streamline

2 Single Horn Growth Separated Flow on Upper Surface Washes Liquid Forward

First Separation Bubble Forms

3 Beginning of Double Horn Growth Separated Flow on Bottom Surface

Second Separation Bubble Forms

4 Fully Developed Double Horn Reverse Flow on Inner Surfaces of Both Horns

Reverse Flow Between Horns

Figure 32-6. Leading Edge Ice Shapes - Formation of Double Horn Ice

AIRFOILS. The accumulation of freezing rain or frozen snow on the upper surface of a wing can cause an even greater effect (than frost) on the lift and drag producing abilities of a wing. In addition, the ice can add a significant amount of weight to the airplane, weight that was not accounted for when computing the takeoff roll, takeoff speed, and initial climb.

F. OTHER SYSTEM EFFECTS. Finally, ice can freeze in the gaps and recesses of the primary and/or secondary flight controls, restricting their movement; freeze over unheated pitot static ports, denying information to the aircrew and systems which need them; and locking brakes after gear retraction so that tires skid and blow out on touchdown. The effects of runway slush and snow on takeoff distance are discussed in Chapter 30.

G. CONTROL OF GROUND ICING. Because of the hazards associated with takeoffs attempted with airplane surfaces contaminated with ice, snow or frost, the FAA has prohibited takeoff with snow, ice or frost adhering to an airplane's wings, propellers, control surfaces, engine inlets, and other critical surfaces. Perhaps the most primitive method of ground removal of snow and ice is to sweep off the loose stuff, and then scrape off the stuff that's frozen to the surface. It's time consuming, especially on large airplanes, but if done carefully, it works.

Another method of preventing the accumulation of frost, freezing rain or snow on an airplane is to keep the airplane protected from the elements until just prior to its use. This can involve simple fabric covers, simple overhead protection (shade hanger), or storing it in a closed hanger (heated or unheated). However, even if one of these methods were used, the airplane would still have to make it from the hanger to the runway. If the airplane was in an unheated hanger whose ambient temperatures were below freezing, with the right temperature and humidity conditions outside, it is possible for frost to form on the airplane after it is removed from the hanger. If the hanger was heated, dry snow falling on the warm surface would melt, and if the airplane remained outside long enough in sub-freezing temperatures, could refreeze both on airfoil surfaces and within the gaps in controls.

Hot air blowers have also been used to melt the ice/snow, and, to some degree, dry the surface. If ambient temperatures are below freezing, however, any liquid water left on the airplane will have the opportunity to refreeze.

The most common method of removing ice or snow from large commercial airplanes is the use of a "de-icing" and/or an "anti-icing fluid". The de-icing fluid is, as its name implies, a method of removing accumulated snow and ice from the surface of an airplane. Anti-icing fluid, however, is used to prevent or delay the accumulation of snow or ice on an already clean airplane. Both use a glycol-based Freezing Point Depressant (FPD) in solution to produce a fluid with a freezing temperature below the ambient temperature and the airplane's surface temperature.

The addition of glycol decreases the freezing point of a water solution when in either the liquid or crystal (ice) phase. As ice melts into the FPD solution, its strength is weakened and the freezing point of the FPD solution will slowly rise toward water's normal freezing temperature. If ambient temperatures are sufficiently low, the water in the solution can re-freeze.

It should be noted that the relatively high speed, low static pressure airflow over a wing at the high angles of attack associated with takeoff and initial climb is accompanied by a significant decrease in ambient temperature. This may cause freezing of water on the wing's upper surface which was just above freezing and was not blown off the wing during the takeoff roll.

Climbing to altitude as well as absorption of additional freezing precipitation into the solution also have the potential to reduce the temperature of the FPD solution to below freezing. Ground de-icing is not intended to provide any protection from ice accumulation once the airplane is in-flight.

H. DEICING AND ANTI-ICING FLUIDS. Although there are various types of fluids used to make the aqueous solutions used for deicing and anti-icing, they can be grouped into two general types, Type I and Type II. Both types are produced to comply with the guidelines provided by the Society of Automotive Engineers (SAE) and the International Standards Organization (ISO). Propylene glycol, diethylene glycol and ethylene glycol are the FPD agents normally used in Types I and II fluids. These fluids can be purchased from the manufacturer either in bulk quantities to be mixed by the user or pre-mixed by the manufacturer. In either case, the actual glycol concentration of the mixture should be tested before it is used.

Although the use of FPD fluids can reduce the

probability of ground ice-induced takeoff accidents, it should only be used when and where appropriate. In addition to the acquisition cost of fluids and equipment, there are also the labor costs of the application and the reclamation costs necessary to protect the environment. The diethylene and ethylene glycol used in deicing and anti-icing fluids is moderately toxic for humans. Ingestion of small amounts of diethylene or ethylene glycol may cause abdominal discomfort and pain, dizziness, and effects on the central nervous system and kidneys. Because it is significantly diluted with water and other additives, it would be difficult for an individual to consume the 3 to 4 ounces of pure glycol necessary for a lethal dose. Inhalation of glycol vapors may cause nose and throat irritations, headaches, nausea, vomiting, and dizziness. Contact can cause irritation to the skin and eyes.

TYPE I FLUIDS. Prior to dilution in a solution, Type I fluids contain a minimum of 80 percent glycol. They have a relatively low viscosity and although they can be used for both deicing and anti-icing, they provide anti-icing protection for only a limited period of time. The procedure generally practiced by many United States air carriers is to ensure that the film remaining after deicing has a freezing temperature of at least 20°F (11°C) below the ambient temperature. This is referred to as a temperature buffer. Canadian and European temperature buffers are similar Type I fluids (10°C or 18°F).

TYPE II FLUIDS. Type II fluids contain a minimum of 50 percent glycol. They also contain "thickening" agents (polymers) which increase the solution viscosity, allowing it to both be applied in a thicker layer than Type I fluids and also remain on the surface longer without running off. They are designed to remain on an airplane's surfaces during ground operation, providing a layer of FPD fluid which melts any frozen contaminants which attempt to collect on the surface. Although United States carriers use a temperature buffer of 20°F for Type II fluids, Canadian and European air carriers use a temperature buffer of only 7°C (13°F). The slightly lower temperature buffer used by the Canadians and European air carriers takes advantage of the slightly slower rate at which Type II is diluted by precipitation (as compared to Type I fluid).

Type II fluids are also useful in providing over night protection when applied prior to spending a cold, wet night on the ramp prior to an early morning departure. In this case, treatment with Type II fluid

will speed the morning's deicing process. Although Type II fluids can be used as both deicing and anti-icing agents, the heating normally associated with deicing degrades their effectiveness. They are more effective than Type I fluids when used as anti-icing fluids.

SAE and ISO Type II fluids are designed for use on airplanes with rotation speeds in excess of 85 knots. Below this airspeed, the fluid's presence on an airfoil's surface can change the airfoil's aerodynamic performance, resulting in characteristics similar to those expected when the airfoil is covered with a layer of frost. Above this speed the shearing action of the airplane's boundary layer overcomes the viscosity of the fluid, allowing it to flow off the wing's surface without degrading the airplane's aerodynamic characteristics. Therefore, like most of the good things in life, FPD solutions have a dark side, and if used improperly, can cause dangerous degradation in an airplane's performance, stability and control. Because of its high viscosity, fully concentrated Type II fluids should not be applied to:
- pitot heads
- static ports
- angle-of-attack sensors
- cockpit windows and the nose of the airplane
- air inlets
- engines
- control surface cavities

MILITARY TYPE I AND TYPE II FLUIDS. Except for the addition of a thickening agent, Military Type I and Military Type II fluids are essentially the same. However, the requirements to which they are certified (material, freezing point, viscosity and plasticity), differ from the requirements specified by the SAE and ISO.

I. DEICING FLUIDS. Deicing fluids are FPD solutions which are sprayed onto the airplane's surface in order to melt and carry off frost and frozen precipitation. Following a thorough de-icing, the airplane's surfaces are returned to a clean configuration by removing any mechanical interferences which are caused by ice, snow or frost. Although deicing with cold solutions is possible, the additional thermal energy available in heated solutions of FPD make it more effective in melting ice, snow and frost.

Once the high temperature fluid melts the frozen ice or snow, it keeps it melted by mixing with the melted ice or snow to form a diluted mixture of water and FPD. The more diluted with liquid water the mixture becomes, the closer its freezing temperature approaches 32°F (0°C). The higher the percentage of alcohol in the original mixture, the more liquid melt water the mixture can accommodate before refreezing. Unfortunately, the greater the percentage of alcohol in the mixture, the greater its cost per gallon and the greater the environmental insult if it is released into the environment. The mixture has approximately the same viscosity as water and will run off the airplane's surfaces with little resistance. The sequence selected in which the airplane is de-iced is designed to minimize the accumulation of liquids in airplane locations where they can re-freeze and cause problems inflight. The thin film of Type I mixture which remains on the aircraft's surface provides little residual effect after the de-icing is completed and freshly falling freezing precipitation is free to start collecting again in a relatively short period of time.

J. ANTI-ICING FLUIDS. Anti-icing fluids are used to protect clean (i.e. deiced) airplane surfaces from the accumulation of frost and frozen precipitation by providing these surfaces with a protective film of FPD solution to delay the reforming of ice, snow or frost. Although anti-icing fluids contain water and FPD, they also contain a thickening agent (a polymer). This thickening agent keeps the fluid from running off the airplane's surface the way Type I fluid does. The presence of a plasticizer in an anti-icing fluid allows the liquid mixture to form a relatively thick, viscous film over the airplane surfaces. This film melts new accumulations of freezing precipitation until either the mixture's freezing point is elevated to the ambient temperature or the film is blown off by the airstream as the airplane takes off. Type II fluids are not normally used to deice airplanes on which ice or snow has already accumulated. It is not normally heated to the temperatures used for Type I fluids prior to application and because of its higher viscosity, requires different equipment from that used for Type 1 fluid. Because Type II fluid adheres to a surface in a relatively thick film, it will continue to melt newly fallen frozen precipitation for some time after its application. The length of time the airplane's surfaces are protected from accumulating newly fallen ice or snow is referred to as the "holdover time". The holdover time is a function of many variables, some of which are listed below:

- airplane surface (inclination, roughness)
- temperature of the airplane's surfaces.
- type of FPD used.
- percentage of FPD in the original mixture.
- temperature of the original mixture.
- method of application.
- thickness of fluid layer.
- rate of application of precipitation.
- nature of the frozen precipitation.
- ambient temperature.
- wind velocity and direction.

Treatment of an airplane's surfaces with anti-icing fluid is not intended to provide any protection from ice accumulation once the airplane is in-flight.

Because of the uncertainties associated with some of the variables used to determine the holdover time, pretakeoff checks are recommended by the FAA when weather conditions dictate. Figures 32-3 and 32-4 show the large variation in holdover times within each of the five weather conditions defined. These variations reflect not only the uncertainty associated with the variables (e.g. rate of application of snow or freezing rain), but the uncertainty associated with the measurement of these variables).

The effects of deicing and anti-icing fluids on the aerodynamic performance of a wing configured for takeoff has been studied by NASA in wind tunnels and during flight tests. These tests have shown that the increased viscosity of Type II fluids allows them to, during certain conditions, remain on the wing during the takeoff and initial climb. While on the wing, Type II fluids cause wing roughening and a general degradation of the wing's lift producing ability. The effect, although measurable at takeoff C_Ls is more pronounced at C_{Lmax}. The loss of C_L is also reported to be more pronounced in wings with leading edge high lift devices deployed for takeoff, especially those with aerodynamic gaps. Measurable increases in drag and changes in the wing's pitching moment were also noted. The magnitude of the loss of lift, the increase in drag and the change in pitching moment were all sensitive to changes in temperature. Colder temperatures increased the magnitude of the change. All of the results were sensitive to the various types of fluid used during the tests. For flaps 5 and a sealed-slat configuration, the worst case loss of Lift at takeoff AOAs was 9% while the decrease in C_{Lmax} was 10%. For a flaps 15, gapped-slat configuration, the worst case loss of Lift at takeoff AOAs was 12% while the decrease in C_{Lmax} was 13%. Test were conducted using the Boeing 737-200ADV airfoil as a baseline and test results cannot, therefore, be applied across the board to all aircraft.

K. INVESTIGATING ACCIDENTS IN WHICH GROUND ICING IS A SUSPECTED FACTOR.

When degraded performance, stability or control of an airplane shortly after takeoff is coupled with freezing or subfreezing temperatures and small temperature dew-point spreads or standing water or slush, the potential effects of frozen contaminants on critical airplane surfaces or equipment should be evaluated. The investigation should address the following areas:

- Inspections used to determine the need for deicing and anti-icing.
- Procedures used to deice and anti-ice the airplane.
- Equipment used to deice and anti-ice the airplane.
- The type of fluid and concentrations in the solution used to deice and anti-ice the airplane.
- The amount of additional frozen contaminants which the airplane could have collected prior to takeoff.
- The criteria used by the ground and flight crews to determine that the airplane was still free of frozen contaminants prior to takeoff.

INSPECTIONS USED TO DETERMINE THE NEED FOR DEICING AND ANTI-ICING.

Some of the factors to be examined in this area include:

- existence of formal procedures.
- adequacy of procedures to detect icing in critical areas.
- visibility of critical areas to include the effects of:
- adequacy of lighting
- viewing angles
- reduced visibility from inside the cabin, due to the effects of Type II fluid on side windows. (Type II fluid has the consistency of honey and degrades visibility through transparencies.)
- training of ground and flight crews performing the inspections.

PROCEDURES USED TO DEICE AND ANTI-ICE THE AIRPLANE.

Some of the factors to be examined in this area include:

- the existence of formal procedures for deicing and anti-icing the airplane.
- compliance with procedures for deicing and anti-icing the airplane to include:

- the sequence followed to deice/anti-ice the various surfaces.
- avoidance of surface areas which should not be exposed to anti-icing fluid (e.g. pitot static ports, cockpit transparencies, air inlets, engines)
- training of ground crews in deicing and anti-icing procedures.
- communication of critical information concerning deicing or anti-icing to the flight crew. This should be transmitted to the pilot in the form of a four element code which includes: the type of fluid used; the percentage of fluid water mixture in the solution; the local time at the beginning of the final deicing or anti-icing step; and the date of the deicing or anti-icing.

THE TYPE OF FLUID AND CONCENTRATIONS IN THE SOLUTION USED TO DEICE AND ANTI-ICE THE AIRPLANE.

Some of the factors to be examined in this area include:

- procedures to ensure the quality of the fluids being used.
- procedures to ensure the accuracy of the mixtures used in the solutions applied to the airplanes.

THE AMOUNT OF ADDITIONAL FROZEN CONTAMINANTS WHICH THE AIRPLANE COULD HAVE COLLECTED PRIOR TO TAKEOFF.

Some of the factors to be examined in this area include:

- the type, temperature and rate of the frozen precipitation accumulating/collecting on the airplane's surfaces.
- the length of time from the beginning of the late deicing/anti-icing procedure performed on the airplane.
- other factors which could affect the collection of additional frozen contaminants on airplane surfaces;
- wind direction and velocity.
- cloud cover/sunlight.
- influence of jet exhausts on melting and freezing of dry snow.
- presence of surface water which could splash/be blown into critical areas.
- use of reverse thrust which could blow contaminants onto or melt dry snow on critical areas.

THE CRITERIA USED BY THE GROUND AND FLIGHT CREWS TO DETERMINE THAT THE AIRPLANE WAS STILL FREE

OF FROZEN CONTAMINANTS PRIOR TO TAKEOFF. Some of the factors to be examined in this area include:

- existence of criteria for determining the need for additional deicing or anti-icing.

- adequacy of criteria for determining the need for additional deicing or anti-icing.

- adequacy of ground and flight crew training on need for additional deicing or anti-icing.

- implementation of procedures for determining the need for additional deicing or anti-icing.

L. NEW TECHNOLOGY. When searching for corrective actions for accidents in which ground icing was a factor, new evolving technologies need to be examined. These efforts to control the risk associated with aircraft ground icing are proceeding along several complementary fronts.

First are the efforts to understand why some wing configurations are more vulnerable to the effects of frost than other configurations. Although it appears that airplanes with "hard" leading edges are more vulnerable than those with high lift devices on their leading edges, not all agree with this hypothesis.

Second are the efforts to ensure that critical surfaces are free of frozen contamination. Type I deicing and Type II anti-icing fluids are being continually improved, as is the understanding of their effectiveness and limitations at various temperatures and under different types and rates of precipitation. Dyes added to Type II fluids which change color as they are diluted with melted precipitation may be able to provide pilots with visual information on not only coverage, but the residual effectiveness of the anti-icing layer. Methods to quantify precipitation rates and types also need to be developed. Technologies which minimize the time from deicing/anti-icing to takeoff by placing the deicing/anti-icing closer to the runways or by minimizing delays due to traffic also provide partial solutions. For especially critical areas, heating devices can also be installed on the surfaces. The effectiveness of inspections is being improved by recognizing the limitations of earlier procedures and effective training of both ground and air crews.

Finally, but equally important are the efforts to ensure that the crew will be notified whenever anti-ice or deice procedures were ineffective and the surfaces remain contaminated. Smart skins or sensors imbedded in critical surfaces will be able to determine both the nature and depth of surface contamination and notify the pilot before the takeoff is attempted. Early in the takeoff roll, air flow sensors will measure the nature of the boundary layer over the wing, sensing changes which indicate the presence of hazardous ground icing. It is easy to blame the aircrew for taking off with ice on the wings. Its much harder, but ultimately more effective, to provide the aircrew with better information on the nature of the hazard and the risks it generates.

3. IN-FLIGHT ICING.

In-flight icing can be divided into two types: structural and engine ice. Structural ice degrades the airplane performance when supercooled water droplets impinge on aircraft surfaces. Ice buildups can the degrade lift production, increase drag, reduce propeller efficiency, increase airplane weight and, if shed by the structure on which it forms, cause damage to systems or structure. Engine ice can degrade thrust or power produced by the power plant by starving it of air.

A. STRUCTURAL ICING. The hazards associated with in-flight structural icing can be divided into two general areas; aerodynamic factors and system operation. Of the two, aerodynamic factors is probably the most obvious. An in-flight airplane is acted on by three forces, the aerodynamic force (which is often divided into its two vector components, lift and drag), thrust, and weight. In-flight icing can and often does affect all three. In addition, in-flight icing can dramatically influence the pitching moments generated by the airflow over an airplane's surfaces. If deice or anti-ice equipment is not available or used, flight into icing conditions can degrade the performance of various airplane systems. Some of the more obvious are the engine and the pitot static systems. In addition, the performance of communication and navigation systems which rely on external antennas can also be influenced by the presence of ice. By stretching the definition of "systems" slightly to include windscreens, you could even include the degradation of the pilot's view of the external environment in this category.

Direct evidence of in-flight icing is extremely volatile. By the time the investigator arrives on scene, the ice will most probably have disappeared due to either natural melting or sublimation, accelerated melting in a post crash fire, or separation from the structure during the ground impact. Because of the fleeting nature of this evidence, investigators need to have a good understanding of when in-flight icing can

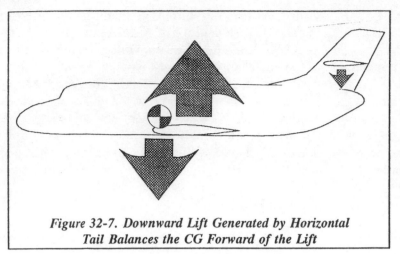

Figure 32-7. Downward Lift Generated by Horizontal Tail Balances the CG Forward of the Lift

occur and how it can effect an airplane and its systems.

Two conditions are necessary for to structural ice to form on the exterior of an inflight airplane. First, there must be liquid moisture in the air. Second, ambient air and aircraft surface temperatures must be below freezing. Although water can remain a liquid as low as minus 40°C, most icing occurs between minus 20°C and 0°C. The lowest temperature at which a Part 25 certified airplane must be tested is minus 22°C, the Standard Day Temperature for just below 19,000 feet. There are two typical environments which satisfy these requirements, freezing rain or drizzle and super cooled droplets of liquid water. The freezing rain or drizzle can most often be found in the clear air of a cold air mass which is below an overriding warm air mass. These conditions can be found along the cold side of

a warm front or occasionally behind a slow moving cold front. Supercooled water droplets are found in clouds, in fact they are clouds. The Supercooled droplets range in size from 5 to 50 microns in size (the diameter of a human hair is about 75 microns) and are associated with cumuloform clouds (which can include billowing stratocumulus clouds) or dense layers of nimbostratus clouds. The larger droplets are associated with the cumuloform clouds while the nimbostratus clouds yield smaller droplets. The size and temperature of the water droplet determines the rate at which it transitions from liquid water to crystalline ice; the smaller and colder the faster the transition, the larger and warmer the slower the transition. The size and temperature of the droplets is also a factor in determining whether the water freezes into

- a relatively rough surfaced, opaque and milky appearing rime ice, or

- a relatively smooth and translucent clear ice, or

- a mixture of the two.

Rime ice is normally associated with small, and very cold droplets which freeze immediately on contact with the airplane surfaces. Rime ice tends to accumulate first at leading edge stagnation points and is primarily confined to the leading edges of the structure. If allowed to build up to any significant amount, the rough shape tends to distort the design shape of aerodynamic structures, greatly altering the intended airflow around the structure. Because of air which is entrained in the ice, rime ice is "milky" in color, has a lower density and is generally weaker than clear ice.

Clear ice is normally associated with larger droplets which are at or just a couple of degrees below freezing. Because these droplets take longer to freeze, they can flow back from the airplane's leading edges as they freeze. This forms a sheet of ice which closely conforms to the original shape of the surface. Because of the larger droplet size and the greater liquid water content associated with the environment in which clear icing is found, the accumulation of clear ice can be very rapid. Since clear ice

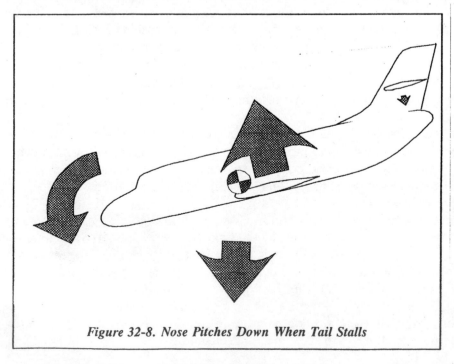

Figure 32-8. Nose Pitches Down When Tail Stalls

does not entrain air, it has a higher density than rime ice and is also stronger.

In between text book rime ice and text book clear ice are a variety of mixed ices. The interesting shapes of these "mixed ice" formations and the shape's aerodynamic effect will be discussed later in this chapter.

The rate at which ice accumulates on an airplane varies with, the size of the liquid water droplets, the amount of liquid water per unit volume of air, the speed at which the airplane surfaces are moving through the air, and the size of the leading edge of the surface moving through the air. Smaller droplets tend to follow the airstream and most pass over the airplane's surfaces. The momentum of larger droplets impedes their movement with the air and they collide with the airplane's surfaces. Parts 23 and 25 certification requirements include droplet sizes from 15 to 50 microns. These requirements are oriented toward liquid moisture in clouds and don't include freezing rain. Rain drops can exceed 1,000 microns in size (one millimeter), over 20 times larger than the largest droplets in the cloud. There are no requirements to certify an airplane in freezing rain. The more droplets in a cubic foot of air, the more will collide with the airplane's surfaces. Liquid water requirements established in Parts 23 and 25 are 0.04 to 0.8 grams per cubic meter for layered or stratiform clouds and 0.2 to 2.7 grams per cubic meter for cumulus clouds. These limits can be exceeded in some cumulonimbus clouds. The faster an airplane moves, the more the volume swept out by the surfaces. This factor has some limits, however. As airplanes move faster, the leading edges of their surfaces are heated by the compression of the air

near the stagnation points. This can heat the surfaces to temperatures where ice cannot form. Finally, thin structure tends to produce the least disturbance of the air in front of the surface, the droplets will have less opportunity to be carried, by the airflow, out of the way of the oncoming surfaces. Ice will therefore collect fastest on the thinnest structures such as wire antennas, temperature probes and pitot tubes; slightly thicker objects next such as struts, horizontal and vertical stabilizers;, and on the thickness structures most slowly including wing and fuselage leading edges, engine nacelles, and propeller leading edges.

There can be synergistic effects where two or more factors combine. For instance, an airplane's propellers are moving much faster than the airplane itself. They are sweeping out a large volume of air and are relatively thin. They contact more water droplets and may therefore accumulate ice faster than other surfaces. Pilots looking out at an easily observable leading edge of a wing may be fooled into believing that the accumulation there is representative of accumulations elsewhere. In fact, considerably more ice may have formed on the leading edge of the horizontal tail, a condition which is more critical than ice on the wings when it comes to landing.

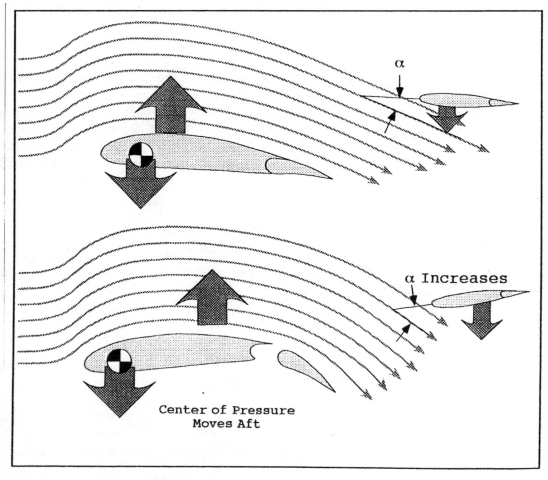

Center of Pressure
Moves Aft

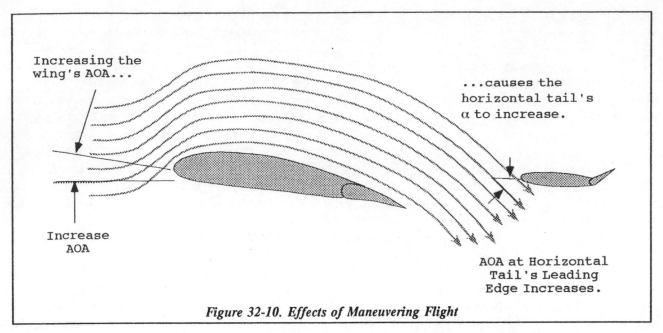

Increasing the
wing's AOA...

...causes the
horizontal tail's
α to increase.

Increase
AOA

AOA at Horizontal
Tail's Leading
Edge Increases.

Figure 32-10. Effects of Maneuvering Flight

When low wing airplanes are flying at high angles of attack, clear ice can collect on lower wing surfaces where it is out of sight of the pilot.

B. HAZARDOUS EFFECTS OF STRUCTURAL INFLIGHT ICING. The aerodynamic effects of in-flight structural icing are primarily a leading edge problem (although freezing rain and refreezing of melted ice running off heated leading edges can occur). The shape of the leading edge ice will depend on the temperature and size of the water droplets and the temperature of the surface onto which they freeze. Examples of these shapes are show in Figure 32-5. Aerodynamically, the double horn shape will have the greatest impact. It can cause the greatest decrease in the coefficient of lift (and therefore stall speed),and the largest increase in parasite drag. This shape can also

cause the stall to be more abrupt. A small increase in angle of attack will produce a large decrease in C_L. It can also cause an asymmetric stall; one wing stalled while the other is not. The single horn formation, which occurs at slightly lower temperatures, and the more classic rime ice formation which occurs at even lower temperatures will, to a lesser degree, cause a reduction in C_{Lmax} and an increase in drag.

The shape of the clear ice which forms at or just below freezing temperatures follows the structure's shape closely and therefore does not have a very pronounced affect on aerodynamics. However, since it is denser and accumulates at a faster rate, it can add significant weight to the airplane, increasing induced drag and slowing the airplane. In addition it can cover antennas and radome, interfering with transmission and reception, and break off in sheets which will be

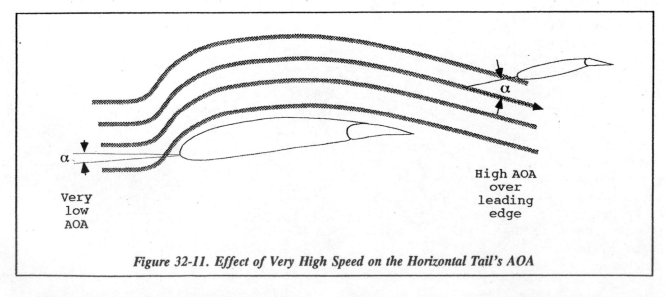

α

α

Very
low
AOA

High AOA
over
leading
edge

Figure 32-11. Effect of Very High Speed on the Horizontal Tail's AOA

carried aft and can cause damage.

As mentioned earlier, ice can accumulate rapidly on propellers, decreasing both propeller efficiency and thrust. In addition to the hazards associated with loss of thrust, it is likely that ice will be shed from the propellers asymmetrically. The resulting vibrations and pounding of the shed ice on the fuselage will at the least be disconcerting to the crew and passengers. At the worst, the vibrations could cause damage to the engine and its mounts.

The formation of structural ice on lifting surfaces can adversely affect an airplane's stability, the most dramatic of which is the stall of the horizontal tail. Lift generated by the horizontal tail (which is in the downward direction) balances the weight of the airplane whose CG is located forward of the wing's center of lift. If the horizontal tail's angle of attack exceeds its stall angle of attack, the horizontal tail will stall, the balance is upset and nose will pitch down. If the nose down pitching is not arrested, the airplane's attitude can quickly reach vertical. If the tail stall occurs while the airplane is making its approach to landing, collision with the ground at an extremely nose low attitude and at a speed well above stall speed is likely.

One condition which can increase the probability of a horizontal tail stall because of leading edge ice is the lowering of wing flaps. When wing flaps are lowered the downwash angle behind the wing will normally increase, which increases the angle of attack of the horizontal stabilizer. Not all airplanes are equally susceptible to tail stalls. The effect of the downwash on the tail is a function of the wing and horizontal tail locations. High wing, low tail configurations are more likely to place the horizontal in the wing's downwash. Low wing, high tail ("T" tail) configurations are less likely to have the horizontal tail in the wing's downwash.

There are other ways the horizontal tail's

angle of attack can be increased beyond its stall angle of attack. If the wing's angle of attack is increased to high C_Ls, its downwash angle will increase, increasing the horizontal tail's angle of attack. Flying at high speeds will also increase the horizontal tail's angle of attack as the airplane flies at lower pitch attitudes. Since a wing with a positive camber will develop a nose down pitching moment which increases as a function of dynamic pressure, the downward lift produced by the tail must steadily increase as speed increases. Thus, three things that might precipitate the stall of a leading edge iced horizontal tail are: lowering wing flaps, maneuvering at high angle of attack, and very high speed flight.

One other factor which influences the hazards associated with the stall of the horizontal tail is the change in the pressure pattern on the horizontal tail when it stalls. The pressure pattern will change so that the aerodynamic pressure over the tail will naturally cause the elevator to want to move toward the trailing edge down position. For hydraulically powered flight control systems which do not allow forces on the control surfaces to cause movement of the control surface, this does not pose any additional problems. The elevator is locked in position by the hydraulic actuator. However, in aircraft with non-powered systems, the pressures on the elevator can be strong enough to over-power the efforts of the pilot, forcing the yoke to the full forward position. Unless the crew is able to return the yoke to the nose up position, safe recovery is unlikely.

Some of the questions which need to be answered following an accident caused by leading edge ice-induced stalling of the horizontal tail include the failure of the crew to either detect and eliminate tail

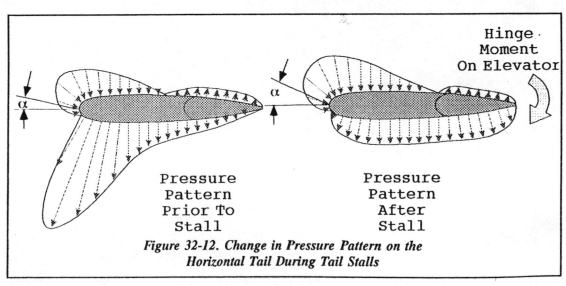

Hinge Moment On Elevator

Pressure Pattern Prior To Stall

Pressure Pattern After Stall

Figure 32-12. Change in Pressure Pattern on the Horizontal Tail During Tail Stalls

ice and the crew's knowledge of the hazards associated with tail icing.

Wing frost and ice on the upper surfaces of swept wing airplanes will increase the thickness and reduce the energy of the boundary layer air as it naturally tends to move outboard. The thickened, low energy boundary near the wing tips will tend to stall sooner than normal. If the wing tips stall before the rest of the wing the center of lift will tend to move forward, causing a nose up pitching moment just when it is not needed; during the stall.

Structural icing has also led to problems with the crews' movement of flight controls. Surface contamination can freeze in locations which prevents surface movement. This can be caused by ice which has formed on the ground, liquid water which collected on the ground but froze in flight or ice which was melted off the leading edge but which flowed back from heated leading edges and froze. Water which freezes inside flight controls can also upset the mass balancing necessary to prevent flutter. One military jet attack airplane experienced flight control jams which occurred when rain water, which accumulated at the bottom of the cockpit, froze and locked the controls.

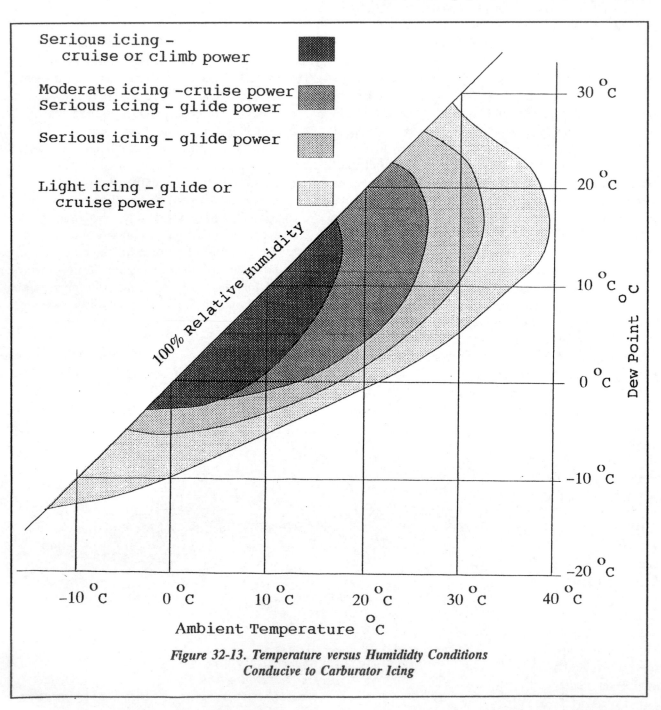

Figure 32-13. Temperature versus Humididty Conditions Conducive to Carburator Icing

Fortunately the pilots were able to break the ice loose by applying pressure to the control stick. In another case, the crew of a commercial jet airliner experienced flight control difficulties when water, which accumulated on the ground, froze while in flight and partially jammed the aileron controls. Wheel brakes can also accumulate water or slush, which can freeze while at higher altitudes. This locks the wheels in position so that they can blow out on touchdown.

C. ENGINE ICE. In addition to the structural type of icing which can destroy the thrust producing lift of propellers, icing can interfere with a reciprocating engine's induction system or damage critical components in a turbo-jet engine. Airplane engine icing is normally divided into two sub-categories: induction icing and intake icing. Induction icing refers to ice which develops within a reciprocating engine's carburetor when cooling associated with the venturi effects and fuel evaporation causes condensation and freezing or deposition. This type of icing occurs most commonly in clear air and at ambient air temperatures well above freezing. It requires neither liquid water in the atmosphere nor freezing ambient temperatures. See also Chapter 10.

Intake icing is a specific type of structural ice which forms on the air inlets for either reciprocating or jet engines. Both types of engine icing can block the engine's air supply, reducing power or thrust available. In addition, inlet ice on jet engines can break off, damaging compressor blades.

Loss of power by a reciprocating engine can be caused when ice forms in the induction system, blocking the source of air to the engine. Although this blockage can occur in atmospheric conditions similar to those which allow structural ice to form, induction icing can also occur in clear air and when temperatures are well above freezing conditions. Impact ice can occur when supercooled water makes contact with components of the engine air induction system which are cooled to below freezing temperatures. This requires the same conditions which foster the growth of structural ice. In fact it is structural ice that forms in areas which block the flow of carburetor air to the engine. This ice can form on the air inlets, air filters, and on the structural components within the induction system on which the supercooled water strikes. Venturi ice forms in parts of the induction system where increases in local velocity cause a drop in local ambient static pressure and an associated drop in temperature. If the drop in temperature is greater than the

temperature-dew point spread, moisture in the air will condense. If the temperature of the surface is below freezing, the moisture can either freeze when it contacts the surface or form directly out of the air as frost. This can occur in the carburetor's venturi or around a partially closed throttle valve. The more the valve is closed, the greater the velocity necessary to keep a constant mass rate of airflow. A fully open throttle valve produces little temperature drop. Maximum temperature drops from the venturi effects are normally small (in the area of 5°C) but can allow ice to form when inlet temperatures are slightly above freezing. A much greater drop in temperature is associated with the evaporation of fuel where it is introduced into the carburetor. The energy necessary to turn the liquid fuel into its gaseous form is absorbed from the air, cooling it as much as 20°-40°C. Again, if the temperature-dew point spread is sufficiently small (and, as the accompanying chart shows, it doesn't have to be very small), water vapor in the air will condense. If the temperature is below freezing, ice will form on structure within the induction system. As the passage for air and the fuel air mixture to the engine becomes smaller, the power output of the engine is decreased. Manifold pressure and, for fixed-pitch prop's, RPM will drop.

The engine can also begin to run rough. As ice continues to grow the opening for passage of the fuel-air mixture can change sufficiently to preclude engine operation and the engine will quit. The evidence concerning the existence of engine icing is primarily circumstantial. Weather forecasts and observations, especially those by pilots flying airplanes with similar engine configurations, may provide evidence concerning the possibility of engine ice. However, the location of the induction system on the engine can effect the temperature of the induction air and variances between engine types as well as engine operating conditions can spell the difference between ice and no ice. When investigating potential induction icing accidents, the availability and use of anti-ice features such as carburetor heat (for venturi, evaporation and impact ice within the induction system) and alternate air sources (for impact ice on induction air inlets and filters) should be examined. The position of cockpit controls and the mechanical devices they operate, need to be determined and documented. Functional checks, where possible, should also be performed. The assistance of an engine technician, experienced in the specific engine type and installation involved is vital. There are also a number of carburetor ice (venturi & evaporation) detection systems on the market. If one of these systems was installed, its operational status and any post crash indications of the existence of ice should be ascertained.

Inflight ice affects jet engines differently than reciprocating engines. For the most part, the problems involve damage to rotating components or disruption of the airflow through the engine and accompanying engine stalls or stagnations. Structural ice which forms ahead of or in the inlets can be shed in slabs which, if they enter the engine, can cause damage to the engine and loss of thrust. In addition, if the slabs disrupt the airflow across the compressor face, compressor stalls which result in engine flame-outs or stagnation can also result. The formation of ice on inlet guide vanes, stators, and compressor blades can result in reduced clearance and interference damage. Post crash evidence may include FOD type damage. Engine stagnations can lead to excessive temperatures in the engine's turbine section, and associated heat damage. Finally,

modern jet engines rely on pitot-static pressure information from the engine inlet. If the sources of this pitot-static pressure information are iced over, engine operation and thrust output can be affected. Evaluation of FDR and CVR data may provide clues concerning engine RPM, thrust produced and operation of deice or anti-ice systems.

When compared to other leading edge surfaces on the airplane, the leading edge of engine inlet ducts and inlet guide vanes are relatively thin and are therefore relatively good ice collectors. In addition, the low static pressures in the engine inlets will be accompanied by lower than ambient temperatures. If small ambient temperature-dew point spreads exist, frost can form on inlet duct surfaces whose surface temperatures are below freezing.

D. OTHER SYSTEM EFFECTS OF IN-FLIGHT ICING. It should be obvious that the effects of in-flight icing go well beyond aerodynamics and thrust. The blockage of the static port of an airplane can cause a multitude of problems. If the pitot pressure is used by the airspeed indicator, an increase in altitude will also cause the indicated airspeed to increase. If the pilot attempts to control the airspeed indication by raising the nose, the airspeed indication will increase further while the actual airspeed is decreasing. The stall and spin of a commercial airliner, despite its experienced crew, was attributed to this series of events. On the other hand, loss of altitude will cause an airspeed indicator to indicate lower than actual speeds if the

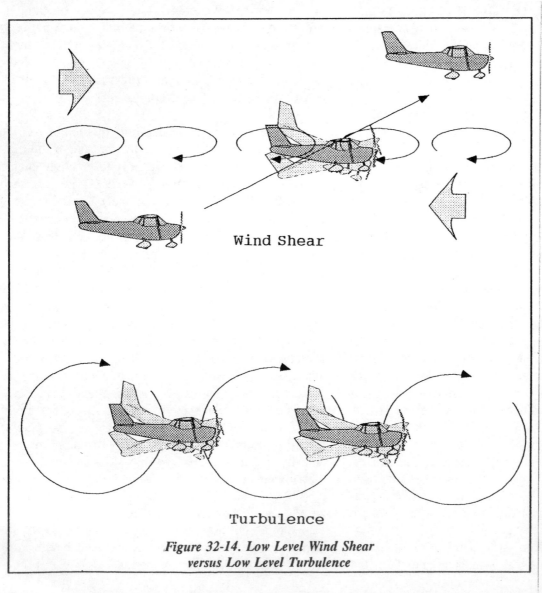

Wind Shear

Turbulence

Figure 32-14. Low Level Wind Shear versus Low Level Turbulence

airspeed indicator's pitot source is blocked with ice. If the pilot attempts to correct the low airspeed indication by lowering the nose, indicated airspeed will decrease further while the airplane's actual speed is increasing. The excessively high airspeeds which result can cause structural failure due to high dynamic pressures, aeroelastic failures (flutter, wing divergence in torsion and control reversal) or over G due to maneuvering loads at very high airspeeds. If the pitot source is used by the airplane's stability augmentation system, erroneous controls may lead to pilot induced oscillations and resulting over G, loss of control and subsequent collision with the ground. The accumulation of ice on communication and navigation antennas can degrade the effectiveness of the transmission or reception of the antennas. The antenna may also fail structurally when the drag created by the ice exceeds the antenna's static strength or the ice creates shapes which are aerodynamically unstable and dynamic loads cause their structural failure. The build-up of leading edge wing ice may also be masked if the autopilot is engaged, providing the small aileron inputs necessary to keep the wings level. If the autopilot is disengaged, either by the pilot or on its own when its authority is exceeded by aerodynamic forces, the airplane may experience a sudden roll-off in the direction of the most contaminated wing. A large aileron input could cause wingtip stall. One theory about a recent accident involved the retraction of the flaps, which would increase the sensitivity of the wing to tip-stall, at approximately the same time that the auto-pilot was disengaged.

E. CONTROL OVER IN-FLIGHT ICING. When investigating accidents involving airplane icing, the investigator needs to examine not only what happened, but why it happened. These are some of the questions that may need to be answered.

- Why did the pilot fly into icing conditions which the airplane was not able to safely penetrate?

- Did the pilot seek a pre-flight weather briefing?

- If a briefing was provided, was it accurate?

- Did the pilot seek or did air traffic control provide updates of significant weather?

- Did the pilot know that ice was accumulating on the critical airplane surfaces?

Another series of questions addresses airplane systems and their ability to detect, prevent the accumulation or eliminate the accumulation of ice.

- Were anti-ice and deice systems functional and effective?

- Did the pilot know how and when to operate anti-ice and deice systems such as airfoil leading edge boots, electrically and engine bleed air heated surfaces, and glycol systems?

- Were the anti-icing or deicing systems installed on the airplane capable of functioning in the icing environment encountered by the airplane?

- Was the airplane flown at a higher or lower than normal angle-of-attack, allowing ice to accumulate on unprotected airfoil surfaces which were aft of leading edge de-icing devices?

Another issue concerning the ability of aircraft to operate in icing conditions is their certification in accordance with the provisions of 14 CFR 25.1419 (for aircraft weighing more than 12,500 lbs.) and 14 CFR 23.1419 (for lighter aircraft). Although some pilots believe otherwise, no aircraft is certified to operate continuously in severe icing.

4. WIND SHEAR.

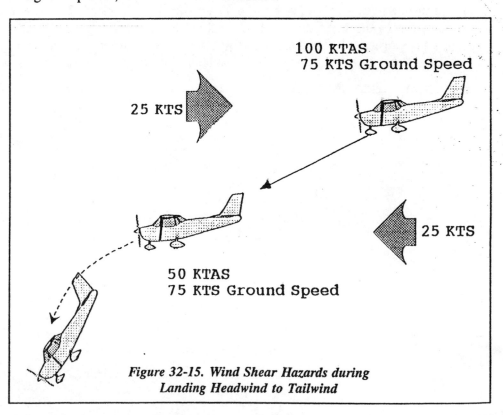

Figure 32-15. Wind Shear Hazards during Landing Headwind to Tailwind

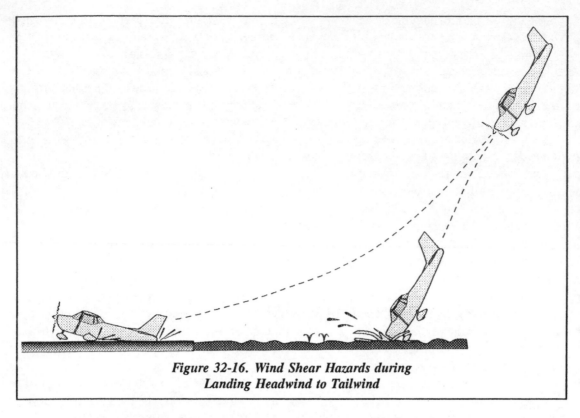

speeds associated with takeoff and landing.

Since landing hazards are normally considered more severe, let's look at them first. In addition, let's keep things simple at first by only addressing horizontal wind shears. If an aircraft flies through a shear involving a relatively rapid decrease of airspeed, the aircraft will tend to lose altitude before equilibrium is re-established. Similarly, if aircraft experiences a sudden horizontal wind shear involving an increase in airspeed, the aircraft will tend to gain altitude before equilibrium is re-established.

Figure 32-16. Wind Shear Hazards during Landing Headwind to Tailwind

Wind shear is a sudden change in either the wind's speed or direction and, at low altitudes, is normally associated with microbursts, frontal passage, sea breeze fronts, temperature inversions, and gust fronts proceeding thunderstorms. Although wind shear induced turbulence at altitude can cause discomfort to passengers, and can, if encountered at excessively high speed, cause structural damage, it is most dangerous when experienced at the low altitudes and at low

Before we get into a full blown discussion of low level wind shear (LLWS), let's clear up the difference between LLWS and turbulence. Both LLWS and turbulence involve a sudden change in wind speed and/or direction. However, turbulence involves repeated changes as the airplane moves through a turbulent air mass. LLWS, on the other hand, involves one change each time the airplane moves from one air mass to another air mass. The differences between the two are important because they influence the tactics used by the pilot to counter the

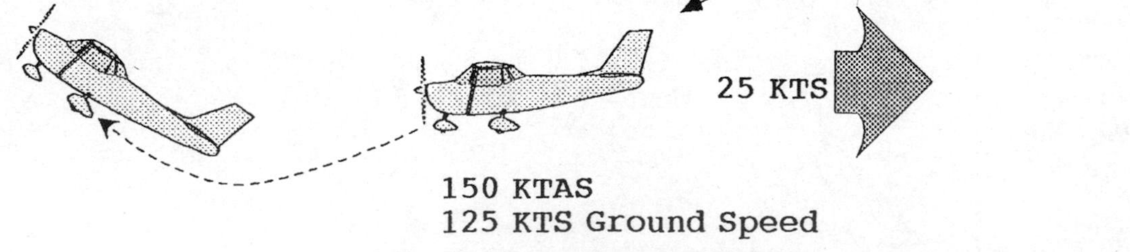

100 KTAS
125 KTS Ground Speed

25 KTS

25 KTS

150 KTAS
125 KTS Ground Speed

Figure 32-17. Wind Shear Hazards during Landing Headwind to Tailwind

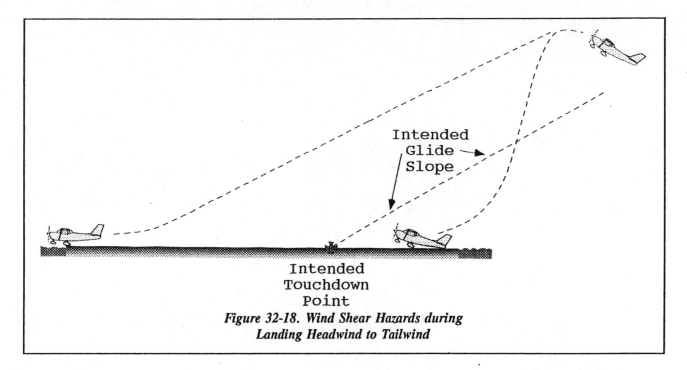

Figure 32-18. Wind Shear Hazards during Landing Headwind to Tailwind

hazards generated by each. Generally, pilots will compensate by carrying extra airspeed when wind shear is forecast. On the other hand, since higher speeds increases the G load imposed by individual gust experienced during turbulence, pilots tend to slow down when encountering turbulence. Of course there are shades of gray when talking about the two. Both involve disruptions of an otherwise uniform airflow

encountered by an airplane, and that's part of the reason why they are confused. The National Weather Service's definition of the LLWSs which are to be included in forecasts is based on a shear of at least 20 knots per 200 feet of altitude. The FAA urges (but doesn't require) pilots to report wind shear encounters while on approach to or departure from airports. FAA guidance suggests that pilots report the gain or loss in

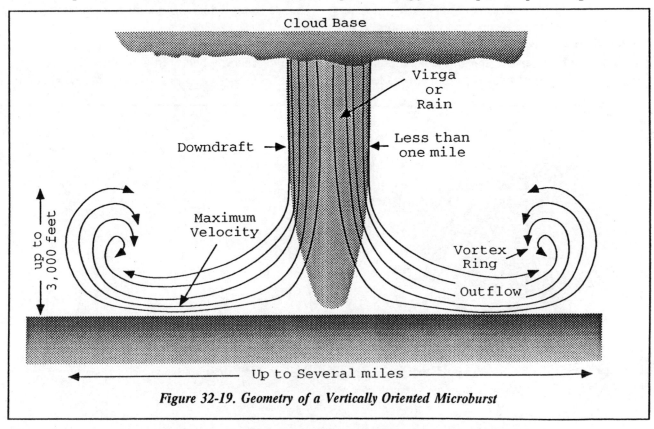

Figure 32-19. Geometry of a Vertically Oriented Microburst

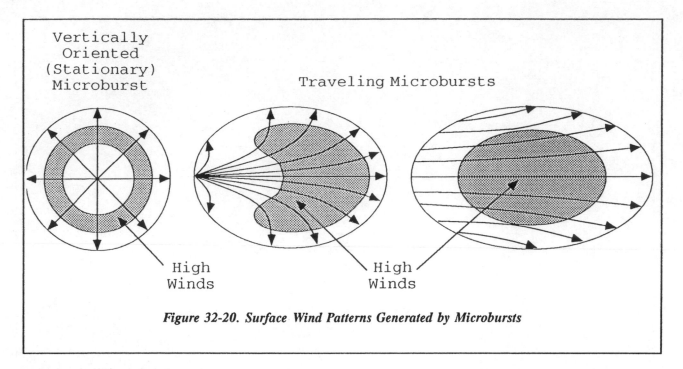

Figure 32-20. Surface Wind Patterns Generated by Microbursts

airspeed as well as the altitude at which it occurred.

LLWS can be divided into two general categories, convective and non-convective. Figure 32-14 shows how convective LLWS can involve both vertical and horizontal shears while non-convective LLWS will normally involve only horizontal shears. Some of the larger airports are equipped with a computerized Low Level Wind Shear Alert System (LLWAS) which was designed to detect the presence of hazardous LLWS. LLWAS does this by automatically comparing the winds measured by various sensors located around the periphery of the airport with the winds near the center

of the field. If the difference between the center sensor and a periphery sensor becomes excessive, the system is designed to notify the tower controller who will, in turn, notify traffic. Airports equipped with LLWASs are identified in the Airport/Facility Directory.

A. WIND SHEAR DURING LANDING. Wind shear during landing can cause problems when a change in the wind speed suddenly causes the aircraft to:

- have insufficient airspeed to prevent it from descending below its intended flight path,

- have excessive airspeed which causes

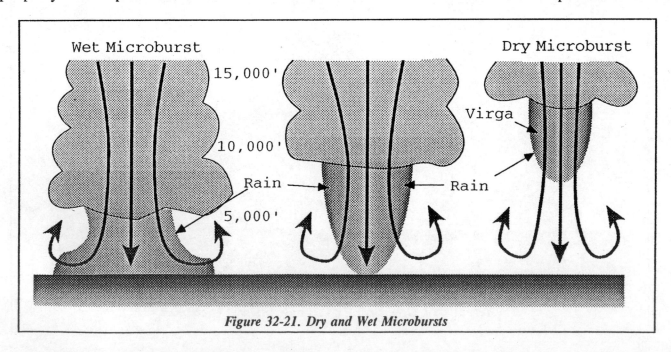

Figure 32-21. Dry and Wet Microbursts

it to climb above intended flight patch, or

-

develop an uncontrollable drift to the left or right.

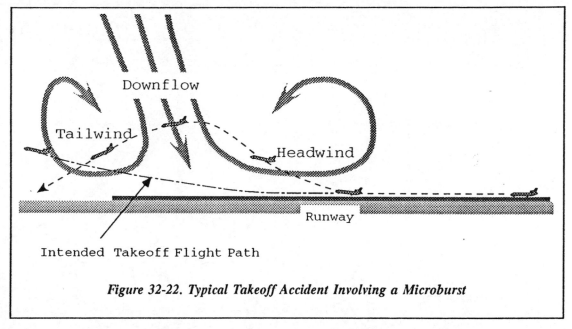

Figure 32-22. Typical Takeoff Accident Involving a Microburst

Of the three, the first (insufficient airspeed) is obviously the most critical. If the ground is too close, the only choice may be between risking a stall while attempting to prevent a descent into the ground and a controlled descent to a point off the runway. In the second and third situations (excessive airspeed or lateral drift), only the pilot who attempts to force the aircraft on to the runway is at risk.

In the first case imagine an aircraft flying a final approach at 100 knots and into a headwind of 25 knots. The aircraft speed over the ground would be 75 knots. If the aircraft were to fly out of the air mass which is moving at 25 knots and into a layer of calm air which is close to the surface (a situation which can occur if there is a temperature inversion) the aircraft airspeed at the moment it entered the calm air would be 75 knots. This could be below the aircraft's stall speed, and immediate corrective action is necessary to prevent a stall. When the aircraft airspeed drops, the nose of the aircraft will naturally tend to drop in an attempt to return to its trimmed airspeed, and the aircraft will tend to drop below the intended glide slope. If the encounter with the wind shear occurs just before the flare, the pilot may not be able to prevent a short or nose low, hard landing. If the loss of airspeed occurred at a higher altitude and the pilot was able to avoid a stall, an increase in thrust would be necessary to offset the increase in drag associated with the lower airspeed. Aircraft on final approach are normally flying at speeds where total drag is close to minimum. A decease in airspeed will then result in an increase in total drag. This situation is especially critical for high by-pass turbojet engines which are slow to respond to requests for increased thrust when operated at low RPM.

One other situation deserves consideration. If the aircraft is flying a coupled approach and the autopilot

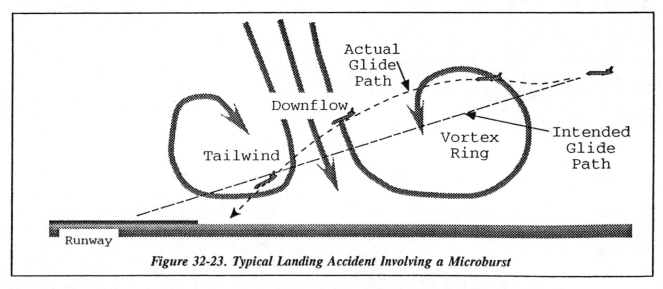

Figure 32-23. Typical Landing Accident Involving a Microburst

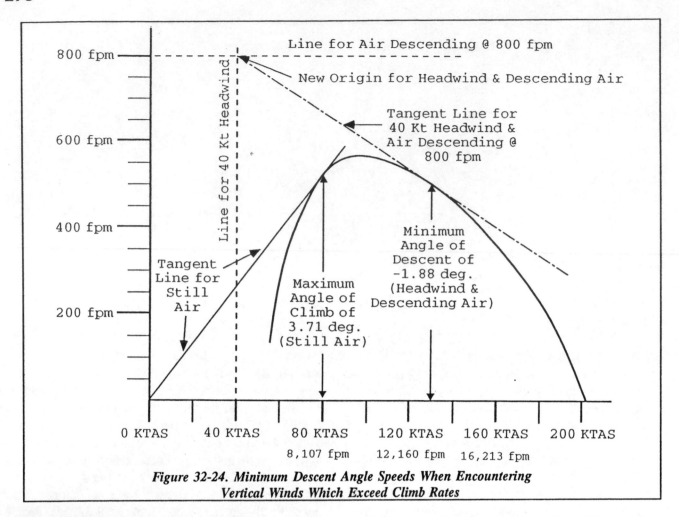

Figure 32-24. Minimum Descent Angle Speeds When Encountering Vertical Winds Which Exceed Climb Rates

overrides the aircraft's natural tendency to nose down in an attempt to maintain a constant glide path, the aircraft speed will rapidly decay through the stall speed. Not good! By the way, the same situation can occur if the initial portion of the final approach is flown in calm air and a 25 knot tail wind is encountered on short final. All of these situations can lead to a short landing.

It is possible that a headwind to tailwind shear results in a long landing and a runway overrun. In this case the aircraft would be holding a higher than normal power setting in order to maintain the recommended glide slope angle while flying into a headwind. When the windshear is encountered and the aircraft both drops below the glide slope and loses airspeed, a large increase in power will be necessary to slow the descent in order to regain the glide slope as well as accelerate to the recommended final approach speed. It is possible that the pilot will overshoot the glide slope and airspeed and end up flying final approach both above the glide slope and above the recommended final approach airspeed.

The questions that needs to be addressed during the investigation of this type of accident include:
        - Should the pilot have been able to recover the aircraft safely?
        - Did the aircraft have sufficient performance to avoid a collision with the terrain?

A detailed analysis of the available airspeeds flown, total drag, aircraft weight, thrust available and estimated wind shear magnitude and direction will be necessary to assess the aircraft performance potential. A discussion of how this can be accomplished is contained in Chapter 29. Coupling the computer generated information with results of pilot-in-the-loop simulations can go a long way toward answering these questions and the development of realistic safety recommendations.

In the second case, the situation is reversed. The aircraft is flying down final at 100 knots in calm air and suddenly flies into an air mass which gives it a headwind of 25 knots. At the moment the aircraft enters the air mass, its speed through the air will be 125 knots and the natural positive static pitch stability

(Chapter 31) will cause the nose to rise. This will place the aircraft high and hot on final and leave the pilot with three choices:

  - go around (safe but embarrassing),

  - continue the approach using a normal glide slope while slowing to normal approach speed (safe if there is adequate dry runway remaining after the longer-than-planned touchdown) or

  - continue the approach by shoving the nose down, pulling the power to idle and hoping that the landing flare can be completed without either a hard or nose wheel first landing (unstabilized approaches are one of the leading causes of landing phase accidents).

In the last case, a sudden increase in cross winds just prior to touchdown can cause the aircraft to weathervane into the wind or drift downwind off the runway. The coupling of directional and lateral stability (Chapter 31) can also cause the aircraft to roll away from the direction of the yaw resulting in wing low touchdowns and wingtip or engine pod contact with the ground. Touchdowns in a crab can cause side loads which exceed the capabilities of the landing gear or loss of directional control after touchdown.

In 1985, the intentional crash of a remotely controlled Boeing 720, and a "demonstration" of the adequacy of anti-misting jet fuel was upset by a low altitude lateral wind shear. When a highly experienced test pilot attempted to salvage the landing rather than execute a go-around, the aircraft hit the ground in a left wing low attitude.

B. WIND SHEAR DURING TAKEOFF. A wind shear which results in a loss of airspeed as the airplane climbs after takeoff can cause problems similar to those experienced during landing. Although the wind shear may have the same results (a crash), the problem facing the pilot is not normally as demanding. The power is already advanced and the aircraft is moving away from the ground. Now, if the airplane encounters a wind shear which results in a loss of airspeed, the crew will not be faced with the delays associated with decision making; increasing thrust or power, raising the nose and cleaning up the airplane configuration. The pilot will still have to ensure that a safe angle of attack is maintained. Some extra terrain clearance may be available and traded-off for airspeed as the aircraft is allowed to descend back through the shear and into the headwind at lower altitudes. However, if the pilot attempts to maintain a climb by increasing the pitch attitude or if the shear occurred at very low altitude with no room to trade a

little altitude for airspeed, a stall or spin crash may be inevitable. Again, a detailed analysis of the airspeeds flown, total drag, aircraft weight, thrust or power available and estimated wind shear magnitude and direction coupled with realistic pilot reactions will be necessary to assess the aircraft performance potential.

5. HEAVY RAIN AND MICROBURSTS.

In addition to the hazards associated with ground and in-flight icing, mother nature has included two other weather related hazards on the list of aeronautical perils; heavy rain and microbursts. Neither is as widespread as icing, but both have accounted for their share of accidents. Although heavy rain and microbursts are often found together and their effects are synergistic (the hazards of each complementing the other), we will, for simplicity, treat them as separate entities.

A. EFFECTS OF HEAVY RAIN. Studies at NASA's Langley facility have shown that heavy rain, in the range of from 100 to 1,000 mm/hr, can reduce C_{Lmax} by 7 to 29% and the stall AOA by 1 to 5 degrees. In addition, drag increases on typical transport aircraft were estimated at 2 to 5%. NASA came to these conclusions by propelling an instrumented NACA 64-210 airfoil along a long track while simulating rainfall with overhead sprinklers. Four related hazards were identified:

  - wing surface roughening by water running off the wings.

    - the water's weight.

    - the water's vertical momentum.

    - the asymmetric application of water on the airplane's wings.

All can be applied simultaneously when an airplane flies into a heavy cloudburst. If this rain is combined with an adverse wind shear (described in the next paragraph), the synergistic effects may exceed the airplane's performance by a wide margin. A layer of rain water which forms on the wings will take a finite length of time, even in the airstream, to flow off the wing. The heavier the rain, the greater the thickness of the film of water on the wings. This film of water can have an effect on the wing's aerodynamic performance which is similar to the effects of a coating of frost, reducing C_{Lmax} by up to 30%, decreasing the stall angle of attack and increasing drag. With total planform areas of over 10,000 ft^2 for large airliners, a layer of water averaging 1/16 inch thick can add over 3,000 lbs to the airplane's weight. In addition to its weight, the rain will impart its momentum on to the airplane's planform. In the vertical plane, the rain's vertical velocity and mass create a momentum which acts on a very large horizontal planform and must be offset by additional lift. The airplane's forward velocity coupled with the airplane's frontal area will provide a target for the falling rain and a momentum change which will show up as additional drag. Finally, the application of rain to the airplane's wings and the degradation of lift will rarely be symmetrical. If one wing stalls while the other is still creating significant lift, the resulting roll could place the airplane in an attitude from which it could not recover before ground impact, a situation most likely when the airplane is operating at high angles of attack and close to the ground.

B. MICROBURSTS. Microbursts are pockets of cold air which are generated in cumuloform or stratocumulus clouds and then descend rapidly to the ground. Although their vertical speed can exceed 45 kts, the speed of the descending column of air can accelerate to over 100 kts as it turns to parallel the earth's surface. Although the phenomena has certainly existed since the beginning of time, it was not understood or even known until the mid '70s. Dr. Theodore Fujita of the University of Chicago and later Dr. John McCarthy of NCAR exposed not only its existence but its probable role in numerous accidents which were blamed on other factors. Prior to that time, takeoff and landing accidents for which no other cause could be found, were sometimes incorrectly blamed on the crew. Since Dr. Fujita's studies, the involvement of microbursts in numerous airplane accidents has lead to increased awareness, pilot recovery techniques and development of both ground based and airborne detection equipment.

The microburst starts as an entrained mass of cool dry air in a cumulo cloud. When the density of the air mass cools sufficiently, it will descend from the cloud in a column which is generally less than one mile in diameter and sometimes accompanied by rain. If the column of cold air and rain descends into dry air, the rain will evaporate, extracting heat from the column of air, cooling it, increasing its density and increasing its rate of descent. If the rain evaporates before it reaches the ground, the microburst is referred to as a "dry" microburst. If, however, the rain accompanies the descending column of air to the ground, the microburst is referred to as a wet microburst. Dry microbursts are more typical of the high Western plains states of the United States while wet microbursts are more typical of the South Eastern states.

As the descending column approaches within a couple of thousand feet of the ground, increasing pressure in front of the column will cause it to spread out and accelerate to speeds which can exceed 100 kts. The exact shape of spreading air will depend on the horizontal component of the column's velocity. If the column is moving straight down, the air mass will take the shape of a horizontal ring with a maximum height of 1,000 to 3,000 feet and a maximum diameter of 2 1/2 miles. Several different potential wind patterns are shown in Figure 32-19.

The ground life of a microburst is normally less than 15 minutes. During the first five minutes the winds will grow to their maximum velocity. If the microburst is moving horizontally across the terrain, the length of time the wind is blowing at a specific ground location may be less than 15 minutes.

One well documented microburst, which touched down at Andrews Air Force Base outside of Washington DC, contained winds which peaked at over 130 kts. A anemometer located alongside the runway showed wind speeds which went from about 12 kts to over 130 kts (that was the anemometer's limit) in less than two minutes. The wind dropped to zero in less than a minute, shifted 180 degrees and then, within one minute, accelerated to 83 kts. In less than 10 minutes the wind speed indicated by the anemometer was back to normal. Although the arrival of the microburst was unannounced, the arrival of the President of the United States, whose plane landed just 10 minutes prior to the arrival of microburst, was announced and the waiting dignitaries and press were treated to quite a show.

Microbursts are a hazard primarily to aircraft at altitudes less than 1,000 feet AGL during their takeoff or landing phases of flight. The scenarios of takeoff and landing accidents are different. During a takeoff, the most hazardous location for a microburst is at or just off the departure end of the runway. As the airplane begins its takeoff roll, it will encounter a headwind associated with the outflow of air from the microburst. This will allow the airplane to takeoff at a lower than normal ground speed. As the airplane climbs into the microburst's down flow, it will experience a loss of airspeed (due to the loss of the headwind) and loss of climb ability (due to its moving into a mass of descending air). An increase in pitch attitude to arrest sinking will keep the airspeed low as the airplane moves into the outflow air which now shows up as a tailwind. The airplane is now set up for a stall or a descent into the terrain. See Figure 32-22 for an illustration of a typical takeoff scenario.

Landing accidents typically involve a microburst sitting between the approach end of the runway and an airplane on final approach. As the airplane enters the microburst's outflow, its speed through the air will momentarily increase and it will probably climb above the glide slope. Attempts to return the airplane to its target airspeed and glide slope will normally involve lowering the nose and decreasing power. If the indicated airspeed is reduced to published approach speeds the speed over the ground will be below approach speeds. Then, in an attempt to regain the glide slope the airplane is descending at a higher than normal rate when it enters the down flow portion of the microburst, it also experiences a decrease in airspeed associated with the loss of the headwind component. Then, with the power back, a higher than normal sink rate and airspeed decreasing, the airplane enters the outflow on the other side of the microburst. The tailwind generated by the outflow can reduce the airspeed to the point where the airplane will stall if level flight is attempted.

The hazards associated with microbursts are now well known. Although recovery techniques have been developed by manufactures and published by the FAA, these techniques are common for all commercial jet transports and are not specifically developed for each different type of airplane. In addition, the techniques do not distinguish between the performance of an airplane when it is close to its maximum gross weight (immediately after takeoff) and at low gross weight (prior to landing). But, even if optimum techniques are employed by the crew following a microburst penetration at low altitude, the airplane still might not have sufficient performance to safely escape the effects of the microburst.

6. MOUNTAIN WAVES.

When high speed winds pass over a single mountain or pass perpendicular to ranges of mountains and through a stable layer of air, large scale turbulent flow is created. It extends both above and below the altitude of the mountain peaks. As the winds pass over the crests of mountains and ridges they will be deflected downward, creating significant down drafts, updrafts and horizontal winds. This turbulent flow can work its way down to the surface and generate the wind shears discussed earlier.

A fast moving rotor-type wave was reported to have been one of the possible factors in a of loss of control accident involving a Boeing 737 attempting a landing at Colorado Springs, Colorado. An early model USAF B-52 searching for turbulent weather east of the Rockies found it just before a horizontal gust ripped off most of the aircraft's vertical tail (Structural design requirements for vertical tails were beefed up as a result of the incident). Each year general aviation aircraft are lost while fighting their way into a headwind in a vain attempt to cross a ridge while caught in down drafts which exceed their climb capability.

There are some common misconceptions associated with the best course of action when caught in a down draft which exceeds the airplane's climb capability.

- Apply full power and increase the pitch attitude until the aircraft is on the verge of a stall.
- Apply full power and hold the airspeed for best angle of climb.
- Apply full power and hold the airspeed for best rate of climb.

Actually, option 1 is a good idea if the pilot knows the aircraft is going to touch-down, even with full power. Touch-down at the lowest possible speed while still under control is a desirable way to crash land. However, at this speed, total drag is above the minimum and that hurts climb performance. Option 2 is good if you can get the aircraft to climb, but is not a good choice if the aircraft is still sinking. Option 3 is good (or bad) for the same reason. The most desirable speed is the one which provides the smallest descent angle. Unfortunately there is no way to calculate this speed with the information provided in the pilot's handbook and the time available while descending toward the ground.

Fortunately, the aircraft accident investigators, while in the comfort their offices, can estimate what was possible. The first step is to obtain and plot the aircraft's full power rate of climb at various airspeeds. You can estimate this yourself on a piece of graph paper (See Chapter 29 for clues), estimate it through flight tests, or get the data from the manufacturer. After you have the graph plotted add a scale below the airspeed which provides equivalent speeds in feet per minute. These values assume zero vertical winds and need to be corrected for estimated vertical and head winds. The correction for the vertical wind (down or we wouldn't have a problem) can be made by drawing a horizontal line (a) across the plot of climb rate at the estimated vertical speed. This line should be above the maximum climb rate (or the airplane should have been able to climb). Next draw a vertical line (b) through the speed of the estimated head wind (a tail wind will be to the left of the abscissa). Next draw a series of lines radiating out of the intersection of the wind speed lines in directions which cause them to intersect the curved line representing the aircraft's full power, still air rates of climb. The slope of these lines will represent the descent slopes for the various airspeeds. The tangent of the descent angle is descent rate (vertical feet per minute) divided by the ground speed (horizontal feet/minute) or vertical feet traveled per horizontal feet traveled. In other words:

$$\tan^{-1}(descent \angle) = \frac{descent\ rate(fpm)}{horizontal\ speed(fpm)} \qquad (1)$$

The idea here is to find the right combination of descent rate and horizontal speed. Once again it is unrealistic to assume that the pilot will be able to make these calculations while facing a ridge line in an aircraft which is losing altitude despite full power. Perhaps, they can give the investigator some clues on what could have been.

7. SUMMARY.

As the saying goes, it doesn't pay to mess with Mother Nature. She has been in business longer than we have and we are just beginning to understand all the tricks she has in store for us.

Basically, accident investigation of these phenomena center around these questions.

1. How did we get into this mess?

2. What went wrong with the equipment designed to prevent this or our procedures designed to avoid this?

3. Once in the mess, what could we have done?

33

High Speed Flight

1. INTRODUCTION.

When an aircraft is flown at speeds above those for which it has been designed, it is being exposed to any number of hazards ranging from structural failure, to degraded control to degraded stability to system failure to engine failure. One of the ways these hazards can be grouped is by the speeds at which they occur (i.e. subsonic, transonic, supersonic and hypersonic). At subsonic speeds we are primarily concerned with the hazards associated with wind gust induced over G, aeroelastic factors and excessive dynamic pressure. We'll leave that area for another day.

When an aircraft designed for subsonic flight ventures into the transonic speed range, we can add to the above list

the effects of compressibility and its effects on stability, controllability and engine performance. As the air-

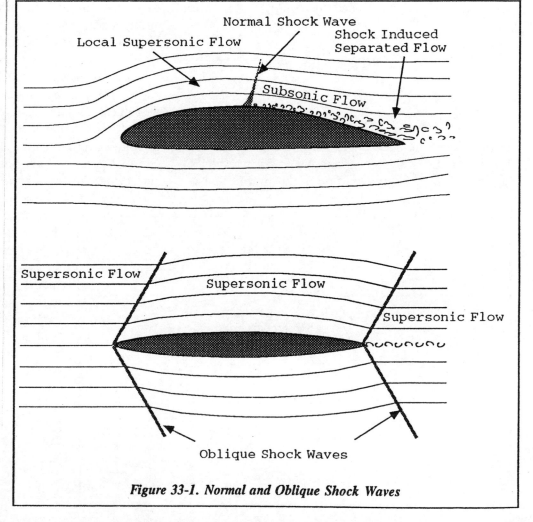

Figure 33-1. Normal and Oblique Shock Waves

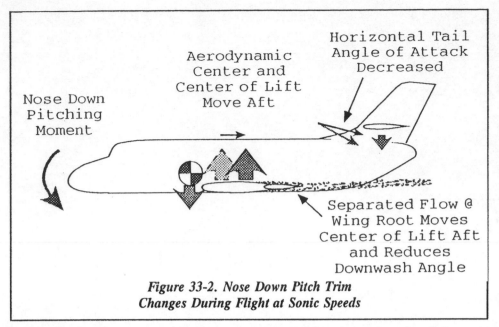

Nose Down Pitching Moment

Aerodynamic Center and Center of Lift Move Aft

Horizontal Tail Angle of Attack Decreased

Separated Flow @ Wing Root Moves Center of Lift Aft and Reduces Downwash Angle

Figure 33-2. Nose Down Pitch Trim Changes During Flight at Sonic Speeds

stagnation temperatures become even greater concerns. For aircraft utilizing computer controlled components, somewhere along the line two questions will come up.

- What is the maximum speed acceptable as an input for programs used by computer controlled systems?

- What will happen if this speed is exceeded?

2. DEFINITIONS AND CONCEPTS.

A. CRITICAL MACH NUMBER. An aircraft's critical Mach number is the minimum free airstream speed (in terms of Mach number) which results in sonic airflow over some portion of the aircraft surfac-

es. In and of itself, there is nothing really "critical" about an aircraft's critical airspeed. When an aircraft flies, the speed of the airflow around the airplane varies from location to location. Obviously, the speed over the top of the wing is faster than the speed under the bottom of the wing. The difference in speed results in the pressure differential which produces lift. Other than the leading edge surfaces, the speed of the air over most of the aircraft surfaces is in fact faster than the speed of the aircraft through the air. As an aircraft accelerates toward the speed of sound, air which is being most accelerated by the passage of the aircraft will reach sonic velocity before the aircraft is moving at Mach 1. This speed, the speed at which some of airflow around the aircraft reaches Mach 1, is referred to as the aircraft's critical Mach number. It is the speed at which shock waves first start forming on the aircraft's surfaces.

B. SHOCK WAVES. Shock waves form when supersonic air encounters some restriction to its movement and then suddenly decelerates, compresses and experiences an increase in static pressure. As an aircraft is accelerated through its critical Mach number "normal" shock waves will typically form on the surface of the top of the wing, often near its root where the wing is thickest. These waves are called "normal" because they form perpendicular (or normal) to the oncoming relative wind. As the aircraft accelerates to supersonic speeds, many of these shock waves will "lean" back away from the oncoming supersonic flow and form "oblique" ("oblique" because they are "oblique" or at an angle to the relative wind), shock waves. At supersonic speeds normal shock waves exist only in front of "blunt" portions of the aircraft, portions whose three dimensional surface is perpendicular to the oncoming relative wind.

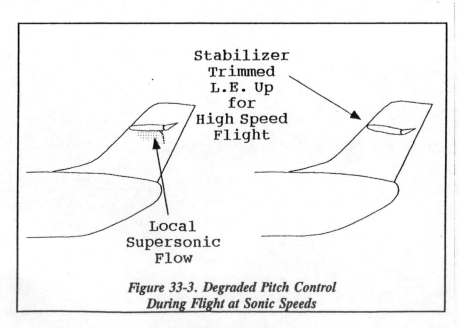

Stabilizer Trimmed L.E. Up for High Speed Flight

Local Supersonic Flow

Figure 33-3. Degraded Pitch Control During Flight at Sonic Speeds

Normal and oblique shocks waves have several similarities and differences. Both involve the abrupt slowing down of supersonic air. In other words both have supersonic air in front of them that rapidly slows down as it moves through the shock wave. In the case of a normal shock wave, the air will slow from supersonic to subsonic speeds. The amount of reduction in speed is represented by Equation (1).

$$M_2 = \frac{1}{M_1} \qquad \textbf{(1)}$$

Where:
M_2 = Mach number behind the shock wave.
M_1 = Mach number in front of the shock wave.

Although an oblique shock wave also slows supersonic flow, it does not slow it to subsonic speeds. The reduction in the airflow's speed as it moves through an oblique shock wave is related to the angle the shock wave forms with the airflow. The larger the angle, the larger the reduction in speed.

Both normal and oblique shock waves involve an abrupt increase in static pressure as the air moves through the shock wave. However, given the same initial Mach number, the pressure rise across a normal shock wave is greater than that across an oblique shock wave. Shock waves which form on the top of wings during transonic flight can cause an additional adverse pressure gradient and premature separation of the airflow behind the wing.

Both normal and oblique shock waves cause additional drag (called wave drag) which is in addition to induced and parasite drag. The increase in drag starts at a speed referred to as the "force (or drag) divergence Mach number" which is slightly greater than the aircraft's critical Mach number. The critical and force divergence Mach numbers are both less than Mach 1.

C. FORCE DIVERGENCE. As an aircraft accelerates into the transonic speed range, the coefficient of drag will increase rapidly and then decrease as the aircraft reaches supersonic speeds. The drag associated with transonic and supersonic flight is referred to as "wave" drag. The speed at which an aircraft first starts to develop wave drag is referred to as its "force divergence" speed (or Mach number). Sweeping an aircraft's wing aft will both increase an aircraft's force divergence Mach number (allowing it to cruise at higher Mach numbers without the penalties associated with wave drag) and decrease the maximum transonic drag coefficient (allowing it to reach supersonic speeds with less thrust). However, at higher supersonic speeds, the coefficient of drag for swept winged aircraft is higher than that for straight winged aircraft.

D. COEFFICIENT OF LIFT. As an aircraft accelerates through the transonic speed range it will experience some decrease in its maximum coefficient of lift. Sweeping an aircraft's wings aft will delay this drop in C_{Lmax} to higher Mach numbers.

E. SHIFT OF AERODYNAMIC CENTER. As an aircraft transitions from subsonic to supersonic flight (or vice versa) a shift in the wing's aerodynamic center will cause a significant change in its pitch trim. When an aircraft is flown at less than its critical Mach

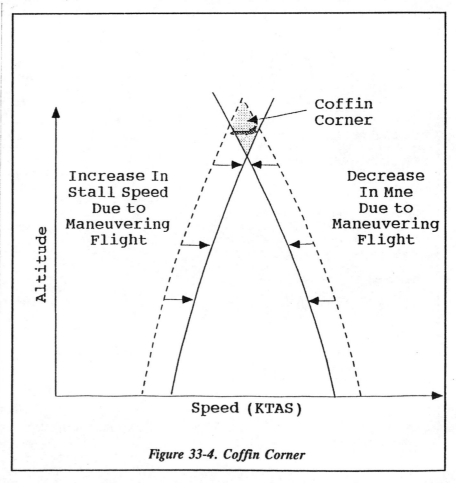

Figure 33-4. Coffin Corner

number, its aerodynamic center will be located at approximately 25% chord. If the aircraft is accelerated to supersonic speeds, the aerodynamic center will shift aftward to the 50% chord location. This aftward shift of the aerodynamic center will provide the average lift vector with a greater moment arm, producing a nose down pitch trim change. As the aircraft slows from supersonic flight to subsonic flight, the aerodynamic center will move forward to its original location, producing a nose up pitch trim change.

3. HAZARDS GENERATED BY EXCESSIVE SPEED.

A. MACH TUCK. As an aircraft exceeds its critical Mach number it will naturally experience a nose down pitch trim change as the aerodynamic center moves aft toward the 50% chord position. In addition, if the thicker portions of the wing root on a swept wing aircraft experience a reduction in C_L (due to the supersonic airflow over the wing) before the outer portions of the wing, the wing's center of lift will move aft, producing an additional nose down pitching moment. A third factor which might be acting to lower the nose can be the reduction in downwash angle caused by the wake resulting from the separated air behind the normal shock wave forming on the upper surface of the wing's root section. The reduction in downwash over the horizontal tail will naturally reduce the downward lift produced by the tail, creating a nose down pitch trim change. Finally, any reduction in the efficiency of the horizontal tail caused by supersonic flight can contribute to a nose down change in the aircraft's pitch trim.

A typical incident involving inadvertent transition would involve an unplanned nose down attitude while at high altitude, high speed cruise. This could be caused by wind shear, turbulence or a system malfunction. Once the nose is down, modern high speed (but designed for subsonic flight) jets will quickly accelerate to transonic speeds. A second problem then often appears. The ability of the crew to decrease the aircraft's pitch attitude can be reduced, allowing airspeed to continue to build while altitude is rapidly lost.

A reduction in pitch control (the ability of the crew to raise or lower the aircraft's nose) is an expected result of flight at sonic

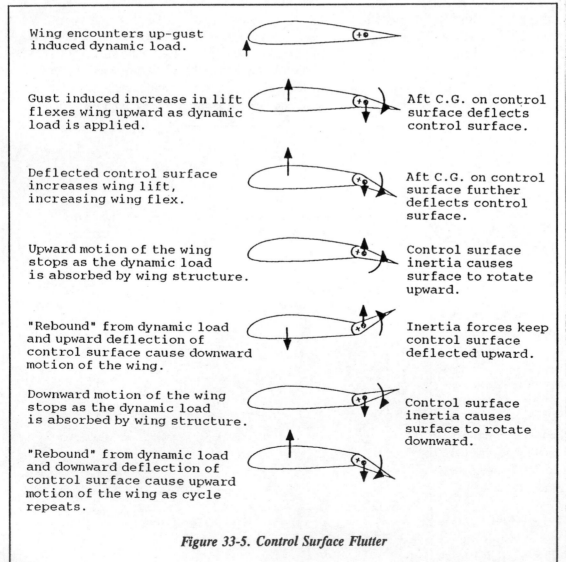

Figure 33-5. Control Surface Flutter

speeds. On aircraft designed to operate at supersonic speeds this represents a reduction in the Gs the aircraft can pull. On aircraft not designed to operate at sonic speeds it can degrade pitch control authority to the point where pull-out from a high speed dive may be difficult or impossible. When the aerodynamic center moves aft the aircraft's pitch stability is naturally increased and its controllability in pitch is reduced as the distance between the aircraft's center of gravity and the lift vector is increased. In addition, if a shock wave forms on the horizontal tail during transonic flight, aircraft equipped with elevators will have their pitch control seriously degraded. This is one disadvantage that is not true for aircraft equipped with horizontal stabilizers which were designed for flight at sonic speeds.

One other factor to consider is the manner in which pitch trim is achieved. If the horizontal stabilizer is moved to achieve a specific pitch trim speed, the stabilizer's leading edge will be up during high speed cruise in order to correct for the natural nose down pitching moment created by higher speed.

All aircraft have a natural tendency to develop a nose down pitching moment as they transition into supersonic flight and also experience a decrease in their ability to pull out of the resulting dive. This can cause impact with the ground or in-flight break-up due to high dynamic pressure, flutter or a combination of factors. If this doesn't happen and the aircraft is able to descend to lower altitudes where the speed of sound is higher (due to the higher air temperatures) and its Mach number therefore lower (assuming a roughly constant dynamic pressure) it will lose some of its nose down trim and regain some of its pitch authority. A pilot-induced over G is now a real possibility. All of these factors need to be examined when investigating high speed in-flight breakups or collisions with the ground.

B. PITCH OSCILLATIONS. If an aircraft decelerates through the transonic range while the pilot is holding a high G load, it is possible for the increase in nose-up trim (as the aerodynamic center moves forward) to cause the aircraft to experience an over G condition. Another form of oscillations can occur as small changes in angle-of-attack cause shock waves to

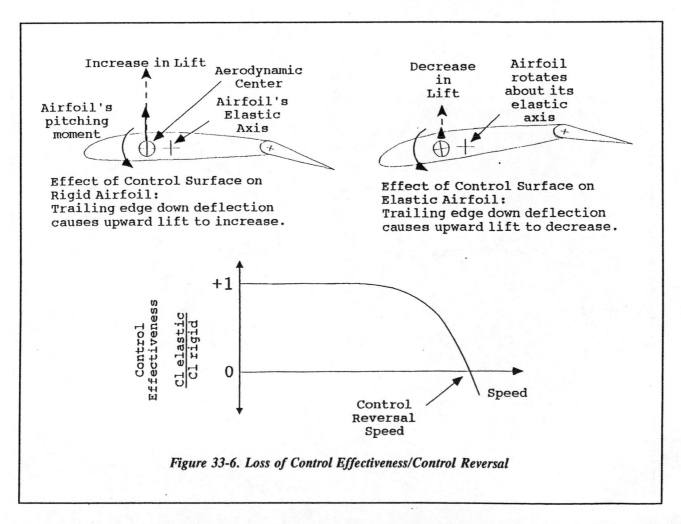

Figure 33-6. Loss of Control Effectiveness/Control Reversal

move back-and-forth across the wing, causing changes in pitch attitude and G load.

C. BUFFET. Separated airflow behind normal shock waves can cause a wake of turbulent air to trail behind the wing. If this wake encounters the aircraft's horizontal tail, it could expose the tail to buffeting airloads for which the tail was not designed. Buffet associated with transonic flight is normally associated with high altitude flight where critical Mach numbers are encountered at lower indicated and true airspeeds.

D. CONTROL SURFACE BUZZ. If separated airflow behind a normal shock wave on a wing's upper surface moves across a reversible (non-powered)

aileron, small symmetrical aileron vibrations can occur. Like buffet, control surface buzz is normally associated with high altitude flight where critical Mach numbers are encountered at lower indicated and true airspeeds.

E. "COFFIN CORNER " OPERATIONS. As an aircraft climbs to high altitudes, its true airspeed for stall will increase. This is due primarily to the decrease in air density at high altitudes. At approximately 36,000 feet pressure altitude, the standard day altitude for the tropopause, the air will have a density of only approximately 30% of that at sea level. The aircraft's critical Mach number, on the other hand, will decease. This is due to lower temperature experi-

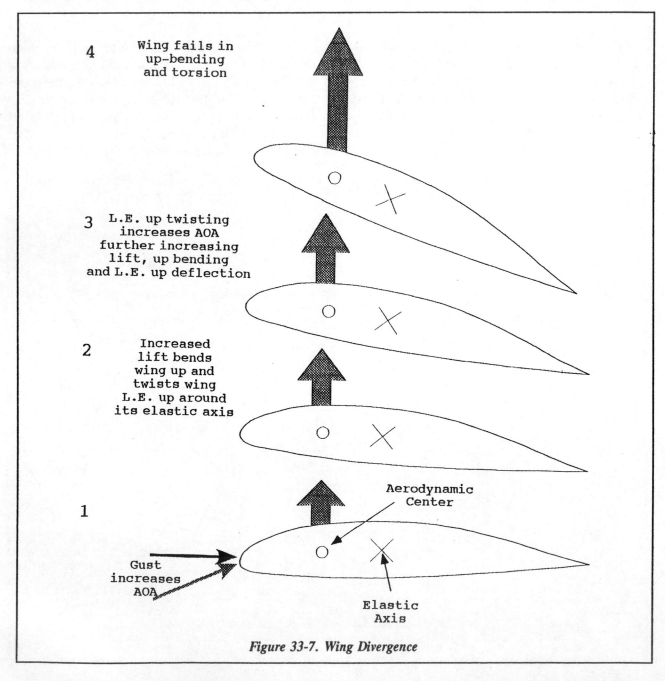

Figure 33-7. Wing Divergence

enced at high altitudes and the direct relationship between the speed of sound in air and air temperature. At 36,000 feet pressure altitude the standard day speed of sound is approximately 574 KTAS while at sea level it is approximately 662 KTAS. An aircraft having a M_{ne} of 0.75 Mach (430 KTAS & 246 KCAS) and a clean stall speed of 180 KCAS, would have an 66 knot range between stall speed and the limiting Mach number. If, however, the aircraft were to experience increased G load (due to either an up-gust or pilot-induced maneuvers) the stall speed would increase as a natural result of G load. For instance, a 1.9 G load (moderate turbulence) would increase the stall speed to 248 KCAS. That's higher than its M_{ne}! The aircraft would stall while exceeding its limiting Mach number. This could cause loss of control and entry into a dive with possible structural overloads. The U-2 is reported to have a limit airspeed of 412 KTAS and a stall speed of 400 KTAS while at its mission altitude. Only 12 knots true margin between stall and overspeed. And that margin decreases every time the airplane turns or hits a bump. The B-47, the first large swept wing jet bomber, consistently operated in this "coffin corner" with only a few knots margin between stall and high speed buffet. A slight gain in altitude would put the airplane into both stall and buffet at the same time. This was a very sporty proposition for the crew as the recovery techniques for one would exacerbate the other. This problem was cured somewhat by the installation of vortex generators on the wings. These essentially raised the coffin corner altitude about 4,000 feet and allowed the plane to operate with a wider margin between stall and buffet.

F. EFFECT ON ENGINE OPERATION. In order to function properly, air entering the compressor section of jet engines has to be subsonic. Entry of supersonic airflow into the engine can cause compressor stalls, engine stagnation and flame out. Therefore, the inlets for aircraft designed to operate in the transonic or supersonic speed ranges must be designed to slow the air entering the engine to subsonic speeds before the air enters the compressor section. Jet aircraft which are operated above their limiting Mach

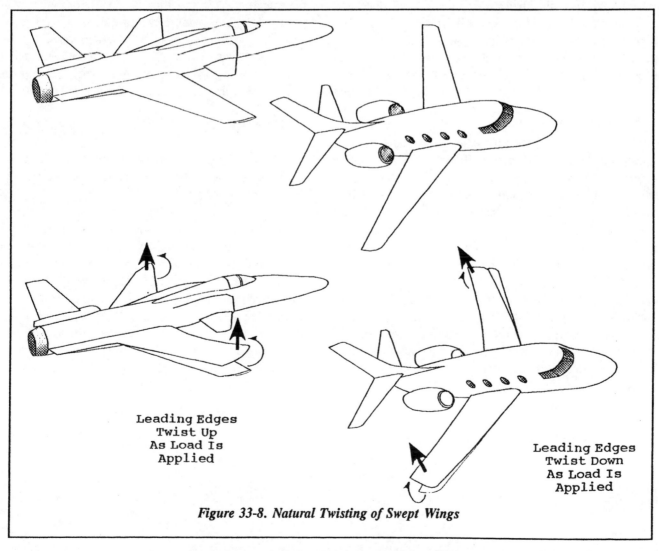

Leading Edges
Twist Up
As Load Is
Applied

Leading Edges
Twist Down
As Load Is
Applied

Figure 33-8. Natural Twisting of Swept Wings

number may therefore experience engine malfunctions associated with the ingestion of supersonic airflows.

G. FLUTTER. Flutter involves a combination of the following three factors:

  1. High dynamic pressure (high density or more likely high speed)

  2. Structural stiffness (the amount of deflection experienced by a load on a structure and its ability to store energy elastically).

  3. Inertia factors (which involve the natural frequency of the structure and the frequency with which the aerodynamic forces are applied.

Flutter is a possibility any time an aircraft flies above its limit airspeed.

H. LOSS OF CONTROL EFFECTIVENESS/-CONTROL REVERSAL. Excessively high speeds or excessively flexible structure can result in degraded control effectiveness or even loss of control. All airplanes are flexible. When exposed to loads, their structure deflects. When a control surface such as an aileron or an elevator is deflected trailing edge down, the wing or horizontal stabilizer to which it is attached will twist leading edge down. When the control surface is deflected leading edge up, the wing or horizontal stabilizer will twist leading edge up. Under normal conditions, the aerodynamic effects of the control surface deflection will far outweigh the aerodynamic effects of the wing or horizontal stabilizer twisting. However, at excessively high speed (above V_{ne}) or after experiencing structural damage, twisting of the wing or horizontal tail can degrade the effect of the control displacement, decreasing roll rate or elevator authority. Since the wing is, relative to the horizontal stabilizer, more flexible in torsion, degraded roll control at excessively high speeds is more common than degraded pitch control. In extreme cases, aileron defections which would normally cause an airplane to roll in one direction, can cause it to roll in the other direction.

This was another quirk of the B-47 bomber. It had an extremely flexible wing with ailerons positioned at the tips. At 440 KIAS, the twisting of the wing exactly matched the force of the ailerons and the roll rate went to zero. Theoretically, above 440 KIAS, you would encounter aileron reversal. No one tried this, as the buffet level at 440K was becoming unbearable. Since this speed varied a few knots between airframes and the operational low level speed for the bomber was 425KIAS, each plane had to be run up to 440K to make sure that some roll authority existed. At its best, the roll rate at 425K was nothing to cheer about.

I. DIVERGENCE. Because of the relationship between a wing's center of lift and its elastic axis, up gusts will tend to increase lift in a manner which also twists the wing leading edge up. The upward twisting of the wing increases its angle of attack, which further increases lift and twisting until either the wing structure absorbs the load or the wing fails. This torsional failure of the wing is referred to as "divergence". Although wings are designed to withstand the torsion loads normally created by lift, when airplanes are flown at excessive speeds, gust loads will be amplified and may cause torsional loads capable of causing wing divergence. The natural twisting caused by lift loads on a forward swept wing design will have a natural tendency toward divergence. On the other hand, the natural twisting of an aft swept wing will naturally tend to minimize the potential for divergence.

4. SUMMARY.

At this writing, not too many civil aircraft are operating in the supersonic range, but we might as well get prepared for it.

There are a lot of civil aircraft operating near the transonic range. An airliner clipping along at Mach 0.8 at 40,000 feet is subject to some of the problems we have been talking about. Remember, the airflow over the surfaces will reach sonic velocity before the aircraft itself. Furthermore, a slight upset or high G situation can easily move the aircraft into a speed range where all sorts of bad things can happen. The investigator needs an understanding of these possibilities.

34

Loads, Stresses and Strain

1. INTRODUCTION.

Those of you having an engineering or scientific background may want to just skim this chapter, especially the first couple of sections. It's pretty basic stuff and it is included for those who haven't been exposed to the notions and concepts associated with loads, stress and strain. In addition, it's easier for us to refer back to this section periodically, rather than repeat the material several times in other chapters. Now, on to the subject at hand.

2. LOADS.

Loads can be subdivided into five general types; tension, compression, shear, bending and torsion. Aircraft structure has to withstand one or more of these loads, and it is important for anyone investigating the structural implications of an aircraft crash to have a firm grasp of what each these loads can do to a structural component. The first three loads, tension, compression, and shear are measured in terms of pounds in the United States aviation community, but you may come across the terms Newtons, Pascals, or

grams if you look far enough. Bending and torsion loads are normally measured in foot-pounds or inch-pounds in the United States, and, as their units of measure imply, are a little more complicated than just plain tension, compression and shear. A bending load is referred to as a bending moment, while a torsion load is referred to as torque (pronounced "tork").

A. TENSION. Refer to the illustration shown in Figure 34-1. The arrows represent two opposing forces acting on a rod with a square cross section. Since the two arrows are in opposite directions, the rod is said to be carrying a "tension" (or "tensile") load. If we could look along any plane perpendicular to the tension load and examine the material on either side of the plane, we would see that the load is trying to pull the material apart. The material in the rod resists this pulling apart. Material having a higher tension strength withstands a larger load before being pulled apart and weaker material lets go at a smaller load. By the way, there always has to be two arrows representing the forces. If there were only one, the rod would be pulled right (or left) off this page. Figure 34-1 could also be an aircraft wing strut, a simple structural component

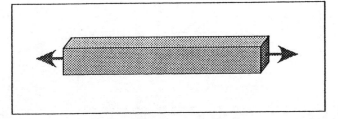

Figure 34-1. Tension Load

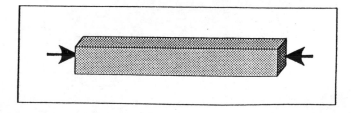

Figure 34-2. Compression Load

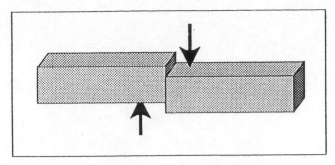

Figure 34-3. Shear Load

which is normally loaded in tension,

B. COMPRESSION. In the illustration shown in Figure 34-2, the arrows are reversed from the directions shown in Figure 34-1 and the rod can be said to be carrying a "compression" (or "compressive") load. The forces are now trying to push the material on

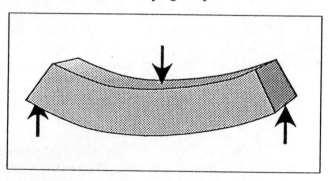

Figure 34-4. Bending

either side of a plane perpendicular to the load into one another. Again, the material in the rod will resist the load up to a point which depends on the compression strength of the material. Figure 34-2 could also be one type of aircraft landing gear, a type which is normally loaded in compression while the aircraft is on the ground.

C. SHEAR. The arrows in the illustration shown in Figure 34-3 are different from those in either Figures 34-1 or 34-2. Although the two arrows (forces) are still in opposite directions, they are no longer aligned with each other. If you consider the material on either side of the plane shown, you can see that the material on one side of the plane is trying to slide along the other; the material on the right side of the plane attempting to move up while the material on the left side of the plane attempts to move down. The resistance of the material is referred to as shear strength and the rod is said to carrying a "shear" load. If you are having trouble visualizing this type of loading, think of a pair of scissors (shears) cutting a piece of paper. Figure 34-3 is analagous to an aircraft rivet. Aircraft structural joints joined by rivets are designed so that under normal aircraft loading, the rivets are loaded in shear (as opposed to tension or compression).

D. BENDING AND BENDING MOMENTS. In order to understand the concept of bending loads, it is first necessary to understand the concept of bending moments. Suppose a 10 foot long beam is embedded into a wall. If a 100 pound load is placed on the end of the beam, the beam will be exposed to a bending moment. The maximum "bending moment" will be created at the base of the beam where it attaches to the wall. The magnitude of this bending moment at the wall can be determined by multiplying the force's magnitude by the shortest distance from the wall to the force's axis. The shortest distance from the wall to the force's axis will also be perpendicular to the force's direction. If that sounds complicated, think of it this way. The direction of the force is straight down and the magnitude of the force in the figure is 100 pounds. Since we are interested in determining the bending moment at a point on the beam where it attaches to the wall, the point is 10 feet away on a line perpendicular to the direction of the force. Multiply this perpendicular distance by the magnitude of the force and "bingo",

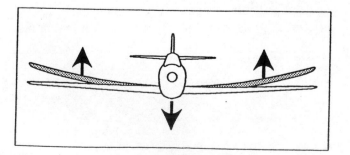

Figure 34-5. Wing Bending

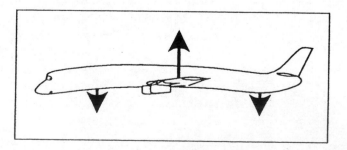

Figure 34-6. Fuselage Bending

you have determined the bending moment:

Force magnitude x Distance perpendicular to the force = Bending Moment or: 100 pounds x 10 feet = 1,000 foot-pounds.

If we wanted to determine the bending moment at a point in the middle of the beam, we could simply take the perpendicular distance from the force to the point about which we are concerned (5 feet) and multiply it by the magnitude of the force (100 pounds). And we then find out that the bending moment at the mid-point of this beam with this load is 500 foot-pounds. Simple? How about the bending moment at the tip of the beam? If the point we are interested in is right under the force, the perpendicular distance must be zero feet and the bending moment must also be zero. This is a very simplistic example and in the real world things are normally a lot more complicated.

When pilots think of bending an aircraft structure, they most probably think of bending the wings; especially when pulling Gs at high gross weights. Figure 34-5 shows a sketch of a fuselage, a wing and the lift vectors representing the lifting force created by the wing. Remembering the example discussed above, it should be obvious that the bending moment for this wing is greatest at the wing root. Even when we replace the single total lift vector by lots of little lift vectors, each representing a small section of the wing (a more realistic assumption), you can see that the greatest bending moment occurs at the wing root.

An airplane's fuselage is also subjected to bending loads. Figure 34-6 shows typical bending loads. In positive G flight both the forward and aft portions of the fuselage are bent downward around the wing. The fuselage forward of the wing is bent down by its weight, while the aft end of a fuselage is bent down by both the weight of the aft portion of the airplane and the airloads created by the horizontal tail.

E. TORSION AND TORQUE. Torsion loads, like bending loads are measured in foot-pounds or inch pounds. Figure 34-7 shows a circular rod which is loaded in torsion. The two vertical forces of 100 pounds are each 10 inches from the center of the rod. The torque developed by each of the two forces is therefore 1,000 inch pounds (100 pounds x 10 inches) and the total torque on the rod is 2,000 inch pounds (2 x 1,000 inch pounds).

Most wings are subjected to torsional loads when

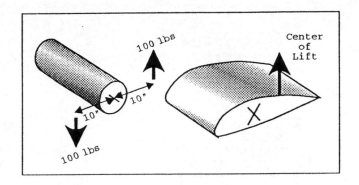

Figure 34-7. Torsion in an Airfoil

they produce lift. When a cambered airfoil creates lift it also produces a torsion load around its aerodynamic center. The relationship of the variables involved determining the magnitude of the torque is shown in Equation 1. The two variables under the direct control of a pilot are airspeed and coefficient of pitching moment about the aerodynamic center. Although there is not much a pilot can do to change either the density of the air the airplane is flying through or the wing's area or its chord, airspeed and wing configuration are routinely changed during flight. The effect of changes in airspeed should be fairly obvious. Increasing speed will increase the pitching moment (torque) about a wing's aerodynamic center while decreasing airspeed will decrease torque imposed on the wing. Increasing an airfoil's camber by deflecting control surfaces (e.g. flaps, ailerons, elevators, etc.) can increase the camber, thereby increasing the coefficient of pitching moment and therefore the pitching (twisting) moment.

$$M_{a.c.} = C_{Ma.c.} \frac{\sigma V^2}{295} SC \qquad \textbf{(1)}$$

Where:

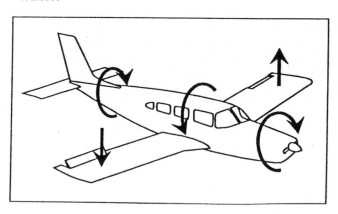

Figure 34-8. Torsion Generated in a Fuselage During Rolling Manuevers

MATERIAL	ULTIMATE TENSILE STRENGTH (ksi) (F_{tu})	YIELD TENSILE STRENGTH (ksi) (F_{ty})	ULTIMATE SHEAR STRENGTH (ksi) (F_{su})
Aluminum 2024-T6	69	57	41
Aluminum 7075-T6 (Clad)	76	67	46
Steel 4340 180 Ksi H.T.	180	163	

Figure 34-9. Examples of Limit (Yield) and Maximum (Ultimate) Stress for Three Common Aircraft Structural Materials

$M_{a.c.}$ = Pitching Moment about the aerodynamic center in foot-pounds.

$C_{Ma.c.}$ = Coefficient of pitching moment about the aerodynamic center.

σ = Ambient air density ratio.

V = True airspeed in knots.

S = Wing area in square feet.

c = Wing chord in feet.

Fuselages can also be subjected to torsion loads. When the wings of an airplane each develop different amounts of lift (e.g. when the ailerons are extended or when one wing stalls in a snap roll), the inertia and air loads created by the empennage will tend to resist the torque created by the differential lift.

3. STRESS.

Stress is a measure of the internal force per unit area generated when some material is subjected to a load. In other words, it's the intensity of internal forces created when external loads cause the material to deform. Some people may argue that deformation (or strain - which will be covered in just a moment) is the result of stress. They're right too. Actually, there can be no stress without strain and no strain without stress; the two are interrelated.

Stress, in the aviation community, is normally expressed in terms of pounds per square inch and abbreviated "psi." In this text we will use the symbol "F" to designate stress. Some other texts may use the symbols f or s.

A. THE CONCEPT OF STRESS. Where we have discussed five types of loads, there are only three types of stress; tension stress, compression stress and shear stress. The tension stress in a material is equal to the tension load divided by the area carrying the load. Compression stress is equal to the compression load divided by the area available to carry the load. Shear stress is calculated the same way. From this you may have deduced that bending and torsion loads may cause tension stress, compression stress or shear stress. You are right, to a degree. Both bending and torsion loads cause the creation of all three types of stress and tension loads create shear stress in addition to the tension stress you might expect.

Let's discuss this some more. Although you can't normally see it, an understanding of stress is very important to the aircraft accident investigator. When discussing the strength of a material, it's the stress the material can withstand before failing that's important. Load will be used to calculate the level of stress within the material, but it's the stress level in the material that determines when and where the material will fail. The maximum stress that standard aircraft structural materials can carry without failure is well established and available to engineers in metals handbooks. You should also remember that the maximum tension stress a material can handle without failure may not be the same as its maximum compression stress which (as you may have guessed) need not be the same as the maximum shear stress. In fact it would be very unusual if all three were equal.

Two levels of stress are important to the aircraft accident investigator: the material's yield stress and its ultimate stress. The yield stress is the stress level at which the material experiences permanent objectionable deformation. Structural engineers design aircraft

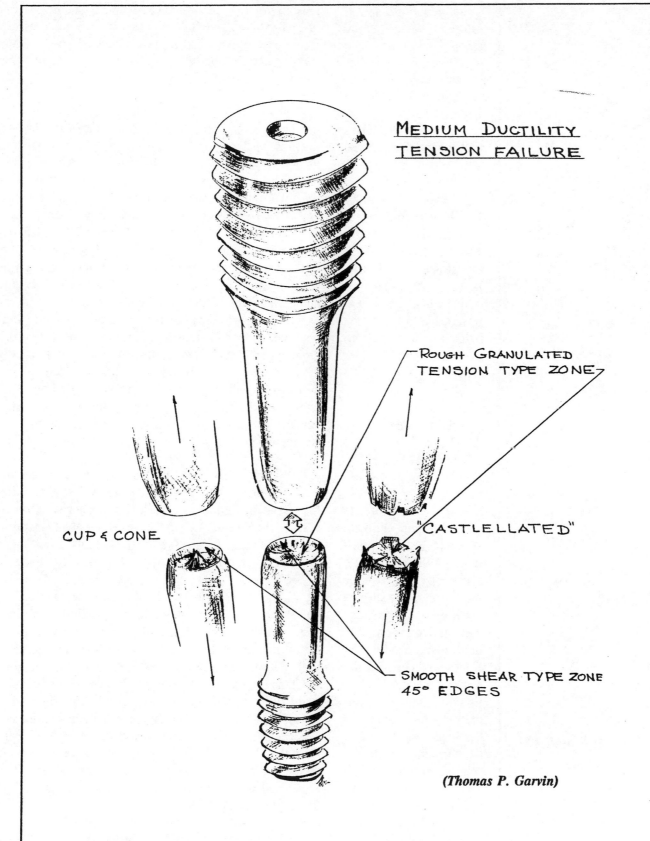

MEDIUM DUCTILITY
TENSION FAILURE

ROUGH GRANULATED
TENSION TYPE ZONE

"CASTLELLATED"

CUP & CONE

SMOOTH SHEAR TYPE ZONE
45° EDGES

(Thomas P. Garvin)

Figure 34-10. Tension Failure in a Material of Medium Ductility

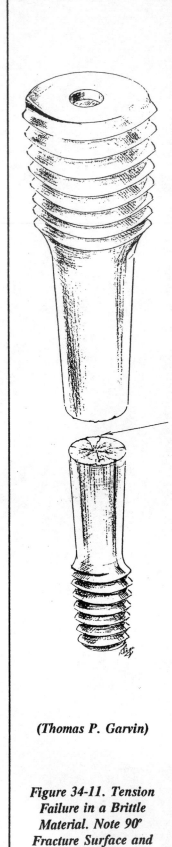

(Thomas P. Garvin)

Figure 34-11. Tension Failure in a Brittle Material. Note 90° Fracture Surface and No Necking Down

structures so that expected load-caused stresses are all below this level. The ultimate stress is the stress level at which the material will fail catastrophically. In common terms that means the material will break in two. A structure which has been exposed to loads which caused stresses to exceed yield levels will bend and deform, but it should not collapse. A structure which has been exposed to loads which caused stresses to exceed ultimate levels will collapse.

B. APPEARANCE OF METAL FRACTURE SURFACES. When load-generated stresses in metal components exceed their ultimate stress level the component will fracture, separating into two or more parts. The surface texture of the fracture area and plastic deformation surrounding the fracture can provide the aircraft accident investigator with clues concerning the nature of the loads which caused the fracture and the relative strength of the material. For instance, if a fractured bolt which is supposed to be made of high strength steel, exhibits fracture evidence normally associated with a lower strength steel, the

reasons for this apparent discrepancy will have to be uncovered and explained. The process by which a fracture occurs and the appearance of the fracture surface and surrounding material is classified as either "ductile" or "brittle."

Since ductile materials tend to be weaker in shear than in tension, they tend to fail along the plane of maximum shear stress and have 45° "shear lips" on their fracture surfaces (see Figure 34-10 for an example). The fracture surface exhibits other evidence of the shearing failure as the crack follows an irregular path following the planes of maximum shear stress. The fracture surface of a ductile material which has failed under a tension load, has been described as silky or fibrous and having a dull appearance and a fine texture. The area surrounding the fracture will exhibit significant plastic deformation, stretching along the axis of the tension load and necking down on the plane perpendicular to the tension load. When looking at this area under a microscope, the surface will normally have a dimpled appearance, showing the plastic deformation of the material as it was pulled apart. A considerable amount of energy (compared to brittle fracture which will be discussed in the next paragraph) will be consumed during this ductile fracture process.

Figure 34-10 shows the way we expect to see the fracture surface of a ductile material when examining the wreckage of an aircraft. However, there are lots of reasons why a fracture surface does not always exhibit the characteristics discussed. We will talk about these reasons in later chapters.

Since brittle materials tend to be weaker in tension than in shear, they tend to fail along the plane of maximum tension stress, along a plane perpendicular to the tension load (see Figure 34-11 for an example). The fracture surface exhibits other evidence of the tension failure, as the crack propagates rapidly with little plastic deformation and therefore little energy consumption. The fracture surface will have a course texture, that is often described as "bright" with points of light reflecting off the sharp edges of the metal grains. The gross plastic deformation normally associated with a plastic fracture is absent, as well as the 45 degree shear lips.

C. STRESSES DEVELOPED BY TENSION LOADS. Referring to Figure 34-12 you can see a 2 inch square rod being subjected to a 100,000 pound load. To determine the tension stress developed in this simple structure, divide the tension load being carried by the structure by the area perpendicular to the tension load. That's the area available to carry the load. In this case that's a pretty simple problem. A 2 inch square has an area of 4 square inches. The load is 100,000 pounds. The calculation is shown in Equation 2.

$$F_t = \frac{P}{A} = \frac{100,000 \; pounds}{4 \; square inches} = 25,000 \; p.s.i. \quad (2)$$

Where:

F$_t$ = Tension stress in pounds per square inch (psi).

P = Tension load in pounds.

A = Area supporting tension load in square inches.

But 25,000 psi of tension stress is not the only stress developed in the rod. Although the tension stress developed by a tension load is referred to as the primary stress, shear is a secondary stress developed by the tension load. Imagine that our 2 inch square rod has been smoothly cut along an angle of 45 degrees to the long axis of the rod. Now assume that the cut surfaces have been coated by a sticky adhesive and the ends rejoined in their original positions. Now, when the 100,000 pound load is applied the adhesive may allow the 45 degree surfaces to slide over one another. That sliding is due to a shear load. Now let's take a look at that same situation by analyzing the vector components of the load acting on that rod.

Referring to Figure 34-12, we can see 100,000 pounds of force acting along the long axis of the rod. The component of the 100,-000 pound force

vector acting perpendicular to the 45 degree plane is about 70,700 pounds. This force is acting on an area of about 5.7 square inches. Which means the tension load is causing a tension stress of 12,500 psi. This turns out to be exactly one-half the tension stress developed along a plane 90 degrees to the axis of the tension load. In other words, the tension stress along the plane 45 degrees to the axis of the tension load has decreased to only 50% of the tension stress on a plane parallel to tension load. If we examined the tension stress on a plane parallel to the tension load, we would discover it had decreased to zero.

By using the same techniques described in the paragraph above, we can show that the component of the 100,000 pound force vector acting parallel to the 45 degree plane (that's the force that is trying to slide the severed surfaces of our rod) is also about 70,700 pounds. The area this force is acting on is also about 5.7 square inches. Which means the shear stress developed by this tension load is 12,500 psi (70,700 lbs divided by 5.7 in^2), and that it is also equal to one half the tension stress developed along a plane perpendicular to the tension load. This shear stress is referred to as a secondary stress. It can be shown that tension stress developed on the plane perpendicular to the tension load is greater than the tension stress developed on any other plane. It can also be shown that the shear stress developed on a plane 45 degrees to the

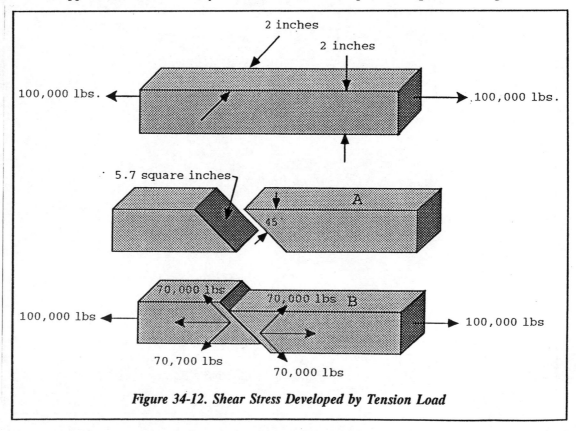

Figure 34-12. Shear Stress Developed by Tension Load

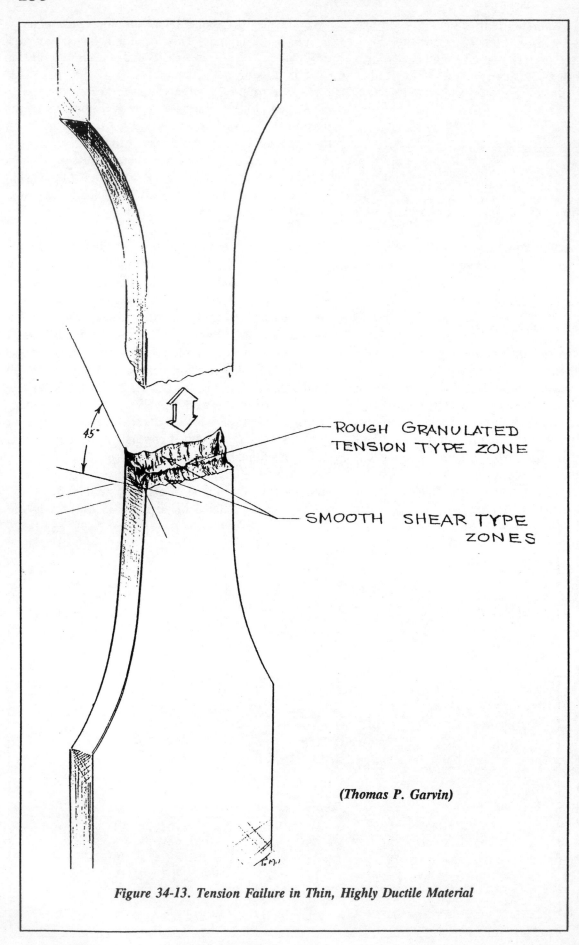

ROUGH GRANULATED
TENSION TYPE ZONE

SMOOTH SHEAR TYPE
ZONES

45°

(Thomas P. Garvin)

Figure 34-13. Tension Failure in Thin, Highly Ductile Material

tension load is greater than the shear stress developed on any other plane.

The bottom line is that a tension load will produce both tension and shear stress. The maximum tension stress will occur on a plane perpendicular to the tension load and will be equal the tension load divided by the cross-sectional area perpendicular to the load. The tension stress is referred to as the "primary" stress. Maximum shear stress will occur on a plane 45 degrees to the tension load and will be equal to one-half the tension stress. The shear stress is referred to as the secondary stress.

D. STRESSES DEVELOPED BY COMPRESSIVE LOADS. When we look at the compression force vectors acting on our 2 inch square rod, we see something very similar to what we saw when we looked at the stresses within the rod loaded in tension. The maximum compression stress is located along the plane oriented 90 degrees to the axis of the load and is equal to the value of the load divided by the

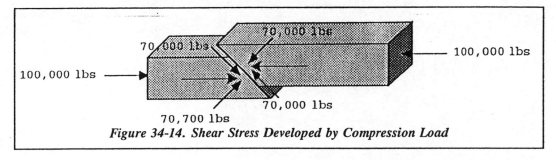

Figure 34-14. Shear Stress Developed by Compression Load

area of the rod perpendicular to the load. Looking at the plane 45 degrees to the axis of the compression load, we see one component of the compression load acting perpendicular to the 45 degree plane (pressing the two planes together in compression) and one component of the compression load acting parallel to the 45 degree plane (attempting to slide to two planes against each other in shear). It can be shown that the maximum compression stress exists on the plane perpendicular to the compression load and that the maximum shear stress exists along planes which are 45 degrees to the compression load. Brittle materials (whose shear strength is lower than their compression strength) will tend to fail along a plane 45 degrees to the compression load when loaded in compression. When relatively ductile material fail due to compressive loads, their failure modes vary considerably. These failure modes will be discussed in greater detail in a later paragraph.

E. STRESSES DEVELOPED BY SHEAR
LOADS. When a simple structure is loaded directly in shear, shear stress will be developed parallel to the shear load and the value of this shear stress will be equal to the shear load divided by the area parallel to the shear load. Figure 34-15 shows a sketch of the shear load on our 2 inch square rod. If the shear load is 100,000 pounds, we can calculate the shear stress to be 25,000 psi. This is called the primary stress developed by the shear load.

It also turns out that in addition to this primary shear stress, shear loads also produce secondary tension and compression stresses. If we examine a small section of the material on the shear plane within the rod and draw arrows to represent the primary shear stress, we can see that additional shear stresses that are equal and opposite shear stress is necessary to keep the material in equilibrium. This situation is shown in Figure 34-16A. We now have four arrows (vectors) representing the shear stresses in the material. Since we are dealing with vectors, we can add any two of these vectors to develop a composite vector

which is equal to the first two.

Let's add the two shear vectors which originate on the upper left hand corner of the square shown in Figure 34-16A. Let's also add the two vectors which originate on the lower right hand corner of the square. We can add the two upper left hand vectors to form a single vector acting up and to the left. We can also add the two lower right hand vectors to form a single vector acting down and to the right. Looking at Figure 34-16B, we can see that the two new vectors are directed away from each other and create a tension stress that is 45 degrees off the original shear stress. It can be shown that this secondary tension stress is equal in magnitude to the original shear stress.

Now let's start over again and add the two shear vectors which originate in the upper right hand corner of the square shown in Figure 34-16A. When we then add together the two vectors in the lower left hand corner of Figure 34-16A, we can see we have created

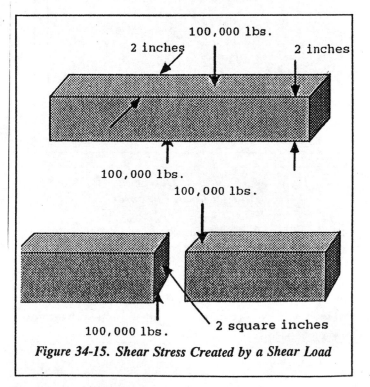

Figure 34-15. Shear Stress Created by a Shear Load

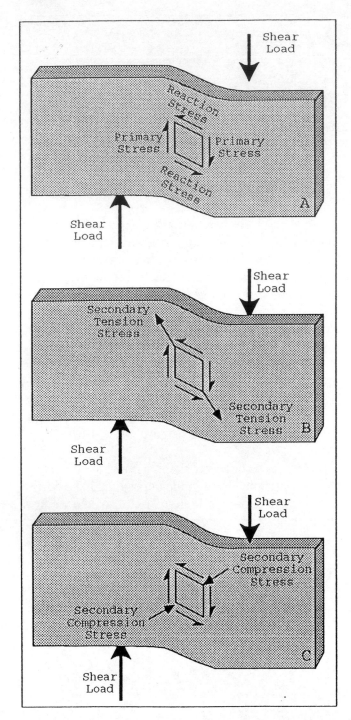

Figure 34-16. Stresses Developed Within a Material Loaded in Shear

two stress vector arrows which are pointing at each other. The primary shear stress can be resolved into a secondary compressive stress which is 45 degrees off the primary shear stress and 90 degrees off the secondary tension stress. Once again, it can be shown that the compression stress is equal in magnitude to the primary shear stress.

The bottom line here is that shear loads produce

primary shear stresses which act in a plane parallel to the shear load and are equal in magnitude to the shear load divided by the area of material resisting the shear load. In addition to the primary shear stress, secondary tension and compression stresses are developed which act at 90 degrees to each other and 45 degrees to the shear stress. The secondary tension and compression stresses are equal in magnitude to the shear stress. With these three different (but equal magnitude) stresses acting on a material when placed under a shear load, the material will fail in a manner which reflects the mechanical property in which it is weakest (tension strength, compression strength or shear strength).

The physical shape of a structural component also plays a strong roll in the failure mode of the structure when overloaded in shear. The effects of physical shape and ductility on the shear failure modes will be discussed later in this chapter.

F. STRESSES DEVELOPED BY BENDING LOADS. Much of the structure in an aircraft is designed to take bending loads. When the aircraft is inflight, the wings bend as their lift supports the fuselage. The fuselage is supported in the middle by the wings but its nose and tail are bent downward buy the weight of the forward fuselage and empennage. Even on the ground bending forces are at work, bending the wings and fuselage down over the supporting landing gear. With all this bending going on it's important for the aircraft accident investigator to understand the types and magnitudes of stresses created when a structural component is exposed to bending loads.

Bending loads create primary stresses of tension and compression. Shear stress is also created as a secondary stress. In order to better understand the stress developed within a structural component that is carrying bending loads, lets perform a simple demonstration. Take a deck of ordinary playing cards, and place the thumbs near the center of the deck while the fingers grasp the edge of the deck. Then bend the deck and examine the cards. You will see that the ends of the cards are no longer in line. They form a stair-step pattern as shown in the top example of Figure 34-17.

Next, take a similar deck of cards that has been modified with a couple of dabs of glue between each card, forming a solid block of stiff paper. Grasp the deck as you did before and bend the deck. It is harder to bend the deck now. A lot harder! If you could bend the deck to the curve of the unglued deck, you would see that edges of the cards were still in line as shown in the middle example of Figure 34-17. If you compared the two bent decks, you would see that the cards on the outside of the curve of the unglued deck had not changed in length. However the cards on the outside of the curve of the glued deck have become longer. The cards were lengthened by tensions stresses created by the bending load.

Next, look at the cards on the inside of the curve of the unglued deck. You guessed it. They have not changed in length. However the cards on the inside of the curve of the glued deck have become shorter. They were shortened by compression stresses. Then, examine the cards in the middle of the both decks. They have not changed in length! They have been exposed to neither tension nor compression stress. Later in this section we will refer to the cards which have been neither stretched nor compressed as being on the "neutral axis", the plane along which neither tension nor compression takes place when a component is "bent".

Finally consider how the sliding of the cards in the un-glued deck was prevented in the glued deck. The sliding of the cards in the glued deck is prevented by the glue binding each card to its partner. That shows the presence of a shear stress between the cards in the center of the glued deck. There are also shear stresses developed which act through the cards in the deck, but this feature cannot be illustrated using this demonstration.

The simple experiment discussed above illustrates several concepts associated with the stresses developed in a structural component which is subjected to bending loads. First, the material on the outside of the curve caused by the bending load is subjected to a tension stress. Second, the material on the inside of the curve caused by a bending load is subjected to a compression stress. Third the material near the center is exposed to neither tension nor compression stresses. Shear stress is also developed in the center of the deck. The bottom example in Figure 34-17 shows the various stresses in the glued deck of cards when it is exposed to bending loads. You can see that there are vertical as well as horizontal shear stresses being developed.

Calculation of the stresses within a structural component subjected to bending loads is more complicated than calculating the stresses in structure being subjected to simple tension, compression and shear

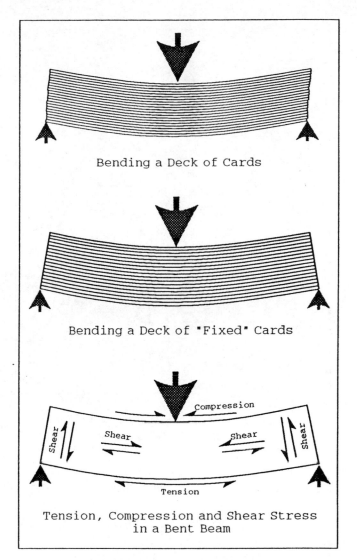

Figure 34-17. Bending

loads. Derivation of Equation 3 is beyond the scope of this text, but will be explained with intuitive reference to the demonstration we just discussed.

$$F_t = F_c = \frac{My}{I} \qquad (3)$$

Where:

F_t = Tension Stress (psi).
F_c = Compression Stress (psi).
M = Bending moment being applied (in.lbs.).
y = Distance from the neutral axis (in).
I = Component's moment of inertia (in⁴).

Equation 3 states that the tension and compression stress developed by a bending moment (M) will vary in direct proportion to the bending load. If the bending moment increases, tension and compression loads will

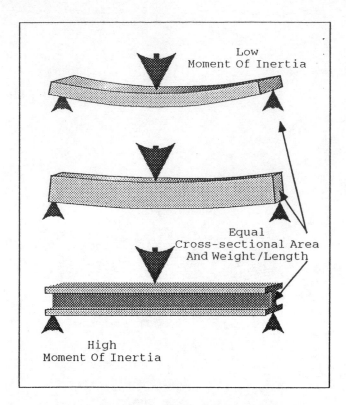

Low
Moment Of Inertia

Equal
Cross-sectional Area
And Weight/Length

High
Moment Of Inertia

Figure 34-18. Strength versus Moment of Inertia
in Beams with Constant Cross Section

also increase. Decreasing the bending moment will decrease tension and compression stresses. That should make sense, increasing the load will normally increase stress. The equation also states that the tension and compression stress varies directly with the distance from the neutral axis (y). Remember that card at the middle of the bent deck of cards? The one that was neither stretched nor compressed? That card was at the deck's neutral axis and was carrying neither a tension nor a compression stress. The further you go from the neutral axis, the higher the stress. In addition, at the neutral axis, tension and compression stresses are zero. The MOMENT OF INERTIA (I) is a mathematical term which is related to the shape and area of an

object, in this case the cross-sectional area of the deck of cards. Generally speaking the larger the area and the further the area is located from the neutral axis, the higher the moment of inertia. The higher the moment of inertia, the lower the maximum tension and compression stress. The higher the moment of inertia, the lower the maximum stress. For instance, suppose a 2 by 4 inch wooden board were stretched between two supports, one end of the board on one support, the other end on the other support. Now suppose a weight were placed in the middle of the board when the 2 by 4 board was placed with the 4 inch side horizontal (see Figure 34-18). It shouldn't surprise anyone if the maximum stress in the board is less if the board were positioned so that the 2 inch side is horizontal. By increasing the moment of inertia, the same size and weight board can carry a larger load. The moment of inertia of the board can be increased even more if some of the wood near the neutral axis is moved toward the upper and lower edges of the board as shown in Figure 34-18, bottom example.

This "I" beam shape increases the efficiency of the board and allows it to carry either a higher load at the same stress or the same load at a lower stress. You can see this concept implemented in not only aircraft structure, but in all sorts of structures.

Another bottom line. The primary stresses created when a structure is loaded in bending are tension and compression. Shear is developed as a secondary stress. Bending loads cause tension stresses to be greatest near the structure's surface on the outside of the curve caused by the bending load. Maximum compression stresses are developed near the surface on the inside of the curve. Bending loads also cause the creation of shear stresses. These shear stresses are greatest at the neutral axis and decrease toward zero near the surfaces where tension and compression stresses are maximized. The average shear stress is equal to the vertical load divided by the area carrying this load.

G. STRESSES DEVELOPED BY TORSIONAL LOADS. When a structural component is placed in torsion, the structure will tend to rotate or flex about a line called the neutral axis. The stresses at the neutral axis are zero. The rotation will be resisted by shear stress (the primary stress) and tension and compression stresses (the secondary stresses) which increase as the distance from the neutral axis increases. Figure 34-19 provides an example of a simple cylindrical shaft which is loaded in torsion. The arrow in the foreground (which is curled in the clockwise direction)

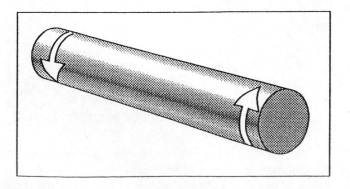

Figure 34-19. Torsion Load on a Cylinder

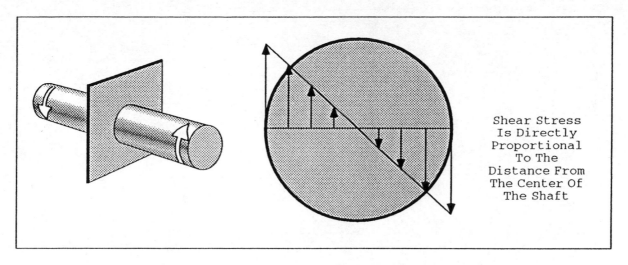

Figure 34-20. Shear Stress on an Cylindrical Shaft Loaded in Torsion

represents a clockwise torsion load which is normally given in inch-pounds or foot-pounds. The arrow in the background (which is curled in the counterclockwise direction) represents the restraining torsion load. Without this restraining load the shaft would be free to rotate.

If we were to cut the shaft shown in Figure 34-19 on a plane perpendicular to the shaft's longitudinal axis and examine the stresses on this plane we would discover that the material was being exposed to a shear stress. The material on opposite sides of the plane are tending to slide across one another. The shear stress will be greatest in the material nearest the outside surface of the shaft. The shear stress at the center (the neutral axis) of the shaft will be zero. The maximum shear stress developed in the shaft will increase as the distance from the shaft's neural axis increases. As mentioned earlier, the shear stress will be maximum when the distance is equal to the radius of the shaft. Once again, the shear stress will be zero when the distance from the neutral axis of the shaft is zero.

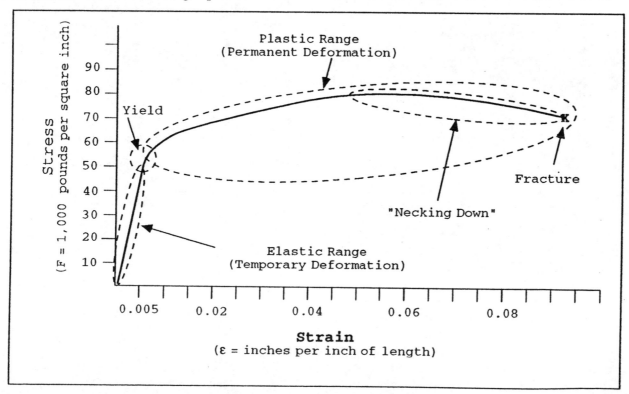

Figure 34-21. Typical Stress/Strain Diagram

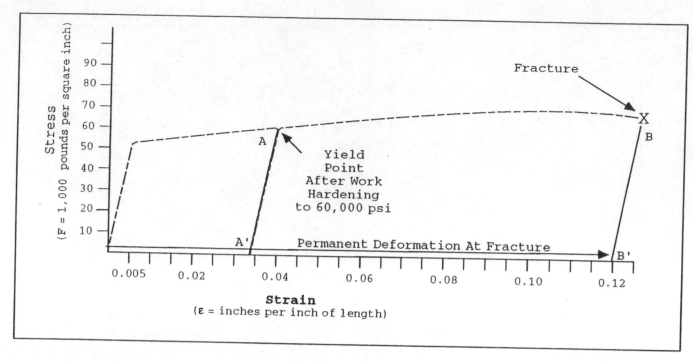

Figure 34-22. Effects of Work Hardening

Let's talk now about the things that can affect the magnitude of the shear stress in a structural component that is being loaded in torsion. First it should be intuitively obvious that the shear stress will increase as the torsion load increases. In fact, every time the torsion load is doubled, the shear stress doubles. Every

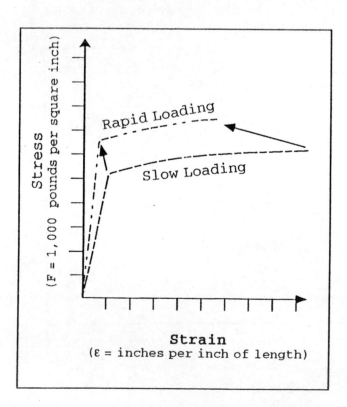

Figure 34-23. Effect of the Speed at which a Load is Applied on the Stress/Strain Curve

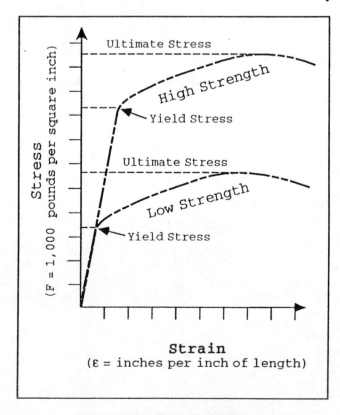

Figure 34-24. High Strength versus Low Strength Materials

time the torsion load is cut in half, the shear load is cut in half. As mentioned earlier the shear stress will increase as the distance from the component's neutral axis increases. One additional factor is involved in determining the shear stresses in the structural component, and that last factor has to do with the component's cross-sectional shape and size. Every two dimensional figure (shape and size) has a parameter called the POLAR MOMENT OF INERTIA. The shear stress will decrease as the polar moment of inertia increases. In fact, the shear stress in a structural component loaded in torsion can be represented by the relationship in Equation 4.

$$F_s = \frac{Tr}{J} \qquad (4)$$

Where:

F_s = Shear stress (psi).
T = Torsional load being applied (in.lbs.).
r = Distance from the neutral axis (in).
J = Component's polar moment of inertia (in⁴).

Let's determine the maximum and minimum shear stresses developed when a 2 inch diameter shaft is loaded with a 100,000 inch-pound torsional load. Since the cross-sectional area of the shaft is circular, and the polar moment of inertia for a solid circular shaft is proportional to the diameter of the shaft's diameter raised to the fourth power (if the diameter of the shaft doubles the polar moment of inertia increases by a factor of 16 - it goes up 16 times!!!) and represented by Equation 5.

$$J_c = \frac{\pi D^4}{32} \qquad (5)$$

Where:

J_c = Polar moment of inertia (circular) (in⁴).
D = Diameter of shaft (in).

Don't worry about how we determined this equation. It's something you will learn (or once learned) in a basic calculus course but you can look up it in a reference book if you really need it. If we substitute the dimensions of our shaft into this equation, we can establish the value of its polar moment of inertia.

$$J_c = \frac{\pi (2\,inches)^4}{32} = 1.57\,inches^4 \qquad (6)$$

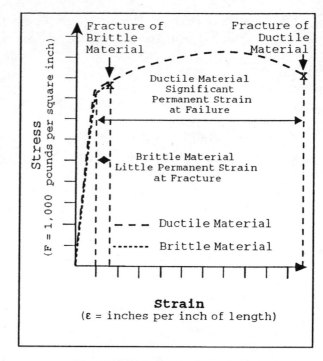

Figure 34-25. Stress/Strain Curves of Ductile and Brittle Materials

Substituting into Equation 4, the shear stress at any distance "r" from the neutral axis (center) of our shaft is equal to:

$$F_s = \frac{Tr}{J} = \frac{100,000\ inch\ pounds \times r\ inches}{1.57\ inches^4}$$
$$= 63,694 \frac{pounds}{inches^3} \times r\ inches \qquad (7)$$

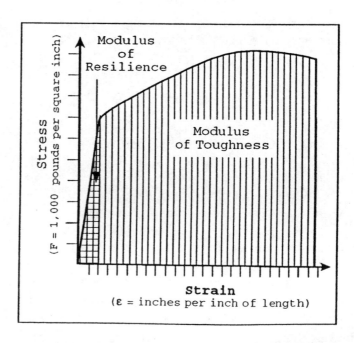

Figure 34-26. Modulus of Resilience and Modulus of Toughness

Thus the shear stress at the surface of the shaft (where r = 1 in.) is equal to 63,700 psi. The shear stress at the center of the shaft (where r = 0 in.) is zero psi. We would expect the shaft to fail first at the surface, where the stress is highest. At a point half way between the center of the shaft and its surface (where r = 0.5 in. the shear stress is equal to 31,850 psi. If you think about it for a second, you should see that the material near the center of the shaft isn't working very hard. It's almost dead weight. In fact, the material within a half-inch of the center of the shaft is carrying much less than one half the load that the material within one-half an inch of the surface is carrying. When it comes to carrying torsion loads, the material near the center of the shaft is not very efficient. Let's see what the designer can do about that.

Suppose we rearranged the material in the shaft so that it was a hollow tube with an inside diameter of 2 inches and an outside diameter of 2.828 inches. The outside diameter of the shaft was specifically selected so that the cross-sectional area of this hollow shaft is exactly the same as the cross-sectional area of the solid shaft. Since their cross-sectional areas are the same, their weight per running foot of shaft is equal. The calculations shown in Equation 8 will show that although the two shafts are of equal weight, the maximum stress developed in the hollow shaft is lower for any given torsional load. That's why we see hollow cylindrical shapes used in aircraft where torsion loads have to be carried.

Now to the proof. The equation for the polar moment of inertia for a hollow circular shaft is:

$$J_c = \frac{\pi (D_o^4 - D_i^4)}{32} = \frac{\pi (2.828 \ inches^4 - 2.0 \ inches^4)}{32} \quad (8)$$
$$= 4.712 \ inches^4$$

Where:
    D_o = Outside diameter of shaft (in).
    D_i = Inside diameter of shaft (in).

When we substitute the dimensions of our hollow shaft into this equation we find out that its polar moment of equation is equal to 4.712"$_4$. That is over twice the value of the polar moment of inertia for the solid circle. If we expose this hollow shaft to the same 100,000 inch-pound torsion load to which we exposed our solid shaft, we can calculate the maximum and minimum shear stress in the hollow shaft.

The maximum shear stress exists on the outside surface of the shaft and is equal to:

$$F_{s_{max}} = \frac{100,000 \ inch \ pounds \ x \ 1.414 \ inches}{4.712 \ inches^4} \quad (9)$$
$$= 30,000 \ p.s.i.$$

The minimum shear stress exists on the inside surface of the shaft and is equal to:

$$F_{s_{min}} = \frac{100,000 \ inch \ pounds \ x \ 1.0 \ inch}{4.712 \ inches^4} = 21,200 \ p.s.i.$$

$$(10)$$

When structural components are exposed to pure torsion loads, it should be obvious that some cross-sectional shapes (in this case the hollow shaft) are more efficient than others. By rearranging the material in the shaft, the maximum stress in the shaft was reduced to less than one-half the original value. This was accomplished by moving the material near the neutral axis (which was not carrying much of the torsion load) to a location further from the neutral axis (where it could carry a much larger portion of the torsion load). In other words, the solid shaft is less efficient than the hollow shaft when carrying torsion loads because in the solid shaft the material in the center is under worked (0 psi), while the material near the shaft's surface is overworked (63,700 psi). In the case of the hollow shaft, all of the material is working at close to the same level of intensity 30,000 psi at the outer surface and 21,200 psi at the inner surface).

These examples are provided to give you some insight as to:

1. How the shear stresses within a structural component vary as a function of the distance from the neutral axis and the structure's polar moment of inertia.

2. Why more expensive hollow structures are sometimes selected over solid structures.

One common misconception should be put to rest. You cannot make a structure stronger by removing material from the center of the structure. In the example we just discussed, all of the material in the original design was used, it was just rearranged.

4. THE CONCEPT OF STRAIN.

"Strain" is a term used to describe the forced

to remember is that you can't have one without the other. Tension and compression stress will produce a strain which is measured as a change in a structure's dimensions. The total change in length is referred to as the total strain and is normally measured in inches, while the change in length per unit length is referred to as the unit strain and is measured in terms of thousandths of an inch per inch of length. A stretching or elongation of a structure's dimension(s) is referred to as a tensile strain while a shortening of a structure is referred to as a compressive strain. Shear stress is accompanied by an angular change in shape. When a circular shaft is loaded in torsion, the angular twisting of one end of the shaft per unit length of the shaft is measured as its angular unit strain. Although the major change in a structure's dimensions is in the same direction as the applied load, the secondary stresses discussed earlier will also cause smaller strains in these directions. Finally, strain can exist in two forms: elastic strain which will disappear as soon as the stress is relaxed and plastic strain which involves permanent deformation.

A. THE STRESS-STRAIN DIAGRAM.

The Stress-Strain Diagram is an extremely useful device which graphically shows the mechanical relationship between stress and strain. This diagram shows the corresponding values of stress and strain from zero stress and strain to the ultimate failure of the samples which were tested. Although these tests are normally conducted to establish the stress/strain relationship of a material (e.g. a specific aluminum alloy, steel or a fiber reinforced plastic) using a simple structural component, the tests can also be performed on a complete structure consisting of several sub-components using a variety of materials. A typical stress strain diagram is shown in Figure 34-21. Some of the important aspects of the diagram are addressed in the following paragraphs. Later on in this section we'll also discuss some specific material properties which are of particular interest to the aircraft accident investigator.

The stress-strain diagram graphically shows the strain which accompanies any specific stress level, and the stress which accompanies any specific strain. In the example shown, tension stress in psi is plotted along

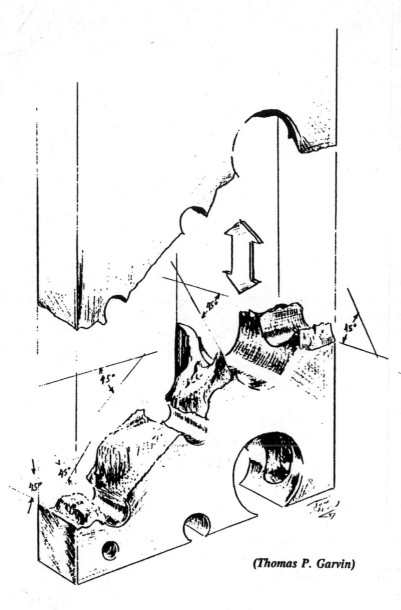

(Thomas P. Garvin)

Figure 34-27. Typical Tension Failure Ductile Aircraft Material

change in one or more of a material's dimensions. For the purpose of this book, we will be referring to strain as the change in dimension (size or angular shape) which is created in the direct response to an applied external force. Strain will not occur without an accompanying stress and stress will not occur without an accompanying strain.

The question, "Which came first, the stress or the strain?" is akin to the question, "Which came first, the chicken or the egg?" Who cares? The important thing

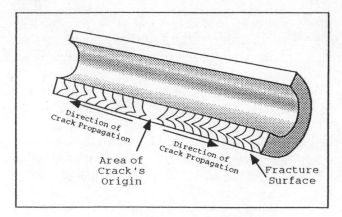

Figure 34-28. Chevron Pattern Marking the Direction of Propagation of a Fast Fracture

the vertical axis while strain in inches of elongation per inch of length is plotted along the horizontal axis. In the example, a stress of 10,000 psi will be accompanied by a strain of 0.001 inches of elongation per inch of length of the component.

If you look at the stress-strain curve in this example, you should be able to identify two distinct regions: the region of relatively low stress-strain where the curve is relatively straight and aligned with the chart's origin and the region of the curve where it curves into a more horizontal attitude, and is no longer aligned with the chart's origin. The first of these two regions is referred to as the material's elastic range while the second region is referred to as the material's plastic range. Before we go on, we need to discuss the differences in the behavior of the material in these two ranges.

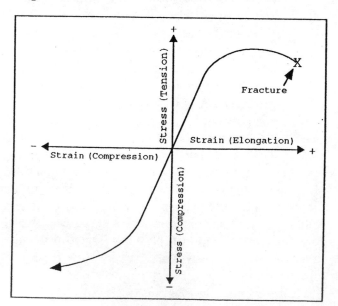

Figure 34-29. Stress/Strain Diagram for a Ductile Material in Block Compression

When our test specimen is first loaded at relatively low stress levels, the specimen's material is said to behave "elastically." When the stress is relaxed, the material returns to its original shape. It is not permanently deformed (in this case permanently stretched). In the example, if the material were stressed to 1,000 psi it would elongate 0.0001 inches per inch of length of the specimen. If it were loaded to 20,000 psi it would elongate to 0.002 inches per inch of specimen. In both of the above cases, the material would return to its original length if the stress were relaxed. In fact, you could load the material up to 50,000 psi (and get 0.005 inches of elongation per inch of specimen length) and the material would still return to its original length when the stress was relaxed. Stress and strain have a "linear" relationship in this "elastic" range. If we double stress, we double the strain. If we double the strain, we double the stress. The slope of the elastic range stress-strain curve is referred to as a material's Modulus of Elasticity. To calculate the Modulus of Elasticity simply divide a stress level by it's accompanying strain.

$$E = \frac{F}{\epsilon} \qquad (11)$$

Where:

E = Modulus of Elasticity (p.s.i.)
F = Stress (p.s.i.)
ϵ = Strain (inches per inch)

For the example shown in Figure 34-21, the Modulus of Elasticity at a stress level of 1,000 psi and an accompanying strain of 0.0001 in./in. is:

$$E = \frac{1,000 \ p.s.i.}{0.0001 \ inches/inch} = 10,000,000 \ p.s.i. \qquad (12)$$

We'll end up with this same answer even if we divide 20,000 psi by 0.002 in/in or 50,000 psi by 0.005 in/in. This is often written as "E = 10 x 10⁶ psi" and verbally communicated as "a modulus of ten million." By the way, a modulus of about ten million is typical of aluminum alloys. Steels are about three times stiffer with typical moduli about 30 million. Some carbon (graphite) fiber reinforced plastic composites are even stiffer with moduli approaching 70,000,000 or seven times stiffer than aluminum and over twice as stiff as steel.

In summary, when the stresses in a material are

within its elastic range, the stress-strain relationship is linear and there will be no permanent deformation when the stresses are relaxed. However, if we expose our test specimen to stresses which are above its "elastic" range, all these things change.

When stress increases into the "plastic" range, the most obvious change on the stress-strain chart is the change in slope of the curve from fairly steep to almost horizontal. In this "plastic" range, a relatively small increase in stress is accompanied by a greater rate of straining (relative to the rate in the elastic range). In the elastic range our test specimen would strain 0.0001 in/in for every 1,000 psi of stress. At 70,000 psi, in the "plastic" range, our specimen will elongate about 0.07 in/in. That's an increase in strain of about 0.001 in/in for each increase of 1,000 psi of stress. That's about 10 times more strain for each 1,000 psi than we saw in the "elastic" range. The material is less "stiff" when stresses are in the plastic range; the material "gives" more for each increase in stress.

Besides being less "stiff" when stresses are in the "plastic" range, material which is exposed to these high levels of stress will not return to their original shape when the stress is relaxed. This change in shape is referred to as permanent plastic deformation. Line A-A' in Figure 34-22 shows how the strain in the test specimen would relax after it had been stressed to 60,000 psi. The distance between the origin of the chart and point A' represents the permanent plastic deformation that would occur if the test specimen were loaded to 60,000 psi and then unloaded.

Let's think about this last comment for a moment. If we loaded our specimen to 60,000 psi it would strain to 0.04 inches per inch of specimen length. If the stress were then relaxed to 0 psi, the strain in the component would relax along line A-A' and the specimen's length would be decreased by 0.006 inches per inch of specimen length. Now suppose we consider this deformed specimen as a "new" specimen. If we load it again, the stress-strain ratio (i.e. the modulus of elasticity) will be the same as the original specimen. However, the material will not go into a plastic range until it reaches a tension stress of 60,000 psi. By loading a ductile material beyond its original yield stress we can "work harden" the material and increase its yield stress. This "work hardening" will not increase the material's ultimate stress nor will it change the material's stiffness, but it will make the material more brittle. Work hardening may also be of some

help in improving the material's resistance to fatigue cracking. We will cover this aspect when we talk about fatigue cracking in Chapter 35.

If, on the other hand, the test specimen were loaded to about 80,000 psi it would probably fail, in this case the tension load caused the test specimen to fracture into two pieces. The stress at which the specimen fractures is referred to as the ultimate stress. Once the specimen has fractured, the stress would be relaxed along line B-B' to zero. The distance from point B' to the origin represents the permanent plastic deformation the will remain after the fracture occurs and the strain is relaxed.

The diagram in Figure 34-21 shows a distinct change in direction at the point marked "Yield Point." The stress corresponding to this point is referred to as the specimen's "yield stress" or "yield strength." If the stress-strain curve doesn't have a distinct change in shape between the elastic and plastic ranges, the point at which a 0.2% permanent deformation occurs is

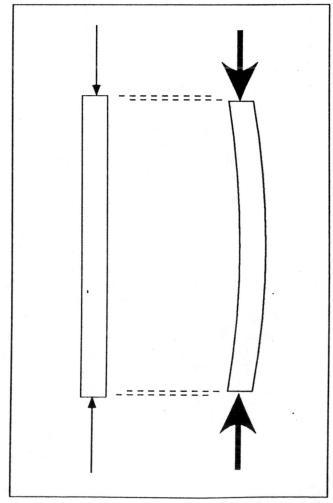

Figure 34-30. Column Carrying a Compression Load

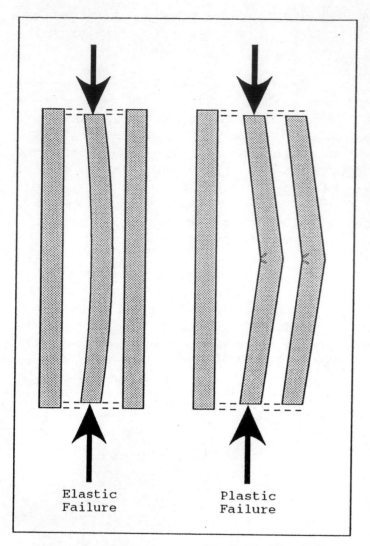

Elastic
Failure

Plastic
Failure

Figure 34-31. Elastic versus Plastic Buckling Failure

One other point on the development of stress strain diagrams. When the load is applied to the test specimen, it is applied slowly. If the load is applied too fast, the shape of the stress-strain curve will change, generally moving up and to the left as shown in Figure 34-23. We will talk more about the effects of the speed of loading on the stresses and straining of aircraft structure in Chapter 35.

B. HIGH STRENGTH VERSUS LOW STRENGTH MATERIAL. When we are talking about strength of structural materials, we are talking about the maximum stress they can withstand prior to failure. In this sense, there are two kinds of failure; failure through excessive deformation and failure by separating into two or more pieces. The first type of strength, strength against excessive deformation, is indicated by a material's yield strength. Beyond this stress, the component will deform rapidly with only minor increases in stress. The second type of strength is the material's ultimate stress, the maximum stress the material can tolerate prior to fracture.

C. DUCTILE VERSUS BRITTLE MATERIAL. Figure 34-25 shows the stress-strain diagrams for two different types of material. The dashed line is for a ductile material while the dotted line is for a brittle material. The difference between the two is the amount of plastic deformation the material experiences before fracture. Brittle materials will have a total strain of less than 5% at fracture while ductile materials will have a total strain of more than 5% at fracture. The amount of plastic deformation exhibited by material A is significantly greater than that exhibited by material B. It should also be obvious that ductility has nothing to do with strength. Ductile material has many advantages over brittle material and is normally preferred over brittle material for structural components.

When looking at the "plastic" portion of a ductile material's stress/strain curve, its initial slope is positive while the latter portion is negative. During the initial phase of the plastic deformation, strain is occurring uniformly across the material as microscopic weaknesses in the material (called dislocations) are worked out. During the latter phase of the plastic deformation, necking down at one specific location occurs, and although the local reduction in cross-sectional area causes an increase in actual stress, the stress based on the original cross-sectional area increases.

referred to as the yield stress. Most aluminum alloys have a yield stress which is based on 0.2% permanent deformation. Since loads which cause stresses to exceed a material's "yield stress" can cause permanent deformation in a structure, and permanent deformation is normally undesirable, designer's normally use the yield stress (as opposed to ultimate stress) when designing structure to carry normal loads. Aircraft structure will therefore be designed to withstand "limit loads" without experiencing more than a 0.2% permanent deformation. In addition, the aircraft structure will be designed to withstand 150% of the limit load and not experience catastrophic structural failure. This higher load, which is at least 150% of the limit load, is referred to as the structure's "ultimate load." We'll get a lot deeper into this subject we talk about "Static Structural Loads" in Chapter 35.

D. MODULUS OF RESILIENCE VERSUS MODULUS OF TOUGHNESS. Mechanical energy can be defined as the product of a force acting over some distance. The area under the stress-strain curve represents the energy required to cause the material to deform or strain. If a material is loaded within its elastic range, the energy required to cause the strain is approximately equal to the area under the stress-strain curve. Since there is no permanent deformation, the material will return to its original shape in the same way a spring returns to its original shape after a load is removed. The maximum amount of energy a material can absorb without permanent deformation is referred to as its modulus of resilience and is graphically shown in Figure 34-26. High resilience is important in structural components such as landing gear and wings which can be exposed to sudden loads.

The maximum amount of energy a material can absorb without fracture is referred to as the material's modulus of toughness and is also graphically shown in Figure 34-26. Toughness is a property that is normally associated with ductile material and, like strength, is a desirable feature in the design of critical structural components because it provides a type of insurance. If a critical structural component is accidentally loaded beyond its design limits, the structure will distort and deform excessively before it finally fractures. Aircraft structure that is supposed to withstand crash loads must be designed to be tough. By absorbing large amounts of energy, "load attenuating" seats, lower fuselage structure and landing gear can reduce the G loads on occupants and maximize the potential for their survival.

E. DEFINITIONS OF TERMS RELATED TO THE STRESS-STRAIN DIAGRAM.

●Actual Fracture Stress - The load at fracture divided by the actual area supporting the load when the material fractures. This is different from the "ultimate stress" because the latter uses the original area supporting the load.

●Elastic Range - The range of stresses or straining during which the material experiences plastic (permanent) deformation.

●Fracture Stress - The maximum primary stress at the instant of fracture.

●Maximum Fracture Strain - The maximum

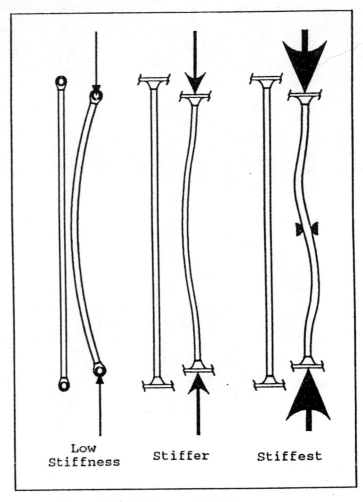

Figure 34-32. Effects of Stiffening on a Long Column's Critical Load

strain at the moment of fracture. Its value will be greater than the Measurable Fracture Strain.

●Measurable Fracture Strain - The permanent strain measured after fracture and after the elastic strain which existed at fracture has been relaxed. It is measured in inches per inch of strain.

●Modulus of Elasticity - The ratio of stress to the unit strain that it will generate while a material is in its elastic range. It is sometimes referred to as "Young's Modulus". It is expressed in terms of psi (or force per unit area) and abbreviated with the symbol "E".

●Modulus of Resilience - The area under the "elastic" portion of the stress-strain curve. It is a measure of the energy that can be tolerated without permanent deformation.

●Modulus of Toughness - The area under the

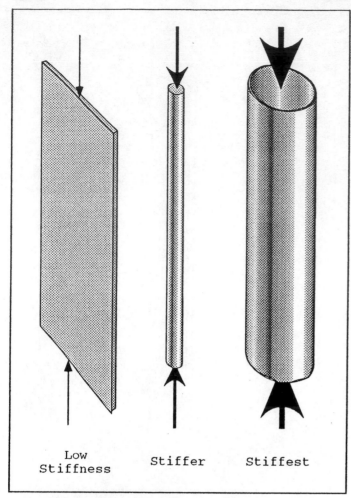

*Figure 34-33. Effects of Radius of Gyration
on a Long Column's Critical Load*

Low
Stiffness   Stiffer   Stiffest

"plastic" portion of the stress-strain curve. It is a measure of the energy that can be absorbed without fracture and is important in evaluating the crashworthiness of a structure.

●Plastic Range - The stress-strain range where plastic deformation will occur.

●Proportional Limit - The greatest stress a material can withstand and still have a constant Modulus of Elasticity. Below this limit, material will return to its original dimensions when unloaded.

●Ultimate Stress - The maximum static stress a material can withstand without failure. It is designated as F_{tu} for tension stress, F_{cu} for compression stress and F_{su} for shear stress.

●Yield Stress - The static stress at which a material will first start to experience permanent objectionable deformation. Above this stress a material

will enter its plastic range. Below this stress it will exhibit elastic behavior. It is designated as F_{ty} for tension stress, F_{cy} for compression stress and F_{sy} for shear stress.

5. FAILURE UNDER TENSION LOADS.

Metal which fractures due to the application of a tension load tends to separate in one of two ways. First, the material can fail by cleavage, a microscopic failure mode associated with separation of the material along the material's grain boundaries without significant plastic deformation. Fracture occurred either at the grain boundary or through the grain itself. When the fracture surface is viewed microscopically, it can be seen that the fracture ran through the material's grain structure. With unaided viewing, the surface appears granular, with sparkles of light reflecting off the surface. Cleavage failure is normally associated with brittle materials. The second type of microscopic failure mode reveals a more plastic fracturing process, with the microscopic material undergoing gross plastic deformation before separation. The surface appears to be dimpled. In locations were the material was simply pulled apart, the dimples are roughly circular in shape with relatively flat faces. In other locations the dimples are elongated, with one side of the long axis being higher than the other. In this case the material was sheared apart, with the dimple's elongation reflecting the direction of the shear load. With unaided viewing the surface appears to have a dull fibrous or silky texture. This fracture surface is associated with failure of ductile materials.

The examination of fracture surfaces to deduce information about the physical properties of the material (e.g. strength and ductility) and the nature of the load which caused the fracture (e.g. load type, direction and rate of onset) is one of the skills required of the aircraft accident investigator. Although the average investigator probably doesn't have the skills and equipment necessary to make a conclusive determination concerning the causes of the fracture, the investigator is responsible for ensuring suspect components get to the appropriate laboratory for expert examination. To do this the investigator must be able to identify unusual fractures and those that could have influenced the accident sequence

A. DUCTILE MATERIAL. The most obvious feature of a tension fracture in a ductile material is the gross plastic deformation in the area surrounding the fracture. The more ductile the material, the more dramatic will be the necking down of the material on either side of the fracture. If you refer to the stress-strain diagram for the typical ductile material under a tension load, you can see that the slope of the stress-strain curve decreases when stresses increase beyond the yield stress.

NOTE: The slope of the stress-strain curve to the right of the yield stress is referred to as the material's tangent modulus. It is a measure of the material's stiffness while in the plastic range. Unlike the modulus of elasticity which has a relatively constant value throughout the elastic range, the tangent modulus is not a constant and normally decreases (stiffness decreases) with increasing strain.

As stress increases from yield stress to ultimate stress, plastic deformation is occurring along the entire component. In addition to the elongation along the axis of the tension load, the component will also experience a slight, but measurable decrease in its cross-sectional area. As it stretches, the material gets skinnier. Once the ultimate stress has been reached, the stress-strain curve appears to have a negative slope until the fracture stress is reached. This apparent decrease in stress is due to the way the stress, which is plotted, is calculated. As the "ultimate stress" in the material is reached, the plastic deformation in the section of the material with the lowest strength will rapidly accelerate. The component's cross-sectional area in this local area of weakness will shrink, causing the actual stress to continue to rise as local strain increases until the component fractures. However, since engineers use the original area to calculate stress, it appears the stress is decreasing when it is actually increasing. Although this may seem strange to the uniniti-ated, it is a logical method for the engineer. What the investigator needs to know is that gross plastic deformation in the form of overall elongation of the component and necking down in the area of the fracture are expected in tension load failures of ductile materials.

In addition to gross plastic deformation in the form of necking down, the fracture

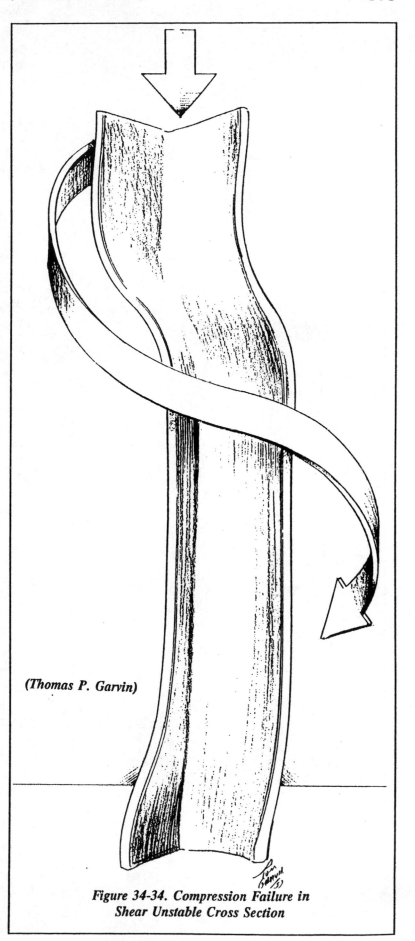

(Thomas P. Garvin)

Figure 34-34. Compression Failure in Shear Unstable Cross Section

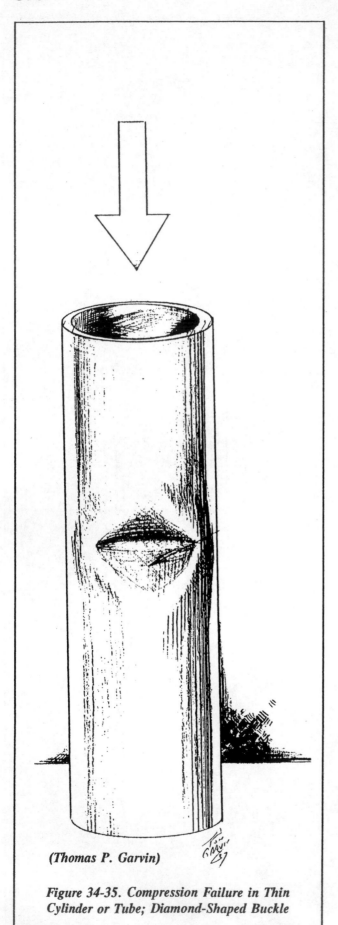

(Thomas P. Garvin)

Figure 34-35. Compression Failure in Thin Cylinder or Tube; Diamond-Shaped Buckle

surface will have some other distinctive features. At a macroscopic level, the surface will have a dull, fibrous appearance. The fracture surface nearest the origin of fracture of the component may have 45 degree "shear lips", whose surface is relatively smooth. The shape of the component and its thickness will affect the appearance
of the fracture surface and comparison to test samples, of known material properties and loading conditions is a valuable aid.

B. BRITTLE MATERIAL. Brittle tension load failures tend to have their fracture surface oriented 90 degrees to the tension load. There is little if any apparent plastic deformation and the cleavage of the material's grains or failure at the grain boundaries will cause the fracture surface to have a bright, granular appearance. When held to the light, pinpoints of light will reflect off the grain surfaces.

One of the characteristics of brittle fractures is the rapid speed at which the crack moves through the material. When compared to plastic fractures, there is little plastic deformation and consumption of energy. This fast moving crack may leave chevron shaped marks on the fracture surface. These marks may have special significance to the investigator since they point back to the origin of the crack. Close examination of the origin may reveal flaws introduced during the manufacture or use of the component.

C. FAST FRACTURE (BRITTLE FAILURE) IN DUCTILE MATERIAL. When stress is rapidly applied to a ductile material, it may fail in a manner which causes its fracture surfaces to more closely resemble that of a brittle material. The "necking down" and 45° "shear lips" commonly associated with ductile failure will be replaced by "granular" fracture surfaces which are perpendicular to the axis of the tension stress. Although this "brittle fracture" in normally ductile material is normally associated with large structures such as, welded steel ships, welded bridges, gas pipelines and pressure vessels, it can show up in aircraft wreckage. The fast fracture will normally occur as a tearing failure, with the leading edge of the tear or crack moving at the speed of sound. Witnesses hearing the noise accompanying this tearing may confuse it with the noise associated with a chemical explosion. The fracture surface often leaves a classic chevron or herringbone pattern marking the direction of the crack propagation. The chevron's "point" back toward the origin of the crack, opening up toward the direction in which the crack is moving.

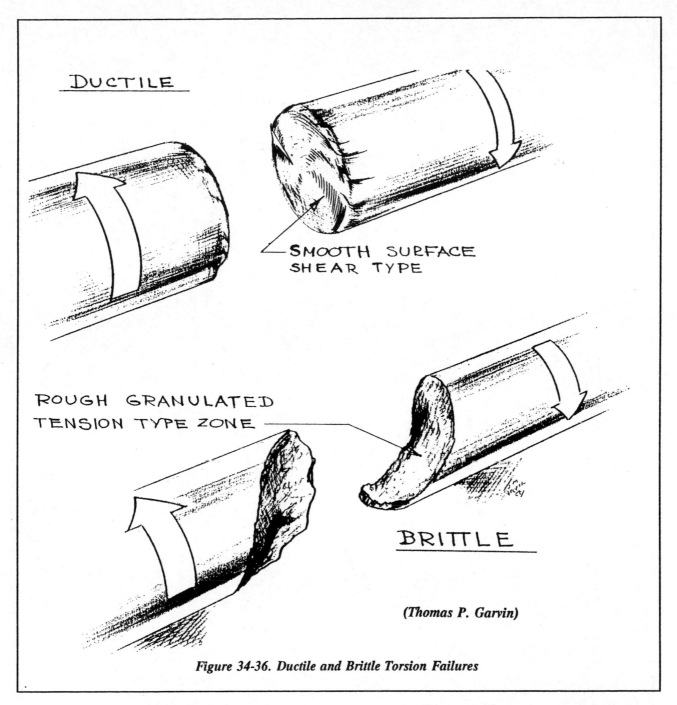

DUCTILE

SMOOTH SURFACE
SHEAR TYPE

ROUGH GRANULATED
TENSION TYPE ZONE

BRITTLE

(Thomas P. Garvin)

Figure 34-36. Ductile and Brittle Torsion Failures

This latter fact can be useful in determining it the crack's origin has indications that fatigue cracking, corrosion, or some other deficient condition existed and precipitated the failure.

Since these fractures are associated with static loads and can normally be traced back to a crack or flaw in the structure, it is important for the investigator to identify and protect these surfaces. A small crack (caused by corrosion an inclusion or even a fatigue crack) can become unstable and grow rapidly in pressure vessels such as a high pressure hydraulic cylinders or landing gear oleo struts. During in-flight break-ups and ground impacts, impact loads can produce many fast fracture surfaces which have nothing to do with the causes of the accident but may provide clues as to where critical failures first occurred and how they progressed.

The investigator must remember that in the "real world" things are neither black nor white. They are just shades of gray. And, although aircraft structure may be classified as either ductile or brittle, we will always be able to find exceptions to the rules we have just established.

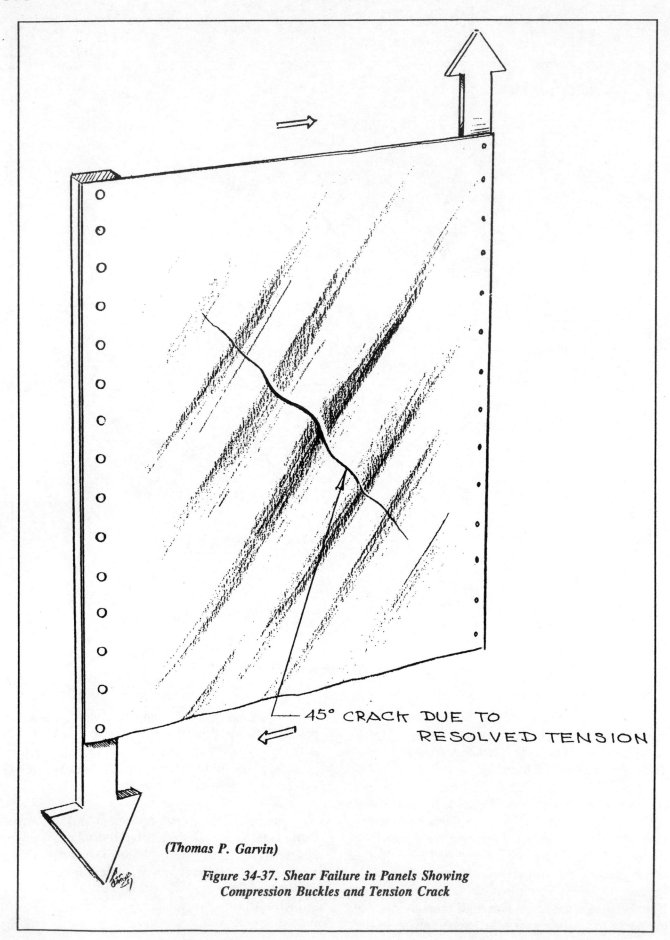

45° CRACK DUE TO
RESOLVED TENSION

(Thomas P. Garvin)

*Figure 34-37. Shear Failure in Panels Showing
Compression Buckles and Tension Crack*

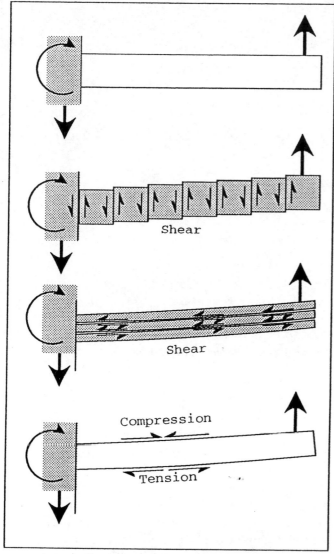

Figure 34-38. Stresses in a Beam Exposed to Bending Loads

Another bottom line. Whenever a fracture surface and associated deformation is not, based on your knowledge of the material, what you expect, you have an apparent anomaly that must be resolved. It is the field investigator's job to identify the fact that may exist and make sure the evidence is preserved as the failed component is transported to the appropriate laboratory. Although the anomaly may be explainable in the context of the accident sequence (e.g. impact loading during a crash, increased ductility due to post crash fire), you need to know why the apparent anomaly exists. On the other hand, the anomaly may be real. The wrong part may have been used or processing error made during the production of the "right" part.

6. FAILURE UNDER COMPRESSION LOADS.

Compression failures are more complicated than tension failures. While the failure load of a component exposed to tension loads is primarily a function of the cohesive strength of the material and the structure's cross-sectional area, the maximum strength of a component loaded in compression is a function of not only the cohesive strength of the material but also its stiffness (modulus of elasticity, cross sectional shape as well as size, the length of the component and the degree to which the ends and other portions of the component are fixed in direction). These factors may be lumped together and referred to as the structure's stability. Short squat structures like a single domino lying face down on a table are very stable. When loaded in compression they get crushed or flattened. Taller structures like a stack of dominos are referred to as columns. A stack of ten or so dominos loaded in compression would be more stable than a stack of fifty. The stability of the structure affects ways the structure can fail. For instance a penny which is placed on a railroad track and run over by a train will exhibit one type of compression failure mode (crushing or flattening) while a flag pole with a 400 pound flag-pole sitter on it will fail in quite another manner (buckling or bending). Although compression stress is calculated in much the same manner as tension stress (i.e. compression load divided by area supporting the load) the stress at which the failure will occur is dependent on those factors we mentioned earlier. For instance long skinny structures which are loaded in compression along their long axis tend to fail at lower stress levels than short squat structures. Structures using materials which have a low modulus of elasticity (low stiffness) tend to fail at lower compression stresses than those that use a material with a higher modulus of elasticity. In the next couple of paragraphs we'll talk about these factors in more detail.

A. BLOCK COMPRESSION. When a short squat structure is loaded in compression we will call it "block compression." When loaded, this structure won't tend to buckle or bend. The materials ability to carry the compression and shear stresses which were discussed earlier will determine its failure mode. When relatively brittle materials (like poured concrete supports) are loaded in compression, they fail in shear along a line about 45 degrees to the load. When ductile material is loaded in block compression, stresses in the material will initially cause the material to strain in much the same manner as when tension loads increase. Strain will at first be elastic (with approximately the same stiffness as the material had when it was loaded in tension) and then progress to a

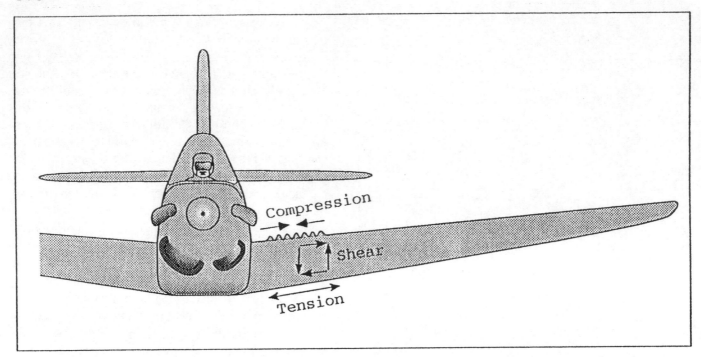

Figure 34-39. Variations of Stresses in a Wing Exposed to Up-Bending Loads

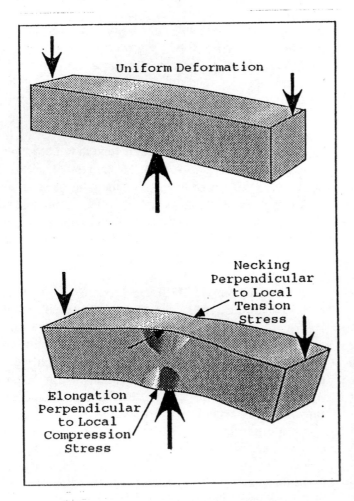

Figure 34-40. Bending Load Failure of a Rectangular Beam made of Ductile Material

plastic range (where the stiffness decreases). But instead of getting longer, the component will get shorter. And instead of getting thinner, it will get thicker, material flowing outward in response to the compression load. As compression stresses increase even further, the stress necessary to keep the material flowing will start to increase, making the material stiffer. When a ductile material is loaded in compression the instability associated with necking just prior to a tension failure doesn't occur. Instead, the material will not have a clearly defined ultimate strength but will instead flow outward until radial cracking in the material occurs. Although block compression failures are not a significant factor in aircraft accident investigation, we often see the local failures which can be explained using the concepts just discussed.

B. COLUMNS LOADED IN COMPRESSION.
A column is a structure which is designed to carry compression loads along an axis which is relatively long when compared to its lateral or cross-sectional dimensions. When a column is exposed to increasing compression loads, the column will react initially by compressing along the longitudinal axis of the column. This straining is similar to the strain in the short block being exposed to compression. However, as loads increase, the column will begin to flex in either bending or twisting. Once flexing begins, the column quickly becomes unstable and its ability to support additional loads rapidly decreases. If the flexing begins while compression stresses are still in the elastic range,

the flexing will occur uniformly over the length of the column and the column will spring back to its original shape after the load is removed. Columns which react in this manner are referred to as long columns. If, however, the compression loads cause the column to become unstable after the compression stresses reach the plastic range, deformation will occur locally (this is called "crippling") and the column will remain deformed after the load is removed. Columns which react in this manner are referred to as short columns. Generally speaking, short columns are "stronger" than long columns. In other words, if two columns are equal in all parameters except length, the longer of the two will fail at the lower load. The long column is less weight efficient. Since high strength to weight ratios are important for aircraft structures, most columnar structures in aircraft are designed as short columns. However, improper assembly or repair, or even damage can cause a short column to become a long column. More about this later.

Determination of the compressive load and stress at which a column will become unstable and fail is much more complicated than determining the tension

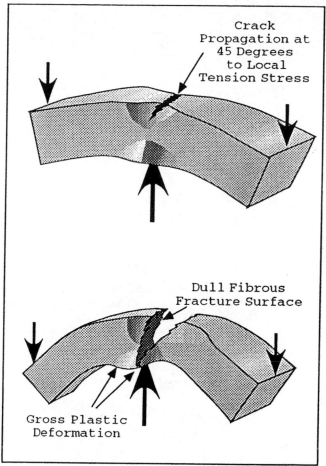

Figure 34-41. Bending Load Failure of a Rectangular Beam made of Ductile Material

load or stress at which a similar structure will fail when loaded in tension and way beyond the scope of this text. However, a general knowledge of these failure modes and the characteristics of the column which determine the failure mode is valuable to the aircraft accident investigator.

The load at which a column becomes "unstable" and fails to carry increasing loads without excessive deformation is referred to as the column's critical load. As mentioned earlier, instability can occur at relatively low loads while the stresses in the column are still within the elastic range and where strain is elastic. Or the instability can occur at higher loads when the stresses are in the plastic range and most of the strain is in the plastic range and deformations permanent. Elastic failures normally occur in "long" columns where the column is relatively long and slender, having a high slenderness ratio. Plastic failures occur in columns in which the slenderness ratio is lower.

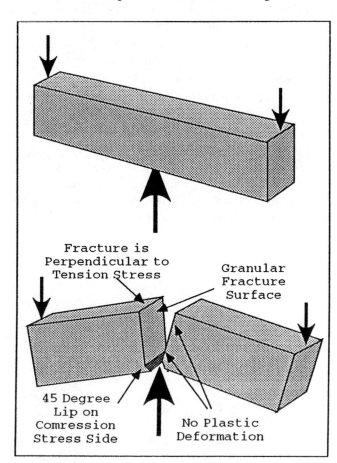

Figure 34-42. Bending Load Failure of a Rectangular Beam made of Brittle Material

C. BUCKLING. Buckling, which is caused by

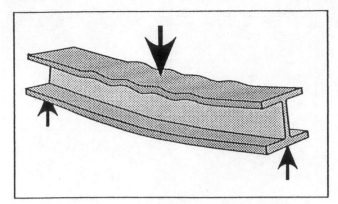

*Figure 34-43. Bending Induced Compression
Buckling in the Spar Cap of an "I" Beam*

compression loads, is a bending or flexing of the structure carrying the compression load. The buckling is in a direction perpendicular to the compression load.

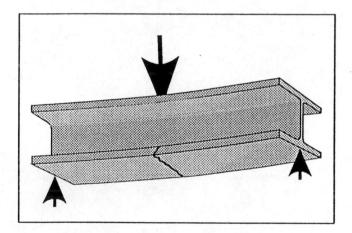

*Figure 34-44. Bending Induced Tension
Failure of the Spar Cap of an "I" Beam*

When buckling occurs, the structure becomes unstable and its ability to carry additional loads will start to decrease, sometimes rapidly. As a structure starts to buckle, some locations on the structure will be exposed to higher stresses than others. Almost all aircraft structure which is loaded in compression has a compression load failure mode which involves buckling.

Elastic Versus Plastic Buckling. Buckling can occur while the highest stresses still remain below the material's yield stress i.e. all of the material in the structure remains in its elastic range. If loads are then decreased below the critical load, the structure will return to its original shape without any permanent deformation. If loading is increased significantly above the critical load, collapse of the column can occur. On the other hand, buckling can also result in local

stresses which exceed the material's yield stress i.e. some of the material in the structure is in the plastic range. This plastic buckling is normally localized and often referred to as "crippling" and the portions of the structure which deformed while stressed in the plastic range will remain deformed after the load is removed. Since, in both cases, the structure failed to carry the load without excessive deformation of the structure, in both cases the structure failed.

Long Columns Vs Short Columns. Whether a column's primary failure mode is elastic (it springs back to its original dimensions when the load is removed) or plastic (it remains deformed after the load is removed) is determined by the column's slenderness ratio. Columns with high slenderness ratios tend to fail first in an elastic mode while columns with a low slenderness ratio tend to fail first in a plastic mode. The maximum load a long column can support without buckling is not so much a function of the compression yield stress of the material used in the column, but rather a function of the column's shape (cross-section and length), supporting structure which increases its stiffness, and the material's modulus of elasticity. On the other hand, a short column will not fail until internal stresses exceed the material's yield stress and are in the plastic range. All other things being equal, short columns fail at higher compressive stresses (and higher compressive loads) than long columns. Since weight is critical in aircraft designs, and short columns are more weight efficient, the majority of aircraft structure supporting compression loads are short columns.

The column's "slenderness ratio" is directly proportional to its effective length and indirectly proportional to its radius of gyration. That means that

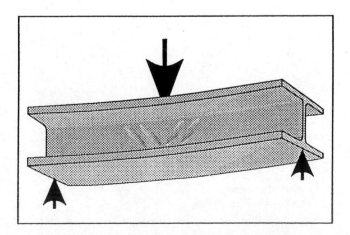

*Figure 34-45. Bending Induced Compression
Buckling in an "I" Beam's Shear Web*

slenderness ratio increases as its effective length increases and decreases as its "radius of gyration" increases. "But," you say, "what do 'effective length' and 'radius of gyration' mean?"

A column's effective length is directly related to its actual length, but is also influenced by design features which stiffen the column. For example, the effective length of a column can be made shorter by preventing the ends of the column from rotating. A column whose ends are fixed and unable to rotate as the column buckles would have an "effective length" that was one fourth the length of a similar column whose ends were free to rotate (see Figure 34-32). The effective length of a column can also be decreased by providing supports along the length of the column which prevent lateral displacement or rotation of the column. Placing restraints which prevent lateral displacement in the center of a column would decrease its effective length by a factor of two. Since the critical column stress of a long column is inversely proportional to its effective length, decreasing the effective length of a column will increase the load at which the column fails while increasing the effective length will decrease the load at which the column fails. This second case (increasing the effective length) is of direct interest to the aircraft accident investigator. Changes (authorized and unauthorized) which remove structure which keeps the ends of the column from rotating or which prevent lateral movement along the

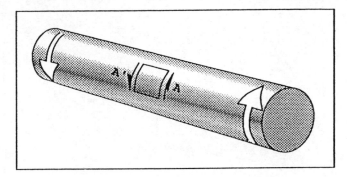

Figure 34-46. Addition of Shear Stresses to Form Tension and Compression Stresses

length of the column, increase the effective length of a column and can significantly reduce the load bearing capabilities of the column. Investigations involving long column failures should involve determination of the adequacy of stiffening supports.

For example, if required supports which stiffen a long thin flight control push rod were not reinstalled after maintenance, the column could lack the strength to move the flight controls. The investigator must keep in mind the fact that, since the stress on the column's material was still in the elastic range, the column will spring back to its original shape after the load is removed and will not exhibit the permanent plastic deformation or fractures normally associated with structural failure. In this case the investigator's physi-

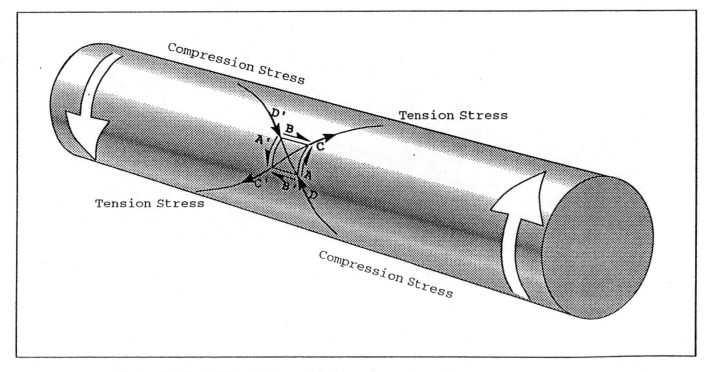

Figure 34-47. Addition of Shear Stresses to Form Tension and Compression Stresses

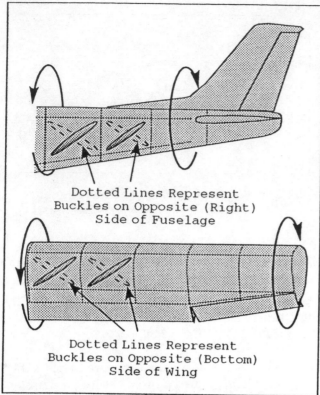

Dotted Lines Represent
Buckles on Opposite (Right)
Side of Fuselage

Dotted Lines Represent
Buckles on Opposite (Bottom)
Side of Wing

*Figure 34-48. Examples of Thin Walled Closed
Cell Structure Carrying Torsional Loads*

The "radius of gyration" is the other factor which affects the column's slenderness ratio and is a mathematical term which is related to the cross-sectional size and shape of the column. An example which compares the radius gyration of two structures with equal cross sectional areas is shown in the Figure 34-33. It is probably intuitively obvious that the rectangular column made of 10 inch by 0.1 inch material will fail before the cylindrical column with a diameter of 1.13 inches. Both columns have equal cross sectional areas, but have different shapes. Since the radius of gyration of the cylindrical column is larger than that of the rectangular column, the cylindrical column is stronger in compression and will fail at a higher load. On the other hand, if the 10 inch by 0.1 inch plate is bent into a circle and the seam welded closed, the radius of gyration of this new hollow cylindrical column will be able to carry an even higher compression load without failure. It is interesting to note that all three of these structures would fail at the same load if loaded in tension.

cal evidence of the failure many be limited to witness marks made when the column flexed beyond its normal limits and contacted surrounding structure.

D. THIN-WALL COLUMNS. Some of the thin, light weight columnar structures used in modern aircraft designs can experience, in addition to pure buckling instability, a combined buckling-torsional instability. Thin walled, light weight columns whose cross-section is "open" (e.g. "L", or "Z" channels) giving them low torsional stiffness, will begin to twist when buckling occurs. If the buckling-twisting occurs

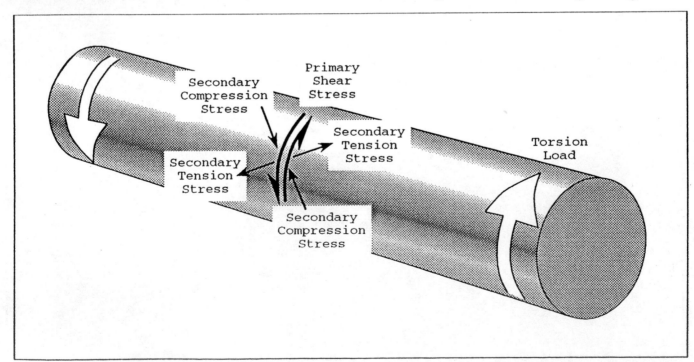

Figure 34-49. Decomposition of Torsion Load into Primary and Secondary Stresses

while the stress in the material is in the plastic range, the deformation could be mistakenly attributed to a torsional load. Careful examination of the structure and an understanding of how the structure will react to various loads, is an integral part of investigation of column failures.

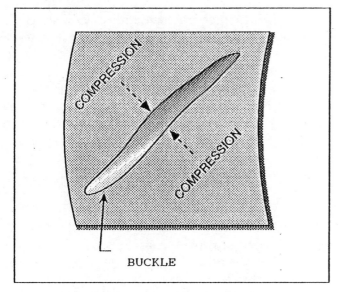

Figure 34-50. Determination of Direction of Torsion Load from Compression Buckles

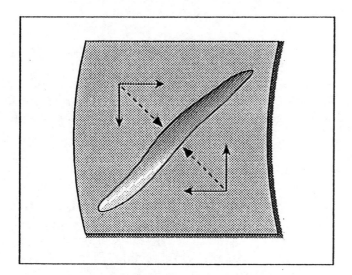

Figure 34-51. Resolution of Compression Loads

E. THIN-WALLED PANELS LOADED IN COMPRESSION. Modern aircraft designs routinely use relatively thin panels to carry compression loads. One example is the upper skin of an aircraft's wing. Our earlier discussions have revealed that, without some added stiffening, a thin panel cannot withstand a

significant compression load. If, however, the edges of the panel are prevented from rotating as compression loads are applied (perhaps by riveting them to other structure, restraining them in a track or even curving the panel into a cylinder), the panel can become an efficient load bearing aircraft structure. When these panels are overloaded in compression, plastic straining of the panel leaves a tell-tale diamond shaped deformation. You can see this phenomena if you carefully press your heel straight down on an empty aluminum can until it just starts to crush. Be careful when trying this. The walls of the can will suddenly become unstable as you apply increasing weight and the load required to cause instability greatly exceeds the load required to collapse a buckled can.

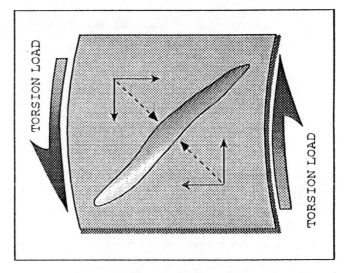

Figure 34-52. Determination of Direction of Torsion Loads

7. FAILURE UNDER SHEAR LOADS.

A. DUCTILE VERSUS BRITTLE MATERI-AL. When loaded in shear, ductile materials tend to fail in a shear mode, with material on either side of the fracture surface showing considerable plastic deformation. The top drawing in Figure 34-36 shows a typical fracture expected when a bolt made of ductile material fails due to shear loads. If on the other hand, the bolt were made of a higher strength and more brittle material, the fracture surface would have the characteristics shown in the bottom drawing of Figure 34-36. If an investigator notes a ductile appearing shear failure of what is supposed to be a high strength and relatively brittle bolt, their "alert" light should start flashing. The ductile characteristics of the bolt could be due to the annealing (e.g. softening) of the bolt during a fire, the improper manufacture of the bolt (e.g. lack of proper heat treatment or hardening), or even the use of a bolt with improper material properties (e.g. bogus part)

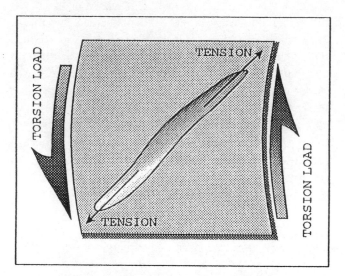

Figure 34-53. Determination of Torsion Load Direction based on Tension Stress

B. SHEAR IN THIN PANELS/SHEETS. When a ductile metal is fabricated into thin metal panels and then loaded in shear, the metal will not have the same failure mode as described in the previous paragraph. When the shear load is applied, shear, compression and tension stresses are developed. Although the maximum shear, compression and tension stress developed are equal, the thin panel is weakest in compression and will fail first in buckling at an angle 45 degrees to the shear load. The thin panel will fail as a long column, 45 degrees to the shear load. See Figure 34-37 for a visual description of this phenomena. If the stresses remain in the elastic range, the panel will return to its original shape after the load is removed. If, however, higher loads cause stresses to increase into the plastic range, permanent deformation will occur, providing the accident investigator with a clue concerning the nature and direction of loading which caused the structural failure of the panel. If shear loads continue to increase after the panel experiences significant buckling, the panel may also fail in tension, with the tension failure occurring 45 degrees to the shear load and 90 degrees to the compression buckling. This is the direction of the maximum tension stress.

8. FAILURE UNDER BENDING LOADS.

Bending loads, like shear loads, develop tension, compression and shear stresses. Remember the deck of cards described earlier in this chapter. The flexing of the cards by pressing down in the center of the deck and pulling up at the edges of the deck is similar to that experienced by a wing supporting an aircraft in flight and essentially the same as that of a simple beam.

Compression and tension stresses in a structural beam being exposed to bending loads behave in much the same way as that deck of cards. The material near the surface on the inside of the curve (the top of the wing while pulling positive Gs) experiences the largest compressive stresses. The material near the surface on the outside of the curve (the bottom of the wing of an airplane pulling positive Gs) experiences the largest tension stresses. The material near the center of the beam experiences lots of shear stress but little or no compression or tension stress.

Let's now look at how we can expect a beam to fail when exposed to excessive bending loads. The material on the inside of the curve will experience compression stress and the outside of the curve will experience tension stress. Both stresses will be highest near the upper and lower surfaces and lowest at the beam's neutral axis. Shear stress, on the other hand, will be zero at the upper and lower surfaces and maximum near the beam's neutral axis.

When a beam is loaded in bending, the resulting straining will cause the cross-sectional area on the inside of the curve to increase as the length of the inside of the beam decreases and the cross-sectional area on the outside of the curve to decrease as the length of the outside of the beam increases.

The cross-sectional shape of a beam exposed to bending loads is important in determining the failure mode of the beam. Since aircraft structure is normally made of relatively ductile material, let's limit our discussion to ductile material. Let's also start with a beam with rectangular cross-section. Since the cross-sectional shape of the beam is rectangular, shear stresses and straining are not that significant and we will ignore them. However, compression and tension stresses and their associated straining may leave signatures which can provide investigators with clues concerning the nature and direction of bending loads. Bending loads which generate only stresses which are in the elastic range will not result in permanent deformations and the investigator will have to rely on witness marks for clues. If, however, compression and tension stresses continue to increase and exceed yield stresses, plastic deformation will cause uniform permanent curvature or "set" over the length of the beam. The direction of the curve indicates the direction of the bending load. If loads increase further, local tension necking will eventually occur on one portion on the outside of the beam and a significant decrease in the cross-sectional area will occur. The necking can be accompanied by a significant increase in the associated cross-sectional area on the inside (compression side) of the curve. If bending loads are not relaxed or continue to increase, tension fracture on the outside of the curve, in the area of the necking can occur. This fracture will progress from the outside surface toward the center of the beam and will be similar to that of a static tension failure in a ductile material (showing lots of 45° surfaces with a dull fibrous appearance). If the bending load continues the fracture can propagate through the center of the beam and through the portion of the beam which was originally in compression and causing a total separation of the beam. This sequence is shown graphically in Figure 34-38. The extensive plastic deformation associated with this process requires a relatively large amount of energy, a desirable characteristic in aircraft structures.

A similar rectangular cross-section beam made of brittle material will have an entirely different failure mode. Since brittle materials have a relatively small plastic range, bending loads which cause tension stresses to exceed the yield stress will not result in much plastic deformation before a fracture on the outside (tension side) of the curve occurs. This fracture surface will be 90° to the original tension stress and have a shiny, granular appearance similar to that of a brittle material experiencing a static tension overload. The crack will normally propagate rapidly through the beam, with little energy absorption, an undesirable characteristic in aircraft structures. As the crack approaches the inside surface of the beam, it direction might change so that it is oriented 45° to the beam's inside surface.

Now let's look at what happens if some other cross-sectional areas are used for a beam carrying a bending load. The "I" beam we discussed earlier is much more weight efficient than the rectangular beam and it (or a similar structure) is therefore often used in aircraft structure applications. At loads where stresses are still in the elastic range deformations will be temporary and the structure will spring back to its original shape as soon as the load is relaxed. However, if components of the beam have a high slenderness

ratio they will behave as long columns for those portions of the beam which are loaded in compression and will start to buckle at stresses much lower than the materials compressive yield stress. Elastic buckling of an "I" beam could occur if the slenderness ratio of either the cap carrying compressive stresses or the shear web was high enough to cause it to behave as a long beam. The edges of the "I" beam's cap could buckle first at the edges of the cap and the shear web could start to develop buckles in a direction perpendicular to the compressive stresses generated as a result of shear stresses. See figure 34-43. If bending loads increase, compression stresses can reach the plastic range and buckling will involve permanent deformations. Examination of the buckles in the caps and shear web on an "I" beam can reveal the nature and direction of loads which caused the failure of the beam.

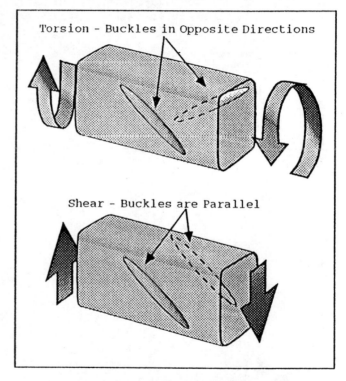

Figure 34-54. Differences in Buckles Caused by Torsion and Shear

At the bottom of the beam tension loads could result in tension stress failure of the lower spar cap. This would show up as a fracture with the classic examples of tension failure: fracture surfaces which are 45° to the local tension stress, necking down of structure adjacent to the fracture surface, and a dull fibrous fracture surface. The fracture surface may propagate into and through the beam's shear web.

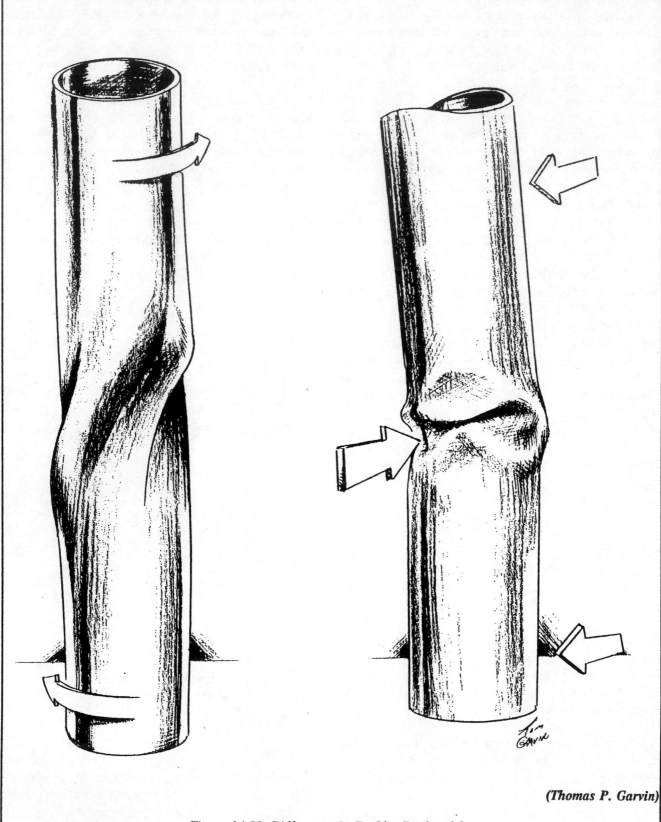

(Thomas P. Garvin)

**Figure 34-55. Differences in Buckles Produced by
Torsion and Bending Loads**

An "I" beam's relatively thin shear web can be vulnerable to shear stress induced by compression buckling near the "I" beam's neutral axis. Figure 34-45 shows the shear web buckling pattern associated with excessively high bending loads.

9. FAILURE UNDER TORSION LOADS.

When a structural component fails due to torsion loads, fracture surfaces and plastic deformation can often provide the investigator with clues concerning the nature and direction of loads which caused the failure. As we have discussed before, structures made of ductile material will have failure modes which are different from those made of brittle materials. Thin walled structures will also behave differently than solid structures.

A. SOLID STRUCTURES. Let's look at solid structures first, starting with a cylindrical shaft loaded in torsion as shown in Figure 34-46. Torsional loads will cause a primary shear stress as the material on the left side of the shaft attempts to slide down across the material on the right side of the shaft (vectors A and A' in Figure 34-46) which is attempting to slide upward. Looking at the diagram of the shear stresses shown on the surface of the shaft, we can see that reactionary stresses are necessary to keep a small unit square of material on the surface of the shaft in equilibrium (vectors B and B' in Figure 34-47). Addition of vectors A and B to form vector C and addition of vectors A' and B' to form vector C' (Figure 34-47) shows how shear stress an be resolved into tension stress. Similarly, addition of vectors A and B' to form vector D and addition of vectors A' and B to form vector D' shows how the same shear stresses can be resolved into compression stresses.

B. DUCTILE MATERIAL. If the cylindrical shaft is made of a ductile material, it will fail in shear. Shear stresses will be highest near the surface of the shaft resulting in significant plastic deformation when shear stresses exceed yield stress. Near the center of the shaft, stresses will remain near zero. Prior to separation of the two parts of the shaft, significant plastic deformation will occur near the surface of the shaft. The appearance of the fracture surface of the shaft will be a smooth surface with helical swirls reflecting the direction of rotation. A small raised tit may exist on one of the fracture surfaces, indicating the neutral axis of the shaft. Because of the large amount of material undergoing gross plastic deformation, the processes of fracturing this ductile, solid, cylindrical shaft will require a relatively large amount of energy, at least when compared to the torsion fracture of a brittle material.

C. BRITTLE MATERIAL. If the cylindrical shaft is made of brittle material, the shaft will fail in tension. By determining the direction of the secondary tension stress, the direction of the tension fracture can be determined (Figure 34-49). In the case we are discussing, the tension stresses exist along a line down and to the right and up and to the left. The fracture will start at the weakest place along the surface of the shaft and proceed in a helical pattern around the shaft until it jumps to the other side. The fracture surface will have a granular appearance, with light reflecting off cleavage grains.

D. THIN WALLED, CLOSED CELL STRUCTURE MADE OF DUCTILE MATERIAL. Aircraft structural designs often take advantage of the weight efficiency of thin walled, closed cell structures to carry torsional loads. One such example is the load bearing skin of modern cantilever wings and fuselages. Another type of structure using this thin walled hollow shaft is the torque tube. Torque tubes are very weight efficient when it comes to carrying torsional loads. Another is the "D" cell spar used in the leading edge of the wing of some fixed wing aircraft and the leading edge of the rotors of some rotary wing aircraft. By the way, if the cell is opened (e.g. a fatigue crack running parallel to the axis of the torsion load) the ability of the structure to carry torsional loads is greatly decreased.

If the wall of the hollow structure is made of ductile material and is also thin enough for the wall to behave as a column when loaded in compression, then the shaft will first fail by buckling when it is overloaded in torsion. The buckles will be at a 45° angle to the torsional load (and the longitudinal axis of the structure) and perpendicular to the compression stress created by the torsion load. If torsion loads continue after severe buckling occurs, tension fractures will occur at a 90° angle to the compression stress developed by the torsion load (which is also perpendicular to the tension stress created by the torsion load and at a 45° angle to the torsion load. If torsion loads continue to increase, tension fractures can continue to grow until complete separation of the structure occurs. All of this makes a whole lot more sense if we simply draw a sketch of the thin walled closed cell structure, the torsional load (be careful of the direction of the torsion) and the shear, tension and compression stresses created by the torsion load. Figure 34-49 shows, in sequence, the torsion load on our structure, the shear (primary) stresses created by the torsion load, the compression (secondary) stresses created by the torsion load, the compression buckling caused by the compression stresses, the tension (secondary) stress created by the torsion load, and the tension fracture caused by the tension stress. These sketches attempt to show the cause and effect sequence involved in torsion failure of this type of structure. Unfortunately, this isn't the way the accident investigator looks at it.

The investigator normally works the sequence in reverse. First the observation is made that the failure involved buckling (and perhaps tension fractures) of a thin walled, closed cell structure. We can then draw two arrows which represent the compression stresses necessary to cause the buckling (Hint: it is perpendicular to the long axis of the buckles. See Figure 34-50.) Next decompose these two vectors into vertical and horizontal components as shown in Figure 34-51. We now have the vectors which represent the shear stresses created by the torsion load. The shear stress arrows (vectors) parallel to the torsion load indicate the direction of the torsion load. This is shown in Figure 34-52. This, believe it or not, is sometimes of concern to the investigator. We can also come up with the same answer by drawing two arrows that represent the tension stress which accompanies the compression stress (Hint: it's perpendicular to the compression arrows and parallel to the long axis of the compression buckles. See Figure 34-53.)

E. DISTINGUISHING BETWEEN FAILURE DUE TO TORSION AND FAILURE DUE TO BENDING OR SHEAR LOADS. There is a potential trap when determining the direction of torsion loads which caused buckling in thin walled, closed cell structure. The buckling may be due to bending or shear and not torsion. The investigator has to first determine if the buckling is due to torsion or bending loads. Buckles due to torsion tend to form a helix around the structure. This is shown in Figure 34-55. If we could look through the structure, the buckles would appear to be in opposite directions. On the other hand, when bending or shear loads cause buckling on opposite sides of the structure, the buckles are parallel to each other. Knowledge of this difference will become important when we start analyzing buckling damage to thin walled, closed cell structures such as wings, box beam wing spars and fuselages. This point is illustrated in Figure 34-54.

10. SUMMARY.

So much for the review. Now that we all understand the concepts of loads, stresses and strains, we are ready to apply those concepts to the investigation of structural failures in aircraft accidents. That's coming up in Chapter 35.

35

Structural Failure

1. INTRODUCTION.

The image of a Sherlock Holmes type of figure, rooting through the still smoking aircraft wreckage, closely examining each piece of broken metal, is an image many people associate with aircraft accident investigation. In fact, although just one part of the aircraft accident investigator's job, the examination of broken structure is a vital part of the investigation. Failure of an aircraft's primary structure and subsequent inflight break-up, is ranked high on the list of risks air crews would rather not face. Failure of a mechanical component which leads to loss of control of the aircraft is not far behind. And even though inflight failure of structural and mechanical components is involved in a relatively small percentage of aircraft accidents, "bent metal" is examined in virtually all aircraft accident investigations. Aircraft accident investigators do spend time letting the broken an twisted metal "talk" to them. Looking at the fracture surfaces, the investigator asks the material if it tore apart slowly, over a long period of time, or suddenly by some inflight overload. Or was the fracture the result of impact forces generated when a perfectly good aircraft was flown into the ground? Was the bent metal deformed by inflight loads, the dynamic effects of an aircraft tearing itself apart in the air or the sudden impact of a collision with the ground? A competent aircraft accident investigator routinely looks at broken and crumpled aircraft components, postulating the magnitude and direction of the forces which caused it to deform or fracture, and integrates this information into potential accident scenarios. There is

a lot to be gained by "talking" to the broken metal that once was a reliable aircraft.

By the way, before we get into this chapter let's establish some definitions. First the word "structure." For now we'll refer to structure as any piece of material that carries a load. In other words, structure is something that is pushed, pulled or twisted at one end and pushes, pulls or twists something at the other end. For variety we'll sometimes use the word "component" in place of "structure." Same meaning.

The next term we need to talk about is "failure," as in "structural failure." When using the term "structural failure," we'll be referring to a failure where the material fails to carry the load it was intended to carry. The pushing, pulling or twisting doesn't get transmitted to its intended destination. "Structure" can "fail" in one of two general ways. First it can "fracture" into two or more pieces. A child may refer to a component which has fractured as "broken." But a structural component also "breaks" or "fails" when its shape is changed so that is can no longer carry its load. In this type of failure the component still is in one piece, but it is bent, stretched or even corroded or worn so that it can no longer do its intended job. Some structural failure modes which may be of interest to the investigator are listed in Appendix D.

The most obvious reason for examining aircraft wreckage is to determine if the strength of the aircraft's primary structure was inadequate to carry normal operational loads. If a structural weakness allowed the

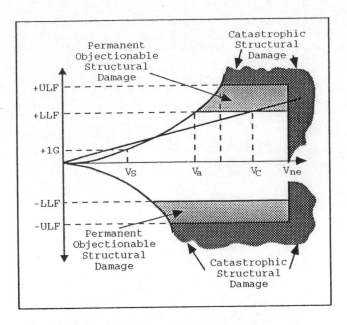

Figure 35-1. Operating Flight Strength Limitations (V-G or V-n) Diagram

structure to fail at lower than expected loads, it is very likely that this weakness exists in other similar aircraft and will, if not corrected, eventually cause other aircraft to crash. If an investigator fails to identify a structural weakness in an aircraft's design, it is just a matter of time before this structural inadequacy is involved in another accident. The consequences of an investigator missing the significance of an unusual fracture surface or bend in a spar can be catastrophic. The most dramatic consequence of a structural failure is inflight break-up. Although design-related inflight break-up has not accounted for a significant proportion of aircraft accidents, the investigator must be sure that a design- related inadequacy was not a factor in the accident. Examination of fracture surfaces, distorted (bent and stretched) metal as well as the marks made as separating parts broke away from the aircraft while it was still in flight, can provide valuable information concerning the reasons for the break-up. In fact, the location of aircraft structural components along the flight path and prior to the primary impact point is one of the most obvious clues of inflight break-up. In this type of accident (inflight break-up), the aircraft structure which separates prior to ground impact can be found at significant distances from each other.

Inflight break-up of general aviation aircraft is normally the result of excessive loads which result from the airplane being flown outside of its certified operating envelope and not the result of inadequate structural strength. One of the most common reasons

for exceeding the envelope is the pilot's attempt to recover from an unusual attitude. This, in turn, is often the result of unplanned flight from VMC into IMC or encounters with violent weather while in IMC. It has also occurred when inaccurate flight instruments caused the pilot(s) to become confused, place the aircraft in a high speed, nose low attitude and then over G the airplane while attempting to recover from a high speed dive. Examination of the wreckage which results from this type of accident (inflight break-up when the airplane is operated outside its normal operating envelope) may contain evidence which is very similar to that encountered when the airplane breaks up in flight due a structurally deficient component. The aircraft accident investigator must, therefore, be alert to evidence which points to structural weaknesses. This chapter attempts to provide many of the clues and knowledge necessary to distinguish between damage caused by pre-existing deficiencies, damage associated with pilot induced overloads and damage generated by crash loads.

Not all structural failures result in the immediate catastrophic inflight break up of an airplane. Many components such as control push rods, cables and bell cranks are used to control or enhance the stability of the aircraft. Other components such as turbine wheels and drive shafts are used to develop or transfer energy and are equally critical to the continued safe flight of the aircraft. Although failure of these structural components may not cause immediate inflight break-up of the aircraft, they may cause the air crew to lose control of the aircraft, and allow the aircraft to collide with the ground. An important job for the individuals responsible for the structural investigation is to also

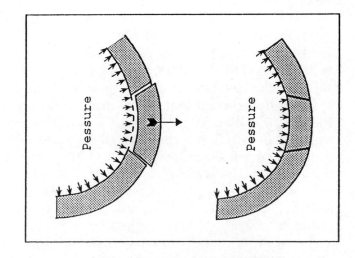

Figure 35-2. Plug Doors Versus Outward Opening Doors

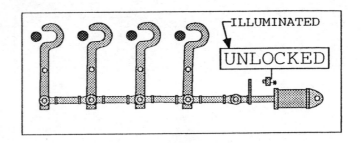

Figure 35-4. Door Lock Warning System Design (Unlocked Position)

identify structural failures which significantly increase the pilots' workload or degrade system operation.

Examination of failed structure may also show that the aircraft's structure did not fail in flight and the failure was probably the result of impact forces generated when the aircraft hit the ground. When a force causes a structural component to fail, the nature of the force (i.e. its direction and the rate at which it was applied) may be recorded in the deformed or fractured material. Fractures created by impact forces tend to look different than those caused by more gradually applied loads. Bending and twisting of structure caused by inflight overloads will differ from the deformations caused by impacts with the ground. A fracture caused by the progressive growth of a fatigue crack will show significant differences from a fracture surface resulting from the fast fracture of a statistically overloaded component.

2. WHY DO AIRCRAFT PARTS BREAK?

There are a lot of reasons why an airplane's structure can fail. Some of the reasons have been inferred in the paragraphs above. The next couple of paragraphs will provide a more detailed explanation of the factors briefly mentioned above, plus several others to tuck away in your "investigation clue bag."

A. OVERLOAD. If inflight loads exceed those for which the part was designed, it shouldn't be a surprise if the part fails to carry the load. All structures, be they bridges, buildings or airplanes, are designed to withstand only specific loads. It is unrealistic to assume that airplanes should be designed and built to withstand any conceivable load it can experience. An airplane that was designed to meet excessive structural requirements would need excessively long runways and would not have the payload carrying capability necessary to make it useful. So, if a struc-

ture is exposed to a load greater that for which it was designed, it will fail structurally, either deforming to the point where it can no longer perform the job for which it was designed, or fracturing into two or more pieces. There are two general reasons for airplane structure to fail: it can either be overloaded or it can be under strength.

Airplane structures are designed to withstand both the loads generated by the air at some maximum airspeed and the loads generated while maneuvering the airplane at some G load. These maximum airspeeds and G loads define an airplane's operating flight strength limitations and are often graphically displayed as a V-n or V-G diagram. Although the V-G diagram will be discussed in greater detail later in this chapter, a brief explanation now will make some of the following comments easier to understand.

Most airplanes can be flown to speeds and "G" loads which can place excessive loads on the aircraft's structure. This can occur when a pilot allows or causes a controllable airplane to fly outside its authorized operating flight strength envelope. It can also occur if

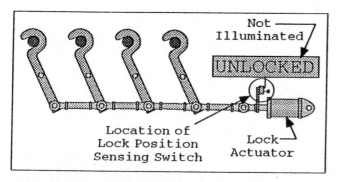

Figure 35-3. Door Lock Warning System Design (Properly Rigged and in Locked Position)

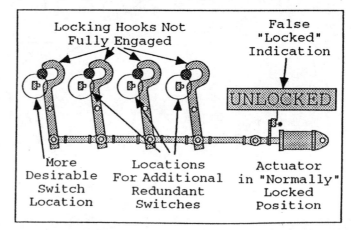

Figure 35-5. Door Lock Warning System Design (Mis-rigged; "False" Locked Indication)

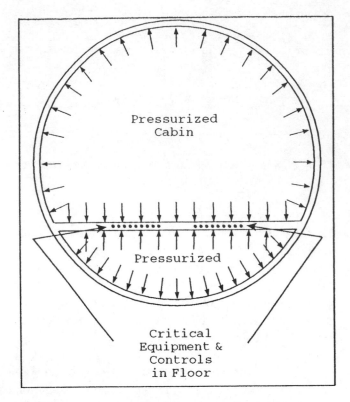

Figure 35-6. Structure Separating Pressurized Compartments and Containing Critical Equipment

the flight control system fails and the pilot loses the ability to keep the airplane within its operating envelope. Figure 35-1 shows the typical operating flight strength limitations for general aviation aircraft and the failures which could be expected if the various limits were exceeded.

If an airplane is flown "through the top" of the envelope by exceeding its "weight/G" limits, the airplane's structure could be bent or broken by lift-

related loads. The section on "Maneuver Loads", later in this chapter, examines this issue in greater detail. If the airplane is flown out the right hand side of the envelope by exceeding the aircraft's limit airspeed, there are several ways the airplane (and its occupants) can get in trouble.

First, aircraft structure that is directly exposed to the onrushing air could be damaged as the dynamic pressure of the airstream is converted to static pressure pressing inward on the structure. We'll return to this subject later in this chapter under the heading of, "Loads due to Dynamic Pressure."

Second, airfoil and control surface deflections caused by the higher than normal air loads can cause twisting, bending and oscillations of the structure. These effects (referred to as "aeroelastic effects") can limit the effectiveness of flight controls or cause structure to fail. "Aeroelastic phenomena" are discussed in greater detail later in this chapter.

Exceeding limit airspeeds in a supersonic aircraft may weaken the aircraft's structure in areas where the dynamic energy in the airstream is converted to heat energy. Avoiding damage to aircraft structure isn't the only reason for keeping airspeed below specified limits. Excessive speed can reduce the airplane's stability, cause the compressor section of a jet engine to stall, and generate unpredictable computer commands when an aircraft is flown at airspeeds for which there are no instructions. See Chapter 33 to get additional details on the hazards associated with high speed flight. The bottom line here is that there are lots of bad things that can happen when an airplane ex-

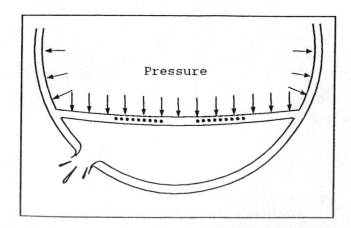

Figure 35-7. Loss of Pressure in One Compartment

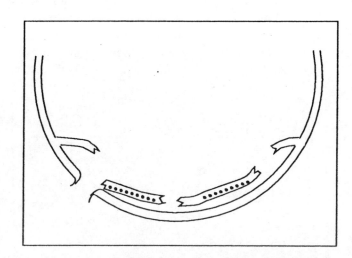

Figure 35-8. Collapse of Separating Structure and Loss of Critical Equipment and Controls

ceeds its Vne (or "redline") airspeed. Since aircraft undergoing flight test exceed the redline airspeed only during specially controlled tests, and then under careful observation, even the designers may not know what will happen if this speed is exceeded in an "operational" aircraft.

Flying the airplane outside its operating limits isn't the only way a pilot can damage the airplane's structure. Hitting the ground at too high a speed can obviously damage the airplane. Structural damage can also result if the pilot lands at speed and weight combinations which exceed the airplane's design limits for landing. By understanding the various forces acting on an airplane and how the airplane's structure reacts to these forces when overloaded, a trained investigator can often interpret structural damage to determine if the structure's failure was induced by the way the airplane was flown. Landing and ground maneuver loads will be covered in a later section of this chapter while crash loads will be discussed in Chapter 36.

Airplane structure has to designed to withstand not only the loads imposed by the pilot, the structure also has to withstand loads the airplane is expected to encounter while flying in its "normal" environment. For example, the pilot of an airplane can expect an airplane's structure to be designed to withstand the air loads created when the airplane encounters wind gusts which can be reasonably expected in flight. The windscreen, engines, leading edge surfaces and all transparencies can be expected to encounter birds. Ice can accumulate on the leading edges of propellers, airfoils, nacelles and nose cones, breaking off periodically to encounter engines and other airplane structure located along its trajectory. Airplanes are designed and certified to be able to withstand reasonable insults from the environment. By the way, what is considered "reasonable" can change over time. For instance, until recently, jet engines where required to withstand the ingestion of a four pound bird without catastrophic failure. As technology improved and the number of people at risk following an engine loss increased, jet engine manufactures have demonstrated the ability of newly developed engines to withstand the impact of 8.5 pound birds.

Airplanes are designed to withstand reasonably expected wind gusts when the airplane is flying at or below the airplane's "rough air speed." Wind gusts can suddenly increase the lift produced by the wing, at best causing a bumpy ride, and at worst causing structural damage or failure. The manner in which

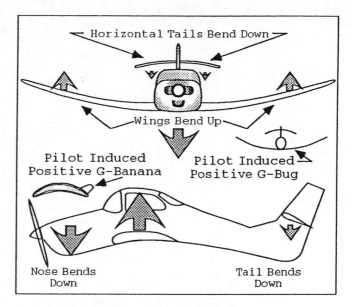

Figure 35-9. G-Bugs and Bananas Caused by Pilot-Induced Positive G Loads

wind gusts create G loads is different from pilot induced G loads. The differences in the load patterns can sometimes be detected in structure which has experienced inflight over G conditions. These differences can provide the accident investigator with valuable clues concerning the cause of the over G. If the cause of the over G condition was gust-induced, the investigator is faced with the problem of determining if the structural damage or failure was due to a gust that exceeded the design criteria of the airplane, if the airplane was flying at a speed above that recommended for rough air, or if the structure was weaker than that necessary to survive the maximum FAA specified gust at the maximum FAA specified airspeed.

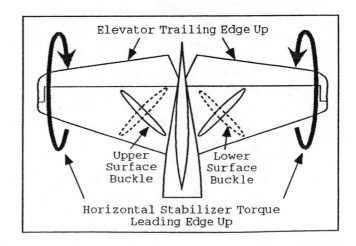

Figure 35-10. Torsional Buckling Resulting from Nose-Up Elevator Inputs

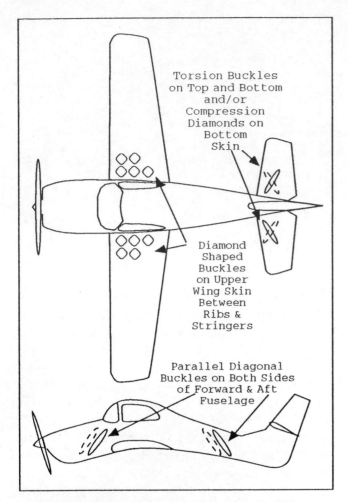

Figure 35-11. Fuselage and Wing Skin Buckling Caused by Pilot-Induced Positive G Maneuvers

B. INADEQUATE STRUCTURAL STRENGTH.

One of the most obvious reasons for the failure of a structural component is that the component lacked the strength to withstand the loads generated when the airplane was flown within its normal operating limits. In other words, the structure was too weak. The potential for weaknesses in the structural design of an airplane immediately starts bells and whistles going off, in the head of an investigator. Questions like, "If this planes structure has a weakness, do all airplanes of a similar design have a similar weakness?" and, "Do all airplanes designed using similar design criteria have a similar structural weakness?" start bouncing around in their heads. If the answer to either of these questions is, "Yes," a systemic problem that affects entire fleets of airplanes exists and the implications can be widespread. There are several reasons why a structural component may be under strength.

• First, the structure could have been doomed to being under strength from the day it was first drawn on the drawing board. The designer may have made an error which was not caught on subsequent analyses or tests. Or the structure may have been produced using substandard materials, improper processes or incorrect fabrication procedures. If repairs were made to the structure after it first left the factory, any of the above errors could have occurred during the repair process.

• Service life issues. It should be obvious that after an airplane leaves the factory, fair wear and tear can cause a general weakening of the airplane's structure. These "service life" issues are normally divided into four sub areas; fatigue cracking, corrosion, wear, and creep. All four are progressive failures which cannot be reversed as the airplane accumulates flight hours or ground-air-ground cycles. And, since structural failure due to a service life issue is normally a systemic problem, the potential problems can reach far beyond the single airplane that crashed. See the section titled, "Service Life Issues" for a more detailed discussion of this subject.

• Exposed to high temperatures. Aircraft structure can be weakened in a relatively short period of time. Exposure to heat can greatly reduce a metal's strength. Exposure of some aluminum alloys to temperatures of 400° F for five minutes can reduce the alloy's strength by 80%. Since structural aluminum is especially sensitive to heat, it is not generally suitable for use in areas where high temperatures are expected. Jet engine hot sections and compressor bleed air lines are made of structure such as stainless steel or titanium alloys which maintain most of their strength in relatively high temperatures. Airplanes which can achieve high supersonic speeds require the leading edge structure to be able to withstand the extremely high temperatures generated at these speeds. Now, if aircraft structure which has been designed for high temperatures is exposed to these high temperatures, no problem! The problem occurs when structure which has not been designed for high temperature operation is exposed to high temperatures. For example, if a critical load bearing component were made of aluminum alloy located near a tube carrying hot, high pressure air from the engine compressor, and the tube failed near the component, the heated component could fail in 1 G flight while it was carrying only a fraction of its design load. In fact, when exposed to inflight fire, load bearing structure can be expected to fail structurally rather than melt. The load bearing structure of an airplane's wing, when exposed to inflight fire, does not melt and fall off. Instead, it is weakened to the point where it can no longer carry the loads generated by flight and fails, normally with an unusual amount of distortion. This does not mean that there will be no melting. Components which are not carrying critical loads can melt and have their droplets carried off by the airstream and be deposited downstream on other components or structure.

3. TYPES OF STRUCTURE.

An airplane's structure is broadly separated into two general areas; primary structure and secondary structure. Primary structures are those parts of the airplane's structure which are necessary for the airplane to safely fly its mission. The parts which make up the airplane's primary structure are the primary load bearing members which carry the aerodynamic forces created by the movement of the airplane through the air and balance them against the thrust of the engines and weight of the airplane and its payload. Primary structure also includes structure which supports the airplane while it is on the ground. If the structural failure of a component could put the safety of an airplane or its passengers in jeopardy, then that part is considered to be part of the airplane's primary structure. Secondary structure is the part of the airplane that is not part of the primary structure and is not necessary for the safe completion of the airplane's mission. Obviously, there is room for interpretation when trying to distinguish between primary and secondary structure.

A. PRIMARY STRUCTURE. The following components are normally considered to be part of an airplane's primary structure.

- Wing structure (including primary and secondary controls necessary for safe flight).

- Fuselage structure carrying flight, ground and cabin pressurization loads.

- Empennage (including primary and secondary controls necessary for safe flight).

- Landing gear structure.

- Engine mounts and supporting structure.

Primary structures can be further sub-categorized as either "critical structural elements" or "principle

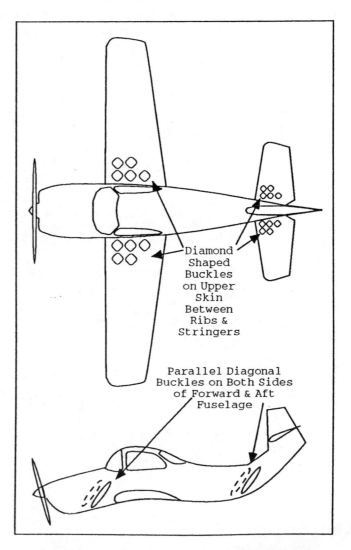

Figure 35-13. Fuselage and Wing Skin Buckling Caused by Gust-Induced Positive G Maneuvers

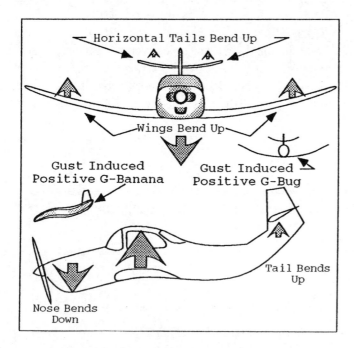

Figure 35-12. G-Bugs and Bananas Caused by Gust-Induced Positive G Loads

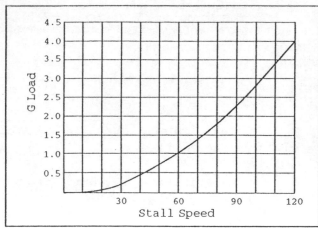

Figure 35-14. Stall Speed Versus G Load

flight-critical loads. And, although these components have failure modes that are similar to those exhibited by primary and secondary structure, they do not fall into either of these categories. Hydraulic pressure lines, drive shafts powering pumps and electric alternators, and gear teeth in transmissions all have failure modes which can give clues concerning the nature of the loads which caused them to fail. The conscientious airplane accident investigator will have a firm understanding of the loads that airplane structure carries and the physical evidence these loads leave behind them when they cause structure to fail.

structural elements". Critical structural elements are those elements whose failure would result in catastrophic failure of the airplane. Principle structural elements are those elements which contribute significantly to carrying flight, ground and pressurization loads, and whose failure could (but not necessarily will) result in catastrophic failure of the airplane.

B. SECONDARY STRUCTURE. The following components are normally considered to be part of an airplane's secondary structure.

- Aerodynamic fairings.

- Tailcones.

- Landing gear doors.

C. OTHER STRUCTURAL COMPONENTS. There are other mechanical components which carry

4. TYPES OF LOADS.

All loads are vectors, and as such have both magnitude and direction. You can change a load by either changing its magnitude (e.g. increasing or decreasing the pounds of force being applied), by changing its direction (applying the force upward instead of down), or by changing both magnitude and direction. In addition, loads take a finite length of time to be imposed. And, although loads can be applied in a very short period of time, they can't be applied or changed instantaneously. This last factor, the fact that loads take time to apply and/or change is important for the airplane accident investigator to understand because a structure's reaction to a load is, to a significant degree, a function of the rate at which the load was applied. In order to better understand this phenomena, we'll separate loads into three general areas; static loads, dynamic loads and repeated loads. For reasons which will shortly become obvious, static loads cannot be dynamic loads and vice versa. However, some types of static loads can be repeated loads and some types of dynamic loads can be repeated loads. Understanding the differences between static, dynamic and

AIRCRAFT WEIGHT	LIMIT LOAD FACTOR AND ULTIMATE LOAD FACTOR (LLF & ULF)	MANEUVER SPEED (V_a)
Above Design Gross Weight	Decreases	Constant
Below Design Gross Weight	Constant	Decreases

Figure 35-15. Maneuver Speed and Load Factors Versus Gross Weight

repeated loads is important because the nature of the load has a lot to do with the failure mode of the structure and the evidence left behind.

A. STATIC LOADS. If a load were applied so slowly that the structure to which the load was being applied was essentially in equilibrium at all times, the load would be considered a static load. As an example, if a diver were suspended over a diving board (feet barely touching the diving board) and then slowly lowered onto the board until the board was fully deflected and carrying the full weight of the diver, the load applied by the diver's weight would be considered a static load. As the diver were lowered onto the board, the board would be deflected steadily downward with no apparent bouncing (oscillations) by the board. Static loads can be either short or long time loads.

SHORT PERIOD STATIC LOADS. If the load were applied for a short period of time (say seconds, minutes or hours) the load would be referred to as a short time static load. Despite the fact that most loads experienced by an aircraft while performing its mission do cause some oscillations in the airplane's structure, these loads are considered static and most structural verification tests are static load tests. (The exceptions would include loads like those experienced by the landing gear during landing.) Back to our hypothetical diver. The diver would be lowered slowly onto the board so that the board supported the diver's entire weight and was fully deflected, remain there for a short period of time, and be lifted slowly from the board. The board would return to its neutral position with no oscillations.

LONG PERIOD STATIC LOADS. If, on the other hand the load was applied for a long time (say weeks, months or years) the load would be considered to be a long time static load. If after being slowly lowered onto the board the diver were left there for let's say a year or two, it wouldn't be too surpassing if, when the diver were finally lifted off the board, the board did not return to its original shape but remained slightly bent. This failure of the board to return to its original shape after a long time static load is referred to as "creep." And, although creep is not normally a concern of the designer of an airplane's structure, creep can be greatly accelerated by the effects of heat. It has been a significant problem for jet engine designers and can cause grief for an aircraft's structure if extremely high temperatures are present.

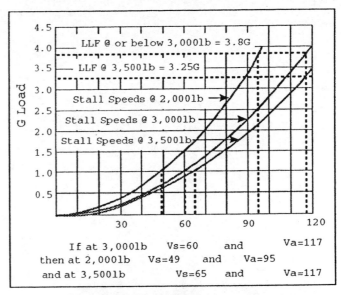

Figure 35-16. Weight Versus Stall Speed Versus Maneuver Speed

B. DYNAMIC LOADS. Dynamic loads occur when loads are applied fast enough to prevent the structure from carrying the load while remaining in equilibrium as the load is being applied. In other words, when the load is applied the structure deflects beyond its equilibrium position, rebounds to less than its equilibrium position, and in a series of ever smaller oscillations converges to its equilibrium position. Dynamic loads can be sub-divided into sudden and impact loads, impact loads being applied much faster than sudden loads.

●**SUDDEN LOADS.** To illustrate a sudden

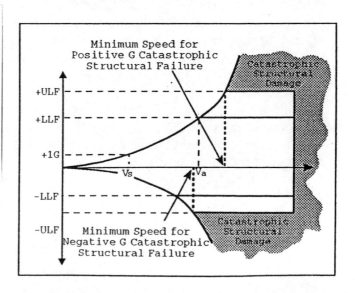

Figure 35-17. Minimum Speed Necessary to Cause a Catastrophic Pilot-Induced Over G

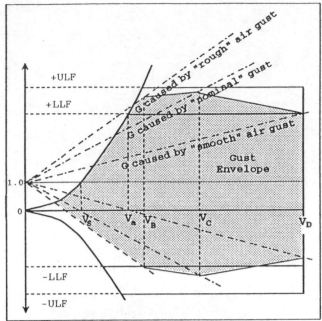

*Figure 35-18. Gust Envelope for Aircraft
Certified under FAR Part 23 or 25*

load, imagine our diver hanging by the hands from a chinning bar with the feet just touching the diving board. If the diver's support were suddenly removed, the weight of the diver would be suddenly applied to the board as the diver dropped. The diving board would deflect past the position of the statically loaded diving board and momentarily stop at a deflection approximately twice that experienced during the static load test. It would then rebound to almost its original position. The board would then begin a series of ever smaller oscillations before finally arriving at the same deflection experienced during the static load test. In addition to experiencing greater deflections than the static load test, the dynamic load will impose greater internal stresses in the diving board. Dynamic load tests are used to test a limited number of airplane structural components such as landing gear.

●IMPACT LOADS. Impact loads are applied at even faster rates. Imagine our diver in a helicopter hovering 1,000 feet over the board. Taking careful aim for the exact end of the board, the diver steps over the

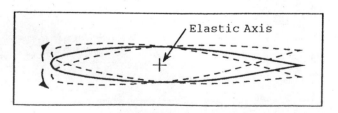

*Figure 35-19. Twisting of an Airfoil
about its Elastic Axis*

side and falls smack onto the action end of the board. Imagine the force created when our diver's weight was applied. It will probably be so fast that the diving board structure between the diver's feet and the supporting structure at the edge of the pool doesn't have a chance to deflect smoothly downward, reacting instead by shattering. Impact loads are normally limited to high speed bird impacts and crash tests. They are important because the internal stresses and deflections associated are so much greater than equivalent static loads or even suddenly applied loads.

C. REPEATED LOADS. Repeated loads are loads that are repeated over and over again. Although the nature of dynamic impact and long time static loads make them unlikely candidates for repeated loads, short time static loads and sudden dynamic loads can be repeated over and over again. During the initial load applications the stresses and deflections experienced are essentially the same as the short time static and sudden dynamic loads. However, the cumulative effects of the repeated loads has real significance to airplane structural designers because repeated loads are one of the necessary ingredients in generating fatigue cracks. If a component undergoes lots of repeated load cycles before it fails due to fatigue cracking, it is said to have experienced "high cycle" fatigue. By lots of cycles, we normally mean hundreds of thousands or millions or tens of millions of cycles. That's a lot. When referring to "low cycle" fatigue, we are referring to components which failed after just hundreds or thousands or tens of thousands of cycles. Obviously, the terms "high cycle" and "low cycle" fatigue are not scientifically precise terms, but are relative terms. We will cover fatigue failure in much greater detail later in this chapter.

5. TYPES OF STRUCTURAL FAILURES.

One of the ways you can look at structural failure is to consider the time it took for the failure to occur. If the failure happened all at once, during the application of a single load, we'll call it an instantaneous failure. If, on the other hand the failure required some period of time to occur, we'll call the failure a progressive failure. In other words, the failure occurred in small increments, building up bit by bit, until the component could no longer do its job. By the way, our discussion in the following paragraphs of this section will be referring to failures in the types of materials traditionally used in aircraft structure. These materials are normally ductile (as opposed to brittle) materials. Additional information concerning their similarities and differences of ductile and brittle material is contained in Chapter 34.

A. INSTANTANEOUS FAILURE. Instantaneous failure can be subdivided into three categories: fractures, where the component separates into two or more parts; plastic deformation, where the component deforms excessively when loaded and doesn't return to its original shape; or elastic deformation, where the component deforms excessively when loaded but does return to its original shape after the load is removed. In this last case the component has "failed" even though the post "failure" component looks exactly like it did before the failure. Its "failure" occurred when deformed excessively under a load, and therefore could not do its job. By the way, if you are having trouble with the terms "elastic deformation" and "plastic deformation", go back to Chapter 34 and review the section on strain. If you're really having trouble, you might want to also look over the sections on tension failure, compression failure, shear failure, bending failure and torsion failure.

B. ELASTIC DEFORMATION/DISTORTION (TEMPORARY SHAPE CHANGE). If we load a structural component to the point were internal stresses cause significant distortion but do not exceed the material's yield stress, the structure will spring back to its original shape after the load is removed. Despite the fact that it returned to its original shape, the structure could have "failed". Suppose a design called for a long slender rod to transmit a "push-pull" force to a hydraulic servo-valve on a hydraulic actuator. Instead of moving the servo value when it was pushed, the rod simply deflected by bending, the rod can be said to have "failed" while internal stresses were still in the elastic range. Another example of elastic failure is the elastic twisting of a long slender wing when an aileron is deflected at high speed. Instead of the deflected ailerons causing the desired rolling motion, the wing twisting can degrade roll rates or even cause roll to occur in the wrong direction. This phenomena, referred to as aileron reversal, is discussed in greater detail later in the section on Aeroelastic Effects. Evidence of this type of structural failure can be very difficult to find in an aircraft wreckage.

C. PLASTIC DEFORMATION/DISTORTION (PERMANENT SHAPE CHANGE). If we load a structural component to the point were internal stresses exceeds the material's yield stress, but do not exceed the material's ultimate stress, the structure will undergo permanent deformation and will not spring back to

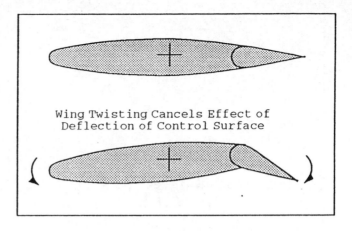

Figure 35-20. Reduction in Control Effectiveness at High Dynamic Pressure

its original shape after the load is removed. Let's go back to that long rod which was supposed to move that hydraulic servo valve. It's been "beefed-up" so that it doesn't bend instead of moving the servo valve. Now suppose that the valve is stuck and resists movement by the rod. If the force applied to the end of the rod creates stresses which exceed the material's yield stress, the rod will deform "plastically" and remain bent after the load is removed. This kind of evidence

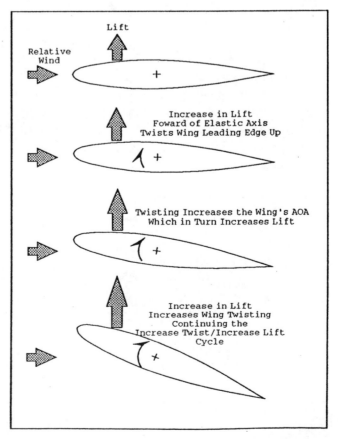

Figure 35-21. Wing Divergence in Torsion

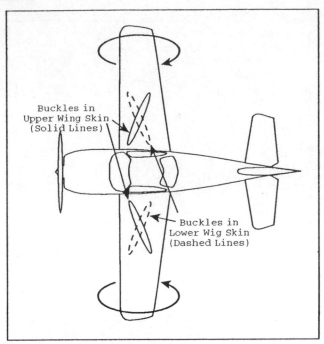

Figure 35-22. Wing Skin Buckling Consistent with Excessive Leading Edge Up Torsion

is a lot easier to uncover. A point to consider; although the rod "failed", it did so only because the servo valve had failed first, and the design of the rod may not be deficient. The buckles in a wing caused by the leading edge of a wing-tip impacting the ground will look different from the buckles associated with up-bending caused by over-G. Examination of the deformed metal in the wreckage will often provide an experienced investigator with lots of information concerning the nature of the load; information that may provide clues concerning events in the mishap sequence.

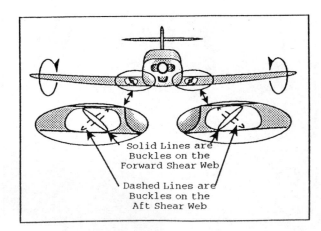

Figure 35-23. Wing Spar Shear Web Buckling Consistent with Excessive Leading Edge Up Torsion

D. FRACTURE. Now let's really crank up the load on a structural component, to the point were internal stresses not only cause significant plastic deformation, they exceed the material's ultimate stress. The structure will then fracture and separate into two or more pieces. In the traditional materials used in aircraft structure (i.e. relatively ductile metals) both the nature of the deformation and the fracture surface can provide valuable clues concerning the loads which caused the failure. For instance, a fracture caused by tension loads will look different from that associated with fatigue cracking. The fracture of a component experiencing a heat induced "creep" will look very different from one failing due to overloads. An experienced eye can detect the origin of a tearing fracture which started as a small fracture in one location and propagated through the remaining structure.

E. PROGRESSIVE FAILURE. Progressive (or "service life") failures are failures that occur over time. The strength of the structure is degraded, some times very slowly, and at other times relatively rapidly. When the remaining strength is no longer adequate to support the imposed loads, the structure will then suddenly fail. Progressive failures are irreversible, the structure will not heal itself. Replacement of the damaged area either through replacement of the entire structural component or the repair with additional structure capable of carrying the loads is necessary. Four general types of progressive failures are covered later in this chapter: fatigue cracking, corrosion, wear and creep. Of the four, the first two are the most important and are becoming more important. As the age of the world's commercial, general aviation and military aviation fleets get progressively older (both in terms of years of life and hours of operation), the original service lives of the aircraft are consistently extended. Although extensions are made only after serious examination of residual strength and structural repair where necessary, there is always the opportunity for human errors in both the estimation of residual strength and the repair of weakened structure.

When a progressive failure appears to be a factor in an mishap, investigators need to fully explore a branch of the investigation which does not normally occur when mishaps involve instantaneous failures; inspections. Because progressive failures occur over time, there is the opportunity to detect the progressive damage before it reaches critical size. In fact, periodic inspections is are an accepted strategy when dealing with the potential for progressive failures.

6. STATIC STRUCTURAL LOADS.

The serious aircraft accident investigator has to have at least a general understanding of the loads that the aircraft's structure is expected to carry. Without this understanding, the investigator will not be able to distinguish between typical inflight overload failures, progressive failures and failures associated with collisions with the ground.

A. LOADS DUE TO DYNAMIC PRESSURE.

Whenever an aircraft flies, it is exposed to the dynamic pressures associated with its passage through the air. The air's pressure on the aircraft will be a function of the ambient static air pressure and the dynamic pressure generated by the aircraft's motion. Where the ambient static pressure is no more than the pressure measured in your basic home barometer, the dynamic pressure is the product of the air's density and the square of the aircraft's true airspeed. An interesting and useful phenomena associated with aviation is that the sum of the static and dynamic pressures is constant in an incompressible airflow. This means that as the speed of the airflow decreases, the static pressure increases. In locations where the airflow completely stagnates, all of the airflow's dynamic pressure will be converted to static pressure. The leading edges of the various aircraft structure take the brunt of this pressure. Although structural design safety margins are normally adequate at certified airspeeds, excessive speed can cause failures due to excessive static pressures. Plexiglas windscreens on some general aviation aircraft have been known to fail during high speed dives. The differential pressure between the inside and outside of an airfoil can also create loads. During one G flight the pressure differential creates the "lift" which supports the aircraft. The differential pressure between the inside and the outside of the aircraft's skin can cause loads on access panels and doors which if improperly fastened can open and be carried off by the airstream. Generally speaking, loads associated with dynamic pressure are a problem only when the aircraft is flown beyond its limit airspeed.

B. LOADS DUE TO INTERNAL PRESSURIZATION.

When aircraft fuselages are pressurized to increase the aircraft's utility at high altitudes, additional internal stresses are created. The maximum tension stress is created along lines parallel to the aircraft's longitudinal axis. Although these stresses are not critical with respect to static loads, they are significant as repeated loads which can cause fatigue cracks to develop in high stress areas. Each time the aircraft

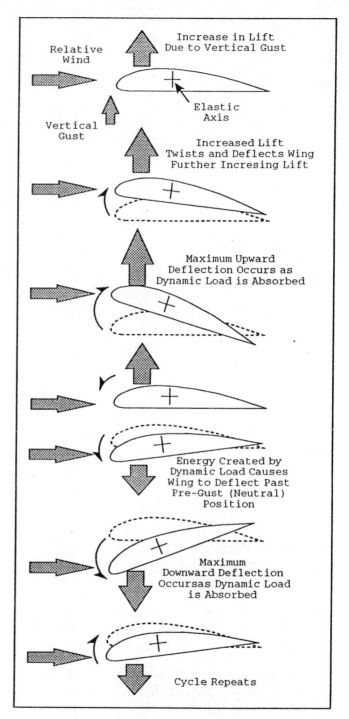

Figure 35-24. Gust Induced Wing Flutter

climbs to altitude and then descents, a fatigue cycle is completed. "Low cycle" fatigue cracking was a factor in the demise of the first commercial jet airliner, the de Havilland "Comet". Concern over this issue still exists in modern airliners which have experienced numerous ground-air-ground cycles. This issue is covered in greater depth later in this chapter under the subject of fatigue.

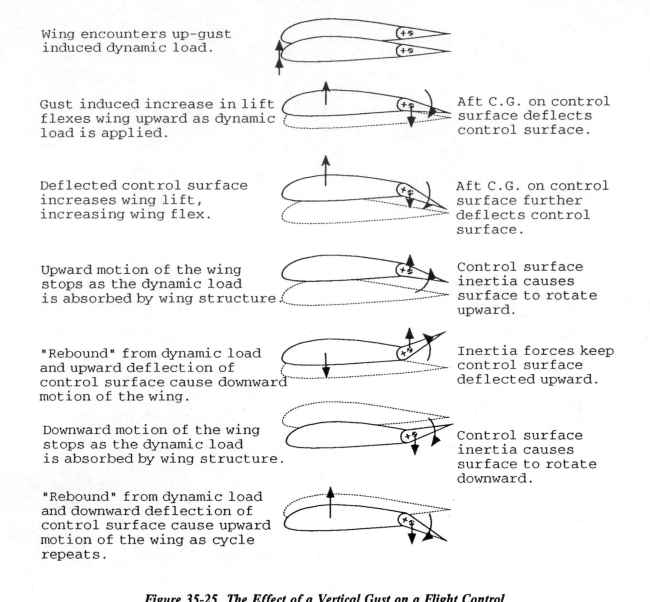

Wing encounters up-gust induced dynamic load.

Gust induced increase in lift flexes wing upward as dynamic load is applied.

Aft C.G. on control surface deflects control surface.

Deflected control surface increases wing lift, increasing wing flex.

Aft C.G. on control surface further deflects control surface.

Upward motion of the wing stops as the dynamic load is absorbed by wing structure.

Control surface inertia causes surface to rotate upward.

"Rebound" from dynamic load and upward deflection of control surface cause downward motion of the wing.

Inertia forces keep control surface deflected upward.

Downward motion of the wing stops as the dynamic load is absorbed by wing structure.

Control surface inertia causes surface to rotate downward.

"Rebound" from dynamic load and downward deflection of control surface cause upward motion of the wing as cycle repeats.

Figure 35-25. The Effect of a Vertical Gust on a Flight Control
with its Center of Gravity Aft of the Hinge

Pressurization loads have been the subject of other types of mishap investigations. Doors to pressurized compartments will naturally open if their locking mechanisms fail. Failures have been attributed to poorly designed locking mechanisms, improper rigging, crew error and normal wear. Although "plug doors" that can only open inward are the obvious answer from a safety perspective, operational requirements sometime dictate other design solutions. The obvious hazard associated with outward opening doors is the ejection of crew/passengers during inflight openings at altitudes.

One factor that has shown up repeatedly in the

inadvertent opening of doors, cargo ramps and canopies of pressurized aircraft, is the less than adequate design of "door unlocked" warning systems. These systems are supposedly designed to alert the crews any time a door, cargo ramp, canopy or other critical opening in the pressure vessel is unlocked. An all too common failure is to warn the crew not of a failure of the system to engage the locks, but instead to warn of only one downstream failure. As an example, a warning system which warns only of the lock actuators failure to move, does not consider upstream failures such as mis-rigging and locking hook stall which will allow the lock's actuator to move but not detect the failure of the locking mechanism's moving into the proper position. This concept obviously has application to all warning systems. Whenever a failure of a warning system is involved in a mishap, the investigator needs to ask the question, "Exactly what was the warning system warning of?"

Other accidental openings (e.g. fatigue failure of the skin, failures of windows, penetrations made by failed engine components) made in the hull of a pressurized airplane have resulted in crew or passengers being ejected at altitude.

Another mishap type involves aircraft with multiple un-vented pressurized compartments. Imagine an aircraft with two or more separate, pressurized compartments. If one compartment is depressurized, the other will maintain its pressure. If, in addition, the structure between the two compartments cannot withstand the load created by the resulting differential pressure, the structure between the two compartments will fail if one side is depressurized. Critical components located in the structure separating the two pressure vessels, can be destroyed and loss of the entire aircraft is possible.

C. MANEUVER LOADS. When an aircraft is intentionally maneuvered by the pilot, predictable loads are imposed on the various structural components. For instance, if the pilot pulls back on the yoke

(or stick) the trailing edge of the elevator will deflect upward, resulting in a downward pressure on the horizontal tail. The horizontal tail will be bent downward. Torsional forces created when the trailing edge of the elevator moves up will cause a nose-up torsional force in the horizontal stabilizer. In addition, the tail of the aircraft will be forced down, causing the down-bending loads in the aft fuselage. The deflection of the elevator will cause the aircraft's tail to move down,

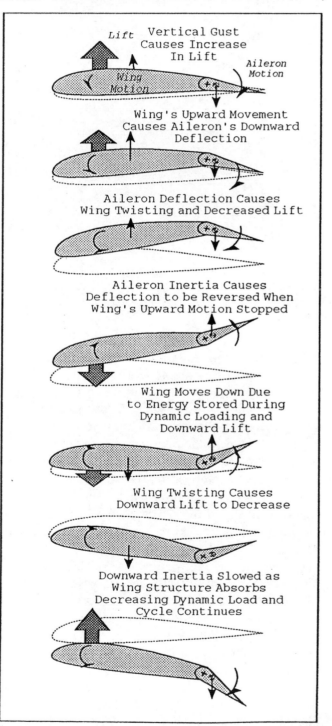

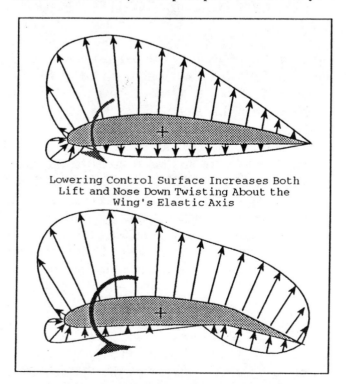

Figure 35-26. Effect of Control Deflection on Airfoil Lift and Twist

Figure 35-27. Airfoil Flutter Induced by Aileron Flutter

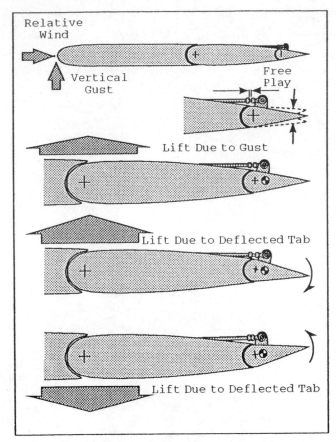

Relative Wind

Vertical Gust

Free Play

Lift Due to Gust

Lift Due to Deflected Tab

Lift Due to Deflected Tab

Figure 35-28. The Effect of a Gust on a Trim Tab with Excessive Free Play in its Control Rod

rotating the wing to a higher angle of attack and increasing the lift produced by the wing. The wing tips will deflect upward as the lifting increases, increasing the bending loads in the wing. Loads in the aircraft's nose will cause the forward fuselage structure to be bent downward. If the pilot maneuvers the aircraft into a negative G maneuver, aircraft structural loads will be reversed. A generalized representation of these bending loads is shown in the Figure 35-9. The effects of torsional loads on the tail is shown in Figure 35-10. Bucking of fuselage skin during pilot induced positive G maneuvers is shown in Figure 35-11. Normally the resulting stresses will remain in the elastic range. If, however, the pilot abruptly applies nose up control at high speeds, plastic deformation and even catastrophic failure resulting from the bending load could occur. The effects of pilot induced negative G loads will be opposite to those shown.

If, however, the aircraft experiences G loads induced by wind gusts, the pattern of bending will change. The wing's angle of attack will increase without being forced down by the tail. In fact, if the gust is strong enough, the horizontal tail can experi-

ence a load which causes it to bend upward. The upload created by the horizontal tail will then cause an up-bending on the aft fuselage. Loads in the forward fuselage will, however, cause down-bending in the forward fuselage. The effects of gust induced positive G gusts are shown in Figures 35-12 and 35-13. The effect of gust-induced negative G loads will be opposite to those shown.

D. LIMIT LOAD VERSUS ULTIMATE LOAD. An aircraft structure is designed to withstand, without permanent objectionable deformation, a specified number of G's while the aircraft is at its design gross weight. In addition, the aircraft must be able to withstand a 50% higher G load without catastrophic structural failure. The number of G's the aircraft must withstand without permanent objectionable damage is referred to as the aircraft's "Limit Load Factor", or LLF. The number of G's the aircraft must withstand without catastrophic structural damage is referred to as the aircraft's "Ultimate Load Factor", or ULF. For example, an aircraft with a design gross weight of 30,000 lbs. and designed to a LLF of +3 G's and an ULF of 4.5 G's should have a wing designed to withstand a "Limit Load" (LL) of 90,000 lbs. without permanent objectionable and an "Ultimate Load" (UL) of 135,000 lbs. without catastrophic structural damage.

E. LIMIT LOAD FACTOR VERSUS GROSS WEIGHT. If an aircraft is loaded beyond its design gross weight (the military does this routinely), its allowable G load would be reduced so as not to exceed the limit load (LL) and ultimate load (UL). For example, if the aircraft described above were loaded to 45,000 lbs. its LLF would have to be reduced to +2 G's and its ULF reduced to 3 G's in order not to exceed the LL and UL. If, on the other hand, the aircraft were loaded to less than its design gross weight, in this case 30,000 lbs., neither the LLF nor the ULF could be increased to meet the LL and UL. That's because structures supporting fixed weights, like engine mounts, have been designed to support these fixed weights at the specified limit load factor without permanent objectionable damage and the specified ultimate load factor without catastrophic structural damage. Thus, if our hypothetical aircraft were loaded to only 20,000 lbs. it would still be limited to +3 G's LLF. This example should give us some ideas concerning how the location of structural damage can vary as a function of the aircraft's gross weight. If an aircraft operating above its design gross weight, initial structural failure of the wing structure

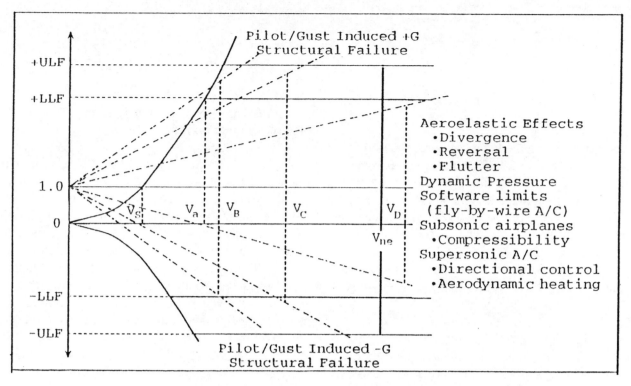

Figure 35-29. V-n Diagram. Structural Overloads
Associated with Exceeding the Operating Limits

is more likely. If, on the other hand, the aircraft is operating at very low gross weights where the wing is not as highly stressed, the initial structural failure during an over G is more likely to occur in structure supporting fixed gross weights.

F. MANEUVER SPEED VERSUS GROSS WEIGHT. The minimum speed at which an aircraft can generate enough lift to generate a G load equal to the aircraft's LLF, or abruptly move the controls to their extreme positions and not damage the aircraft, is referred to as the aircraft's "maneuver speed". Another way of thinking of the maneuver speed is the speed at which the line representing the stall speed a various G loads and line representing the LLF intersect.

If an aircraft is flying at a speed slower than the maneuver speed, the aircraft will stall before it can hurt itself. If the aircraft is flying faster than its maneuver speed and the pilot hauls full back on the pitch control, the aircraft is likely to get permanently bent. Full aft controls at just above the maneuver speed can generate G loads which exceed the LLF. Equation 1 provides an estimate of maneuver speed.

$$V_a = V_s \sqrt{LLF} \qquad (1)$$

Where:

$V_a =$ Maneuver speed
$V_s =$ One G stall speed
LLF = Limit load factor

All of these speeds and load factors are given at the aircraft design gross weight. When an aircraft is operated above its design gross weight its LLF must be reduced so as not to exceed the aircraft's limit load. When the aircraft is flown at weights below the design gross weight, the LLF cannot be increased because of the load limits established by structure supporting fixed gross weights. As an aircraft's weight is decreased below its design gross weight, its maneuver speed decreases but the LLF remains constant. As an aircraft's weight is increased above its design gross weight, its LLF decreases but its maneuver speed remains constant. These relationships are depicted in the matrix shown in Figure 35-15. Equation 2 shows how maneuver speed changes as weight is reduced. Equation 3 shows how LLF changes as weight is increased.

The maneuver speed at the lower gross weights

$$V_{a2} = V_{a1}\sqrt{\frac{W_2}{DGW}} \qquad (2)$$

$$LLF_2 = LLF_1\left(\frac{DGW}{W_2}\right) \qquad (3)$$

decreases because, at the lower weights, the stall speed at all G loads decreases. The reason the maneuver speed remains constant at the higher gross weights is a little more complicated. The stall speed for the higher gross weights increases for all G loads and the stall curve is shifted to the right (higher stall speeds). At the same time the LLF is decreasing. The mathematics works out so that the maneuver speed remains constant as the LLF decreases at the higher weights. Trust us.

The significance to the mishap investigator is simply this. At low gross weights, weights significantly below the design gross weight, it is possible to over G the aircraft at speeds below the "published" maneuver speed. If the aircraft is over G'd, damage is more likely in structure supporting fixed weights. When an aircraft is flown at weights above its design gross weight, it can be over G'd at G loads below its published LLF. However, in this case the damage is more likely in the structure that is carrying the extra weight.

During the above discussions we centered on maneuver speed, limit load factors and less than catastrophic damage to the aircraft. But, if speeds above Va are necessary to bend the aircraft, how much speed is necessary to rip the aircraft apart? The answer is the minimum speed at which you can pull enough G's to exceed the aircraft's ultimate load factor. This speed shows up on the V-G diagram as the intersection of the stall speed curve and the line representing the ultimate load factor.

G. GUST LOADS. When an aircraft runs into a gust of wind the airflow over the wing will change, changing both the aerodynamic force and pitching moment created by the wing. As a result of the encounter with the gust the aircraft will be accelerated up or down, experiencing a change in G loading. This change in G loading will be referred to as ΔG. The total G load on the aircraft during the excursion would be represented by the equation $G = 1 + \Delta G$. ΔG can be a positive or a negative value. The magnitude of the change in G loading imposed by the gust (ΔG) can be

estimated by using Equation 4.

$$\Delta G = \frac{0.115m\sqrt{\sigma}\,V_e\,(KU)}{W/S} \qquad (4)$$

Although Equation 4 may look a bit terrifying, it's really quite logical.

●m = the slope of the wing's lift curve. Remember what that means? Yeah, how many units of C_L do you get for each degree increase in AOA. Theoretical airfoils, the kind that are straight, infinitely long and have no wing-tips have a CL versus AOA slope of about 0.1 before airflow separation begins to occur. If you hit a gust with one of these things, and the gust increases the wing's AOA by one degree, the C_L goes up by 0.1. So who cares? Well, real wings have wing-tips and they are sometimes short and stubby (low aspect ratio) and/or are swept back. All of these things tilt the lift curve back, decreasing its slope. Now, if a short, stubby, swept back wing gets hit by a gust which increases its AOA by one degree, the CL may only increase by 0.05, half of what we had before. And, checking the equation, you should be able to see that if m is decreased, ΔG is also decreased. The bottom line is that straight, high aspect ratio wings are more sensitive to wind gusts. Low aspect ratio, swept wing aircraft are less sensitive to wind gusts. An aircraft with a low aspect ratio or swept wing might run through a gust and call it "moderate", while a high aspect ratio or straight wing aircraft encountering the same gust could get "severe" results.

●σ = density ratio. If the air is more dense you can expect it to pack a greater wallop. If the gust is encountered at high altitude, all other things being equal, the ΔG caused by the gust will be less. At 40,000 feet this can reduce the impact of the gust by a factor of almost four.

●V_e = equivalent airspeed. Equivalent airspeed is what your airspeed indicator would be showing if it didn't have to contend with position and compressibility errors. In other words, it is pretty much a measure of dynamic pressure. If the dynamic pressure of the air passing over the wing is higher, you can expect to get hit with a bigger bump. The faster of two otherwise equal aircraft running through the same gust will report the more severe jolt.

●U = Represents the vertical speed of the gust. The FAA requires

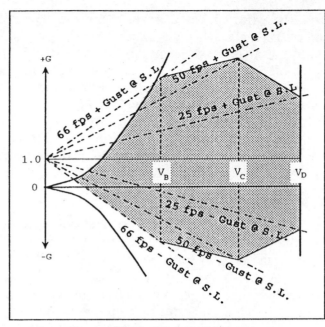

*Figure 35-30. Structural Tolerance
to Vertical Gusts*

aircraft to be structurally capable of withstanding a gust as high as ±66 fps at the aircraft's Design Speed for Maximum Gust Intensity (V_B), as high as ±50 fps at the aircraft's Design Cruise Speed (V_C), and as high as ±25 fps at the aircraft's Design Dive Speed (V_D).

• K = the "Gust Alleviation Factor". Since it takes some time for an aircraft to fly though a gust, the gust doesn't act on the whole aircraft at once. This delaying factor attenuates the effects of the gust and is a function of the aircraft design.

• W/S = the aircraft's wing loading. Since this term is on the bottom of the DG equation, high wing loadings mean a smoother flight. Lower wing loadings mean a larger ΔG for a given gust intensity. Wing loading can be changed by changing the wing's configuration (use of Fowler flaps or sweeping the wing) or by changing the weight. For most aircraft, changes in weight have the highest potential to change wing loading. Although effect of wing loading may superficially appear to mean that the heavier an aircraft is, the least likely it is to be over G'd by a wind gust, this is not necessarily true. More on this later.

H. DESIGN REQUIREMENTS. Since horizontal gusts do not cause significant changes in the G load experienced by an airplane, the design requirements associated with gust-induced loads are concerned with the effects of vertical positive (up) and negative (down) gusts. FARs 23 and 25 address the gust

magnitudes that an airplane's structure must withstand without damage. Referring to the equation for the ΔG, you can see that the effects of a strong gust can be offset by sufficiently low airplane speed, while at high speeds the effects of a given gust magnitude will be more severe. Thus the maximum gust an airplane has to withstand at low speeds is higher than the maximum gust it has to withstand at high speeds. When flying at their "design speed for maximum gust intensity" (V_B) between sea level and 20,000 feet, airplanes certificated under parts 23 and 25 have to withstand a vertical gust of 66 feet per second (fps) without structural damage. When flying faster than this speed, an airplane's ability to withstand severe vertical gusts is limited to lesser gusts. When flying at their "design cruising speed" (V_C), airplanes certified under parts 23 and 25 have to withstand a vertical gust of 50 fps without damage. This speed is often referred to as an airplane's "vee never operate" (V_{NO}) speed and is shown on the airspeed indicator as the top of the green arc. Pilots are cautioned to fly at speeds above V_{NO} only in smooth air. At an airplane's "design dive speed" (V_D) the airplane's structure is required to withstand, without damage, a vertical gust load of only 25 fps. Above 20,000 feet the maximum gust magnitudes for V_B, V_C, and V_D decrease linearly to 38 fps, 25 fps, and 12.5 fps respectively, at 50,000 feet.

I. FACTORS AFFECTING GUST LOADS. Figure 35-18 is representative of the aircraft only at its design gross weight (DGW). As an aircraft's weight decreases below the design gross weight, or for that matter, increases above the design gross weight, its response to vertical G loads changes. Referring to Equation 4, you can see that if the aircraft's weight is decreased and all other factors remain constant, the G load induced by a given vertical gust (ΔG) will increase. Similarly, if the aircraft's weight is increased above it design gross weight, the G load induced by that given vertical gust will decrease and the total G load on the airplane (1 + ΔG) will decrease toward a limit value of 1 G. It sounds as if it is safer (or at least smoother) to be at heavy gross weights when flying through turbulence, but life's not quite that simple. An airplane's tolerance to G loads will change as its weight changes. Remembering the earlier, the LLF will be constant below the aircraft's design gross weight, but will decrease toward a limit value of zero as gross weight increases. At low gross weights, it's the structure supporting fixed weights (e.g. engine mounts) that generally limit the G-induced loads. At high gross weights, the wings become the structurally limiting factor. These relationships can be seen in

Figure 35-18. At the low gross weights shown on the left side of the diagram, a given vertical gust will most likely overstress the structure supporting fixed weights. On the right side of the diagram the same vertical gust is more likely to overstress the wings. In the center of the diagram, at weights near the airplane's design gross weight, the airplane would experience overstress in neither the structure supporting fixed weights nor in the wings. The initial point of structural failure of an airplane being exposed to excessive G loads will, therefore, be a function of the airplane's gross weight at the time of the overstress.

Finally, it should be kept in mind that the FAA gust load requirements only apply to an aircraft in one G flight. They do not apply to aircraft in maneuvering flight. Thus, if an airplane in a level 60° bank turn runs into a 50 fps gust while flying at V_C, the airplane is likely to be over G'd.

J. GROUND LOADS. Ground loads are imposed on an airplane's structure during landing, braking, turning and in general while rolling over the ground. These loads can be structurally significant, especially when landing gear are attached to the wings, and can be the determining factor in the strength requirements of a structure. An airplane's landing gear and the structure to which it transmits loads are designed to dissipate the kinetic energy associated with the vertical velocity at touchdown as well as longitudinal and lateral loads associated with ground maneuvers.

K. LANDING LOADS. FAR 23 specifies the touchdown attitudes and maximum sink rates landing gear structure must withstand for aircraft weighing less than 12,500 pounds. The descent rate at touchdown is based on a formula involving the airplane's wing loading and residual lift at touchdown, but must be at least 7 fps, but no more than 10 fps. FAR 25 specifies the sink rate at touchdown for transport type airplanes as 10 fps when at the airplane's design landing weight and 6 fps at its design takeoff weight. Compliance with these criteria must be verified through drop tests. Since 10 fps translates to 600 feet-per-minute (fpm), touchdowns at these rates would certainly be classified as "hard" landings. Some military aircraft are designed to even tougher standards. USAF fighter and attack airplanes are designed to withstand landings at 18.5 fps (1110 fpm) while the USN requires the design of similar aircraft to withstand 22.5 fps (1350 fpm). It should be noted that the kinetic energy associated with landings is proportional to the square of the descent rate. Thus, USN fighter and attack airplanes have to absorb approximately five times the energy (i.e. $(2.25)^2$) of a similar airplane certified to FAR 23 or 25. While FAA certified transport type airplanes are designed to survive a drop test from as much as 18.7 inches at their design landing weight, USN fighter and attack aircraft have to be drop tested from over seven feet. That's a really hard landing. Also, remember that this is a one-time, dynamic-load test and frequent landings at these sink rates are beyond the scope of the design requirements.

In addition to the vertical energy associated with a landing, the landing gear and associated structure has to withstand spin-up loads which force it aftward, spring-back loads which force it forward and lateral loads which result from touchdowns in a crabbed attitude or with a side drift. Spin-up loads are due to the fact that when an airplane touches down the wheels are not turning. Within a few feet of touchdown a considerable drag force is generated, spinning the wheels up to a rotational velocity appropriate for the airplane's ground speed. During the interval from initial touchdown to the time the wheels are fully spun-up, the wheels are skidding. That's where all the rubber on the touch-down zone of the runway comes from. The spin up load is a significant design consideration. This load is often resolved into a separate drag strut which sometimes features an oleo device which attenuates peak loads by allowing exaggerated "stroking" of the gear in an aftward direction. Since this is a dynamic load, there will be oscillations. As the wheel spins up, the spin-up load will drop off and the landing gear will spring-back in a forward direction. The strong cyclic loading associated with spin-up and spring-back loads may generate a requirement for periodic overhaul and/or non-destructive inspections of the landing gear structure.

7. VIBRATION AND AEROELASTIC PHENOMENA.

Aircraft structure is relatively flexible. At least when compared to buildings, bridges and cars. You can often see aircraft structure deforming. When lifting off the ground the wing tips of a modern airliner will noticeably deflect upward. Strong turbulence will cause these same wing tips to bounce. Helicopter rotor blades flex noticeably when producing lift. These deflections are just the tip of the iceberg. Most inflight deflections of aircraft structure are not observable to the unaided eye. For example, torsional loads such as those developed when ailerons are deflected, will cause a wing to twist opposite to the direction in which the aileron is deflected. Although it is possible to decrease these deflections by increasing the aircraft's structural strength, the penalty would be increased expense and weight, and decreased payload, range, and other performance parameters. Instead, aircraft are designed so that although flight related loads do cause bending and twisting, the bending and twisting does not have unacceptable consequences. The trick is to design the structure to be just stiff enough so that the deflections resulting from flight loads are not excessive and do not degrade the aircraft's controllability, stability or strength. Three interrelated factors are involved: the forces attempting to defect the aircraft's structure; the stiffness of the structure resisting the force; and the difference between the load's frequency and the structure's natural frequency.

The difference between the load's natural frequency and the structure's natural frequency can be important source of repeated loads and fatigue cracking. The deflection in a structure which is being exposed to a dynamic load will have its defection amplified or dampened by a factor which is a function of the ratio of the loads frequency and the natural frequency of the structure being loaded. These relationship are shown in Equations 5 and 6.

$$d_{dynamic} = (d_{static})(Dmf) \qquad \textbf{(5)}$$

Where:

d = deflection
Dmf = Dynamic Magnification Factor

$$Dmf = \frac{1}{1-(f_1/f_2)^2} \qquad \textbf{(6)}$$

Where:

f_1 = Load Frequency
f_2 = Structure Natural Frequency

By examining Equations 5 and 6, you can see that as the load's frequency approaches that of the structure, the Dmf will become increasingly large. As the Dmf becomes larger, $d_{dynamic}$ becomes increasingly large. And, as $d_{dynamic}$ becomes larger, the stress in the structure becomes increasingly large. If the dynamic deflection is sufficiently large, the associated strain will also be large and sudden failure of the structure could occur. Even if the deflection is not large enough to cause sudden failure, the repeated deflections could cause low cycle fatigue cracking and progressive failure. The natural frequency of the structure in some

aircraft has resulted in the prohibition of prolonged cruising with the engines operating in a specified RPM range. Unauthorized shortening of the diameter of metal propellers has also caused the propeller's natural frequency (or a harmonic) to move into the engine's normal RPM operating range and subsequent fatigue cracking failure of the propeller.

A. CONTROL REVERSAL. When a control surface is deflected into the airstream, the total airfoil's camber is changed thereby changing the airfoil's $C_{Ma.c.}$ The resulting change in the pitching moment created by the airfoil will be airfoil nose-down if the control surface is deflected down and airfoil nose-up if the control surface is deflected up. If the airplane is operating within its approved operational limits and the torsional stiffness of the airfoil's structure has not been compromised, the airfoil's structure will resist the torsional force without objectionable twisting. However, if twisting does occur, the effectiveness of the control input will be at best reduced and under worst case conditions could result in an aircraft response in a direction (pitch, roll or yaw) opposite to that intended.

Let's look at an example. If an aileron is deflected downward, it's generally because the pilot (or autopilot) wants to raise that wing. The downward deflected aileron will normally increase the wing's camber, producing more lift and raising the wing. If, however, the nose down pitching moment (also caused by increasing the camber) causes the wing to twist downward, the lift produced can be less than expected. In fact, if the nose-down twisting of the wing is very excessive, lift can actually decrease and the wing will drop. This will probably result in loss of aircraft control.

There are two factors which can cause reversal. First, the wing may not be stiff enough. When a torsional load is applied to a wing tip it will twist about its elastic axis, which is normally somewhere close to 50% chord. Although modern aircraft certified by the FAA should not suffer from control reversal, structural damage, improper repairs or even weakening of the structure due to overheating could reduce its stiffness and increase its vulnerability to control reversal.

The second factor involves excessive airloads. If an airplane is operated at speeds above V_{NE} the torsional loads developed by control deflections may result in excessive airfoil twisting, degraded control and, if speeds are sufficiently high, control reversal. Figure 35-20 shows how control effectiveness can be degraded by excessive speed. Remember that the pitching moment created by an airfoil varies directly with the square of the airplane's speed. Although aileron reversal was a problem with some high speed, thin winged aircraft (most notably the B-47), the problem has been mostly solved by either using spoilers (which don't produce the objectionable pitching moments), or by using ailerons near the tip for only low speed operations when the flaps are deployed and smaller inboard ailerons for high speed operations. Although loss of some aileron control effectiveness was discovered during the flight test of one modern fighter/attack aircraft, structural stiffening and modification of leading edge devises corrected the problem before the aircraft went into production. Reversal is not a common problem, and if it occurs it will probably be due to in-service structural weakening or excessive speed. If reversal did occur, and torsional strain did not reach into the plastic range, there is likely to be little direct physical evidence in the wreckage. Discovery of damage which both existed prior to ground impact and which could have degraded torsional stiffness is one exception. Another possible source is flight data recorder information which shows rolling or pitching opposite to the direction of control surface displacements.

B. DIVERGENCE. Divergence, like reversal, finds its source in either insufficient torsional stiffness or excessive dynamic pressure. Unlike reversal, which may provide some warning through decreased control effectiveness, divergence will result in sudden and catastrophic torsional wing failure. Consider the relationship between the wing's aerodynamic center, located at approximately 25% chord when the airstream is subsonic, and the wing's elastic axis, which is often located just forward of 50% chord. If the wing is exposed to an upward vertical gust, lift will increase, increasing the nose up torsional force on the wing. This will tend to twist the wing to a higher incidence angle, further increasing the wing's angle of attack, lift and the nose up torque. If the wing lacks sufficient torsional stiffness, or if dynamic pressure is excessive, this process of more lift causing more twist which causes more lift which causes...well you get the idea. The wing is literally twisted up and off the airplane. Although the FAA certification process should identify any design weaknesses which could result in divergence, failure to maintain the torsional stiffness of the wing or operation at speeds in excess of the airplane's limiting airspeed could result in wing divergence.

Evidence of divergence includes indications of inflight break-up with the wing separating first. In addition to evidence of overload in bending, the wing would show evidence of excessive leading-edge-up torsion.

C. FLUTTER. Although flutter is more complicated than either of the two other aeroelastic phenomena, it is actually much more common as a source of airplane accidents and incidents. We left it to last not because it was the least important, but because it is the most difficult to understand. By first explaining divergence and control reversal we hope you will be better able to understand the concept involved in flutter. Flutter requires one more ingredient be added to the issues of inadequate torsional stiffness and excessive dynamic pressure associated with control reversal and wing divergence. Inertia. Under the right combination of these three factors, self-excited oscillations of any airfoil (i.e. control surfaces, wings, stabilizers or propeller blades) can result in repeated cyclic loads and deflections of the airfoils.

If divergent flutter occurs the loads can rapidly build to levels sufficient to cause structural failure of the airfoil and in some cases the destruction of the airplane. Even if the loads are not severe enough to cause immediate overload failure, the cyclic stresses can cause low cycle fatigue. Repeated loads can also drive the control surfaces to displacements which can prevent the pilot's control of the airplane. Cockpit vibrations can be severe enough to prevent the reading of instruments. Bad stuff, this thing called flutter. It can rip an airplane to pieces, it can make it uncontrollable, or both. Let's see how it can happen.

Remember that, in addition to the factors of low stiffness and/or high aerodynamic loads, we need to add just the right inertia to produce flutter. Inertia is a function of an item's total mass and the distribution of the mass. Inertia helps determine the natural frequency of a structure and any change in its total mass or distribution can change its frequency. If a cyclic load is applied at or close to a structure's natural frequency, and the load is a sufficiently large displacement (straining) of the structure, the stage is set for flutter.

Let's look at an example. Referring to Figure 35-24, you can see that our sample wing has an aerodynamic center at approximately 25% chord and an elastic axis just forward of 50% chord. Assume that a sudden gust causes the wing to bend and twist upward similar to the manner we saw in wing divergence. Only this time, the wing is stiff enough to withstand the static load. However, gusts are a dynamic load and the wing will both bend and twist past the displacement it would have achieved under a static load. As a result it will spring back towards its original (pre-gust) displacement. If the variation in aerodynamic loads as the wing bends and twists, the resistance to bending and twisting and the bending and torsional inertia are just right (or wrong if you're the pilot), the wing will be displaced downward in bending and twisting only to spring upward again.

After being excited by the gust, the cyclic twisting and bending can be self sustaining, maintaining a roughly constant amplitude or even experience a steady growth in amplitude. Once these cycles begin, the only way to stop them is to: change the inertia of the structure (that's impossible to do in the air); increase the structure's stiffness (attempting to freeze the stick/yolk and rudders in a constant location may help with a non-powered flight control system); or quickly slowing down (the test pilot's option when encountering flutter on a new airplane).

Improperly balanced flight controls and looseness in the control linkages can also induce flutter. Consider the flight control on the airfoil shown in Figure 35-25. The flight control's center of gravity is located aft of its hinge line. If the airplane flies into an up-gust and experiences an upward acceleration, the control surface will naturally want to move down.

First let's look at a situation where only the control surface flutters. When the control surface moves down, the low pressure area on the top of the wing increases near the trailing edge of the wing and a high pressure area forms under the control surface. Both the low pressure on top and the high pressure below tend to force the control surface back up toward its original position. If the control surface's center of gravity is too large or is too far aft of the hinge line, its inertia can move it up past its original position. Again, if the variation in aerodynamic pressure as the control surface deflects, the resistance of the control linkage to uncommanded control surface displacements and the inertia of the control surface's mass are just right, the control surface will be displaced downward only to spring upward and downward again. The forces created by this action can exceed the ability of the pilot to control them and loss of airplane control could result. The loads may also be large enough to cause the control's surfaces to separate from the aircraft due directly to overload or indirectly due to low cycle fatigue.

Now, let's look at how a fluttering control surface can set the airfoil to which it is attached to fluttering. As the control surface is deflected downward by the upward movement of the airfoil, the airfoil will create more lift, magnifying the upward acceleration. If the control's linkage is loose (or if the force on the surface causes excessive straining of the linkage), the control surface will be further displaced, accelerating the airfoil's upward movement. Propelled by a dynamic load, the airfoil will overshoot the displacement associated with a static load and spring back toward its original position. As the airfoil reverses direction, the inertia of the control surface will cause it to move to an up position, now driving the airfoil downward. If the frequency of the fluttering aileron is close to that of the airfoil's structure (or a harmonic) the dynamic magnification factor will be high and the dynamic deflections of the airfoil can grow. Once more, if the aerodynamic pressures developed by the deflected control surface, the resistance of the control linkage to uncommanded control surface displacements and the inertia of the control surface's mass are just right, the airfoil will be driven into a series of self sustaining bending and twisting oscillations.

Figure 35-28 shows what can happen if there is excessive looseness or "play" in the flight control linkage. In this case, a change in the airflow causes a trim tab with excessive play in its connecting rod to deflect. The resulting change in pressure on its surface causes it to be deflected in the opposite direction. The proper combination of inertia, airload and stiffness can result in a fluttering trim tab. If the frequency of this flutter matches the natural frequency of the control surface, it, too, could begin to flutter. If not, the airplane will probably just lose its trim tab.

The examples discussed are just that; examples. There are other flutter modes, too numerous to discuss in this text. Just remember that flutter depends on the integration of three different factors: aerodynamic loads, structural stiffness and inertia. Since FARs dictate that an airplane design's resistance to flutter be verified through both analyses and test, why do we find examples of flutter-induced accidents? First, airplanes are flight tested at speeds up to their design dive speed (V_D), but no promises are made above this speed. In fact, because of the normally stringent requirement surrounding the expansion of the flight envelop up to V_D, V_{NE} is normally set as a percentage of V_D. Improper or imprudent handing of an airplane can result in V_D being exceeded and aerodynamic loads imposed well in excess of those which the design can accommodate. Second, the design's adequacy is verified with the control surface's total mass and mass distribution in accordance with the approved design. Ice which forms in the trailing edge of a control surface when rainwater collects and freezes or incorrect structural repairs can both alter the control surface's mass and its center of gravity, changing its inertia characteristics. Even an improperly applied paint job can place the inertia in the critical area. Some aircraft procedures mandate that flight controls be "balanced" after every paint job or their "top" airspeeds be severely constrained. Finally, the stiffness of control surfaces and airfoils can be degraded to the point where flutter can be induced. Both primary and secondary control surfaces have a maximum amount of "play", which if, exceeded, can induce flutter. Progressive failures such as fatigue cracking, corrosion and wear can reduce a structure's stiffness, as can improper repair or even improper assembly of the original structure.

Although the issues just discussed will direct you toward the root causes of flutter, how about some clues you can use at the crash site to determine if flutter should be considered as an event in the mishap sequence.

Look for evidence of loss of control, separation of control surfaces or airfoils prior to ground impact, repeated cyclic bending and/or torsional loads and low cycle fatigue involving stress/strain well into the plastic range. Although evidence of excessive pre-mishap speeds may be hard or impossible to find in the smoking hole, flight data recorders and Air Traffic Control radar tapes may provide clues. And, although loose flight control linkages and excessive paint may be masked by the impact and fire damage, a review of maintenance records and procedures can provide background evidence of conditions which could have contributed to flutter.

8. V-N DIAGRAM.

The V-n diagram is a useful device to explain operating limits of an airplane in terms of aerodynamic stall, load factor and airspeed. The left hand side of the airplane's operating envelope is defined by the airspeeds at which the airplane will stall at various G

loads. This line represents Equation 7.

$$V_s = V_{s1}\sqrt{G} \qquad (7)$$

Where:
   V_s = Stall speed being calculated.
   V_{s1} = Stall speed at 1 G.
   G = G load at which V_s is being calculated.

The pilot cannot maneuver the airplane to G loads above this line because of the airplane's aerodynamic limitations (it stalls).

The Ultimate Load Factor (ULF) and Limit Load Factor (LLF) lines define the upper limit G to which a pilot can maneuver the airplane without experiencing structural failure. The airplane is aerodynamically capable of generating G loads above these limits, but can suffer permanent objectionable structural damage if operated above its LLF and catastrophic structural damage if operated above its ULF. The speed associated with the intersection of the aerodynamic limit line and the LLF line represents the minimum speed at which an airplane can generate sufficient G loads to damage the airplane. When operated below this speed, the airplane will stall before it generates sufficient G load to cause permanent objectionable damage to its structure. This speed is referred to as the airplane's Design Maneuvering Speed (V_A). V_A may not be less than the airplane's one G, flaps up stall speed. The speed associated with the intersection of the aerodynamic limit line and the ULF line represents the minimum speed at which an airplane can generate sufficient G loads to cause catastrophic structural damage to the airplane. When operated below this speed, the airplane will stall before it generates sufficient G load to cause catastrophic damage to its structure. Similar limits exist for an airplane's negative limit and ultimate load factors. These speeds are important to an investigation which is focusing on an inflight break-up. If the speed at which the failure occurred was within these limits, then pre-existing damage or other factors were involved.

The right side of the airplane's operating envelope is defined by its limit airspeed. If an airplane is operated above this speed, any number of bad things can happen. From a structural standpoint, the higher than intended dynamic pressures can cause aeroelastic problems (divergence, control reversal and flutter) and high static pressures near stagnation points.

An airplane's designed-in structural tolerance to

vertical wind gusts is shown in Figure 35-30. The straight lines originating at one G and zero knots show the total G load imposed on an airplane's structure by vertical wind gusts of 66 feet per second (fps), 50 fps and 25 fps, as a function of the airplane's equivalent airspeed. Referring back to Equation 4, you can see that the ΔG imposed by a given vertical gust varies directly with the airplane's speed. An airplane's structure is designed to withstand, without damage, a vertical wind gust of 66 fps when flying at or below its Design Speed for Maximum Gust Intensity (V_B), 50 fps when at or below its Design Cruising Speed (V_C) and 25 fps when at or below its Design Dive Speed (V_D). Gust intensities are gradually reduced when airplanes are operated above 20,000 feet.

9. SERVICE LIFE (PROGRESSIVE FAILURE) ISSUES.

When a baby is born, the genes in their body determine a maximum calendar life to which the adult can be expected to live. Baring premature demise due to accident or illness and given proper maintenance of the body through nutrition and exercise, the body should live out that life. Mechanical structures also have an expected life, which is referred to as the structure's "service life." A mechanical structure's service life is not so much a function of its calendar age but is more a function of the number and magnitude of loads to which the structure is exposed and the environment in which it was used during its life. The

most common service life issues addressed by the aircraft design engineer are fatigue, corrosion, wear and creep. Of the four, fatigue and corrosion are arguably the most important for aircraft, because of the increasing age of the world's aircraft population. For general aviation aircraft which often sit unattended, unused and unwashed for long periods of time, corrosion is normally of greater concern. For commercial airplanes which are in the air, making money, the ground-air-ground cycles can pile up quickly and fatigue cracking becomes a greater concern and has been directly involved in a number of dramatic airplane crashes. In 1994 approximately 50% of the U.S. Air Force's aircraft were more than 18 years old. The U.S. civil airline fleet had about 33% of the aircraft at over 20 years old. The general aviation fleet, deprived of their source of "new blood" with the cessation of production of most types during the 1980s, is even older. Fatigue and corrosion is an increasing concern in this "aging" fleet of aircraft. As components approach the end of their service life, failure rates increase. This also increases the importance of inspections designed to catch impending failure and repair or replace excessively aged parts.

10. FATIGUE.

Structural failures induced by fatigue are the most serious because they are the common mode of fracture of modern mechanical equipment and because they can occur during an interval that is considered to be within the structure's normal service life. Like high altitude hypoxia, fatigue cracks are insidious and provide little or no warning of impending failure.

The human is a very resilient structure with respect to fatigue. Even when worked within a inch or two of life, and then given adequate rest, food and lots

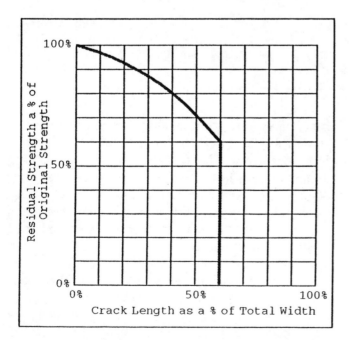

Figure 35-31. An Example of Residual Strength as a Function of Crack Length

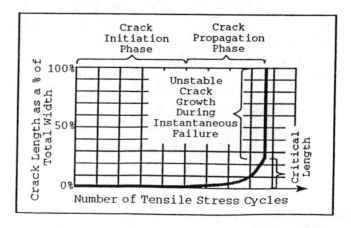

Figure 35-32. Phases of Fatigue Crack Growth

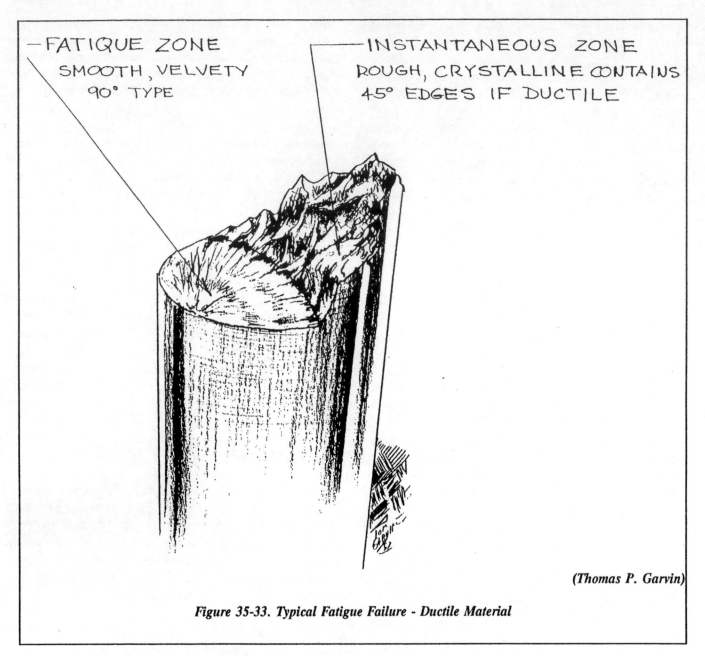

FATIQUE ZONE
SMOOTH, VELVETY
90° TYPE

INSTANTANEOUS ZONE
ROUGH, CRYSTALLINE CONTAINS
45° EDGES IF DUCTILE

(Thomas P. Garvin)

Figure 35-33. Typical Fatigue Failure - Ductile Material

of water, the mind and body can spring back to good as new, perhaps even better. With respect to humans, the effects of physical and mental fatigue are reversible. With respect to metal structures things are different. The microscopic and macroscopic damage occurring during metal fatigue are not reversible. Not only can't the damage be rectified, but, given the same kind of usage, the rate at which the damage occurs is constantly accelerating. Things are getting "worser faster."

Where aging of the human is primarily associated with the calendar, aging of a mechanical structure (with respect to fatigue) is primarily related to the number and intensity of the load cycles the structure

has been exposed to. If an aircraft is placed in proper storage for, let's say ten years, very little, if any, of its fatigue life will be used up. On the other hand, if a plane is kept in constant service, it can use up its "fatigue life" while it is still relatively "new". Thus, when pilots buy used airplanes they are interested in not only its year of manufacture, but the number of flight hours on the airframe. For pressurized airplanes, the number of ground-air-ground cycles (number of missions) is of even greater importance. Like humans who "fail" at well below their "normal" level of performance when fatigued, structural components which suffer from fatigue cracks can fail well before they reach their "limit loads". Not long ago, five wing sections, which had been removed from various

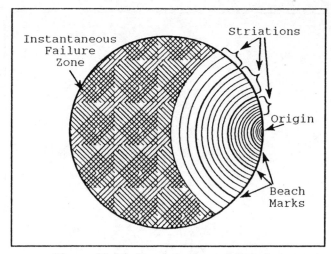

Figure 35-34. Beachmarks and Striations

aircraft because they had reached the end of their "service life", underwent static load tests to determine if they still met their original strength requirements (i.e. no permanent objectionable deformation at 100%

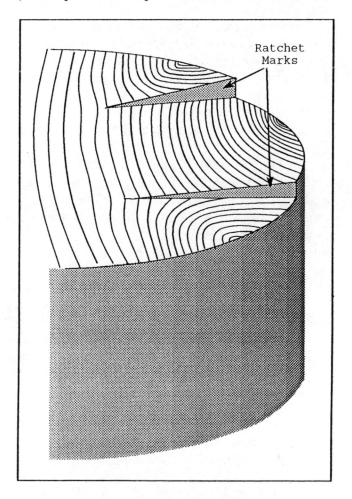

Figure 35-35. Beach Marks, Striations and Rachet Marks

of their limit load and no catastrophic failure prior to reaching 150% of their limit load). Two of the wings failed catastrophically prior to reaching their limit load. Two failed catastrophically before reaching their 150% of limit (ultimate load). Only one wing lived up to its design strength requirements and catastrophically failed at a load exceeding 150% of its limit load. Fatigue cracks can cause premature failures.

A. GROWTH OF FATIGUE CRACKS. Fatigue crack-induced failures go though three different crack growth phases. The first phase, often referred to as the crack initiation phase, involves the creation of many microscopic cracks as the ductile material slips along shear planes, 45° to the tensile stress. Detection of microscopic cracks require magnifications of greater than 25-50 times, and generally can be detected only through destructive testing. During the second phase of fatigue crack growth, often referred to as the crack propagation phase, macroscopic crack(s) develop along a plane perpendicular to the local tensile stress. Macroscopic cracks require magnifications of less than 25-50 times, and generally can be detected with the use of various non-destructive testing techniques. These cracks grow in short spurts as stress increases during tension stress cycles.

As the crack grows in size, the stress intensity along the crack tip increases while the structure remaining to carry the load decreases. During this second phase, crack growth is said to be "stable" or self-stopping. The third stage is set when the crack reaches a "critical length"; the length at which the uncracked structure cannot support the applied load. During the third and final phase crack growth becomes "unstable" or self- sustaining, with the crack speeding through the remaining material at sonic speeds.

The last two phases of crack growth are docu-

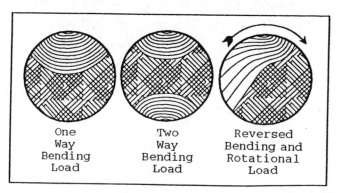

Figure 35-36. Beach Marks and Striations Resulting from Cyclic Bending Loads

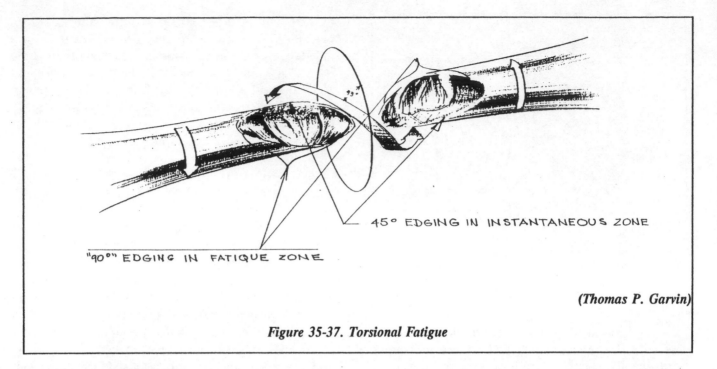

45° EDGING IN INSTANTANEOUS ZONE

"90°" EDGING IN FATIQUE ZONE

(Thomas P. Garvin)

Figure 35-37. Torsional Fatigue

mented in the physical features of the fatigue fracture surface. There will be at least one portion of the fracture surface which shows two of the classic characteristics of a brittle fracture; little to no plastic deformation and a fracture surface perpendicular to the local tensile stress. However, the true brittle failure's granular surface with its bright shiny facets will be replaced by a smooth, velvety surface, often with extremely closely spaced concentric rings. This portion of the fracture surface documents the crack propagation phase. The third phase of the crack growth will show up as either a fast (brittle) fracture surface, with its rough, granular surface (often with chevrons pointing back to its origin at the fatigue crack), or a ductile failure with its 45° fracture surfaces, necking down and dull, fibrous texture. Thus, the surface of a fatigue fracture will show two distinct areas, one documenting the growth of the crack(s) and one documenting the instantaneous failure. If the fatigue cracks originated in more than one location and grew to significant size, there will be more than two distinct surfaces.

B. BEACHMARKS AND STRIATIONS. Just as the growth rings within the trunk of a tree document the trees annual growth, fatigue cracks document their growth pattern during the crack's propagation phase. Unaided visual examination of the crack surface will often reveal small, ring shaped ridges which extend across the fracture surface made by the advancing fatigue crack. These "Beachmarks" are unique to fatigue fractures and go by many other names such as stop marks, clamshell marks or arrest marks. Beachm-

arks may be visible to the naked eye, but under certain conditions may require magnification of 10 to 20 times for positive identification. Although beachmarks are related to the individual fatigue crack advances, they actually document interruptions in the crack's progress and are caused by changes in loading which result in work hardening of the crack tip and pauses in the load cycles which allow corrosion of the material at the tip of the fatigue crack. Therefore, if a component were exposed, without interruption, to uniform cyclic loads it may not show any beachmarks. This situation is typical of some laboratory experiments. Microscopic examination of a fatigue fracture surface (magnification of more than 20 times) will show that between two beachmarks there may be thousands of smaller, microscopic ridges. These microscopic ridges identify the step-like progress of the advancing fatigue crack and can provide the investigator valuable information on when the crack started and its size during any inspections that were performed prior to the component's failure. These microscopic cracks are referred to as "striations". Since striations are exceeding small ridges on the fracture surface, abrasive actions such as fitting two surfaces together to see it they "fit" can cause irreversible damage to the surface, erasing the ridges forever. If you have even the slightest suspicion that a fracture was caused by a fatigue crack, protect it from both additional physical and chemical damage.

Ratchet marks is another term used to describe the surface features of a fatigue fracture. A single large fatigue crack may be traceable to origins at several smaller fatigue cracks. If several small fatigue cracks are growing in roughly the same plane and are adjacent to each other, they will eventually overlap. When overlapping occurs, the metal between the two cracks will fail, allowing the two small cracks to join up into a single larger crack, with a perpendicular ridge showing where the two were joined.

The type of loading a fractured component was exposed to can often be identified by examining the beachmarks and striations on a fatigue fracture surface. While cyclic tension loads can result in a fatigue crack starting anywhere on the circumference of a solid shaft, one way cyclic bending will always have the fatigue crack starting on the surface experiencing the maximum tension stress. Two way cyclic bending can have fatigue cracks starting on both of the sides which are in tension. However, because the crack initiation phases may be different on the two sides, the size of the crack propagation zones on the two sides may also be different. Rotational bending loads are created when a bent shaft is rotated. Imagine a shaft which is loaded in bending, then being rotated. As the shaft rotates, the location on the circumference being exposed to the maximum tension stress changes, moving opposite to the direction of rotation. In this way, the cyclic tension stress necessary for fatigue moves in a way which results in the unique pattern shown in Figure 35-37.

Cyclic torsional loads which produce fatigue cracks will produce an unusual fracture surface. Since torsional loads on a cylinder result in maximum tension stresses along a spiral 45° to the shaft's longitudinal axis, fatigue cracks will develop along this 45° spiral. If the shaft has no significant stress concentrations, the crack will form in a direction similar to that formed when a solid shaft made of brittle material is overloaded in torsion. The exception will be the fatigue crack's fracture surface which will show beachmarks. If, however, there is a circumferential stress concentration, ratchet marks will connect a series of 45° fatigue cracks.

Reversed cyclic torsion loads result in two sets of cyclic tension stresses oriented 90° to each other and therefore will tend to have two sets of fatigue cracks which are 90° to each other. The net result is often two 45° spiral cracks, originating in an "X'ed" pattern.

The intensity of a load can alter the beachmark and striation patterns on the surface of a fatigue fracture. In addition to increasing the distance a crack advances during each stress cycle and therefore the

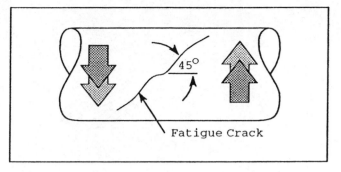

Figure 35-38. Cyclic Torsion Load-Induced Fatigue Crack in a Solid Shaft

spacing between the ridges of adjacent striations, the higher stress levels will decrease the size of a "critical" crack. In other words, the size of the instantaneous failure zone will increase as load intensity increases and decrease as load intensity decreases.

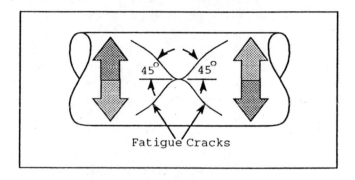

Figure 35-39. Reversed Cyclic Torsion Load-Induced Fatigue Crack in a Solid Chaft

Stress concentrations can effect the growth of fatigue cracks, and therefore influence the shape of striations and beachmarks. For instance, if a stress concentration existed around the surface of a solid shaft being exposed to one-way cyclic bending, the

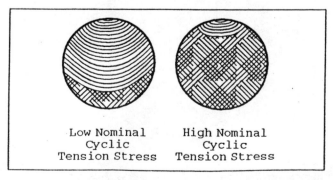

Figure 35-40. Effect of Stress Intensity on Beach Marks and Striations

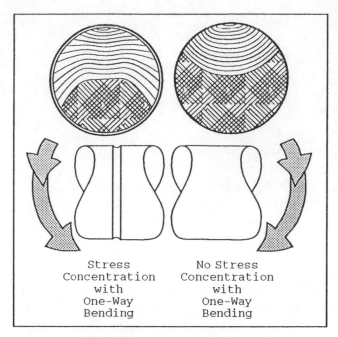

*Figure 35-41. Effect of Stress Concentrations
on Beach Marks and Striations*

concave growth pattern associated with one-way bending would be replaced by a pattern which became convex. This is due to the stress concentration at the surface which caused the fatigue crack to advance further during each stress cycle.

Four conditions are necessary for fatigue cracks to develop and grow: a material which is prone to fatigue cracking, tension stress; the stress must, at least locally, reach the plastic range; and the stress must vary cyclically in its intensity. Notice that time, in and of itself, is not included in this list, although time is required to accumulate the cyclic stress cycles necessary to create fatigue cracks and then make them grow.

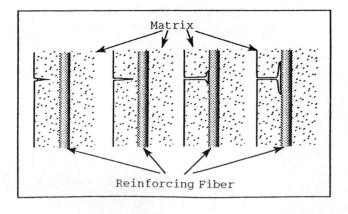

*Figure 35-42. Crack Stopping
in Fiber Reinforced Plastics*

Let's look at these four factors, keeping in mind that like chicken noodle soup (which can't be made without chicken and noodles), fatigue cracks can't be made without these four ingredients. Eliminate one and you have eliminated fatigue cracking.

The first requirement for fatigue cracking is that the material must be prone to fatigue cracking. Not all materials are equally vulnerable to fatigue cracking. Brittle materials tend to fracture suddenly long before developing significant fatigue cracks. The fibers in fiber reinforced composite materials tend stop fatigue cracks before they develop any significant length. Even within the various materials which can develop fatigue cracks there are a wide ranges in "fracture toughness", a material's natural ability to resist stress in an already cracked material. We measure fracture toughness by multiplying stress times the square root of the crack length. Materials with a high fracture toughness allow a crack to grow only when exposed to a high combination of stress and existing crack length. In materials with high toughness, the fracture energy (measured in terms of lbs/in) can also be measured. An interesting feature of fracture energy is that it tends to rise as a crack grows; a phenomena that helps explain why fatigue cracks are self-stopping and tend to grow in small, stable steps until the uncracked structure is insufficient to prevent unstable crack growth.

Materials such as fiber reinforced plastics (one of several types of composite materials) have a natural propensity to stop fatigue cracks. In the illustration shown in Figure 35-42 you can see how a crack moving through the matrix material will eventually (soon rather than later) encounter a fiber. The crack will not propagate into this new medium but will instead divert itself along rather than through the fiber. As you will see later in this chapter this "blunting" of the crack's tip will reduce the stress intensity and stop its growth.

One of the questions an investigator must ask when researching the circumstances surrounding a fatigue crack-induced failure is the choice of materials. "Did it have adequate fracture toughness?"

The second requirement for fatigue cracking is that the structure must be experiencing tension stress. Although tension stress is the primary stress when a structural component is loaded in tension, tension stress will exist as a secondary or component stress when a component is loaded in shear, bending or

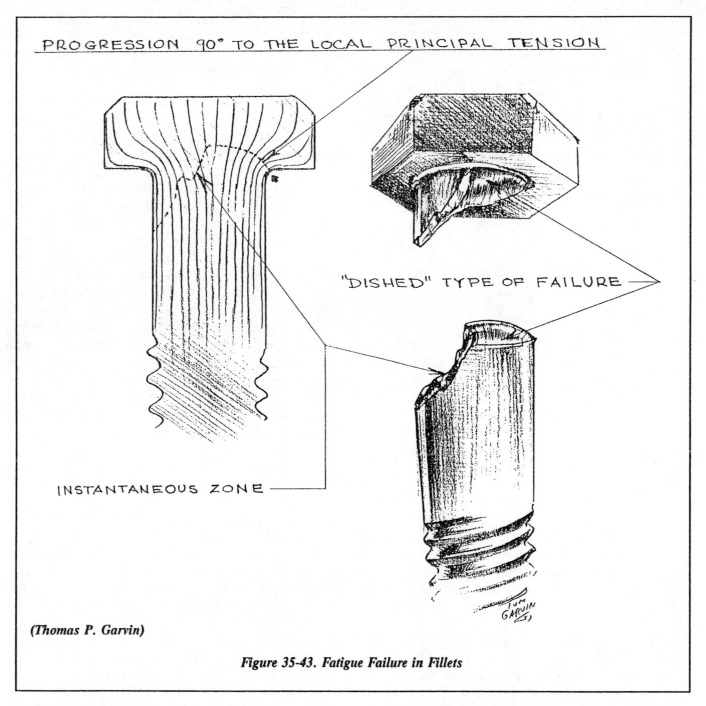

(Thomas P. Garvin)

Figure 35-43. Fatigue Failure in Fillets

torsion. As mentioned earlier, the fatigue crack will propagate perpendicular to the local tension stress. It should be pointed out that the local tension stress may not be parallel to the applied tension load. Discontinuities in the structure geometry can cause local tension stress to be diverted or "flow" around corners and cutouts. For example, in the structure shown in Figure 35-43, the lines in the structure representing the "flow" of tension stress will be the same as streamlines representing the flow of an airstream around obstructions. When using streamlines to represent an airstream, lines which get bunched closer together represents an increase in the speed of the local air-

stream. Similarly, when the stress lines get bunched closer together the magnitude of the stress is increasing. Thus, the magnitude of the tension stress will be higher. With the geometry shown in Figure 35-43, a fatigue crack forming at point A would grow along the curved path shown by the dotted line, remaining perpendicular to the local tension stress.

The third factor which must occur is that stress must, at least locally, reach the plastic range. It would appear that a simple solution is to ensure that the nominal (average) stress in structure is kept in the elastic range, below its proportional limit, but life isn't

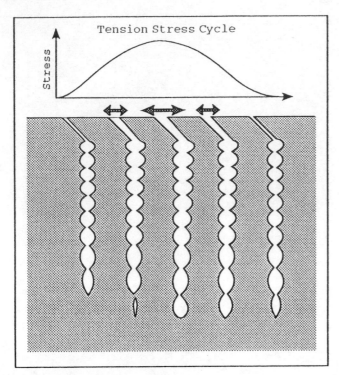

Figure 35-44. Fatigue Crack Growth During a Tension Stress Cycle

so simple. Through our discussion so far, we have assumed that tension stress is equally distributed across the area carrying the load. A quick peek back to Figure 35-43 should reveal that, like the old song goes, "It ain't necessarily so!". As the tension stress lines flow around the corner from the left to the right, they tend to get closest together near the radius, while near the structure's center the stresses lines are further apart. The stress has been "concentrated" near the radius, while the structure's center has been "off-loaded." So while the average or nominal stress across the area B-B' may be within the material's elastic range, the material near the surface could be well into its plastic range. Just how stress can be "concentrated" to many times its nominal value will be discussed later in this chapter. For now let's leave it with the notion that, it is not hard to achieve "local plastic stress".

The fourth factor which must occur for fatigue cracking to exist is that stress must vary cyclically in its intensity. Fatigue cracks grow only during tension stress cycles. Although the exact grow mechanism is still debated (its hard to get an eyeball into the tip of these cracks to see exactly what's going on), one theory is shown in Figure 35-44. As the tension stress increases, the material in front of the crack tip fails in shear, allowing the crack tip to advance, then blunting the crack tip and stopping its growth. When the

tension stress is relaxed, tip crack closes, returning the crack tip to its original pointed edge. In the next section we will see how this sharp crack tip significantly increases stress concentrations.

Fatigue cracks and failures are often characterized by the environment which in which they were created. Large, complex structures involving drilling, machining and welding, (e.g. airplanes) contain cracks when they roll out of the production hangar. Although the avoidance of these cracks is one of the objectives of any production line, inspection techniques can only detect cracks which exceed specific lengths. In addition, inspections cost money and extend the time required to build an aircraft. As a result, aircraft manufacturers assume that cracks smaller than the lengths which can be detected during planned inspections exist in various parts of the structure and that crack growth will start in the crack propagation phase. Issues which focus on the propagation phase take priority. On the other hand, for some small, individual components such as drive shafts, gears and axles, a crack-free surface is often assumed, which leads to the assumption that fatigue cracks must go through the lengthy crack initiation phase.

Fatigue cracking is also categorized by the stresses which initiate and/or propagate the crack. If stresses are high enough to reach the plastic range and result in general yielding of the material, failures can occur in less than approximately 10^4 cycles. This is referred to as "low cycle" fatigue. The number of cycles to failure is relatively low because the stress level is relatively high (e.g. above yield). On the other hand, if nominal stress levels can be kept below the plastic range, the number of cycles to failure will be higher. The boundary between low cycle and high cycle fatigue is somewhat arbitrary, and is primarily determined by whether or not nominal stress levels have extended into the plastic range. While high cycle fatigue is normally associated with low amplitude, high frequency vibrations which occur over a long period of time, low cycle fatigue is often characterized by much lower frequency loads with higher stress levels. Abnormal operations such as flutter and out-of-balance engine operations can also cause high amplitude oscillatory loads which can lead to low cycle fatigue failure.

When the cyclic stresses which cause fatigue are the result of residual thermal stress and not mechanical stress, the fatigue is referred to as "thermal fatigue". When a component heats and cools, the internal metal rarely does so at a uniform rate. Some parts heat faster than others. When cooling, the metal that cools last will have a residual tension stress. Thermal fatigue cracks will grow each time the component goes through a heating/cooling cycle. The crack propagation rate can be increased when the oxidized metal at the crack tip (formed when the crack is held open at elevated temperatures) acts as a wedge when the crack closes.

C. FACTORS WHICH INFLUENCE FATIGUE CRACK GROWTH.

The fatigue life of structures undergoing cyclic loads is predicted as a function of the expected stress cycles the structure is expected to experience and the environment in which the load cycles are applied. As mentioned earlier, fatigue cracks go through three different phases of growth. Changes which result in differences between the actual load cycles and those predicted can both shorten the crack initiation phase, speeding the arrival of macroscopic cracks and shorten the length of the crack propagation phase, speeding the crack's growth

to critical length. Aircraft accident investigators need to have some understanding of the factors which can shorten the fatigue life of a structure.

Stress concentrations result when changes in the geometry of a structure cause local stresses to exceed the nominal stress. For example a structure with a one-half inch by ten inches cross-section would have a nominal stress of 20,000 psi if uniformly loaded in tension to 100,000 lbs. (See Figure 35-45). If, however, the structure was actually one-half inch by eleven inches with a one inch hole in the center, (see Figure 35-45) the stress at the edge of the hole could actually, if the material remained in its elastic range, increase to 60,000 psi, while the nominal (average) stress remained at 20,000 psi.

The ratio of the actual tension stress to the nominal tension stress is referred to as the elastic stress concentration factor (K_t) and is determined by the structure's geometry. Another way of expressing this relationship is through the Equation 8.

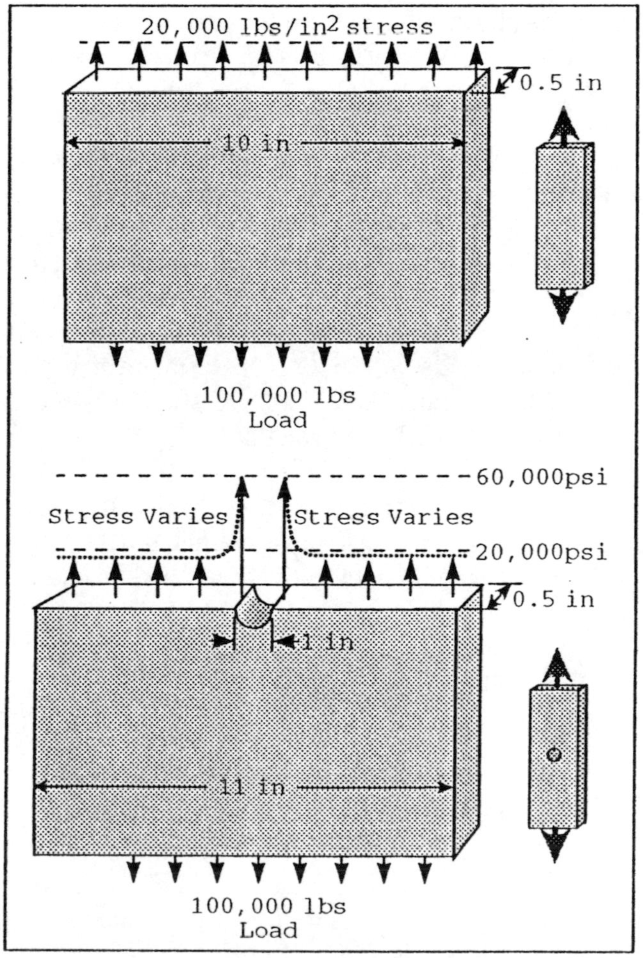

Figure 35-45. Stress Concentration Due to a Hole in a Plate

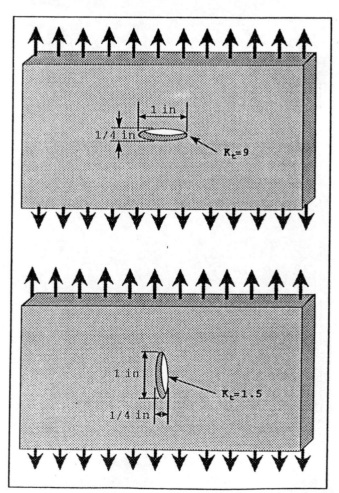

Figure 35-46. Effect of Hole Orientation on Elastic Stress Due to Elliptical Holes

$$\sigma_{max} = K_t\sigma_{nominal} \qquad (8)$$

Where:

σ_{max} = Maximum local tension stress.
K_t = Elastic stress concentration factor.
σ = Nominal tension stress.

Any time there is a change in the cross-sectional area carrying a tension stress, there will be an appropriate K_t. The general equation for the maximum elastic stress concentrations caused by elliptical holes in wide panels is represented by Equation 9.

$$K_t = 1 + 2\left(\frac{a}{b}\right) \qquad (9)$$

Where:

K_t = Elastic stress concentration.

a = Hole dimension perpendicular to the tension stress.

b = Hole dimension parallel to the tension stress.

For a perfectly round hole, then, the dimensions "a" and "b" are equal, K_t is equal to 3 and the orientation of the load creating the stress is not important.

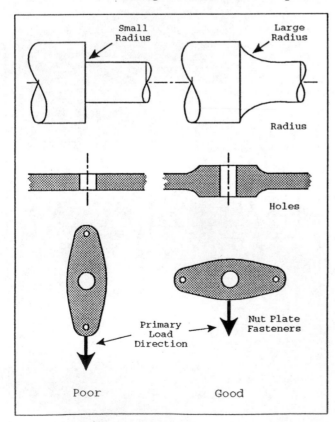

Figure 35-47. Examples of the Reductions of Stress Concentrations Through Design

This is not the case for elliptical holes. Suppose we change the dimensions of the hole to an ellipse whose dimension "a" remains at one inch but whose dimension "b" is reduced to 1/4 inch. The elastic stress concentration factor would increase to 9 (K_t=9) and the maximum local tension stress would increase to 160,000 psi. On the other hand, if we increased the "b" dimension to 4 inches while holding the "a" dimension at 1 inch, the elastic stress concentration factor be would only 1.5 and the maximum local tension stress would be only 30,000 psi. From this little exercise you should be able to see that the orientation of load to the specific geometry has a critical influence on the value of K_t.

It should be obvious by now that the elastic stress concentrations generated by a component are a function of the design's basic geometry. Examples of how a component's geometry can exacerbate or mitigate elastic stress concentrations are shown in Figure 35-46.

Manufacturing techniques used to fabricate a component can influence elastic stress concentrations. Some of the ways elastic stress concentrations can introduced during design are listed below:

● Machining or grinding tool marks perpendicular to local tension stress produced by applied loads. Marks which are parallel to tension stresses do not significantly increase elastic stress concentrations.

● Burrs in drilled holes. Use of dull drill bits and failure to de-burr holes can result in notches in holes which greatly increase the elastic stress concentrations caused by the original hole.

● Failure to prepare holes to the fit of fasteners which will be inserted into the holes. If a fastener with a fillet radius between its head and shank were inserted into a hole in which a chamfer or transition radius had not been machined, the hole's sharp edge would cut into the fastener's transition radius, creating an additional stress concentration in a area which already has a stress concentration.

● Machining errors which mis-align connecting surfaces. Jogs in surfaces when the machining or grinding of a part fails to align adjacent surfaces will produce a stress concentration just the same as a part which was designed with a sharp corner.

● Placement of Holes so that they Increase

Stress Concentrations. Holes which are drilled too close together or are placed around existing stress concentrations can unnecessarily increase stress concentrations.

●Use of broaching or punching to produce holes. Both of these techniques produce higher elastic stress.

●Use of electrical discharge and electro-chemical machining which create rough surfaces on fatigue critical components. Both of these techniques produce rough surfaces which should, if exposed to significant cyclic loads, receive surface treatments which smooth the surface.

●Loose fasteners. Loose fasteners can alter the load paths intended by the designer, possibly defeating design features which would have reduced elastic stress concentrations.

●Improperly bonded surfaces. Failure of bonded surfaces intended to divert the load path from fasteners (such as rivets) can result increased elastic stress concentrations at the riveted joint.

●Identification of parts using techniques which create elastic stress concentrations. Marking methods which remove metal, create rough surfaces or deform metal and can create stress concentrations. Some examples are: impression stamping, mechanical engraving, electric etching and acid etching.

●Failure to use or mis-use various surface preparation techniques. Surface preparation can be as simple as providing a smooth surface finish and painting or can involve more complex processes associated with the surface hardening and/or the introduction of favorable residual stresses. They include, but are not limited to; polishing, painting, plating, shot peening, cold working, carburizing, nitriding and cyaniding. Surface preparation techniques will be discussed in greater detail later in this chapter.

●High stresses in the short transverse direction of extrusions, forgings and heavy plate stock. Although these types of problems don't really increase stress concentrations, they do degrade a material's ability to resist the effects of stress concentrations, shorten the initiation phase of a fatigue crack's life cycle and speed the arrival of macroscopic cracks and the propagation phase.

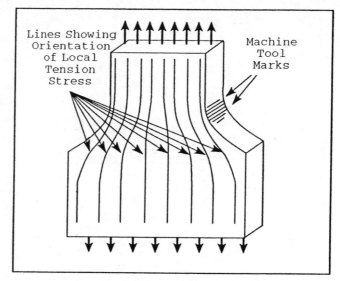

Figure 35-48. Machine/Tool Marks Perpendicular to Local Tension Stress

Just as elastic stress concentrations can be introduced during the design and manufacture process, these concentrations can also be introduced into the structure after the airplane has left the factory and entered service. In addition to most of the list shown above, you can add corrosion, wear, improper repairs and damage as possible sources of elastic stress concentrations.

While elastic stress concentrations influence microscopic crack growth during the initiation phase of a fatigue crack's life cycle, stress intensity is one of the primary factors determining crack growth during the propagation phase of a fatigue crack's life cycle. Stress intensity will depend on the crack's geometry (in terms of the ratio of crack length to width of the

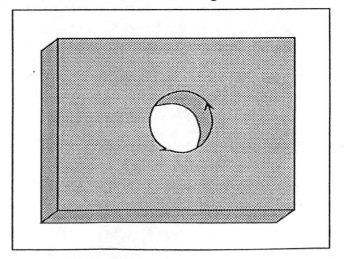

Figure 35-49. Burrs in Drilled Hole

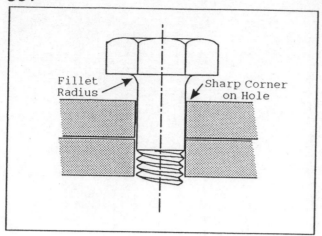

*Figure 35-50. Sharp Corners on Holes Into Which
a Bolt with a Fillet Radius will be Inserted*

original structure), the crack's size (the larger the crack the greater the intensity) and the applied stress (the greater the applied stress, the greater stress intensity).

Since a fatigue crack grows by a minute amount during each stress cycle, it should not be surprising that variations in the cycle should affect crack growth. The nature of the cyclic loads which cause fatigue cracking can be described in a number of ways. Since these variations of load patterns can effect the rate at which cracking is initiated and propagates, the investigator should have some knowledge of their effects. Some of the more important factors used to describe a cyclic stress cycle, and their effects are briefly listed in Figure 35-54.

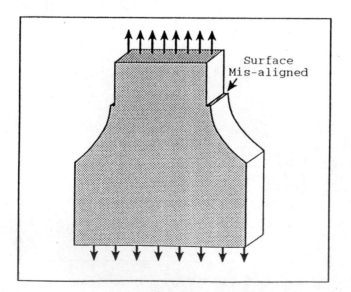

*Figure 35-51. Machining errors Which
Mis-align Attaching Surfaces*

• Maximum stress - The highest algebraic stress value experienced during the stress cycle. Tension stress is considered positive while compression stress is considered negative. s_{max}

• Minimum stress - The lowest algebraic stress value experienced during the stress cycle. Tension stress is considered positive while compression stress is considered negative. s_{min}

• Stress range - The algebraic difference between the maximum stress and the minimum stress experienced during the stress cycle. $s_r = s_{max} - s_{min}$

• Alternating or variable stress - One-half of the stress range. $s_a = s_r/2$

• Mean stress - The algebraic mean of the maximum and minimum stress experienced during the stress cycle. $s_m = (s_{max} + s_{min})/2$

• Stress ratio - The algebraic ratio of the minimum stress to the maximum stress. If the maximum tension and compression stresses are equal, the stress ratio, R = -1. $R = s_{min}/s_{max}$.

It should be pointed out that the growth rate of a fatigue crack is not determined only by the cycle's maximum stress and the number of cycles per unit time. Changes in any of the cyclic stress cycle parameters described above can change the cracks growth rate. In addition, there are other factors which can influence the growth rate. For instance, cyclic stresses which gradually increase from a minimum to a maximum and then drop to a minimum before starting over again will produce fatigue cracks which have a faster growth rate than cracks growing in response to stress cycles which start high and then gradually drop before repeating the cycle as shown in Figure 35-55.

The reason for this phenomena is the work hardening of the crack tip during each period of crack growth. As the crack tip expands, the material in front of the crack undergoes plastic straining, increasing its yield stress. If each subsequent cycle has a higher amplitude tension stress, the crack can continue to grow. If, however, each subsequent cycle has a lower amplitude, the crack may not grow until the amplitude increases at the start of the next cycle. In the extreme, the average rate fatigue crack growth may be slowed by periodically increasing the maximum stress of a single cycle to the point where substantial plastic straining of the crack tip occurs and the crack tip is work hardened to a much higher yield stress. After the "overload", the rate of crack growth will be temporarily halted or slowed to a rate lower than that occurring before the "overload". This is due to the crack having to work its way through a larger zone of high yield stress material. Although this practice of "retarding" crack growth by overloading may be safe in the controlled and low risk environment of the laboratory, asking aircrew members to periodically go out and pull just enough G's to put the aircraft just over its limit G load, so that the growth of fatigue cracks can be temporarily halted, would probably win the dumb idea of the year award.

D. THE S-N CURVE. The standard method of displaying the fatigue strength of a component is the stress-number curve, or more commonly the "S-N" curve. This chart plots the number of load cycles required to cause a fatigue crack-induced fracture when the component is exposed to constant amplitude cycles at various loads or stresses. The curve shown in Figure 35-57 shows the S-N diagram for a hypothetical component. The vertical axis represents the cyclic tensile stress or load being applied to the component. The horizontal axis represents the number of cycles the component withstood before it failed due to fatigue cracking. Because of the large number of cycles normally necessary to cause a fatigue failure, this axis is plotted on a logarithmic scale.

To build the curve, a large number, perhaps hundreds, of duplicate components are tested to failure at various stress or load levels. The stress/load level at which the component was cycled and the number of cycles to failure are plotted on the chart. Some components were loaded to failure in a single cycle. The various stresses or loads at which they fractured are shown on the vertical axis. The fact that they did not all fail at exactly the same level reflects the natural variation in apparently identical components. Another group of components were loaded at only 90% of their estimated ultimate strength. From the chart below you can see that there was a very significant variation in the number of cycles necessary to initiate and propagate the crack to its critical size. When loaded to only 80% of its static strength, the number of cycles required to fail the component increases dramatically. However, the range of the number of cycles to failure also increased, even more dramatically. As the cyclic loads maximum stress decreases, the number of cycles required to cause a fatigue crack-induced failure steadily increases. It should be obvious that predicting

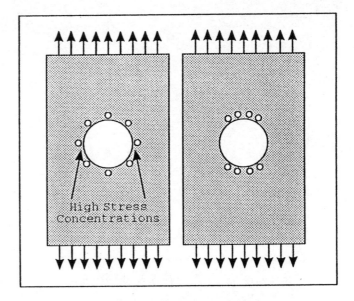

Figure 35-52. Placement of Holes Which Unnecessarily Increase Stress Concentrations

the number of load cycles a component can withstand at a given maximum stress is significantly less accurate than predicting the static load at which it will fail. Statistical methods are used to develop confidence limits were we can say that a component stands, for example, a 90% chance of withstanding 100,000 cycles at 70% of its ultimate strength.

Fatigue limit. In some materials, the number of load cycles necessary to cause fatigue failure becomes infinitely large when the stress level drops below a certain value. This stress level is referred to as a

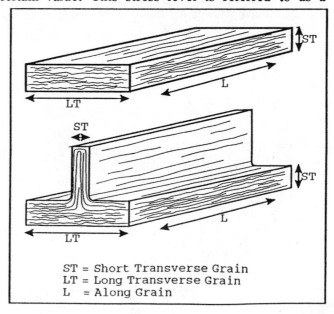

ST = Short Transverse Grain
LT = Long Transverse Grain
L = Along Grain

Figure 35-53. Short Transverse and Long Transverse Directions

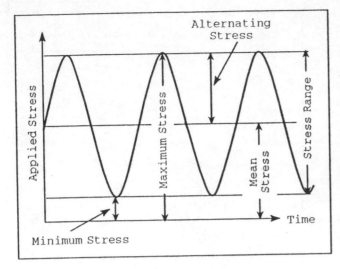

Figure 35-54. Cyclic Stress Definitions

material's endurance or fatigue limit. Although alloys of ferrous metals often exhibit an endurance limit, non-ferrous alloys such as those of aluminum and magnesium do not exhibit an endurance limit and will fail if exposed to a sufficiently high number of stress cycles. The S-N curve for the lower component represented in Figure 35-59 has a fatigue limit of 10,000 psi while the upper component represented in Figure 35-59 does not have a fatigue limit.

Obviously, the service life of a component can be increased if the cyclic loads to which it is being exposed can be kept to a minimum. However, in aircraft, where weight is always critical, this cannot always be done. Therefore a means of tracking how much of a critical component's fatigue life has been used up is necessary. In this way, fatigue critical

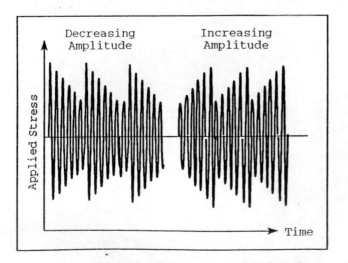

Figure 35-55. Stress Cycles of Decreasing Amplitude Versus Stress Cycles of Increasing Amplitude

components can be replaced before they fail. The simplified version of Miner's Theorem, explained in the next paragraph, should help the investigator understand how this can be done. The effect of changes in stress concentrations and stress ratios are shown in Equation 10.

Miner's Theorem. There are several methods which can be used to monitor the fatigue life of a component or a system so that critical components can be repaired or replaced before a fatigue crack grows to a critical size. One of the more simple methods, called Miner's Theorem, is analogous to using up the money in a checking account. If a component's fatigue life is thought of as money in a checking account, it should be obvious that the account can be emptied with a few major withdrawals, lots of small withdrawals or a medium number of medium sized withdrawals. Or, more likely in the real world, a very small number of large withdrawals, a larger number of medium withdrawals and the largest number of small withdrawals. However, regardless of how the money is withdrawn, its total value can't exceed the money in the bank. The S-N curve can be used in a similar manner to determine how much of a component's fatigue life has been withdrawn and how much is left.

First, the percentage of the component's fatigue life consumed at each stress level is calculated. This is done by dividing the actual (or predicted) fatigue cycles for each stress level by the total number of cycles the component could experience at that stress level before it can be expected to fail due to fatigue cracking. This is the percentage of the fatigue life consumed at that stress level. Next the percentages of fatigue life consumed at each stress level are computed and summed. The sum is the percentage of the fatigue live consumed. To find the percentage of the fatigue life remaining, simply subtract the percentage of the fatigue life consumed from one. These relationships are shown in Equation 10.

$$D_1 = \frac{n_1}{N_1} \quad \therefore \quad D_n = \frac{n_n}{N_n} \quad \therefore \quad D_T = \Sigma D_1 + D_2 + \ldots D_n \qquad \textbf{(10)}$$

where:

D_n = Percentage of fatigue life consumed at stress level n.

n_n = Actual or predicted cycles at stress level n.

N_n = Cycles to failure at stress level n.

D_T = Total percentage of fatigue life consumed.

In the example shown in Equations 11-15, the S-N curve shows a component with a fatigue life of

10,000 cycle at 80% of its static strength, OR 100,000 cycles at 60% of its static strength, OR 1,000,000 cycles at 40% of its static strength. In this example we will assume that cyclic loads will only exist at or close to these levels. In the real world we would probably use more stress levels. With these assumptions we can calculate the remaining fatigue live of a component which has experienced 2,500 cycles at 80% strength, 30,000 cycles at 60% strength and 350,000 cycles at 40% strength by using these equations. Stress level one will be 80% strength, stress level two will be 60% strength, and stress level three will be 40% strength. Then, n_1 = 2,500 cycles, N_1 = 10,000 cycles and the percentage of the fatigue life consumed during cycles to 80% strength (stress level one) is:

$$D_1 = \frac{n_1}{N_1} = \frac{2,500\, cycles}{10,000\, cycles} = 25\% \quad (11)$$

Likewise, n_2 = 30,000 cycles, N_2 = 100,000 cycles, and the percentage of the fatigue life consumed during cycles to 60% strength (stress level two) is:

$$D_2 = \frac{n_2}{N_2} = \frac{30,000\, cycles}{100,000\, cycles} = 30\% \quad (12)$$

Likewise, n_3 = 350,000 cycles, N_3 = 1,000,000 cycles, and the percentage of the fatigue life consumed during cycles to 40% strength (stress level three) is:

$$D_3 = \frac{n_3}{N_3} = \frac{350,000\, cycles}{1,000,000\, cycles} = 35\% \quad (13)$$

And the percentage of the total fatigue life consumed during the 382,500 stress cycles would be:

$$D_T = \Sigma D_1 + D_2 + D_3 = 25\% + 30\% + 35\% = 90\% \quad (14)$$

The percentage of the components fatigue life remaining would be:

$$1 - D_T = 1 - 90\% = 10\% \quad (15)$$

This 10 percent could be used up in several different ways. 250 cycles at 80% strength, or 3,000

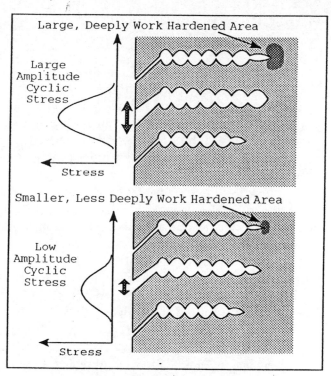

Figure 35-56. Work Hardening of a Fatigue Crack Tip During Crack Growth

cycles at 60% strength, or 35,000 cycles at 40% strength would do it. Or lots of other different combinations.

Finally, it should be remembered an aircraft should not be expected to fall apart when its fatigue life is used up. Because of the large scatter in the fatigue life of components, conservative estimates are normally used. In addition, if adequate inspections have identified critical fatigue cracks and have been tracking their growth, the decision to replace cracked components before they reach critical size or retire the aircraft should be an economic decision. However, errors do occur and fatigue cracks do grow unnoticed

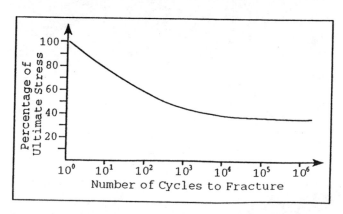

Figure 35-57. Typical S-N Curve

to critical size as their fatigue life is used up and fatigue crack-induced accidents do occur.

E. INSPECTION AS A CONTROL AGAINST FATIGUE FAILURES.

One of the baseline defenses against catastrophic fatigue failures is periodic inspections which detect fatigue cracks and allow their repair before the cracks become a safety issue. Several questions need to be answered in order to ensure that fatigue cracks are detected before they approach a critical length. Where do we look, when do we start to look, how often do we look, how do we look, and has anything changed since we answered the first four questions? If a fatigue crack-induced structural failure is involved in an accident causal sequence, then one or more of the above questions was not answered correctly. It's the investigator's job to find out which of the five questions were blown and why.

●Where do we look? During a modern aircraft's design, the loads on individual structural components are analyzed and tested to ensure that they have sufficient fatigue life to withstand the planned mission load cycles. Potentially critical components are analytically identified and watched during subsystem and full-scale fatigue tests. During structural qualification tests, one airframe is exposed to the number and amplitude cyclic load cycles anticipated during the aircraft's life cycle. During frequent inspections, critical components are checked for the presence of fatigue cracks and their rate of growth. When unanticipated cracks are occasionally discovered, they should be critically evaluated and countermeasures developed. After the aircraft's safe fatigue life has been demonstrated, "lead-the-fleet" aircraft are often used to accumulate "real" load cycles while undergoing close surveillance. But, humans are human and humans make mistakes which occasionally result in accidents. Then it is the investigator's job to find out if we were looking in wrong place and if so why.

●When do we start looking? In uncracked structures, microscopic cracks must develop before macroscopic cracks form and grow. The number of cycles a component will tolerate before developing macroscopic cracks is a function of the components geometry and the stress concentrations it develops, the environment in which it operated and the nature of the stress cycles. Uncracked structures may go for a significant length of time before macroscopic cracks develop and grow to a length which can be detected during non-destructive inspections. But, since it is virtually impossible to produce a large, complex structure without small crack-like flaws, it is generally assumed that small undetectable cracks already exist in the newly manufactured components and that they will grow at rates which can be predicted. The secret is to start the looking in time to detect the cracks before they grow to critical size.

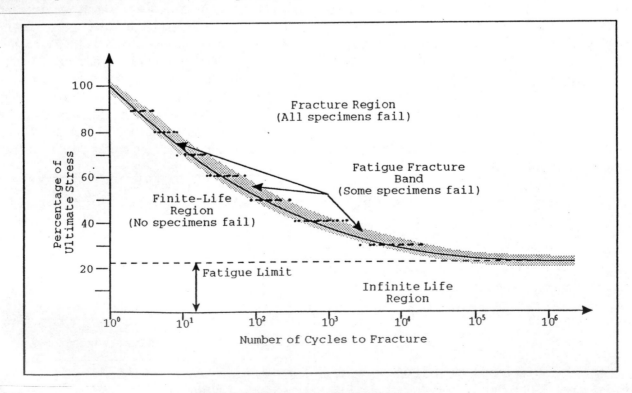

Figure 35-58. Constructing an S-N Curve

●How often do we look? As a crack grows larger it is easier to detect. Therefore there may be some economic incentive to delay expensive inspections until the probability of detecting a crack is sufficiently large. However, as a fatigue crack grows, both the residual strength of the structure and its structural safety factor are reduced. The critical crack length, which provides the minimum residual strength, can be calculated using fracture mechanics theory. For every inspection interval there can also be calculated a maximum permissible crack length, which will grow to the critical crack length during the inspection interval. For each design, the ability of various inspection methods to detect cracks shorter than a certain length will become questionable. Simply put, for each inspection technique, as used on a specific design, there is a certain sized crack (or smaller) which stands a reasonable chance of being missed. If a crack is smaller than this size, then the chances of detecting the crack become even smaller. This crack would then be the smallest crack we can reasonably expect to discover using a specific technique. The number of cycles it takes for the crack to grow from this length, at which it can first be detected to its maximum permissible length is the maximum inspection interval.

●How do we look? If the growth of a fatigue crack to critical length and the subsequent failure of a critical structural component was involved in an accident causal sequence, then it is the investigator's responsibility to identify any deficiencies in the inspection techniques and procedures. There are six general approaches used to detect the presence and criticality of fatigue cracks. These techniques are used during both the manufacture and the operational use of aircraft. In addition to examining the use/misuse of these inspection techniques, these techniques should be considered as tools which can also assist the investigator during the investigation process. These techniques are briefly discussed later in this chapter.

F. CONTROLLING FATIGUE CRACK FAILURES.
The primary control of hazards associated with fatigue cracking in aircraft is the achievement of damage tolerant designs. All damage tolerant designs contain at least four features: testing which establishes the point in an aircraft's service life where fatigue cracking can be expected (establishing the duration of the crack initiation phase); testing to determine how fast existing fatigue cracks will grow (establishing the duration of the crack propagation phase); demonstrations showing that the cracked structure will have sufficient residual strength; and demonstrating that

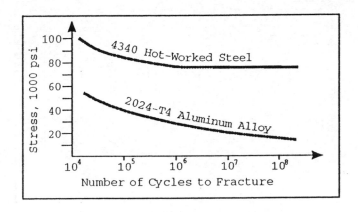

Figure 35-59. Examples of Components With and Without Fatigue Limits

potentially critical damage can be detected before it becomes critical. FAA 25.571 and its associated advisory circulars contain the FAA's approach to achieving damage-tolerant designs for commercial airplanes. The FAA considers a damage tolerant design to be one that can experience serious fatigue cracking, corrosion, or other accidental damage to the aircraft's structure and still have the remaining structure contain sufficient residual strength to carry reasonable loads without failure or excessive structural deformation until the damage is detected. A damage tolerance evaluation should include analyses and tests to ensure that:

●loading spectrums used in the evaluation are those expected during the airplane's service life.

●structure contains multiple load paths to ensure adequate residual strength.

●crack stoppers are appropriately used to control crack growth.

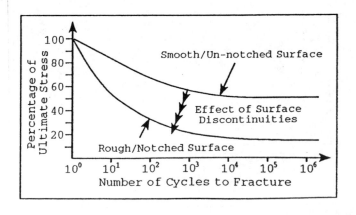

Figure 35-60. Effect of Surface Discontinuities on the S-N Curve

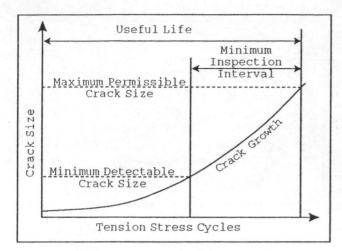

Figure 35-61. Maximum Inspection Interval as a Function of Cycles Required to Grow to Critical Length

●materials which provide a controlled, slow rate of crack propagation are used when appropriate.

●the design allows inspections which provide a high probability that a failure of a critical structural component will be detected before strength has dropped below acceptable levels.

●critical structural components can be repaired or replaced.

●limit the probability of concurrent multiple damage sites which could contribute to a common fracture path.

While the damage tolerance concept includes consideration of fail-safe designs, it is different from

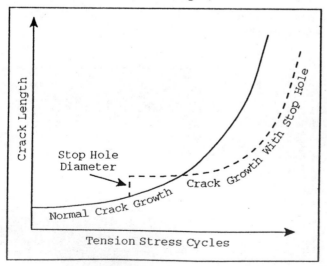

Figure 35-62. Effect of Stop Drilling on Fatigue Crack Propagation

the fail-safe concept. Fail-safe designs attempt to ensure that catastrophic failures are not probable after a fatigue failure or obvious partial failure of a single, principle structural member.

Fatigue crack damaged structure must be either repaired or replaced before the crack grows to a critical length. Improper repair or replacement can actually weaken a structure's fatigue strength. For example, "stop drilling" a fatigue crack will not stop its growth. If properly accomplished it will slow its overall growth rate. If improperly accomplished it could actually increase the rate of crack growth. The effect of stop drilling on the growth rate of a fatigue crack is shown in Figure 35-62. The figure shows how drilling at four different locations affects the crack's propagation. At location "A" the crack's tip has not been drilled out and the rate is not changed. At point "B" the crack tip has been removed, but the material just ahead of the tip, which has already used up some of its crack initiation phase, will start to grow another crack in short order. At point "C" the tip of the crack, as well as some of the material just in front of it, has been included in the drilled out material. However the stress concentration caused by the crack/hole combination will cause the crack initiation phase to be shorter than originally. At point "D" the back edge of the drilled hole is just touching the original crack tip. Again, the crack's propagation stops until the stress concentration caused by crack/hole combination, which is higher than when the hole is drilled at point "C", starts a new fatigue crack. Changes in the size of the stop drill hole will also have an effect on the stress concentration for the "repaired" crack.

The improper use of splice plates when repairing fatigue cracks can also generate new stress concentrations which generate new problems. Holes used for fasteners which hold the splice plates in place and abrupt changes in the load path through the splice plate are two examples shown in Figure 35-63. Patches using a Boron epoxy composite have been used to repair cracks in a way that does not create the above mentioned problems.

Despite the emphasis on the avoidance of stress concentrations during the design of critical components and structure, excessive emphasis on ease of manufacture, low weight and other factors occasionally result in designs which naturally produce stress concentrations. Some examples of desirable and undesirable designs are shown in Figure 35-64.

Since fatigue cracking requires tension stress, it is sometimes possible to avoid fatigue cracks by creating a favorable (compression) residual stress. This is done by placing the material on surface fatigue vulnerable locations in compression. Now, before tension stresses can be created near a component's surface, the compression stress must first be overcome. Shot peening was first used by rail car manufacturers over a century ago when they ran into fatigue problems with rail car axles. Hardening the surface of steels through carburizing, nitriding or cyaniding not only protects the surface, it also creates a residual compressive stress near the material's surface. Bolts which are rolled (in a machine called a header) have superior fatigue strength to those which are cut on a lathe. Not only are the sharp corners and burrs associated with machining avoided, the material in the thread roots and the head/shaft radius is in compression. Compressive residual stresses in these two areas are especially important since they are the locations where fatigue cracks normally start.

Design features can also be used to reduce the amplitude of cyclic tensile stresses, one of the necessary ingredients in making fatigue soup. For example, you might expect that head bolts, which attach a reciprocating engine's cylinders to its crankcase, experience high cyclic tension loads each time a cylinder fires. However, when the head bolts are torqued to specification values, the tension load created in the bolts also creates a compression load on the opposing faces of the cylinder head and the crankcase. This compressive force must be totally relaxed before the tension stress in the head bolt can increase. Cyclic tension stress has been replaced by a high, but static stress. Fatigue is avoided, if the bolts are kept torqued (tensioned) to the proper values. Too much torque and bolt will fail in tension, too little torque and the bolt will fail in fatigue cracking.

G. SPECIAL CONCERNS FOR INVESTIGATIONS INVOLVING FATIGUE CRACK FAILURES.

The first problem facing an investigator involved in a accident involving a fatigue crack-induced failure is realizing that fatigue cracking is involved. The investigator may hope to stumble across the "golden BB" in the thousands of fracture surfaces scattered through the wreckage, but success is more likely by using a systematic approach of attempting to identify which fractures occurred first and closely examining those few for evidence of fatigue. If fracture surfaces include evidence of possible fatigue, they must be carefully preserved until delivered to a compe-

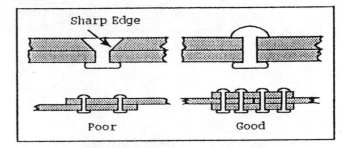

Figure 35-63. Stress Concentrations Caused by Faulty Repairs

tent laboratory. Once fatigue is confirmed, the real investigative work begins. What were the root causes which allowed a fatigue crack to develop and then propagate until it caused a catastrophic failure? The following paragraphs can be used as a sort of checklist, identifying potential links in the chain of events leading to the fatigue failure. Earlier in this section we mentioned that fatigue cracking required four things. Fatigue cracking which propagates to a catastrophic failure requires even more. The investigator's job is to identify all the deficiencies which allowed the accident to occur, so that actions which prevent or minimize the recurrence of similar accidents can be taken.

1. Is the basic design prone to fatigue cracking?

• Does the design's geometry avoid creation of stress concentrations?

• Do the manufacturing techniques minimize stress concentrations?

• Does the material selected have the highest practical fracture toughness?

• Do manufacturing techniques create favorable stress concentrations?

2. Did manufacturing or assembly errors induce stress concentrations or defeat fatigue avoidance features?

• Was the correct material used?

• Were surfaces machined to the appropriate dimensions/finish?

• Were subassemblies properly mated?

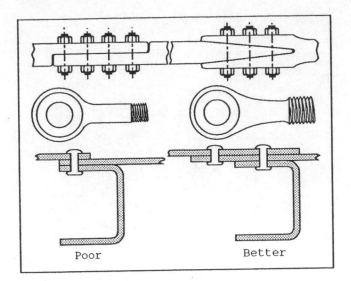

Figure 35-64. Examples of Designs Which Magnify or Minimize Stress Concentrations

• Were tool marks avoided on critical surfaces?

3. Did usage induce stress concentrations or encourage crack growth?

• Did changes in the load spectrum change expected crack growth rates?

• Did operations, maintenance or repair induce stress concentrations?

• Did operations, maintenance or repair lessen fatigue avoidance features?

• Did other service life issues degrade fatigue avoidance features?

4. Why did inspections fail to detect cracks before they reached a critical length?

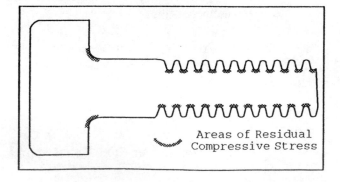

Figure 35-65. Residual Compressive Stress in Rolled Bolts

• Was the fatigue fracture area identified as a fatigue critical area?

• Were inspections specified for the failed component?

• Were the inspections performed? As specified?

• Should the flaw have been detected at the last inspection?

• Was the specified inspection technique adequate?

• Was the specified inspection interval adequate?

• Was performance of the inspection adequate?

11. CORROSION.

Corrosion is the natural disintegration of a material as it is attacked by one or more substances in its environment. During the refining process, energy is added to metal ores and other raw materials in order to produce the mechanical properties necessary in structural components. Mother Nature, the great equalizer, doesn't like variances in energy levels and sets to work trying to bring the material back to the low energy levels existing in the products of corrosion. In aircraft structural components, Mother Nature's attack will reduce the strength and ductility of components, turning strong metals into weak metallic oxides, hydroxides or sulfates. These compounds, if not removed from the structure, can exacerbate the problem by providing an environment which is ever more favorable to continued corrosion.

Although the attack will take one of two general types, chemical attack or electrochemical, there are many different specific forms of corrosion. The reaction between the metals and the environment is also accelerated by physical factors such as high temperatures or stress. Although new and more corrosion resistant materials are constantly being developed, the environments into which aircraft are being flown are also becoming more demanding. Eliminating all potential for corrosion is not a realistic goal. Instead the solutions lie in one of the four following objectives: designs which minimize corrosion potential; regular cleaning and protection of vulnerable areas thereby minimizing the environment's opportunity to attack; regular inspections which detect corrosion at an early stage, before significant damage has been done; and repair of corroded components before they cause a safety hazard. Failure of any one of these elements could lead to an accident.

A. FORMS OF CORROSION. Corrosion of metals can be the result of direct chemical attack by reactive substances in the environment or by electrochemical reactions when the various elements in a structure combine with the elements in the environment to form a battery. In both cases the result is the same; disintegration of load bearing metal structure into non-load bearing flakes or dust.

CHEMICAL VERSUS ELECTROCHEMICAL ATTACK. Direct chemical attack involves the reaction between a metal structure and some chemical agent. If you pour a corrosive acid on a metal wing, the acid and the metal will react to form new and undesirable compounds. Of course we don't normally pour acid on a wing, but iron will rust in the presence of atmospheric moisture and an aluminum surface will oxidize when exposed to the air. During electrochemical attack, the aircraft structure and the environment form a battery containing an anode, a cathode, an electrolyte and a conductive path between the anode and cathode. Electrons flow from the anode to the cathode while metal particles on the surface of the anode are transformed into ions. These ions react with the electrolyte to form corrosion products. The process results in the steady disintegration of the anode while the cathode remains relatively unchanged. This is essentially the same process found in the dry cell battery powering your flashlight. And, like your flashlight, removal of any one of the elements in the circuit (anode, conductor, cathode or electrolyte) will interrupt the circuit and stop the corrosion. Easier said than done for an operational aircraft.

UNIFORM ATTACK VERSUS LOCALIZED ATTACK. Corrosion can be uniformly spread over large areas or localized in well defined locations. In the case of uniform attack the damage will have penetrated to an equal depth in all areas and loss of strength is measured as a percentage of the thickness of the structure that has been lost. Uniform corrosion is normally due to direct chemical attack. Typical examples include the reaction of exposed metal surfaces with airborne chlorine or sulfur compounds or oxygen. Heat can accelerate the process and oxygen rich engine exhausts are a frequent cause of uniform

attack. Fumes of chemical sources such as batteries are another common source of uniform attack. Localized attack is normally the result of electrochemical reactions and are categorized as either pitting or selective attack.

PITTING. Only very small areas of a metal's surface will show the effects of pitting. The small holes called "pits" will be randomly located across the metal,s surface and may be accompanied by the products of corrosion such as a powdery residue. Although the pits may appear to have damaged only a small percentage of the surface, they penetrate deeply in a branching manner, causing loss of strength and ductility which is way out of proportion to the metal's

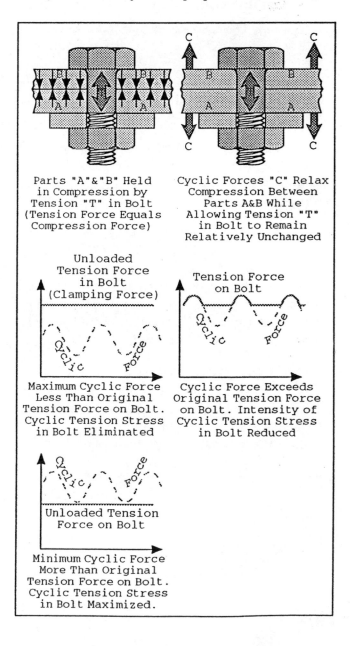

Figure 35-66. An Example of Avoidance of Cyclic Tension Stress

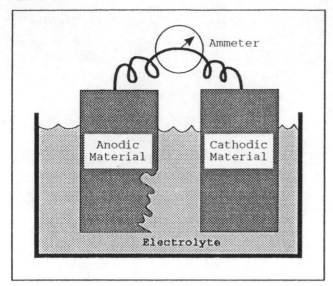

Figure 35-67. Schematic of a Galvanic Cell

surface appearance. Pitting starts with the chemical action of acids or alkali solutions or saline solutions on unprotected metal surfaces. The attack is the result of galvanic action of dissimilar metals on the surface of an alloy or solutions containing concentrations of dissolved oxygen or metal ions. Pitting corrosion of aluminum and magnesium alloys produces gray or white corrosion products on the material's surface.

SELECTIVE ATTACK. When corrosive actions seem to favor one part of a component or assembly above another, the corrosion is called selective. This type of corrosion may start off as pitting and then take on a new character once it enters the subsurface areas. Or it may start from the edges of plates or sheets where moisture can collects.

INTERGRANULAR CORROSION. The primary type of selective attack is intergranular corrosion. Since grain boundaries often are rich in small particles of dissimilar alloying metals, these boundaries are less corrosion resistant than nearby

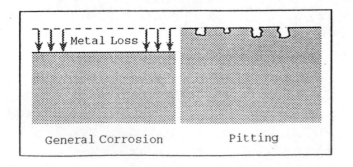

Figure 35-68. Pitting

areas. This type of attack, therefore, first centers on the grain boundaries within a metal component before consuming the grains themselves. Like pitting, the damage from this form of attack causes a loss of strength and ductility which is out of proportion to the amount of metal that is corroded. The potential for intergranular corrosion can be minimized by plating or cladding. Alclad is a common name for a composite sheet, plate, tube, wire, etc. onto which a thin layer of high purity aluminum has been metallurgically bonded. The protective layer is anodic with respect to the core, providing electrolytic protection as well as a mechanical barrier to the environment. Anodizing 2000 and 7000 series aluminum provides a relatively hard and impervious barrier to the environment.

EXFOLIATION. Exfoliation is a form of intergranular corrosion whose progress can go undetected until all structural integrity is lost. The grain boundaries attacked by this type of corrosion are normally the flattened/elongated grains of extruded or rolled metals. Exfoliation normally starts its attack along an edge of the metal which is both concealed from view, exposes many more grain boundaries per unit area than the surface, and has the potential to collect moisture and contaminants. A fastener hole with a slightly loose rivet and machined edges is a choice location. From there the corrosion can move undetected along the flattened or elongated grain boundaries. The grain's shape will naturally restrict the progress toward the surface and encourage its progress parallel to the surface. The formation of corrosive products will eventually cause a swelling which may be the first clue that the component is corroding. The eventual "leafing-out" of corroded sections of metal give this type of corrosion its name. By the time it "leafs-out" it will probably be too late for repair and replacement of the component may be the only recourse. Sealing the hidden edges of and holes in extruded and rolled metal components along with the traditional cleaning and drying are the primary preventive actions. NDI techniques which go beyond ordinary visual inspections (e.g. ultra-sound, eddy current and x-ray) will be necessary to detect its occurrence.

GALVANIC CORROSION. When electrochemical corrosion occurs at a level involving two different components (or different parts of a component) as the anodes and cathodes, it is often referred to a galvanic corrosion. Although electrochemically it is exactly the same as the intergranular and exfoliation corrosion discussed above, now we are talking about the reaction between an aluminum hinge and a steel hinge pin, or the graphite in a graphite fiber reinforced plastic skin and an aluminum fitting. The further apart the materials are in the galvanic series, the faster the corrosion will propagate. When two different metals from different groups in the galvanic series (shown in Figure 35-71) are in contact with each other and both are in the presence of an electrolyte, the most anodic metal will corrode. In the case of aluminum alloy and steel, it will be the aluminum which corrodes the fastest, while the steel may suffer little to no damage. Another factor in the rate of corrosion during galvanic corrosion is the mass of the anode versus the mass of the cathode. If the mass of the anode is much less than the mass of the cathode, the anode will disintegrate much more rapidly than if the reverse situation exists. For example an aluminum fastener in a steel structure will corrode much faster than an aluminum plate with a steel fastener. (See Figure 35-72.) A cadmium plated steel bolt in an aluminum plate will cause less corrosion than an unplated steel bolt because aluminum and cadmium and steel are in the same group. The steel cannot corrode the cadmium because the plating prevents the electrolyte's contact with the steel.

FILIFORM CORROSION. A random, thread-like form of corrosion than occurs on the surface of metal skins, under protective coatings (typically polyurethane paint) is referred to as filiform corrosion. In itself, it is not very destructive, but it may provide a foothold for other types of corrosion. Defective primer coats are normally the immediate source of this problem.

B. CORROSION AND MECHANICAL FACTORS. The damaging effects of corrosion can be accelerated by a number of factors that go beyond the direct chemical and electrochemical factors discussed so far in this section. Residual stresses within a component and/or static or cyclic stresses applied to a component can combine with a corrosive environment to launch a more aggressive attack, causing deterioration of the component at a rate faster than that experienced solely by corrosion. Combined attacks covered in this section include stress-corrosion cracking, corrosion fatigue and fretting corrosion. In addition the effects of improper heat treatment are also included.

STRESS-CORROSION CRACKING. When metal components are subjected to long term static tensile stress (either residual or externally applied) while in a corrosive environment, they may suddenly fracture without any change in the applied

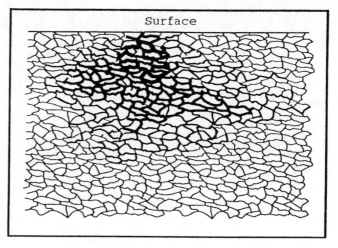

Figure 35-69. Intergranular Corrosion (Starting at a Pit)

load. Like fatigue, the cracks develop in two stages, initiation and propagation. During the initiation phase, protective films on the surface of the component are broken-down and an electrochemical attack on the component begins. During the propagation phase the electrochemical attack moves into crack surfaces, focusing on the crack tip or apex. As a result the crack tip corrodes faster than the sides of the crack. Outside the cracks there is little evidence of the corrosive attack. The cracks will propagate between the grains (intergranular) or through the grains (transgranular),

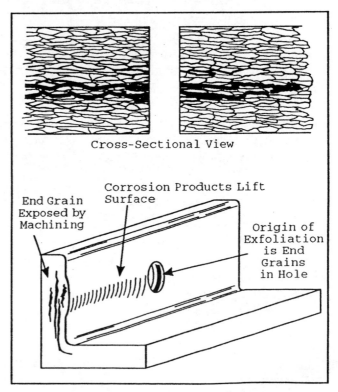

Figure 35-70. Exfoliation Corrosion

resulting in a fracture that appears brittle, even though the metal is ductile. Although the cracks will normally grow perpendicular to the local tensile stress, the alloy type, changes in alloy composition and the environment will influence crack growth. Cracks will continue to grow until they either reach a critical length when the remaining metal fails in a sudden brittle fracture or until the static stress which is causing the growth is relieved. Steel fasteners which are driven with excessive interference press fit into aluminum structures are an example of how the conditions for stress-corrosion can be set-up. Stress-corrosion can be mistaken for other types of fractures. For instance, differences in the corrosion's penetration rate can cause a beachmark-like pattern which can be initially mistaken for fatigue cracking. Stress-corrosion cracking can also be confused with hydrogen-embrittlement cracking. Although final determination of the exact cause of the crack will be made after a detailed laboratory analysis, it is still up to the investigator to preserve the evidence and get it to the laboratory. Prevention of stress-corrosion cracking will center on either minimization of the corrosive environment or static tension stress. For example, use of shot peening or surface rolling to develop residual compressive stress on the surface will increase a component's resistance to stress-corrosion cracking.

MOST ANODIC
Pure Magnesium
AZ31A Magnesium
Zinc
7075-T Clad Aluminum
5056 Aluminum
Pure Aluminum
2024 Clad Aluminum
Cadmium
2024 Aluminum
Steel
Iron
Lead
Chromium
Brass & Bronze
Copper
Stainless Steel
Titanium
Monel
Silver
Nickel
Inconel
Gold
MOST CATHODIC

Figure 35-71. Electromotive Grouping of Some Dissimilar Metals

cause a stress concentration which is constantly increasing in value. Eventually, the cyclic stresses which originally caused the failure of the protective finish will cause a fatigue crack to develop out of the pit. By the way, there will likely be more than one pit and a fatigue crack could be developed by each one. As they grow they will tend to join up at ratchet marks. In other cases, the fatigue crack will develop first and then provide a "home" for the corrosion, drawing the corrodent in by capillary action. Because of the weakening effects of corrosive attack at the crack tip, the crack will propagate faster than it would in a corrosion-free environment. The existence of a unforecast corrosive environment could make the service life estimate, established during a damage-tolerance evaluation, extremely optimistic. The time related effects of corrosion often make the identification of the origin of the fatigue crack easier. However, corrosion of the fracture surface and the associated striations can make a detailed analysis of propagation rates more difficult. Once the corrosion fatigue fracture is discovered, the accident investigator must take special precautions to clean and protect the fracture from additional corrosive damage. Recommended corrective actions could include: eliminate or reduce corrosion potential; use of a more corrosion resistant material; reduce the magnitude or frequency of cyclic tensile stress; introduce or increase the residual surface compression stress.

CORROSION FATIGUE CRACKING. Simultaneous exposure to a corrosive environment and the cyclic tension stresses necessary for fatigue cracking can result in corrosion fatigue. Like almost all fatigue cracks, corrosion-fatigue cracking will start on the surface, often in an area where continued metal straining causes the protective finish to break down. In this case, corrosion fatigue will have shorter crack initiation and crack propagation phases than straight fatigue cracking in a corrosive free environment. The corrosive agent will typically cause pitting which, as it grows, will

FRETTING CORROSION. Fretting involves the unintentional, low amplitude rubbing/movement of two highly loaded surfaces against one another. This rubbing/movement causes abrasive wear. Normally the fit between the two surfaces will allow oxygen or some other corrosive agent to enter the area of fretting and attack the unprotected surfaces. The mechanical fretting of the surfaces is then joined with chemical corrosion on the same surfaces to create what is called "fretting corrosion." Since attempts to reduce movement by tightening the fit between the two surfaces is rarely successful, application of a suitable lubricant between the two surfaces will minimize wear and can provide some protection against direct chemical attack.

IMPROPER HEAT TREATMENT. Heat treatment is one of the elements available while attempting to control corrosion. For example, proper heat treatment of a clad aluminum alloy plate or sheet will cause the cladding's grain structure to be such that minimizes its susceptibility to intergranular corrosion. Improper heat treatment may allow corrosion to penetrate the cladding and attack the lower level vulnerable aluminum alloy. Forging and extrusions of copper and zinc-based aluminum alloys which, if improperly quenched after heat treatment, can allow non-uniform areas of the parent and alloying metal. As a result, they can become more vulnerable to intergranular corrosion.

HYDROGEN EMBRITTLEMENT. Another adverse effect of improper heat treatment is hydrogen embrittlement. Although not really corrosion, it is lumped into this discussion because we couldn't find a better place to put it! During acid etching, electrolytic plating processes and arc welding, hydrogen can penetrate into and remain in the processed metal (typically steel) component. The presence of the hydrogen in the metal can alter its physical and mechanical properties, reduce its ductility or induce internal cracking. In some cases the internal pressure caused by the hydrogen can cause blisters just below the metal's surface. Fracture surfaces will resemble brittle fractures with intergranular cracks with lots of branches. The fracture surfaces can be confused with those caused by stress-corrosion cracking. Proper heat treatment by prolonged "baking" of the processed metal at elevated temperatures will allow the hydrogen to diffuse harmlessly out of the metal. If a missed "baking" is the reason for a component's fracture due to hydrogen embrittlement cracking, it is very likely that the entire lot of components which were "baked" with the defective component are also defective.

C. CORROSION CONTROL. Although all corrosion control efforts will have several common factors, the efforts directed toward direct chemical attack will vary slightly from those used for electrochemical attack. Common factors involve choice of materials which are least vulnerable to corrosion, isolation of reactive materials, cleaning, inspection and repair. Differences focus on the anodic material,

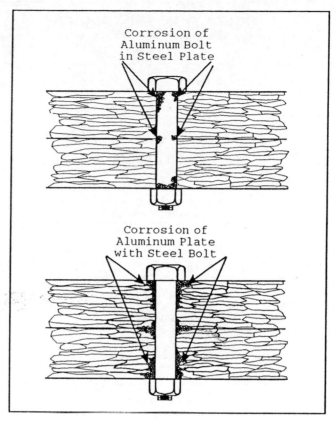

Figure 35-72. Galvanic Corrosion During Contact of Aluminum and Steel

cathodic material, electrolyte and conductive path combination necessary for electrochemical corrosion.

●**MATERIAL SELECTION.** Since not all structural materials are equally resistant to corrosion, the materials used in modern aircraft should be chosen to provide maximum practical protection from both the direct chemical and the electrochemical attacks expected in their operational environments. For example, 7000 series aluminum is generally more susceptible to stress corrosion cracking than 2000 series aluminum. The threat of galvanic corrosion should be considered wherever dissimilar metals make contact.

●**MATERIAL ISOLATION.** Aluminum components with inadequate corrosion resistance can be isolated from electrolytes or dissimilar metals by cladding, surface conversion techniques (e.g. anodize, alodine), zinc chromate primer and organic finishes. Cadmium plating, phosphate coatings and organic finishes can improve low-alloy steel's corrosion resistance.

●**PREVENTION OF ELECTROLYTE FORMATION.** Since all types of electrochemical

CORROSION PRONE AREAS
Base of Bulkheads
Battery Compartments
Battery Vent Openings
Bilge Areas
Cooling Air Vents
Engine/APU Exhaust Areas
External Hinges
Galleys
Joints in External Skin
Landing Gear
Lavatories
Loose Rivets/Fasteners
Water Entrapment Areas
Wheel Wells

Figure 35-73. Corrosion Prone Areas

corrosion require moisture for the electrolyte and moisture can also serve as a medium for corrosive agents, unwanted moisture is a primary source of corrosion problems in aircraft. As a result, keeping metallic surfaces dry is a primary ingredient in any effective corrosion control program. The SR-71 is reported to have had few structural corrosion problems. The structure gets so hot during flight that moisture can't exist. After landing, it remains hot enough to drive off any moisture that falls on it before it can be tucked into its hangar. The structure of the normal high flying jet is not so lucky. It gets rained on. It gets sufficiently cold at altitude so that water vapor can condense and collect within structure after landing. In humid climates these aircraft seem to sweat after landing. Aircraft need to be sealed against the entry of water and ventilation or drainage provided wherever water can accumulate. The other half of the electrolyte equation is the corrosive agent in the form of acids, alkalis or salts. These agents can be supplied by industrial atmospheres, marine atmospheres, the aircraft itself (e.g. products of combustion, leakage of corrosive fluids or gases) or the cargo. Routine cleaning of the aircraft, especially areas of known contaminate accumulation is a necessary component of

keeping it dry.

●INSPECTION AND REPAIR. Despite our best efforts corrosion will occur. Regular inspections of corrosion prone areas (see Figure 35-73) must be conducted and accepted procedures used to remove any corrosion found or replace structure damaged beyond acceptable limits.

D. INVESTIGATION OF ACCIDENTS INVOLVING CORROSION. When corrosion-caused failure of a structural component is involved as a causal factor in an accident sequence, the issue of the adequacy of corrosion control efforts must be addressed. Some of the questions which should be answered are listed below.

●Were the materials used in the design unnecessarily vulnerable to the direct chemical attack by the operational environment or likely to create the potential for galvanic corrosion?

●Was the corroded component adequately isolated from the environment and/or dissimilar materials with the potential for galvanic corrosion?

●Was the corroded area provided with adequate drainage and ventilation?

●Was the corroded area accessible and identified for regular inspection and cleaning?

●Was the corroded component regularly inspected and cleaned of corrosive contaminates and provided with adequate drainage of water accumulations?

●Were damaged corrosion protection finishes, platings and sealants repaired or replaced when appropriate?

●Was all corrosion removed using approved procedures and tools?

●Were specified corrosion control features (alodine, sealants, etc.) replaced after corrosion removal/repair?

12. WEAR.

The slow, removal of material from the surface of a component by mechanical action is referred to as wear. Although in almost all cases wear is undesirable, wear during "break-in" on new or overhauled equipment is often a necessary ingredient in establishing proper operation and long service life. The wear we will be discussing is the bad kind, the kind that leads to premature failure and breakdown.

We will discuss five different types of wear; abrasive wear, adhesive wear, chafing/fretting/false brinelling, erosive wear, and contact stress fatigue. Although the last is actually a form of fatigue, it is included in this section because it involves wear forces and the failure products normally look more like the results of wear than the other fatigue failures we have discussed.

A. ABRASIVE WEAR. Abrasive wear occurs when small abrasive particles (contaminants) cut into and remove material from the surfaces of the two components which are held together while moving. When abrasive wear occurs, one of the questions the investigator must answer concerns the origin of the particles. What was the composition of the contaminating particles? That means fluid samples and help from a laboratory. Where did the particles come from? If a lubricant was used to separate and cool the surfaces, did the lubricant bring in the particles? What is the status of up-stream filters? You did save them, didn't you? Has this happened before? Can we expect this kind of contamination in the future on other aircraft? Do we have to improve the filtering system or its service to prevent recurrence? How about the surfaces that were worn? Were they of the specified hardness? If the surfaces were soft, were they inappropriately vulnerable to the particles? Back to the lab! When was the last time the worn parts were replaced? Maybe they just wore out due to failure to replace them at the appropriate interval!

B. ADHESIVE WEAR. Adhesive wear occurs when the microscopic projections of the surfaces of the two components which are sliding across each other make contact, weld together and break-off. You should recognize that what started off as adhesive wear can quickly become something akin to abrasive wear. But this time the abrasive's origin is/are the components themselves. A good laboratory should be able to help us out here. If a lubricant was supposed to keep the two surfaces apart, the direction of our investigation will take a path different from that used for abrasive wear. Now we need to know why the lubricant didn't do its job. Was the surface lubricated? Was the supply exhausted? Did the pump fail? Were the passages blocked? Were the lubricating properties destroyed

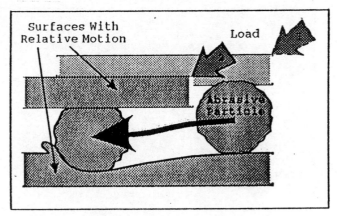

Figure 35-74. Abrasive Wear

when the oil was overheated? Was the proper lubricant used? Was the lubricant inadequate for the task it was asked to perform? Other questions could be directed toward the finish on the two surfaces. Were one or both excessively rough, allowing high points on one to penetrate the lubricant and make contact with the other surface? These are the type of questions we would ask when investigating an abrasive wear failure.

During the "break-in" phase we mentioned earlier, we normally expect some wear. This is adhesive wear where the mating surfaces settle into each other as acceptable high points on opposing surfaces are worn down. Lubricants are normally changed more frequently eliminating wear-generated particle debris. Different types of lubricant may also be specified during the break-in phase. This type of wear is not considered damaging. The fairly smooth, shallow groves referred to as "scoring" are the least severe form of damaging adhesive wear. "Galling" is a more serious form of adhesive wear. The inadequate lubrication has allowed greater surface contact pressures and high spots on the surface make contact, weld and then pull off pieces of significant portions of the opposing surfaces. If allowed to continue, the friction welding of the two surfaces will become more and more extensive. Eventually the welding will become extensive enough to stop the motion of the surfaces and the components will seize. "Seizure" is the most severe form of adhesive wear.

C. CHAFING/FRETTING/FALSE BRINELLING. Chafing and fretting both refer to the wear associated with the rubbing of two surfaces. Since microwelding occurs in both chafing and fretting, both are similar to adhesive wear. Chafing differs from adhesive wear in that during chafing, the contact between the two surfaces occurs when the two parts are unintentionally moved and come in contact. During

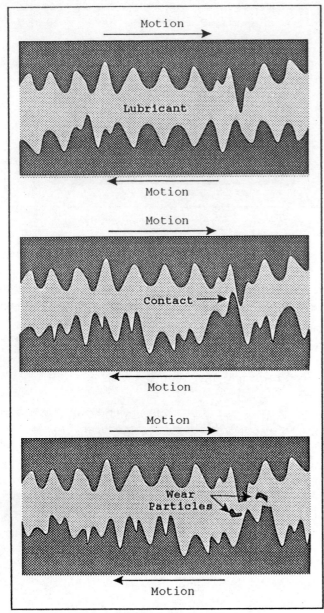

Figure 35-75. Adhesive Wear

Brinelling is an impression in a hard surface created when a hard or sharp surface is forced, under heavy pressure, into a softer surface. False brinelling is a type of fretting induced wear, where bearings which normally rotate when a system is in operation, slide on the surface. Over an extended period of time, this sliding will wear grooves which resemble true brinelling.

D. EROSIVE WEAR. Erosive wear is similar to abrasive wear in that foreign particles are cutting tiny chunks out of a surface. It's slightly different from abrasive wear in that the abrasive particles gain their penetrating energy by a fluid that is carrying them along. Sand could constitute the foreign particles and the airflow over the aircraft or through the engine could be the fluid. The sustained cutting of the sand can remove surface coatings and in severe situations change the shape of turbine blades and vanes.

E. CONTACT STRESS FATIGUE. The wear modes discussed above are all associated with some sort of sliding action. The wear due to contact stress fatigue is somewhat different. This type of wear shows up as cavities or pits in a surface and occur during two different situations. Subsurface origin, surface origin and subcase origin fatigue all develop as a result of stresses created while two surfaces roll against each other while under compressive load. Figure 35-76 shows the type of deformations occurring when two rolling surfaces are forced together. Parts subject to this type of failure include rolling and sliding types of bearings and gears. The cavities formed in all three cases form stress concentrations which can be the source of additional fatigue cracking and the component's catastrophic failure. The material removed from the cavity will normally be hard and brittle. When released it can serve as an abrasive agent, causing abrasive wear to not only the component which generates the particles, but all components sharing the same lubricant. During cavitation fatigue a metal surface is exposed to cyclic compressive forces created by a liquid. Again small pits are developed and the particles generated are released to cause additional problems.

●SUBSURFACE ORIGIN FATIGUE. This type of contact stress fatigue shows up as relatively small pits on the surface of hardened steel components used in anti-friction devices such as ball, roller, and needle bearings and roller cams. The pits are the result of high shear stresses just below the contact surfaces when rolling occurs during high compression forces.

adhesive wear, the lubricant which was supposed to ensure separation typically fails to do the job. In addition, the end result of chafing is normally the wearing through of the softer material, separating a wire or cable into two pieces or wearing a hole in a tube or pipe.

Fretting occurs due to low amplitude rubbing of two surfaces which are held together under pressure. Again, it is similar to adhesive wear in that microscopic surface projections make contact, friction weld and then break-off. Its difference from adhesive wear lies with the sliding action associated with adhesive wear while fretting involves no net movement, the surfaces are only involved in minute, cyclic deflections.

When combined with tiny inclusions in the metal (despite the use of vacuum melt steels) fatigue cracks develop below and the surface. The cracks propagate parallel and perpendicular to the surface. When a small volume of metal below the surface is surrounded by cracks, it is released from the surface leaving a pit with a flat bottom and near vertical sides. The sides of the wall will quickly break down if (as is normally the case) operation continues. Failure can be precipitated by exposing the parts to higher than normal loads, stresses and rotational speeds or just using it beyond its recommended life.

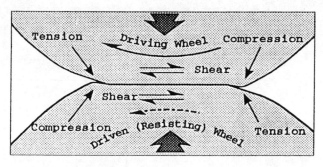

Figure 35-76. Deformation Caused When Two Rolling Surfaces are Held Together in Compression

●SURFACE ORIGIN FATIGUE. If the rolling while under compressive force is accompanied by a sliding action, as when one gear is driving another, the contact stress fatigue cracks tend to originate at the surface and are called surface origin fatigue. These cracks penetrate into the surface diagonally, causing the eventual separation of an arrowhead-shaped particle. The point of the arrowhead will point in the direction of rotation. But, like the vertical sided hole generated by subsurface origin fatigue, the arrowhead will be quickly destroyed if operation continues. Gear teeth are a primary target for surface origin fatigue. The pits formed can lead to bending fatigue failure of the gear teeth. Surface origin fatigue can also occur in the portions of reciprocating engine connecting rod and crankshaft bearings onto which the power stroke forces are directed. Maintenance of an oil film which prevents contact between the two surfaces will prevent the shear stresses necessary to cause this type of surface contact fatigue.

●SUBCASE ORIGIN FATIGUE. Case-hardened rolling/sliding surfaces which are also subjected to compressive forces (such as gear teeth and some roller mechanisms) can be suddenly destroyed by subcase origin fatigue failures. These failures are sometimes referred to as "spalling" and occur in a manner very similar to that experienced in subsurface origin fatigue. The difference is that the fatigue cracks (which are parallel to the surface) form below the case-hardened steel at a depth much greater than that seen in subsurface origin fatigue. In subsurface origin fatigue the cracks will be only a few thousandths of an inch below the surface and the particles released will be very small. However, subcase origin fatigue cracks will release particles 0.040 inches or more deep, causing very large stress concentrations and producing very large particles for abrasive wear. Subcase origin fatigue failures are the result of the use of subcase materials with inadequate shear strength or operation

at excessive loads. If the cause was inadequate shear strength then that condition probably exists in other parts from the same lot.

●CAVITATION FATIGUE. For aircraft, cavitation fatigue is only a potential problem for components in direct contact with low pressure liquids and whose surfaces are vibrating or moving through the liquid. At low pressures, small cavities or negative pressure "bubbles" form as the component's surface moves away from the liquid. As the process reverses (i.e. the component's surface and the fluid move toward each other) the bubbles collapse violently in an implosion. In a situations involving cavitation fatigue, this cycle is completed in microseconds and the compressive stresses involved can be several thousands of pounds per square inch. When the geometry of the system causes the implosions to repeatedly occur in the same location, the requirement for contact stress can occur and pits will form, adding abrasives to the fluid and creating stress concentrations in the component. Cavitation can occur in aircraft fuel and hydraulic

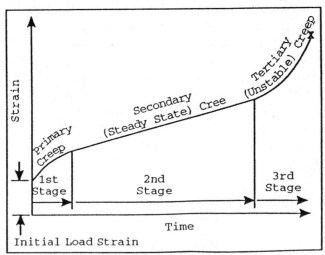

Figure 35-77. The Three Stages of Creep

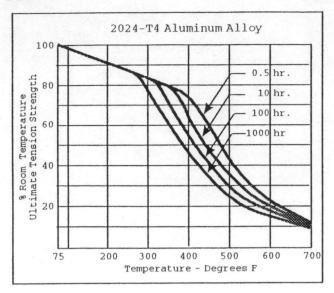

Figure 35-78. Loss of Strength in 2024-T4 Aluminum Alloy at Elevated Temperatures

pumps when system pressures drop to unacceptably low values.

13. CREEP.

Creep (sometimes referred to as stress rupture) is the slow, gradual, plastic deformation of materials exposed to long-term loads. It occurs even when structures are loaded below their elastic limit. Creep will cause prestressed fasteners to steadily lose their preload over time. If not periodically checked and adjusted for creep, the effectiveness of preloading used to minimize tension stress cycles will be reduced and fatigue cracks can develop. High temperatures will increase the rate of plastic strain and allow failures at lower stress levels and in a shorter period of time. Creep-caused structural failure is not a common reason for structural failure, which itself is not a common cause for aircraft accidents. However, it shows up often enough to warrant some knowledge by the aircraft accident investigator. Although creep of high speed rotating components in jet engine turbine sections was a relatively frequent cause for engine failures, creep-caused failure of modern jet engines is rare. The military suffered a rash of non-catastrophic failures when some older engines were asked to go the extra mile during the Persian Gulf War.

A. STAGES OF CREEP. Creep usually occurs in the three distinct stages shown in Figure 35-77, which shows permanent plastic strain as a function of time. Immediately after the initial elastic strain associ-

ated with an applied stress, the first stage of creep (called primary creep) will occur at a steadily decreasing rate. Eventually the creep rate will stabilize and the second stage or secondary creep will be established. The rate of deformation as a function of time will remain essentially constant during this stage. During the third or tertiary stage the rate of deformation will start to increase steadily, becoming unstable. During the latter stages of creep, internal, microscopic defects form and grow into visible pores in the grain boundaries. Fissures will open up where the grain boundaries intersect, then combining with other fissures to form multiple cracks. When the cracks cause sufficient loss of residual strength the component will suddenly fail.

B. EFFECTS OF OVERHEAT. Overheating can precipitate creep-related failures of exhaust valves in reciprocating engines. Short-term creep can occur in aircraft structures when the structural metal is heated above its normal operating temperature by fire or overheat (e.g. jet engine bleed-air leak) conditions.

Consider a normal category general aviation aircraft structure which should withstand at least a 3.8 G load without experiencing permanent objectionable structural deformation and a 5.7 G load without experiencing catastrophic structural failure. A reduction in strength of 74% could lead to permanent objectionable deformation if the aircraft was flying at just 1 G. A reduction in strength of 82% could lead to catastrophic failure if the aircraft was flying at just 1 G. That may sound like a pretty significant reduction in strength, and it is. But, if you look at Figure 35-78 you can see that at 525 °F 2024-T86 aluminum has lost 73% of its ultimate strength. At 600 °F it has lost 84% of its ultimate strength. Both in-flight fires and jet engine bleed air can get a lot hotter than that. Aircraft structures carrying flight loads will fail in short-term creep long before they reach their melting temperature of 1100 °F to 1200 °F. These failures will exhibit gross stretching and twisting and necking down. Major structural components like spar caps can be distorted like a pulled bar of warm taffy. Its the aluminum that's not carrying a load that melts and gets blown back as a spray with the airstream when exposed to uncontained inflight fire.

Probably the most frequent location of crash wreckage showing evidence of creep failure is the crash site itself. Especially if there has been a post crash fire. Before aluminum structure melts, it will lose strength to the point where it can no longer support its own weight. It will then give way to gravity, drooping down toward the ground. As it droops, the metal will stretch with cracks opening up on the tension side of the bend.

13. NON-DESTRUCTIVE INSPECTION TECHNIQUES.

Non-Destructive Inspection (NDI) refers to inspection techniques which will not do significant harm to the object being inspected. It is sometimes referred to as a Non-Destructive Evaluation (NDE) or Non-Destructive Testing (NDT). They're all the same. Although only six specific techniques will be discussed in some detail and several others mentioned in passing, there are many, many more. NDI started in the factory. Many of the new developments in NDI technology are also oriented toward the factory were an endless stream of parts are conveyed down a line and even the final products are funneled to a single local before being packaged off to the customers. The need was to evaluate the adequacy of the product before it was shipped to hopefully satisfied customers. However, the need to evaluate the structural adequacy of airplanes periodically during their lifetime has lead to the modification of these techniques to satisfy field requirements. Now we can check the adequacy of aircraft in the field without having to load it to failure. There are limitations as well as advantages to each of these techniques. Some are good for this and some are good for that. The aircraft accident investigator has to have knowledge of these techniques in order to evaluate the adequacy of NDI performed (or not performed) prior to a service life-related accident. In addition, some NDI techniques can provide the investigator valuable assistance when evaluating crash evidence.

A. VISUAL INSPECTION. This is the simplest and most common of the non-destructive inspection processes and uses the good old Mark I eyeball. Many inspections for fatigue cracks simply call for a visual inspection. The eye can be aided through illumination (a flashlight held at an angle to the surface can help pinpoint cracks), magnification (magnification of ten times using a lens is a low cost and in many cases efficient method of detecting cracks), and remote viewing (which allows viewing of surfaces in locations where the head simply cannot take the eyeball). Mirrors are the simplest of these remote viewing devices. In addition to assisting the non-destructive inspection technician, mechanics mirrors are often used by accident investigators to examine and photograph locations to which their head and/or camera are

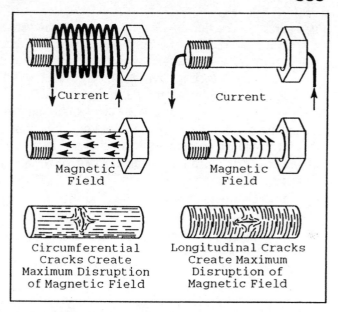

Figure 35-79. Detection of Longitudinal and Circumfrential Cracks Using Magnetic Particles

denied access. Borescopes provide the opportunity to get the eyeball to even more remote locations. While the older style borescopes provided only a straight line-of-sight (typically through an inspection access hole in a turbine engine), magnification and illumination, modern borescopes can bend around corners, creeping into previously inaccessible locations. These devices can use either fiber optics which are viewed through an eyepiece or remote video cameras which are viewed on a cathode ray tube. Obviously, when used as an inspection tool they can only be used to detect surface cracks, distortions and corrosion. In an investigative role, borescopes (and similar devices) can get the investigator's eye to locations which could be viewed only after parts are moved or disassembled, exposing them to the potential of addition post-crash damage or movement.

B. DYE PENETRANTS. Dye penetrant inspections are used to detect small surface cracks and discontinuities which may not be visible during strictly visual inspections. The process makes cracks which are open to the surface easier to see by increasing the crack's conspicuity, making it appear bigger and in a bright color which contrasts with the metal color. The technique, although simple in concept, is somewhat lengthy. After a component is thoroughly cleaned, it is then covered with a colored liquid (penetrant) which has the characteristic of being readily absorbed into surface cracks. The component's surface is again thoroughly cleaned, removing all surface penetrant before a "developer" liquid is applied to the component's surface. The developer draws some of the penetrant from the crack to the surface, making the crack appear larger. Many penetrants include a phosphorescent material which, when exposed to ultra-violet ("black") light, glows in the dark, further assisting in the visual identification of surface cracks. Although the dye penetrant technique requires neither the sophisticated equipment nor the sophisticated technical skills of the some of the NDI techniques described later in this section, it does have several limitations. First, it can detect only surface cracks. Subsurface cracks cannot be detected and even surface cracks whose aperture has been obscured may go undetected. For instance, if a surface finish such as un-removed paint covers the crack's opening, the penetrant will not be able to enter the crack. Contaminants can also clog the crack, preventing the penetrant's entry. For these reasons, a thorough cleaning of the component is necessary to ensure the inspection's reliability. Smearing the aperture of a crack can also deny the penetrant's access to the crack, defeating the purpose of the process. Smearing can occur when "stop drilling" fails to eliminate the tip of a fatigue crack or when surface wear works metal over or into the cracks aperture. Dye penetrant inspections can be automated for high volume operations using relatively immobile equipment. This process typically includes dipping components in a series of cleaner, penetrant, and developer vats with appropriate forced drying prior to inspection under an ultra-violet light in a dark room. On the other hand, dye penetrant kits also come in spray cans which can be carried to the ramp and used on the aircraft. Naturally, dye penetrant inspections can only be used in applications where the inspector can gain access to the surface being inspected.

C. MAGNETIC PARTICLE INSPECTION.
Magnetic particle inspections provide another way to assist the eye by increasing the conspicuity of a surface crack (or subsurface crack which is very close to the surface). Although the process requires more specialized equipment than the dye penetrant process, it makes the crack even more conspicuous if properly used. The process makes use of the fact that when a magnetic field is induced in a component made of ferro-magnetic material, surface (and near-surface) cracks will alter the component's magnetic field. If magnetic particles are placed on a magnetized surface they will align themselves along the magnetic field, this pattern showing any variations caused by the cracks. Finally, if the magnetic particles are phosphorescent, and viewed in a darkened room under an ultra-violet light, the pattern around cracks will be more apparent. It should be obvious that the orientation of the crack to the magnetic field is important. Cracks parallel to the magnetic field will cause minimum alterations in the magnetic field while cracks perpendicular to the magnetic field cause maximum alterations. Therefore, if the cracks being looked for could have orientations which are perpendicular to each other, the part will have to have (at different times) magnetic fields which are perpendicular to each other. Figure 35-79 shows how magnetic fields which are longitudinal and circumferential are necessary to check for both longitudinal and circumferential cracks in a tube. Magnetic particle inspections, obviously, can only be performed on ferro-magnetic materials, materials which can be magnetized. Once magnetized, virtually all aircraft components must be de-magnetized. In addition, magnetic particle inspections normally require disassembly of components.

D. EDDY CURRENT INSPECTION.
Eddy current is the first technique we have discussed which does not require direct viewing of the crack. This techniques involves the use of a probe to both generate an electro-magnetic field and sense and evaluate the "eddy-current" generated in the material being inspected. The presence of both surface and near surface cracks in the material will alter the shape of the eddy-current and the magnetic field it generates. The variations in the magnetic field are then evaluated electronically to provide information regarding structural deviation from the component used to calibrate the equipment. This can be as simple as a twitch on a meter's needle. More advanced models use a computer to provide a cathode-ray-tube presentation of the shape of the eddy-current in the material being inspected. This technique requires the calibration of the equipment for the specific design being inspected and the size of crack being searched. Typically, a sample component with a crack of known dimensions is used to calibrate the equipment. A template is also typically used to guide the probe along a pre-established path. Once the equipment is set up, the inspection is relatively routine and inspection of an area such as a rivet hole may only take a minute allowing the inspection of a large number of similar locations in a relatively short period of time.

However, the highly repetitious task of inspecting hundreds of rivet holes during shift after shift can lead to human factor problems associated with boredom, fatigue and human errors. Like the inspection techniques previously described, eddy-current requires direct access to the area being inspected. In addition, only material which conducts electricity can be inspected using this technique, no non-conductors or semi-conductors allowed. One additional use of the eddy current technique, one that is especially of interest to investigators working fire accidents, is the use of eddy currents to evaluate the extent of fire damage to aircraft structure. Eddy current techniques can be used to evaluate the heat treatment of metals. Metal components and structure which have been exposed to abnormally high temperatures during accidents will be re-heat treated. The extent and severity of the fire can be estimated by determining the new heat treatment of the structure. Some of the problems that can defeat the effectiveness of an eddy current inspection are: failure to thoroughly clean the area to be inspected: failure to warm-up and accurately calibrate the equipment; human factor problems associated with the implementation of boring, repetitive tasks; and use of the technique in areas or for equipment for which it was not designed.

E. ULTRA-SONIC INSPECTION.

Ultra-sonic inspections make use of high-frequency sound to find surface and subsurface defects. High-frequency sound waves are generated by a transducer and then beamed through the part being inspected. Either the reflected waves or the remnants of waves which penetrated the part are then measured with a receiver and electronically evaluated. Typically the evaluation is done through the use of an oscilloscope which presents a signal representing the sound energy recorded by the receiver. For large, repetitive jobs, the signals can be used to draw a picture showing the flaws and discontinuities in the part through which the sound traveled. There are two different ways the sound waves can be applied to and retrieved from the part being inspected; immersion of the part into a fluid which carries the sound waves to and from the part and direct contact inspection where the transducer and receiver are in direct contact with the part. The immersion technique requires large, bulky equipment which is not very mobile, but can be automated, saving labor hours for large components which are being inspected on a regular basis and eliminating many of the human factors problems. The direct contact technique is much more mobile, allowing use in the field on the aircraft or major fabrications. In addition to detecting flaws such as cracks, ultra-sound can also be used to determine the thickness of components and search for evidence of corrosion or excessive wear. Discontinuities as small as 5% of the total thickness can be measured. However, the orientation of the flaw to the sound wave is important. If the flaw is oriented in a way so that it does not reflect or block a significant amount of sound energy, the flaw can go undetected. Orientation of the transducer and receiver is important to the success of the inspection. Like eddy current inspection, ultra sonic inspections require precise set-up and calibration of the equipment using parts with known defects.

F. RADIOGRAPHIC INSPECTION.

In its simplest form, radiographic inspection equipment and procedures are not much different from those used to X-ray your leg the last time it was broken. Very short wave electromagnetic radiation (X-rays or Gamma rays) are generated and directed through the part being inspected and toward unexposed radiographic film. The variations in the exposure of the developed film can be interrupted to determine variations in the density of the part being inspected. Rays passing though cracks, flaws, voids and corroded areas will not be attenuated as much as rays passing through sound material. These un-attenuated ray will expose the film to a greater degree, showing up on the developed film as a darker area, while rays which are attenuated by sound structure will show up as a lighter area. To the untrained eye, cracks, flaws, voids and corrosion may appear to be just another shadow on the film. You may remember that when the doctor clipped the x-ray film showing your shattered leg to the light board, the doctor probably called in another doctor to get a second opinion. You may also remember the pain, shortly before that moment when the nurse manipulated your leg around the x-ray table, trying to ensure that the "pictures" would show all of the cracks. Orientation of the x-rays so as to illuminate the discontinuities and proper interpretation of the film are therefore important aspects in ensuring the thoroughness of the inspection. The x-ray and film process is slow. The film has to be developed, denying the inspector immediate feedback on either the soundness of the structure or the thoroughness of the inspection. Is the part good or bad? Was the inspection valid or invalid? This problem can be solved by replacing the film with a screen which generates electromagnetic signals which can be transmitted to and displayed on a cathode ray tube, providing the inspector with immediate feedback on the structure's fitness and adequacy of the inspection. Various electronic devices can be used to save the pictures for future reference. Going one step further, tomography techniques can be used to take a large number of computer controlled pictures from different angles and locations. Each of these pictures show a "slice" of the structure from a slightly different angle. This information is provided to a computer which in turn builds a 3-dimensional picture of the part's surface and internal structure. The computer will then generate, at the command of the operator, any cross-section of the structure. This computer generated picture can be displayed on a cathode ray tube, and can be changed at the flick of a keyboard to show other angles. Color displays can be used to enhance the variations in the intensity of the rays penetrating the part, simplifying the inspectors task and increasing the inspection's reliability. Naturally, as we go from the relatively simple generator and film to the full-blown tomography unit, we lose mobility and increase expense. Simple X-ray generators and film packages are now fairly compact and have been used in the field on aircraft for many years. The tomography unit requires an elaborate safety room with thick concrete walls, doors that meet high security prison standards and computer equipment that could bankrupt some small nations. And, although they can take some pretty hefty components, you would be hard pressed to get a wing from even the smallest airplane into one of these things. Radiographic inspection is the first method we have discussed that presents significant, life threatening hazards to humans. Failure to follow safety precautions can ruin your day; perhaps your life. These precautions can interfere with other tasks, especially when they are performed in the field on aircraft. Use of excessively high power settings can result in radiation penetration to distances further than those intended. Exposure for longer than necessary time periods can cause cumulative damage. These hazards are insidious, providing no immediate indications of their presence or the damage they are doing to the body. The risks generated in radiographic inspections are widely understood, but rely heavily on the humans performing the inspections for control. Despite the disadvantages of costs higher than other inspection techniques, highly qualified personnel and safety hazards, x-rays are often selected as the preferred method of inspection because of their ability to "look through" the structure and the "hard copy" record of the results. Radiographic inspection also provides a valuable tool for the inspector who would like to "look inside" a component before testing or disassembling a component. If an inspector suspects that a component's failure may have started or sustained an accident sequence, the prudent inspector will have the internal status of the component established through radiographic inspection before exposing it to the risk of additional damage during testing or movement during disassembly.

G. NEW DEVELOPMENTS IN NON-DE-STRUCTIVE INSPECTION TECHNIQUES.

Because of the need for reliable methods of detecting ever smaller flaws in aircraft structure, many new and innovative methods for detecting these flaws are being developed. Some of these new techniques are briefly described below.

- Holographic Analysis. Holographic images of perfect specimens are superimposed over a laser lit image of individual specimens. The individual specimens may be loaded to identify its reaction under stress.

- Pulsed Thermography. Measures variations in temperature or heat flux after a specimen is heated. Debonds, cracks, voids, corrosion and other flaws are detected because their heat emission characteristics differ from the parent material.

- Shearography. Two laser illuminated images of a specimen are evaluated to measure changes in motion in a stressed specimen. Flaws show up as out-of-plane motion on the specimen.

- Non-invasive Magnetic Inspection. Ferromagnetic components can be inspected using a highly portable device which basically uses the concepts of magnetic particle detection with out the particles and the bulky, messy equipment.

- Ion Microtomography. This technique is reported to provide images which are 1,000 times more detailed than X-ray CAT scans by passing computer-positioned beams of high-energy ions through a sample and then using the computer to construct a two dimensional slice and three dimensional images of the sample.

- Robotics. These are devices which, using any of several of the techniques described above, automatically perform the routine, tedious inspection tasks. The NTSB has identified the human as a weak link in most "field" NDI operations. The human is just not very reliable when it comes to boring, repetitive tasks. The robot would do the routine work, calling in the human only when there is a decision to be made.

14. SUMMARY.

Too much on structural failure? Perhaps. At least when the structural engineer baffles you with some technical jargon on why something broke, you'll have a place to go to find out what he was trying to say.

36

Crash Survivability

1. INTRODUCTION.

One of the issues which must be addressed in any aircraft accident in which a person was injured is the question of crash survivability. What actually caused the injury? Did the injury have to occur? Could something have been done to eliminate or reduce the severity of the injury? In some aircraft accidents the impact forces are not high enough to preclude the survival of some or all of the occupants. A shallow angle impact on unobstructed terrain at general aviation aircraft approach speeds should be survivable. Even the cart-wheeling crash of a wide-bodied jumbo jet allowed a significant number of crew and passengers to survive. In other apparently less severe impacts, the survival rate was lower. The aircraft accident investigator must be prepared to identify the various factors which influence the potential survivability of a crash. This chapter will discuss five factors which play a major roll in determining if the occupants of an aircraft crash will survive. It should be realized, obviously, that not all crashes are potentially survivable. Given enough speed at impact, a high enough impact angle, or the right combination of the two, it is not realistic to expect survival. This chapter will identify and discuss the factors which have on occasion been ignored during the design or modification of aircraft and have contributed to serious or fatal injuries in an otherwise survivable aircraft crash.

A potentially survivable aircraft crash is one during which the deceleration forces transmitted to the occupants did not exceed human tolerances and in which the structure surrounding the occupants remained substantially intact. (This assumes, of course, that a post-crash fire did not inhibit survival.) This chapter will also show how deceleration forces can be calculated and related to human tolerances.

In the early 1960s the United States Army Transportation Research Command launched a series of studies to examine the factors which determined the chances of crew and passenger survival during helicopter and light fixed wing aircraft crashes. Using the results of studies concerning crash forces, structural collapse or failure, crash fires and crash related injuries, design features which degraded the chances of survival and increased the risk of injury were identified. Design features which improved the odds of crash survival were then developed and tested. Effective design solutions to specific crashworthiness issues were then consolidated and published in Crash Survival Design Guide (1967). The results of these efforts were amazing. A crashworthy fuel system retrofitted into Army UH-1 helicopters all but eliminated thermal induced fatalities and injuries. In addition to these fuel systems which resist rupture, puncture and tearing, seats and associated restraint systems have been designed to significantly minimize crew and passenger injury due to the high G loads associated with crash impacts. The latest generation of Army helicopters demonstrated a significantly lower crash survival injury or fatality rate than the previous generations. The design requirements which lead to this dramatic improvement in safety have been periodically updated and are currently documented in US Army's Aircraft

Crash Survival Design Guide USAAVSCOM TR 89-D-22B. This document has grown into the five volumes listed below and no attempt will be made in this text to go into the detail contained in these volumes. Instead, this chapter will summarize the issues of crash survival, providing the investigator with the general knowledge necessary to identify potential breaches of good crash survival design concepts.

US Army Aircraft Crash Survival Design Guide

Volume I - Design Criteria and Checklists

Volume II - Aircraft Design Crash Impact Conditions and Human Tolerance

Volume III - Aircraft Structural Crash Resistance

Volume IV - Aircraft Seats, Restraints, Litters, and Cockpit/Cabin Delethalization

Volume V - Aircraft Postcrash Survival

2. CREEP METHOD OF INVESTIGATION.

Investigation of potentially survivable aircraft crashes becomes a little easier if a systematic approach is taken, breaking a complicated series of events into smaller, more digestible bites. One such approach uses five aircraft design factors which control the chances of crew and passenger survival during a crash. The mnemonic for these factors is **CREEP.**

CONTAINER
RESTRAINT
ENERGY ABSORPTION
ENVIRONMENT
POST CRASH FACTORS

The first four of the five CREEP factors relate to the dynamic portion of the crash itself. The portion of the crash sequence which starts when the aircraft first makes contact with the ground and continues until the aircraft comes to its final resting position. These four factors are concerned with the initial and any subsequent impacts with the terrain, the associated deceleration forces acting on the aircraft and its occupants, and the deformation and dislocation of aircraft structure and its contents.

The fifth factor relates to the occupant's attempts to egress the aircraft before suffering additional injuries not directly resulting from the dynamic portion of the crash and most often involving post-crash fire..

A. CONTAINER. In order to survive a crash it is first necessary to provide a "living space" for the occupants during the dynamic portion of the crash. If this space is crushed or punctured, the chances of survival fall drastically.

It didn't take long for the founders of the flying services to realize that a cockpit designed to withstand a 40G longitudinal load was a desirable feature in military aircraft. Lindbergh had all of the heavy stuff (engine and fuel) in the "Spirit of St. Louis" in front of him so that if he had to make a sudden stop he wouldn't become the meat in an aluminum and steel sandwich. Modern crop dusters followed that idea when they placed their cockpits behind the "dust" hopper. Today, structural engineers can predict how and where an aircraft's structure will fail during a potentially survivable crash. Structure surrounding crew members and passengers should be designed to thwart the collapse of living space and avoid its penetration by objects which could cause injuries to occupants. Aircraft components such as landing gear actuators and supports should not be designed so that their crash-related failure modes allow the dynamic forces of the crash to drive these components through the crew or passenger cabin.

One of the investigative techniques used to address this issue starts with the determination of the origin of injuries experienced during the dynamic portion of the crash. If avoidable design features allowed the violation of living space and associated injuries, then corrective actions such as structural redesign or revised maintenance actions may be appropriate.

When investigating the living space available to the occupants, the investigator must recognize that the apparent living space observable in the post crash wreckage may be deceiving. The casual observer can be fooled into believing that the occupants had MORE living space than they actually had. This illusion is due to the fact that a significant portion of the deformation experienced during the dynamic portion of the crash is "elastic"; it springs back and is not directly observable in the post crash wreckage. To the uninitiated, the wreckage may appear to have had living space available to the occupants, but their injuries may show that the space had collapsed during the crash. The elastic rebound of the structure may explain the apparent inconsistencies. If you have had the opportunity to observe the high speed films of aircraft crash tests performed at NASA's Langley facility, you were able to see first hand how an aircraft's structure can collapse to a lethal volume for an instant an than expand to an apparently livable space.

Another aspect of living space to be evaluated is roll-over protection for low or mid-winged small aircraft. The problem is that unless there is specific roll-over protection designed into the airplane, an airplane which manages to get over onto its back during a botched landing or takeoff can put the occupant's head at considerable risk. High wing aircraft provide lots of structure over the pilot's head. But many low wing homebuilts provide only Plexiglas. There is one report of a small, homebuilt jet which got onto its back while sliding down the runway. Although head injuries were fatal to the pilot, the crash was survivable and the airplane was repaired to fly again. Adequate roll-over protection could have allowed the original pilot to make the first flight after the repair.

The federal requirements for the protection of occupants during survivable crashes in general aviation fixed wing aircraft are contained in FAR Part 23.561. This paragraph defines the minimum crash inertia loads, along the various aircraft axes, which the aircraft structure must withstand.

The federal requirements for the restraint of occupants and items of mass within the cockpit and cabin of Transport Category Airplanes is contained in FAR Part 25. Ultimate inertia force requirements are specified in the Emergency Landing Conditions section: paragraph 25.561. This paragraph defines the minimum crash inertia loads, along the various aircraft axes, which the aircraft structure must withstand.

B. RESTRAINT. If the occupants have been provided with adequate living space, the next series of questions deal with the restraint of the crew and passengers and the equipment and components around them. Occupants of any moving vehicle must be protected from injuring collisions within the vehicle; being thrown against the sides of the living space or having objects such as cargo or equipment thrown at them. The strength of all restraints should be sufficient to prevent injury at the force levels which can be expected during the most severe, but survivable, crash. The aircraft accident investigator must examine all restraint system failures to determine if their failure contributed to injuries experienced by the crew or passengers. These failures should include not only mechanical failures of the restraint system, but also failures of the restraint system to do its job properly. One way to look at restraint systems is use the "tie-down chain" concept and look at all parts of the system. A typical tie-down chain might include the

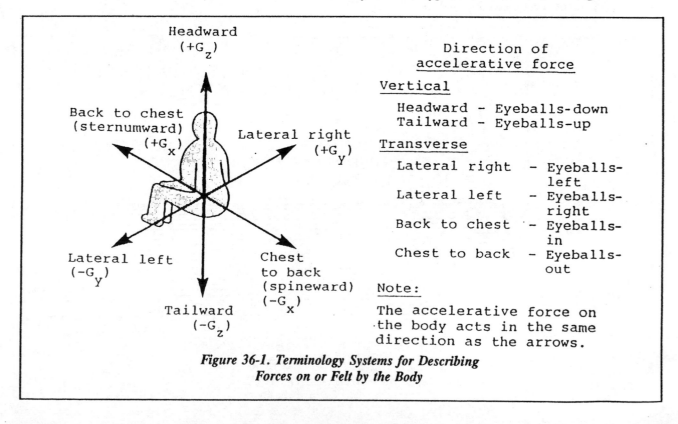

Figure 36-1. Terminology Systems for Describing Forces on or Felt by the Body

floor, the seat rails, the seat attachment fittings, the seat, the belt and harness attachment fittings, the belt and harness webbing and the buckles.

Some of the issues which are of special interest to restraint systems are: varying strength requirements along the aircraft's three axes, the influence of dynamic loading (as opposed to static loading), uneven sharing of loads when multiple restraints are used, the propensity of the restraint system to injure the occupant, and the ease of release of the restrain system in a post-crash environment.

When discussing human restraint systems the normal assumption will be that the occupant is being restrained while in the forward facing position. While aft facing seating will be mentioned briefly, side facing seating will not be discussed.

When restraining humans it is important that the restraint system design does not contribute to the injuries while preventing unwanted movement of the occupant. The geometry of the restraint is critical. The lap belt in a system preventing forward movement needs to transmit loads to the portion of the body which is best able to withstand theses loads; the pelvic joint. Belts which provide restraint above the joint put excessive loads on stomach and other internal organs. Belts which provide restraint below the pelvic joint are likely to allow submarining especially during impacts with a significant vertical component. During the submarining process the occupant slides under the belt, suffering additional injuries due to both the lack of restraint and the process of being squeezed through the gap between the belt and the seat.

Shoulder harnesses are important because they limit the upper body torso's ability to rotate forward and down during decelerations. This motion should be avoided because it causes dynamic loads to the upper body which can exceed the load at the body's center of gravity. Also, the occupant's flailing envelope is enlarged and it increases the likelihood of head injuries which, even if not fatal or even serious in the long term, could degrade the occupant's ability escape unassisted from the aircraft.

In order to minimize back injuries, shoulder harness restraint should be applied at an angle between horizontal and approximately 25 degrees upward. Shoulder harnesses which extend over the shoulder and then downward behind the seat may increase the likelihood of back injuries and submarining. When designed or installed in this manner, crash related tension loads on the shoulder harness place additional loads on the spine, and any negative G (upward) loads are carried through the shoulders to the spine instead of the though the pelvic joint. In addition, the tension load in the harness created by a forward deceleration also creates a downward force on the torso which can increase the potential for submarining.

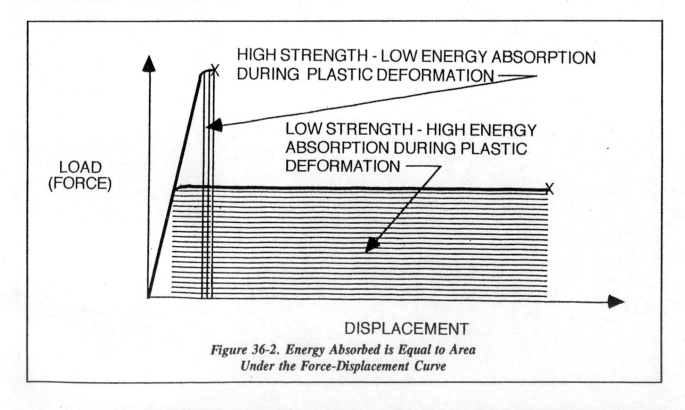

Figure 36-2. Energy Absorbed is Equal to Area
Under the Force-Displacement Curve

Before leaving the lap belt and shoulder harness, two more areas need to be touched on; crotch straps and web width. Although use of a lap belt tie-down (crotch strap) is not normally seen on other than aerobatic aircraft, they can provide additional protection on any aircraft (or fast moving vehicle). The primary purpose of the this device is not to restrain the body at a point between the legs, but to resist the upward movement of the lap-belt caused by the dual action of the upward pulling shoulder harness and the dynamic reaction of the body as it responds to forward/vertical decelerations by attempting to submarine under the lap belt.

Use of a restraint webbing with a large width is desirable from a safety perspective because it spreads the restraining load over a wider area. Narrower belts and harnesses are more comfortable to wear and their associated hardware is lighter. The minimum recommended widths for forward facing seats are 2.5 inches for lap belts with a desirable width of 4 inches in the center abdominal area and 2 inches for shoulder straps.

When discussing restraint systems, a brief review of the coordinate axis system used by structural engineers and its relationship to crashworthiness requirements is in order. Crash loads generated in response to crash generated declarations are three dimensional vectors. Structural requirements to withstand these loads are expressed in terms of longitudinal

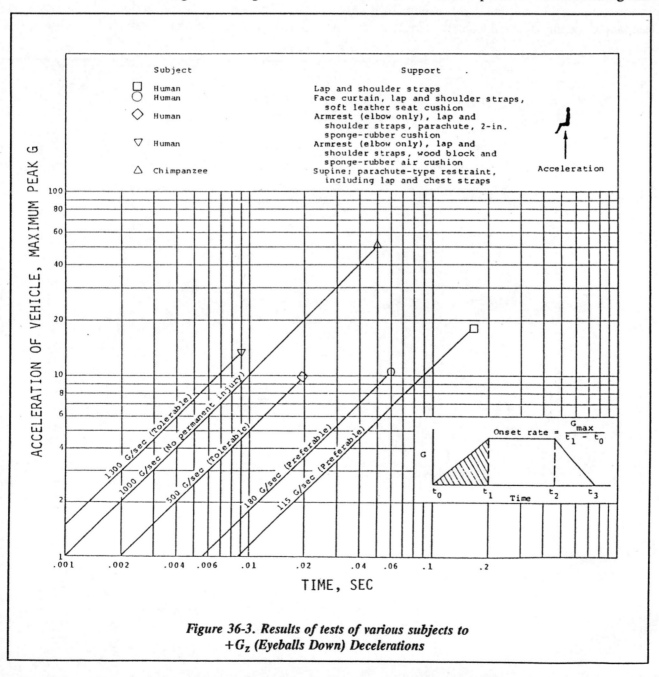

Figure 36-3. Results of tests of various subjects to
+G$_z$ (Eyeballs Down) Decelerations

(X or parallel to the fuselage) axis, lateral (Y or parallel to the wing) axis and vertical (Z or up-and-down) axis requirements. A discussion of loads along these axes can get a little confusing, so many investigators have the habit of describing forces in terms of which way the victim's eyeballs move. The X axis is eyeballs in or out; Y axis is right or left; and the Z axis is up or down. Figure 36-1 explains this and correlates all the different ways of describing forces.

It should be obvious that the magnitude of crash related deceleration loads are not equal along all these axes. The strength requirements of restraint systems should also be tailored to their orientation in the aircraft. In other words, a seat which meets or exceeds crashworthiness requirements when oriented so that the occupant is facing forward, may not meet minimum requirements if the direction the seat is facing is rotated 180 or 90 degrees. It should be obvious, therefore, that the crash investigator should be alert to unauthorized modifications which could degrade the restraint system's effectiveness.

The federal requirements for the restraint of occupants and items of mass within the cabin of general aviation fixed wing aircraft are contained in FAR Part 23. This defines the requirements for seat and restraint systems in terms of occupant weight and G pulse size and shape. Similar requirements for Transport Category Airplanes are contained in FAR Part 25.

The structural connections of the restraining devices are also worthy of review by the investigator.

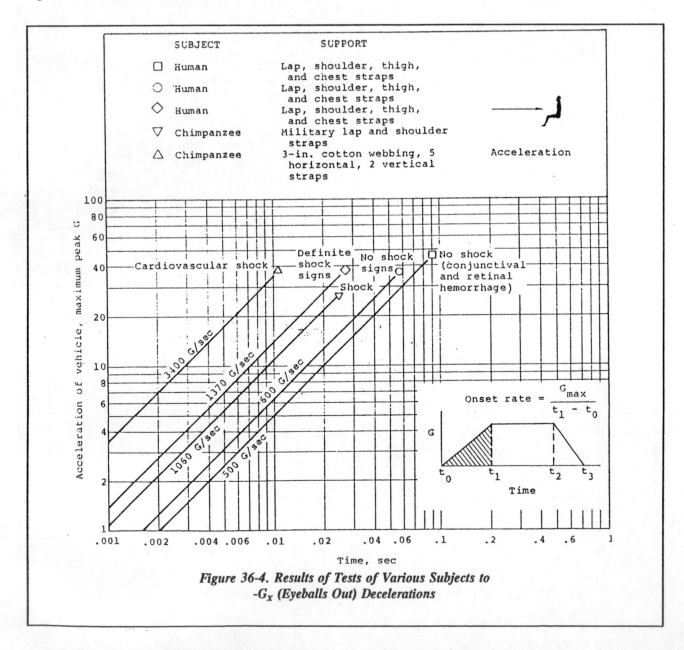

Figure 36-4. Results of Tests of Various Subjects to
-G_x (Eyeballs Out) Decelerations

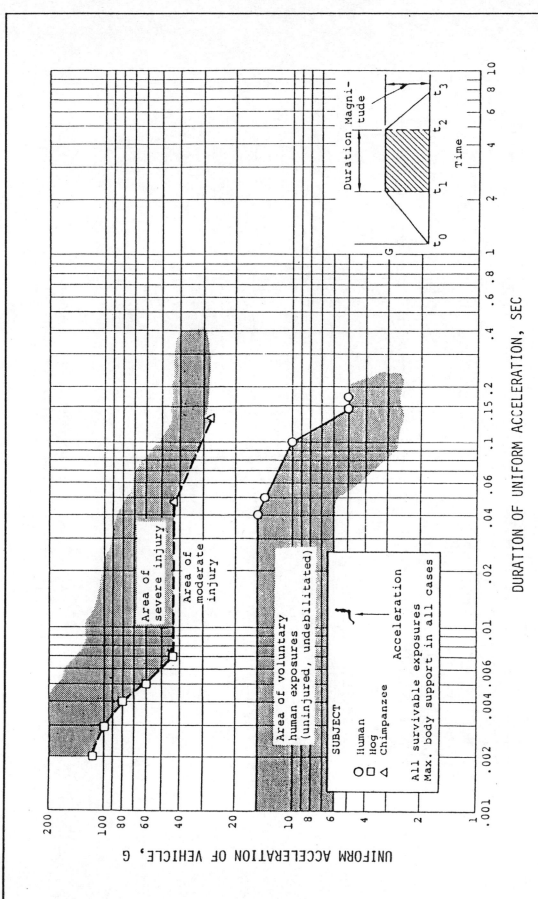

*Figure 36-5. Duration and Magnitude of +G$_z$ (Eyeballs Down)
Decelerations Endured by Various Subjects*

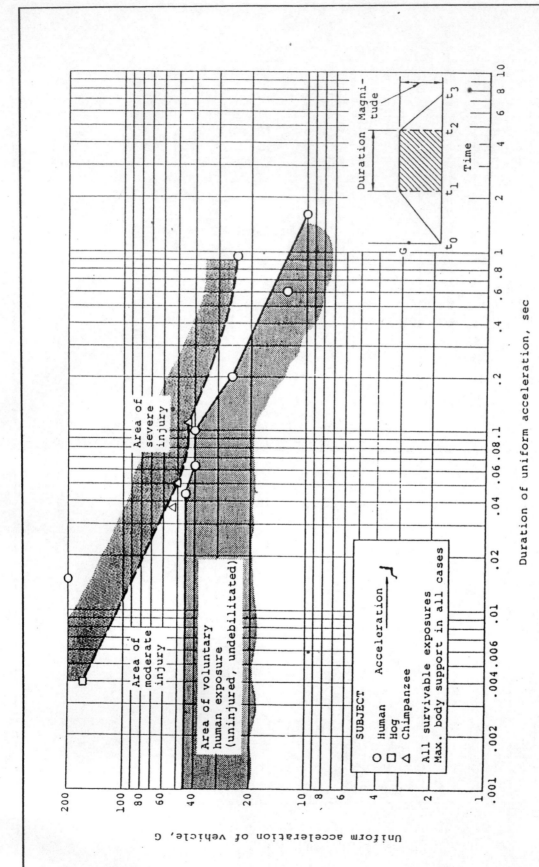

Figure 36-6. Duration and Magnitude of -G$_x$ (Eyeballs Out) Decelerations Endured by Various Subjects

This should include bolts, rivets, and welds as well as the structure through which the restraint system is attached to the aircraft structure. Loss of integrity of the structure to which the restraint system is attached is just as hazardous as loss of integrity of the restrain system itself. The structural adequacy of the webbing itself, the radius of metal fittings around which the webbing is wrapped and the stitching which secures joints in the webbing can be reasons for failure of the personnel restraint system. The anchorage point for the lap belt and shoulder harness can successfully be located on either the seat or the airframe. If the airframe is used and a load attenuating seat is used (see the section of energy absorption for discussion of load attenuating devices) the design must accommodate the movement of the seat expected during a crash. Finally, since the seat is normally an integral part of the restraint system, failure of the attachment points which secure the seat to the floor should also be investigated. Seats which pop out of their tracks or fittings when the aircraft's floor predictably buckles during a survivable crash can directly contribute to injuries.

One last comment before leaving the subject of restraint systems. The occupant should be able to release themselves from the system after it has been exposed to the loads expected during a survivable crash and in any attitude in which the occupant is suspended. There is, for example, a type of seat belt release found in automobiles (but not aircraft) that cannot be released if the occupant is hanging upside down from it. You don't suppose anyone would ever install that type of belt on his airplane do you?

C. ENERGY ABSORPTION. The deceleration forces created during a crash may be high enough to cause fatal or serious injuries, even if a safe living space, adequate crew and passenger and restraint and a delethalized flailing envelope are provided. Since the crew and passengers' bodies are not rigidly attached to the airframe, the design of the aircraft structure and seats may cause the acceleration forces experienced by the crew and passengers to be either amplified or attenuated.   A soft deep seat cushion can greatly amplify the vertical Gs experienced by someone sitting in the seat. Similarly, a deep seat cushion that deforms only at high loads, absorbing energy as it gives, can greatly reduce the vertical crash loads to which a seat occupant is subjected.

Energy absorbing structure can also be placed between the crew and passengers and the point(s) of impact, reducing the chances of injury and increasing the probability of survival. On the other hand, rigid structure located between the point of impact and the crew/passengers can transmit the high decelerations forces at the point of the impact directly to the passengers. Seats can also be designed to reduce crash related injuries, deforming gradually while "stroking" over relatively large distances to absorb energy which would otherwise have to be absorbed by the occupants. A word of caution - if occupant restraint systems are anchored to the aircraft structure and not a load attenuating seat, the restraint system must be designed to accommodate the movement of the seat. Seats which simply collapse when exposed to crash loads, allow the occupant to fall to the floor which is likely to be already moving toward them at a high speed. In addition, crushable structure between an occupant and the impact point also tends to protect the container from being penetrated during the impact sequence.

Energy absorption may be calculated by graphing the force (load, stress) involved versus the displacement or plastic deformation of the structure. The energy absorbed is equal to the area under the curve during deformation. This is shown in Figure 36-2. If the structure is high strength material, such as carbon fiber composites, which doesn't deform very much, there will be little energy absorbed and most of the forces will be transmitted to the occupants. If, on the other hand, the structure is designed to deform in a controlled manner, a considerable amount of energy can be absorbed.

Investigators should investigate the cause of G load induced injuries. High G loads may have been avoidable through better design. Any existing load attenuating system performance may have been degraded by the manner in which the system was used or maintained.

D. ENVIRONMENT. Hopefully, the designers will build a secure box around the crew and passengers and secure them to it. Although we may be able to restrain the torso, it is normally impractical to secure the head and limbs of the crew and passengers. The volume through which the unrestrained extremities can be expected to move should have been delethalized to the maximum degree possible. Obstructions which could cause injuries should have been either removed from within the flailing envelope or padded to reduce the severity or probability of injury.

E. POST-CRASH CONDITIONS. All too

frequently, crew or passengers survive the dynamic portion of a crash, only to suffer additional injuries or death when they are unable to safely exit the aircraft in a timely manner. The two primary factors in the causation of fatalities during otherwise survivable crashes is post crash fire and inability to quickly exit the damaged aircraft.

Fire is far and away the most significant post crash hazard. Although fire can kill and injure directly through heat, the toxic fumes and smoke produced when material in the aircraft interior burn are more often the direct cause of death.

Control of fire, therefore, is one of the key issues in crash survival. The first order of precedence is to prevent post crash fires. As mentioned earlier, the US Army has done a lot to prevent crash-related fires. Intelligent designs can place fuel lines and containers in the least vulnerable locations so that a structure which is expected to collapse or fail during a crash will not cause a flammable fluid-carrying line to rupture.

The second step is to provide materials in the aircraft cabin which do not react unfavorably to heat. Notice that we did not say, "do not react to heat." Unless we are willing to furnish our cabin interiors with fire brick, any candidate material chosen will have some reaction if exposed to enough heat. Some will burn. Others will melt. Still others will char and produce thick smoke. Some may decompose into a toxic chemical. An organic material (wool, for example) will almost always produce hydrogen cyanide when it decomposes.

There is no perfect material. It appears that the worst material to put in the cabin is the one that produces thick smoke which inhibits both vision and breathing and, therefore, escape. The one that decomposes to a toxic chemical may be acceptable because the concentrations will not reach a lethal level for several minutes.

This brings up a point about carry-on luggage. Let's suppose that we have a plane with a capacity of 100 passengers. Each passenger brings luggage on board that weighs 10 lbs per passenger. As a frequent flyer you think that's a little low, but that is the figure the FAA allows to be used in estimating weight and balance. This gives us 1000 lbs of who-knows-what per plane which is going to contribute, one way or another, to the egress problem if there is an accident.

That's probably more than the weight of all the cabin furnishings including seat cushions. The knowledge that we are willing to allow that much uncontrolled material on board takes a lot of the fun out of cabin safety design.

From the investigator's point of view, it is always wise to assume that there were things in the cabin which really didn't belong there and probably didn't help during the crash.

The availability of the means to rapidly exit the aircraft after the crash is critical in any crash in which fire or submersion is possible. This means that occupants must be able to free themselves from their restraint systems, expeditiously locate and move to the exits, and then operate the exits. Pre-crash knowledge of exit availability and operation, the ease of locating the exits and the ability of the occupants to operate the exits all are factors which must be addressed during the investigation of a potentially survivable accident.

Failure to inform occupants of emergency egress procedures, exits which are difficult to locate during actual crash conditions, exits whose operation violates normal behavioral stereotypes and exits whose design makes their operation vulnerable to the loads and associated damage expected during impact are all factors needing review during the investigation.

Design of airplane exits is predicated on the normal parked attitude and configuration. Obviously, this is not always the case. Sometimes the occupants will have to exit from an airplane that is in an abnormal attitude and perhaps in a very unusual configuration. Although Part 125 airplanes have specific emergency exit requirements levied on them, many general aviation airplanes have only one exit which can be easily jammed if the airplane ends up inverted. Many homebuilts are just as bad or worse. In any case, it's the crash investigator's responsibility to look into any factor which impedes egress to the point where injuries result.

The federal requirements for Emergency Evacuation, Emergency Exits, Emergency Exit Marking and Emergency Access for General Aviation Aircraft are contained FAR Parts 23 and 25.

3. CRASH SURVIVAL CALCULATIONS.

This section complements Chapter 5 on Wreckage Distribution. There are at least two approaches to the problem of calculating impact forces. One method treats it as a problem in energy transfer which gets a little messy and is beyond the scope of this text. An easier method, one that can be used by the field investigator, is to consider the kinematics of the crash.

First, crash survivability is a function of both the peak G experienced and the duration of the G. The human body can withstand a really big jolt for a very short period of time. Figures 36-3 and 36-4 show some of the basic limitations in the eyeballs down and eyeballs out direction.

Second, if we are going to calculate the load actually experienced, we need to calculate both the average peak G and the time duration. Can we do this? Yes, if we can establish the impact velocity in both the vertical and horizontal direction and the respective deceleration distances. That was the substance of Chapter 5 and we defined deceleration distance (either vertically or horizontally) as the sum of the depth (or length) of the impact crater and the aircraft structure "crush distance;" the amount it deformed.

It might be helpful to discuss how this really works before explaining how it can be done in the field. There has been a lot of research on impact forces on both automobiles and airplanes. Some years ago, NASA became the owner of several aircraft structures which they crashed under very controlled and instrumented conditions.

If you hook an oscilloscope up to a plane that is about to crash and display the G load with respect to time, you see that the impact is not smooth. The impact is a series of ragged peaks for short durations of time. If there is more than one impact, these irregular peaks continue until the plane has come to a stop.

Each of these oscilloscope displays has a characteristic shape based on the nature of the impact. Some, for example, approximate a rectangle. This would be characteristic of (say) a gear up landing. The G load is established at touchdown and is constant until the plane comes to a stop, then it drops to zero. Others approximate a triangle. As the plane crashes, the G load starts at zero and builds to a maximum. As the plane decelerates, the G load goes back to zero.

There are other possible pulse shapes. An airplane that flew into the vertical side of a mountain would have a pulse shape that resembled a right triangle. The aircraft would feel the max G load as it hit the mountain and the Gs would drop rapidly to zero as the rest of the structure came to a stop. Another shape might be described as a half sine wave (which most of us would call a semi-circle) describing an impact where both the G load and the velocity are constantly changing throughout the impact.

We never have a real accident with enough recorded data to tell us anything about the pulse shape, the peak G or the impulse time. Nevertheless, by knowing the circumstances of the accident, we can pick a pulse shape that would most closely describe the accident. Since this involves some guess work, the solution is to pick the pulse shapes that represent the maximum and minimum pulses that could have existed.

In the case of a general aviation accident, the minimum pulse is always going to be represented by a rectangle. That's as gentle as it ever gets. The maximum pulse (assuming that the plane didn't fly into the side of a mountain) is the equilateral triangle. The correct answer should be between those two extremes. For an air transport accident, the minimum pulse would still be a rectangle, but the maximum pulse might more closely resemble a trapezoid which is really a triangle split by a rectangle which represents a period of constant deceleration.

Anyway, we know (from research) that there is a relationship between the area under the pulse shape we have selected, the velocity of the aircraft and the distance through which that velocity was reduced. Thus we can derive some formulas that, for any selected pulse shape, will allow us to calculate peak G and duration based on velocity and stopping distance.

Is this accurate? Not bad! The method has been verified by matching the calculations on a research crash with the data recorded and calculated by a computer. Besides. You are going to calculate both "best case" and "worst case" solutions. How wrong can you be?

Equations 1 through 6 are for rectangular and triangular pulses, which are the ones you will use most often. In all equations, it is assumed that there are a series of impacts and the aircraft is still traveling at a measurable speed as it finishes the first impact. If that's true, then the reduction in velocity is expressed as $V_1 - V_2$. It is that velocity that is absorbed through the deceleration distance S. If the velocity went to zero in the first place, just use the impact velocity.

As discussed in Chapter 5, we can't deal with the actual velocity of the aircraft. We must resolve this into the horizontal and vertical velocities because all of our tables are based on horizontal and vertical Gs. Equation 1 is the horizontal G calculation for a rectangular pulse.

$$G_H = \frac{V_{H_1}^2 - V_{H_2}^2}{2g\ S_H} \qquad (1)$$

Where:

G_H = G load in the horizontal direction

V_{H1} = Entering velocity in the horizontal direction (FPS)

V_{H2} = Exit velocity in the horizontal direction (FPS)

g   = Acceleration due to gravity: 32.2 ft/sec^2

S_v = Horizontal stopping distance (Feet)

Equation 2 is the vertical G calculation for a rectangular pulse. Notice that there is only one velocity component as the vertical velocity is assumed to go to zero after the initial impact.

$$G_v = \frac{V_v^2}{2g\ S_v} \qquad (2)$$

Where:

G_v = G load in the vertical direction

V_v = Velocity in the vertical direction (FPS)

g   = Acceleration due to gravity: 32.2 ft/sec^2

S_v = Vertical stopping distance (Feet)

Equation 3 is the horizontal G calculation for a triangular pulse.

$$G_H = \frac{V_{H_1}^2 - V_{H_2}^2}{g\ S_H} \qquad (3)$$

Equation 4 is the vertical G calculation for a triangular pulse.

$$G_v = \frac{V_v^2}{g\ S_v} \qquad (4)$$

The vertical pulse duration for the rectangular

pulse is shown in Equation 5.

$$t_v = \frac{(V_{1_v} - V_{2_v})}{g\ G_v} \qquad (5)$$

Equation 6 shows the vertical pulse duration for the triangular pulse.

$$t_v = \frac{2(V_{v_1} - V_{v_2})}{g\ G_v} \qquad (6)$$

The equations for the horizontal pulse durations are the same except they use horizontal values.

As stated, you can assume that the initial vertical velocity went to zero during the initial impact. Horizontally, assuming that the horizontal velocity went totally to zero during the initial impact would be true only if the plane stopped right there and did not bounce or continue on past the impact crater. Keep in mind that we are trying to relate G to the reduction of a certain velocity through a certain distance.

Now what? Now that you have a range of G forces that were present in the impact, you go to a chart such as the ones in Figures 36-5 and 36-6 to see whether the impact was reasonably survivable or not. Armed with that knowledge, you consult with the physician who examined the injuries and see if you agree that this was or was not a survivable accident. See also Chapter 24. If it was--and they didn't--you need to find out why using the CREEP method described earlier in this chapter.

4. SUMMARY.

A well-designed aircraft can absorb a lot of impact energy before the forces reach the occupants. If the occupant's tie-down chain, which includes the seat belt, shoulder harness, attachment fittings, seat, seat attachment fittings and floor, is breached, then the occupant receives no benefit from the energy absorption capabilities of the structure. The occupant continues forward at the initial impact velocity and eventually absorbs all the impact forces with his body. A survivable accident suddenly becomes unsurvivable. Use of the techniques described in this chapter allows the investigator to determine whether or not the occupants should have survived the impact.

PART 1V

Analysis, Reports and Investigation Management

We're all done with formulas and mathematics. From here on, we can sit it a comfortable room and drink coffee and not get our hands dirty. It is tempting to say that the hard part is over. All that is left is the paperwork.

Actually, the hard part is not over. The paperwork is the hard part. All we have done so far is collect the data and establish the facts. Now we know what happened; not why it happened. Here, you might want to go back to Chapter 1 and review the comments on findings, causes and recommendations.

If we did it right, we've got a lot of data and we have a certain amount of confidence in it. We have probably formed some tentative conclusions, but we still have to show how all the facts of this puzzle fit together.

Inevitably, we are going to deal with facts which don't exactly support our conclusions and may even oppose them. If we are satisfied with the way the facts were gathered, we can't ignore them merely because we don't like them. We must either adjust our conclusions or admit that there are other possible alternatives. The nature of this business is that there are almost always other possible alternatives.

This process of sorting out the facts is loosely referred to as "analysis." The term "Analysis" is defined as "the separation of a thing into the parts or elements of which it is composed" or "the determination of causes from results."

There are a number of different analytical techniques available and they all have one thing in common; none of them are perfect. Some are basically data management systems which leave it to the investigator to figure out what it all means. Some force consideration of alternatives. Some force the investigator to move inexorably from result to cause. Some can be used to analyze the entire accident; some just a portion of it.

We offer several different techniques and leave it to the investigator to select the one most useful at the time.

Regarding report writing, it is absolutely true that the investigation is no better than the report of it. Nothing will come of the investigation unless the report convinces someone, somewhere that something ought to be done.

Unfortunately, investigators tend to be technically oriented and they view report writing as an unpleasant chore. This is sometimes reflected in accident reports where an otherwise excellent investigation must be considered a failure because the report was so poorly written.

A separate chapter addresses the problem of

managing the investigation. In a large accident involving a lot of investigators, this is a full time job for the IIC. It can even be a problem for the Group Chairman or the individual investigator. This subject is covered in this part, because it is so closely related to report writing. Organizing the investigation and outlining the report are essentially the same process.

37

Analytical Techniques

1. INTRODUCTION.

Someone once said, "If you can't draw a picture of it, you don't really understand it." You might say the same thing about an accident. If you can't make a drawing or diagram showing how the various factors involved in the mishap sequence are tied together, you don't really understand what happened and why.

There are many different analytical techniques that can be used to model an accident. Some emphasize the cause and effect relationship (e.g. fault tree analysis) while others emphasize the timing of the events in the sequence (e.g. multi-linear-event-sequencing). There are five general reasons for using an analytical approach to accident investigation.

1. First, it provides a framework around which to organize the data or evidence that you collect during the investigation.

2. Second, it assists in ensuring that the investigation follows a logical path that ties the various chains of the accident sequence together. Missing links can then be identified, allowing the investigation to focus on these areas, and search for additional evidence.

3. The third factor is primarily associated with management of larger investigations, where a number of individuals, or teams are involved. Integration of individual and team efforts is vital. The management of this larger effort becomes easier by having a model of potential accident sequences showing how the evidence supports or contradicts the various hypotheses.

4. Fourth, for many professional investigators the most unpleasant part of an investigation is writing the report. This task becomes easier when a model of the most likely accident sequence is available as an outline to guide the writing. When properly applied, the model can be used to ensure that the investigator's conclusions and recommendations are supported by facts and appropriate analyses. If the report is being written by a number of different teams, a visual model of the sequence become even more important. Without an understanding of how each fits into the big picture, the final report can be so disjointed that it will make even the best investigation look bad.

5. Finally, a "picture" which clearly shows the sequence of events can be useful in communicating your conclusions to the officials who have the authority to make the changes you are recommending. These "pictures" can be used to clearly display the investigation team's logic and can be used in briefings to higher authorities or in the report to supplement the narrative.

As mentioned earlier, (Chapter 4) the initial thrust of aircraft accident investigation is toward data collection. The data comes from a number of sources including witnesses, records and the wreckage; and it accumulates rapidly. A problem confronting all investigators is the problem of correlating these data

DATA	SOURCE
Time	ATC Radar or FDR
Weather	Weather Analysis
Airspeed	ATC Radar or FDR
Altitude	ATC Radar or FDR
Heading	FDR or Calculated
Rate of Climb/Descent	FDR or Calculated
Significant Terrain	Plotted on Map
Eyewitness Locations	Statements
A/G Transmissions	A/G Transcription
Cockpit Conversation	CVR
Location of Other Aircraft	ATC Radar
Significant Events	FDR, Witnesses
Navaids	Plotted on Map
Correct Approach/Departure	Plotted on Map
Terrain Elevations	Plotted on Map

Figure 37-1. Flight Path Diagrams

and organizing it into some logical form that helps explain the accident.

To begin with, it is a good idea to develop some plans on how the data will be managed. One method is to bring a set of pre-tabbed file folders to the investigation and establish a small filing system as the investigation progresses. You can assume that certain types of data will always be collected (weather information, witness statements, weight and balance figures, and so on) and you can establish a pretty accurate filing system that will work for most accidents. You would, of course, bring some blank folders for information you hadn't expected to collect.

There are two benefits to such a system. First, you won't lose or misplace documents or data you have collected. Second, when you begin your analysis, you can rapidly locate pertinent documents.

As we advance technologically, the notebook or laptop computer is emerging as a useful tool to bring

to the investigation. We can use it to store reference information, store collected data and run basic mathematical calculations for us.

Once the data is collected, there are a number of analytical techniques available to the investigator. All have certain advantages and disadvantages. These can be discussed under two broad headings; data management methods and cause resolution methods.

2. DATA MANAGEMENT METHODS.

These methods are meant to correlate or portray data from a number of different sources against a common base. They do not necessarily lead to cause resolution, but they allow the investigator to see the interaction of various events that were occurring at about the same time or place. Some provide a visual picture of what was going on and are useful devices for explaining the sequence of events in a complex accident.

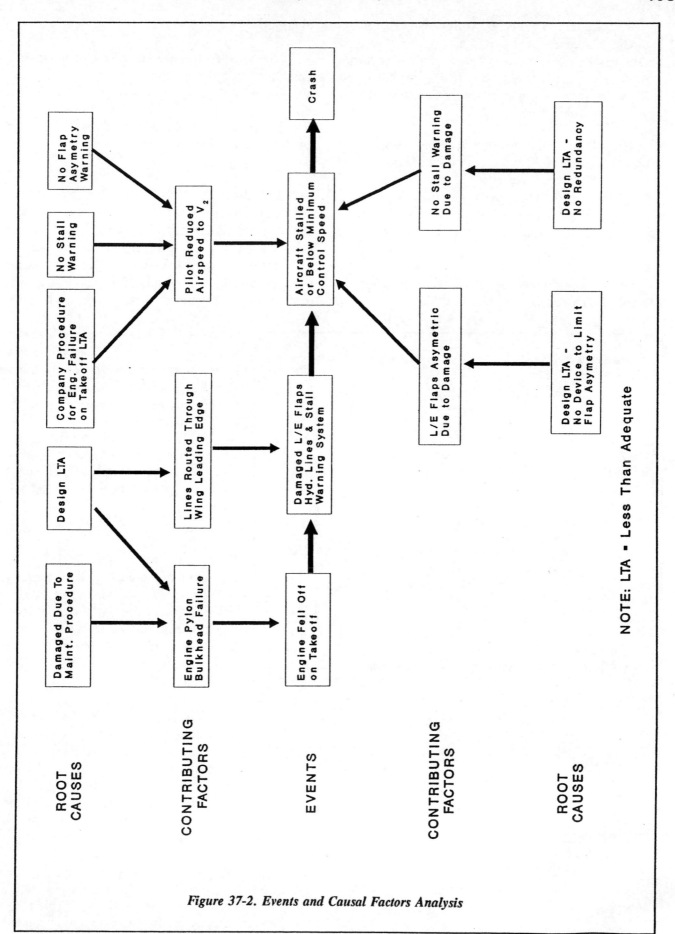

Figure 37-2. Events and Causal Factors Analysis

A. FLIGHT PATH DIAGRAMS. The Flight Path Diagram is frequently used as a means of plotting a number of factors from different sources against a plot of the aircraft's flight path. If time is added, this correlates the factors against a common base of both time and location.

The Flight Path Diagram itself is usually constructed from the ATC radar data (See Chapter 19.) If this is not available, it can sometimes be reconstructed from FDR information. The flight path is generally plotted on a large scale map; preferably one that shows elevations and cultural features (See Chapters 7 and 26.)

Once the flight path is established, a considerable amount of data can be annotated on the diagram. See Figure 37-1.

The benefit of this method is that it allows the investigator to see what the plane was doing and what was being said at any given time or point in the accident sequence. It is, of course, only a two-dimensional depiction. Sometimes it is necessary to supplement the Flight Path Diagram with a vertical depiction.

B. MATRIX ANALYSIS. This is a method of organizing data from different sources that ought to be the same; but isn't. If, for example, you decide to sample a number of identical aircraft (or parts) for certain conditions, you may find that there is some variance among the samples. Matrix analysis is a method of studying these variations.

The classic application of matrix analysis is on witness statements. If you have a large number of eye witnesses, you will undoubtedly discover that there is considerable disagreement among them on what happened. One way to cope with this is to build a matrix consisting of the witness names (or identification) listed down the left side of the chart and the variances in their statements listed across the top.

Then, each witness' observations on those variances are plotted by his name. By summing the totals at the bottom, you develop an idea of where the consensus lies on each of the variances. Assuming that consensus to be true, you can sum the totals to the right side of the matrix and determine the most and least reliable witnesses.

Obviously, this is not absolutely accurate, but it works surprisingly well if there are enough witnesses who saw the accident from approximately the same vantage point. The same thing could be done with your sampling of aircraft mentioned earlier. List the aircraft down the left column and the variances in condition across the top. At the bottom, you determine the most common and least common condition and at the right you determine the best and worst aircraft. This type of analysis could be done on a computer

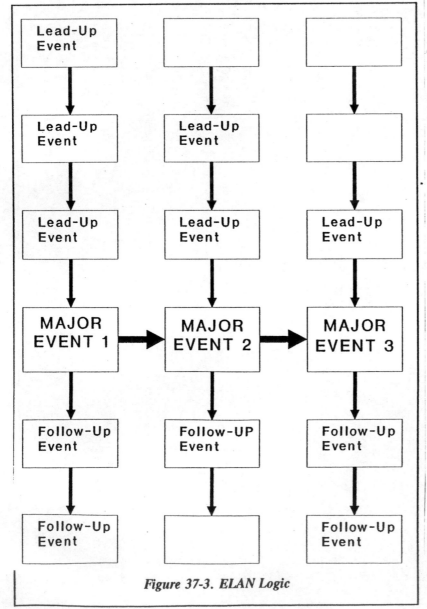

Figure 37-3. ELAN Logic

using spreadsheet software.

C. SIMULTANEOUS TIME AND EVENTS PROCESSING (STEP). This technique plots all known events of the accident against an expanded time line. The time line is drawn at the bottom of a large sheet of paper and expanded so that one foot of paper equals about one minute of time. Wherever possible, events are plotted in a manner depicting the actual time it took for the event to occur. A comment by the Captain which took four seconds, for example, is plotted showing a duration of four seconds on the STEP chart.

Not only are CVR and A/G transmission statements plotted, but all FDR data is plotted along with known points at which the aircraft entered and exited various weather conditions. This permits the investigator to examine the total event sequence in terms of what events were occurring simultaneously and which may have masked or interfered with other events. In a way, this is similar to a flight path diagram except that the

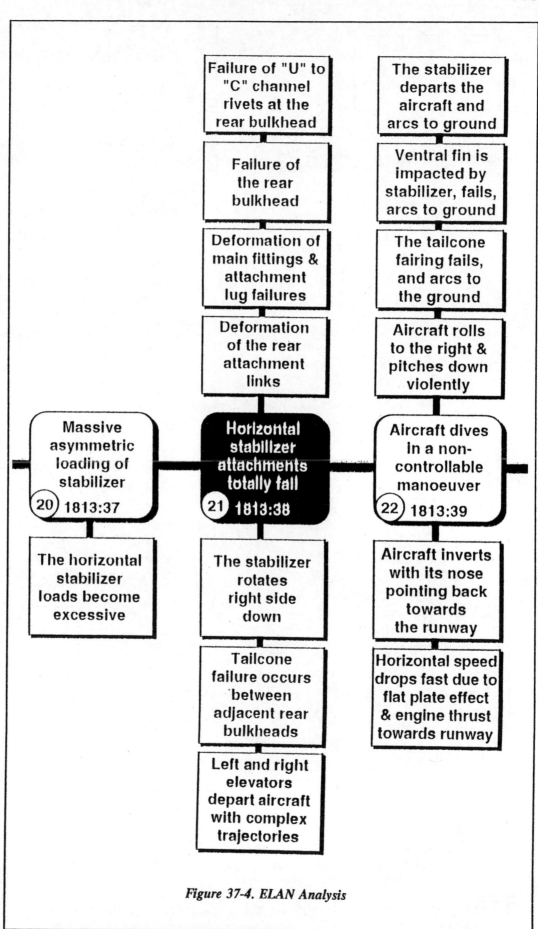

Figure 37-4. ELAN Analysis

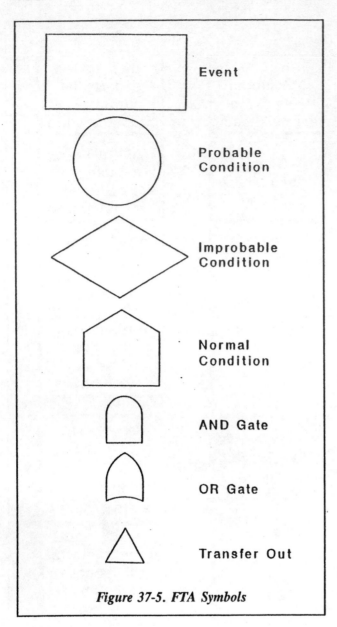

Figure 37-5. FTA Symbols

Event

Probable Condition

Improbable Condition

Normal Condition

AND Gate

OR Gate

Transfer Out

baseline is time instead of location.

3. CAUSE RESOLUTION METHODS.

These techniques are oriented less to assembling all the data and more toward organizing the significant elements into a cause and effect relationship. The result is almost always a diagram which may depict events either in a linear fashion or a network (tree) fashion. Some of them are excellent methods of diagramming the reasoning the investigators used to arrive at various conclusions.

A. EVENTS AND CAUSAL FACTORS ANALYSIS. This method is fairly simple and has been widely used. It is easily taught to a group of investigators and it has the advantage of forcing logical thinking about an accident. The investigators are required to resolve the significant events of an accident one at a time. Once a particular event is resolved, the group proceeds to the next event and is not allowed to back up to an already resolved event.

The analysis is best conducted on a large sheet of paper or a blackboard. Five horizontal lines are established and labeled EVENTS (center line), CONTRIBUTING FACTORS (lines immediately above and below the center line) and ROOT CAUSES (top and bottom lines.)

Figure 37-2 is an example of such an analysis based on the DC10 accident at Chicago O'Hare where an engine fell off the aircraft on takeoff.

The analysis starts on the center line where the

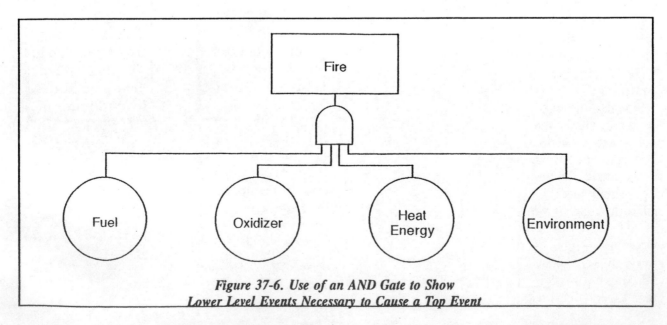

*Figure 37-6. Use of an AND Gate to Show
Lower Level Events Necessary to Cause a Top Event*

basic facts of the accident are listed sequentially. These are strictly facts without any cause and effect relationship. It is important that the group agree that these basic events accurately describe the accident.

Once the event line is established, the group considers the events one at a time and lists the factors that precipitated that event on the lines above and below the center line. Once they agree on these contributing factors, the group examines each contributing factor and lists the root causes for each of those factors on the top or bottom lines. It is likely that some factors will have more than one cause and some will be related to previous events. In some cases, it is possible to carry the analysis another step up or down using a sixth or seventh line.

Once causes are identified for each contributing factor, the group moves on to the next event. When all of the events have been analyzed, the causes of the accident appear on the top and bottom lines and the group has a written diagram showing their logic in arriving at those causes.

B. EVENT LINK ANALYSIS NETWORK (ELAN). This analytical technique is a development of Accident & Investigation Research, Inc. See Bibliography. This method has many of the characteristics of EVENTS AND CAUSAL FACTORS ANALYSIS. The center MAJOR EVENT line is established first and these factual events are broken into the smallest possible discrete steps. For convenience, the events are numbered and correlated with a time line.

Other events are termed either LEAD-UP EVENTS or FOLLOW-UP EVENTS. Lead-up events occurred prior to and resulted in the corresponding major event and are listed sequentially above that major event. Follow-up events occurred after the major event and resulted from it. These are listed sequentially below the major event. In most cases, the bottom follow-up event has continuity with the top lead-up event for the next major event. This logic is shown in Figure 37-3.

This method of analysis lends itself to computerized depiction and it can be refined by color- or shape-coding various events. While this method does not demand cause and effect relationships throughout, it does account for more events and more data. Figure 37-4 shows four major events from an ELAN analysis.

C. FAULT TREE ANALYSIS (FTA). There are

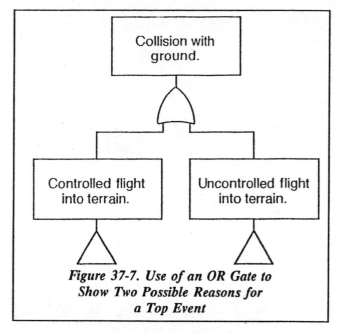

Figure 37-7. Use of an OR Gate to Show Two Possible Reasons for a Top Event

a number of tree-shaped analytical methods. They all start with a single event at the top and proceed downward through layers of precipitating events. As the number of events grow, the shape of the diagram starts to look like a Christmas tree (or the root system of any tree.) In theory, the root causes are found at the bottom or at the lower end of each tree branch.

All "tree" analytical techniques are exercises in deductive reasoning. You start with the accident itself, or a portion of it, and work through other known or reasoned events to the root causes. This is the opposite of an inductive reasoning technique, of which there are many, where you work from a known event (cause) to the results of that event (accident.) Some of the

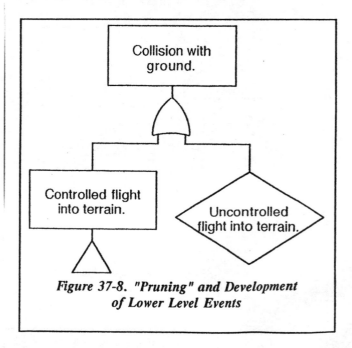

Figure 37-8. "Pruning" and Development of Lower Level Events

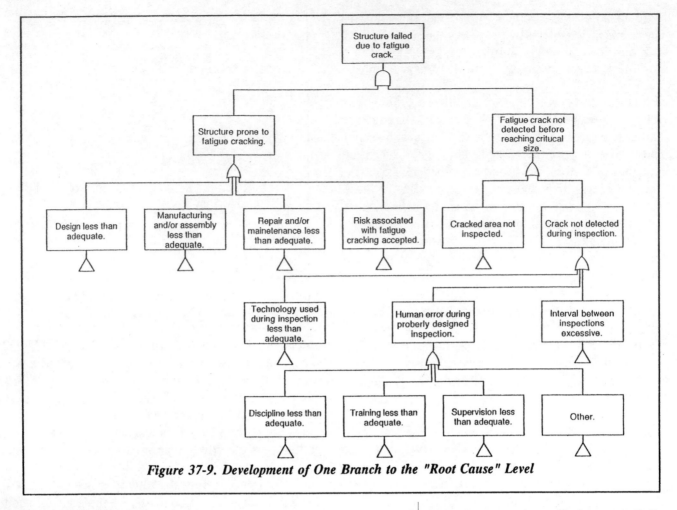

Figure 37-9. Development of One Branch to the "Root Cause" Level

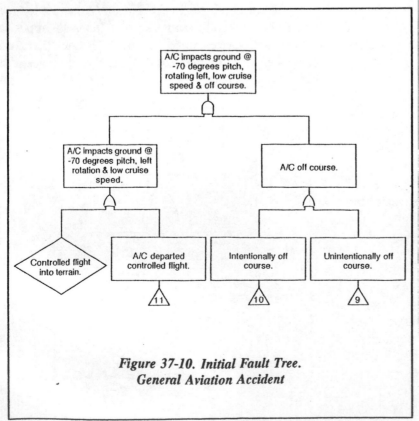

Figure 37-10. Initial Fault Tree.
General Aviation Accident

inductive reasoning techniques are Failure Mode and Effects Analysis (FMEA), Event Tree Analysis and Sneak Circuit Analysis. Although these can be useful in examining cause and effect relationships, they are not commonly used in aircraft accident investigation and will not be discussed here.

Back to FTA. Since Fault Tree is basically a logic technique, it is not difficult to learn or use. It can be used either for the entire accident or a small part of it. It has the advantage of forcing consideration of all possible events and causes, but the disadvantage of taking up a lot of room. Fault Trees are usually spread across several sheets of paper and referencing the various parts to the whole is something of an art. The symbols normally used with it are shown in Figure 37-5.

A rectangle indicates an event that can be further analyzed. The top event (the accident) is a rectangle. The diamond is used to depict an event that either can't be further analyzed because of lack of information or shouldn't be further analyzed because the evidence developed during the investigation rejects that possibility. The circle depicts an event that cannot be analyzed further. It is the root cause of that branch of the tree. These "root causes" should be the focus of the investigator's attention when developing recommendations. They are what is wrong and what needs to be fixed. Recommendations which are directed toward eliminating or limiting the adverse impacts of root causes are most effective, while recommendations which are directed toward the effects of root causes are less effective.

Sometimes, an event which represents a normal or acceptable condition must be listed, or the tree makes no sense. If, for example, "Hypoxia" is shown in a rectangle, it would be wise to list "High Altitude" as a normal condition that must exist in order for hypoxia to occur. These conditions are shown in a house-shaped symbol.

AND gates are placed between levels of the tree when all the events immediately beneath the gate must occur in order for the event above it to occur. This implies redundancy in the system in that more than one event must occur to get through the gate to the next higher level. An example of an "and" gate can be shown by considering the events necessary to result in a fire.

OR gates are used where only a single event must occur to make the next higher event happen. In traditional system safety analyses this implies a possible single failure mode in the system. When a fault tree is used as an investigative tool, the "or" gate means something a little different. Since OR means it could be "this or that", an "OR" gate means we are not sure which branch of the tree is true. During the initial phase of the investigation there may be lots of "ORs".

As the investigation progresses and more evidence is uncovered and analyzed, many of the branches coming out of the "OR" gates should be "pruned". Hopefully, as the investigation comes to a close, all of the "OR" gates will have been eliminated.

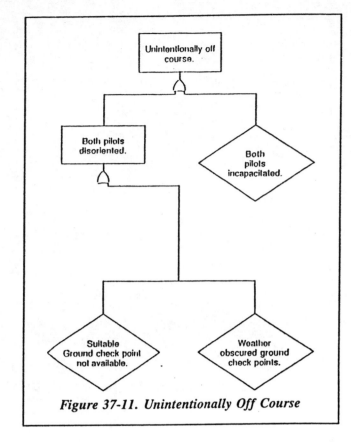

Figure 37-11. Unintentionally Off Course

Proof of elimination of each one of the branches is an important part of the investigation and should be thoroughly documented in the report. If some "OR" gates still exist after all the available evidence has been thoroughly examined and sources of new evidence have been exhausted, then the report should discuss the reasons why the investigation could not decide on which of the branches under the "OR" gate was most

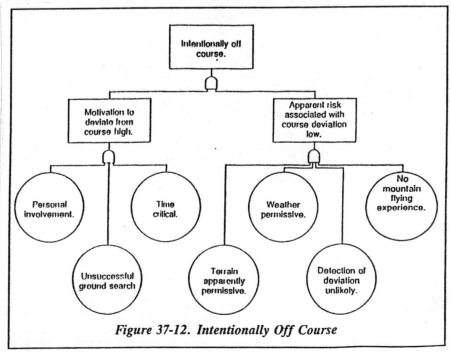

Figure 37-12. Intentionally Off Course

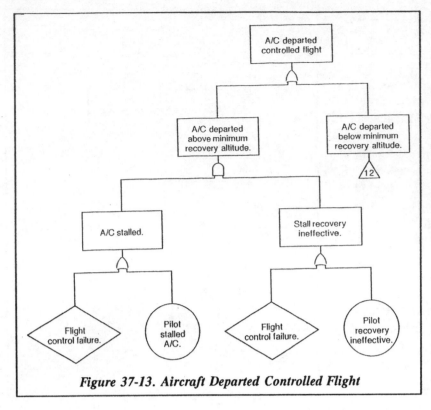

Figure 37-13. Aircraft Departed Controlled Flight

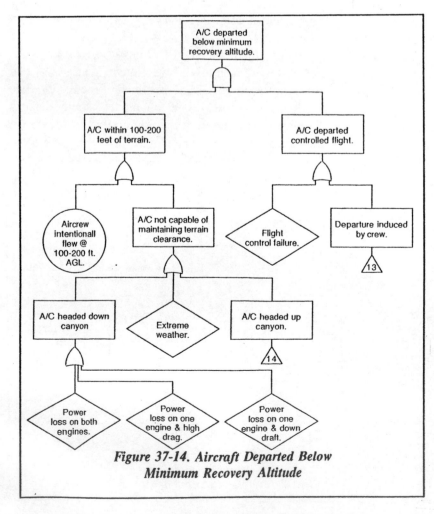

Figure 37-14. Aircraft Departed Below Minimum Recovery Altitude

probable.

Transfer symbols (triangles) are methods of relating the branch drawn on one page to the tree on a preceding page. In the following discussion, there is only one tree depicted, but it is split among several figures. The triangles show how they all connect.

Once the "tree" is completed it can be used as guide in writing the analysis portion of the report. If a chronological (first events first) approach is used, simply start at the bottom of the tree at the root cause which occurred first and work your way up, jumping from branch to branch as the various causal and resulting events took place. On the other hand, if the decision is made to address each of the various events in sequence, the fault tree works equally well for this.

In the aircraft industry, FTA is most frequently used to analyze hardware failures. Since all we are really dealing with is a logic and reasoning technique, it doesn't have to be that way. To illustrate the process, we'll examine a general aviation accident and show how FTA can be used to explain it.

This particular accident involved a single engine airplane that crashed in a wooded box canyon several miles off the filed route of flight. Examination of the wreckage showed an impact about 70 degrees nose down with a slight rotation to the left. The engine was operating at a high power setting and there was no indication of any system malfunction.

The initial fault tree is shown in Figure 37-10. The initial site survey would provide this kind of information. Notice that "Controlled Flight into Terrain" is shown in a diamond. Intentionally diving into the ground was, at this time, not considered a likely possibility. Since the impact evidence does not support CFIT, this will not be pursued further. The other three events at the third level will be analyzed separately.

Unintentionally Off Course. This event is analyzed in Figure 37-11. Notice the use of triangles to show where this connects to Figure 37-10. All branches of this event end in triangles for reasons developed during the investigation. The wreckage was at high altitude in mountainous terrain while the filed flight path was over flat terrain. Weather was VFR and it is considered unlikely that the pilots would not have noticed that they were not where they belonged. There were plenty of visible check points available including a large mountain. Dual incapacitation is a remote possibility. Hypoxia is considered unlikely since the altitude at the impact site was only 7000 feet.

Intentionally Off Course. This emerges as a more likely possibility and is analyzed in Figure 37-12. Interviews with friends of the pilots indicated that neither

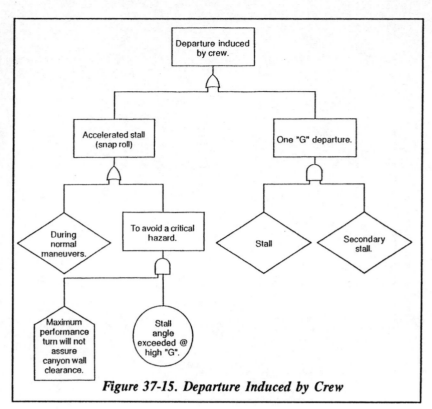

Figure 37-15. Departure Induced by Crew

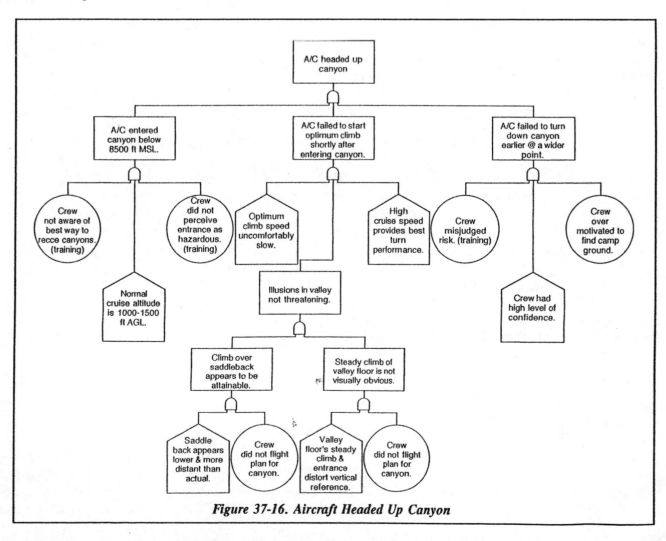

Figure 37-16. Aircraft Headed Up Canyon

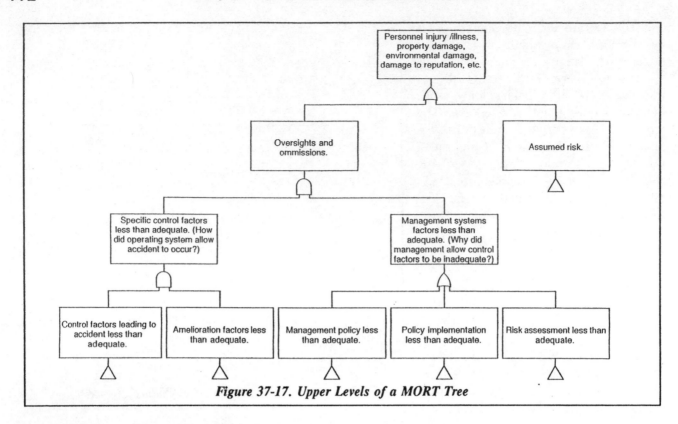

Figure 37-17. Upper Levels of a MORT Tree

had any particular experience with mountain flying and both were interested in locating suitable sites for a future camping trip. Survey flights into the canyon showed some common visual illusions. It looked like it had an exit at the top end and the relatively steep incline was not apparent from the air. All of these branches end in circles. If this is what occurred, these are root causes.

Aircraft Departed Controlled Flight. Figure 37-13 shows the analysis of this branch. Flight control failure is shown in a diamond as the wreckage evidence does not support this. While it must be acknowledged as a possibility, there is no reason to analyze it further.

Aircraft Departed Controlled Flight Below Recovery Altitude. This possibility is analyzed in Figure 37-14. All engine power loss possibilities are shown in diamonds because the wreckage evidence does not support power loss. Since the weather was excellent and not a factor, this is also shown in a diamond. If the aircraft departed controlled flight (stalled) below recovery altitude, it is likely that it was being intentionally flown below stall recovery altitude in the first place. This is shown in a circle.

Departure Induced by Pilots. This is shown in Figure 37-15. This analysis introduces the possibility that the plane was stalled during a maximum perfor-

mance turn. Notice that the fact that the plane probably could not be turned around in that location is shown as a normal event or condition.

Aircraft Headed Up Canyon. The last event to be analyzed is shown in Figure 37-16. This accounts for the pilots' lack of experience in mountain flying and the visual illusions present. Some illusions are shown in diamonds as there is not enough information available to further analyze them.

Taking the analysis as a whole, the most likely causal factors emerge as circles. Notice that in some cases ("No Mountain Flying Experience", Figure 37-16 satisfies both sides of the "A/C Entered Canyon Below 3500ft" AND gate) the same factor is shown on both sides of an AND gate. In effect, this eliminates the redundancy of an AND gate because either event listed below it will get you through the gate.

Some possible causal sequences lead to dead-ends: they go nowhere when they are blocked by an AND gate and there is nothing in the opposite branch that will satisfy the gate.

Notice that the tree forces you to consider all possibilities however unlikely. This is beneficial because it answers the common post-investigation question, "Why didn't you consider ____?" (Answer:

"We did. Here's where we considered it and this is why we didn't pursue it. ")

If you ever use this technique, save the results. You will find the tree handy for winning future arguments.

D. MANAGEMENT OVERSIGHT RISK TREE (MORT).

This technique resembles Fault Tree, but is somewhat more complicated. It applies a pre-designed logic tree (similar to a fault tree) to an accident and forces examination of both the physical equipment and the operational and management inadequacies. The focus of the technique is on management and system weaknesses. It is a difficult technique to apply without formal instruction. There are books available on the subject.

There are a couple of aspects of the MORT philosophy that do, however, deserve mention. This philosophy is shown at the uppermost levels of the tree and centers on the management and system weaknesses mentioned earlier. If you look at the diagram shown in Figure 37-17, which is the tipity top of the MORT tree, you can see the MORT philosophy attributes losses associated with accidents to EITHER "oversights or omissions" OR "assumed risk". In other words, accidents occur because people screw up OR because the risk associated with the mishap was assumed.

Look at the right side of this first branch we find that MORT uses a very specific definition of "assumed risk". "Assumed risk" must have been accepted by an appropriate level of authority after being identified, analyzed and, to the maximum degree, quantitatively evaluated. This is a very important concept. Many "accidents" involve factors which were fully understood prior to the "accident" and a conscious decision made not to do anything about it. Corrective actions may have cost too much, may have degraded performance excessively, or may have required an excessive length of time to implement. In other words, people knew it was going to happen, they just didn't know when or where.

Looking at the left side of the first branch, we can see that when "oversights and omissions" result in "system" deficiencies, these deficiencies exist in not only the operational system, they also exist in the management system which allowed the operational deficiencies to exist. In other words, when an accident is the result of "oversights and omissions", the inves-

tigation must focus on both the deficiencies in the operational system "AND" the management deficiencies which allowed the operational deficiencies to exist. While the left hand branch below the oversights and omissions "AND" gate examines "what happened", the right hand branch examines why it was allowed to happen. Remember that the tree shown below represents only the very top levels of a MORT tree. There are over 1500 potential events to be evaluated in a full MORT analysis.

Finally, the use of "LTAs" deserves mention. You won't find any causal factors identified in a MORT analysis. Instead you will find only LTA, which stands for "Less Than Adequate". When an accident report points out performance which was "less than adequate", it is less offensive and harder to defend against, than performance which was "deficient" and a "cause" of the accident. LTA still implies that changes are necessary to improve performance.

4. SUMMARY.

It must be said that formal analysis techniques are not widely used in aircraft accident investigation. One reason is that they are not well understood by all investigators.

Another is that by the time the investigator has spent a few weeks gathering data, he may have already made up his mind about the accident and doesn't feel that a formal analysis is necessary.

A third reason reason is what we might call "habit." In the old days, the investigator collected all the data, reached a conclusion and then wrote a report that would justify it. This is the "backward" system of report-writing. Start with the conclusion and then make the report support it. That's still going on, by the way, and the astute reader of accident reports can usually spot one that has been written in that fashion.

This reluctance to use analytical techniques ignores that fact that there are other people to be convinced besides yourself. Maybe you do understand the accident, but how are you going to convince others that your conclusions are correct if you can't show them how you arrived at those conclusions?

Most analytical techniques are not particularly difficult. They can be made difficult by adding in some probability calculations; applying them to a part of the accident that doesn't need analyzing; or trying to carry the analysis to too many levels. At that point, the analysis takes on a life of its own and we forget that it is only a tool and it must be of some value to the investigation.

There is no one-size-fits-all analysis technique. There are many available and all have their advantages and disadvantages. You should always select the one that suits your needs on that particular accident.

Don't feel that you have to analyze the entire accident. An airline accident involving an inflight cabin fire and subsequent emergency landing had a lot of ramifications besides the fire itself. An analysis of just the causes of the fire (source of fuel, ignition, etc.) was very useful.

Remember that any analytical technique is meant to help you, the investigator, correlate the information you have developed or focus your thinking on accident causes. Coincidentally, most techniques are also useful in explaining the accident to others or educating people who do not have a technical background in investigation.

38

Aircraft Accident Reports

1. INTRODUCTION.

The product of any aircraft accident investigation is always a report. It can be said, in fact, that the investigation is no better than the report. A superbly written report cannot do much to overcome a bad investigation, but a poor report can definitely ruin a good investigation.

2. TYPES OF REPORTS.

A. NARRATIVE REPORTS. These are the most common reports and generally follow the facts-analysis-conclusions format. They are difficult to write and their quality is sometimes inconsistent. Since factual data is scattered throughout these reports, they do not lend themselves well to computerized data collection systems and subsequent analysis. Nevertheless, this is the only type of report that analyzes and explains the accident.

B. DATA COLLECTION REPORTS. These go with computer systems. They are designed to collect data about the accident in a logical and consistent manner. These days, both narrative reports and data reports are commonly used for each accident.

3. NARRATIVE REPORTS - ICAO FORMAT.

The format suggested in Annex 13 to the ICAO Convention is used commonly throughout the world.

A. TITLE. Operator, manufacturer, model, nationality, registration number, place and date.

B. SYNOPSIS. Notification of accident, identification of investigation authority, accredited representation, investigation organization, report release authority, date of publication and a brief resume of the circumstances leading to the accident.

1. FACTUAL INFORMATION
    1.1. History of flight.
    1.2. Injuries to persons.
    1.3. Damage to aircraft.
    1.4. Other damage.
    1.5. Personnel information.
    1.6. Aircraft information.
    1.7. Meteorological information.
    1.8. Aids to navigation.
    1.9. Communications.
    1.10. Aerodrome information.
    1.11. Flight recorders.
    1.12. Wreckage and impact information.
    1.13. Medical and pathological information.
    1.14. Fire.
    1.15. Survival aspects.
    1.16. Tests and research.
    1.17. Additional information.
    1.18. Useful or effective investigation techniques.

2. ANALYSIS.

3. CONCLUSIONS.

4. SAFETY RECOMMENDATIONS.

APPENDICES. As appropriate. These are not specified in the ICAO format. Here is a list of appendices commonly found in reports.

A-1. Communications transcripts.

A-2. Flight Data Recorder readouts.

A-3. Flight crew information and records.

A-4. Flight plan and weight and balance information.

A-5. Technical and engineering investigation reports.

A-6. Pertinent pages from manuals and handbooks.

A-7. Pertinent maintenance records.

A-8. Releasable statements and testimony.

A-9. Maps and diagrams.

A-10. Photographs.

4. REPORT CONSTRUCTION.

Keep in mind that aircraft accident reports are seldom written by a single individual. They are the product of the group and they are written in separate pieces and assembled later. In order for this to work, the group must have a good understanding of what the final product will look like and how their part will fit into it. If not, they will each do it their own way and the results will be difficult to knit into a coherent report.

A. FORMAT. The format for the report should be known before the investigation even starts. Most investigative agencies use a standard report format and all their investigators are familiar with it. If none is available, use the ICAO format. Brief all members of the investigative team on what format is to be used.

B. OUTLINE. The report format is a fairly broad "generic" format that will fit any aircraft accident. A particular accident, obviously, will have unique features or aspects that need to be included in the report. Referring to the ICAO format listed above, some of the topics under FACTUAL INFORMATION won't be needed for every accident. Some may require additional topics; perhaps one on wreckage recovery, for example. Additionally, you'll notice that no topics are specified for the ANALYSIS section. Obviously, subtopics will be needed based on the circumstances of the accident.

This section can and should be outlined fairly early in the investigation. Three basic topics that fit into almost all report analysis sections are OPERATIONAL FACTORS, MAINTENANCE AND MATERIAL FACTORS, and HUMAN FACTORS. Under those major headings, it is not difficult to list the subtopics that should be analyzed for that accident.

C. APPENDICES. Likewise, the appendices are usually predictable. While you may not yet have the information that goes in a particular appendix, you know its coming and you know you need it.

Early in the investigation, you should have a draft outline of the report including the analysis section and the appendices. Post this where everyone can see it or print it and distribute copies of it. It is important for investigators to know how the report will come together and where their part fits. This outline is also useful as an investigation management tool. See Chapter 39.

D. TOPIC OUTLINES. Now that you have specified the format and constructed an outline, go one step further and specify how the individual sections should be written. This is very helpful to investigators who have an excellent technical background, but aren't particularly good at writing. It is also helpful to you in that the sections written by individual investigators will be organized more or less alike and will be easy to edit into the final report.

One useful system is called the F.A.C. method. The first paragraph in each section or subsection deals with the Facts. What was investigated in this area and what was observed? (Note: Here we are not talking about the facts of the accident. We are talking about a factual description of what the investigator did to investigate this subject.) The second paragraph covers the Analysis of those facts. What did the investigator determine from his investigation? The third paragraph is a short one and covers the Conclusions about that subject. It either was or was not a factor in the accident. To illustrate this, suppose that one of the investigators is examining flight crew qualifications. Here's a simplified version of what his report might look like.

CREW QUALIFICATIONS.

(Facts) All flight crew records were examined including certificates, log books, initial qualification records, training records and flight check records.

(Analysis) Based on an examination of those records, both pilots were qualified and current in the aircraft.

(Conclusion) Flight crew qualifications were not a factor in this accident. (Or, perhaps they weren't qualified and it was a factor in the accident.)

Most investigators find it fairly easy to write in that general style and format. Here we also see the benefits of selecting the appendices early. If, for example, the investigator wanted to include a copy of the pilot's Flight Check Record, he would be able to refer the reader to the correct appendix.

E. EDITING. The individual mini-reports prepared by the investigators should be accepted and approved in draft and then turned over to the report editor. If the individual parts are all submitted in a common format, the task of editing for coherency and continuity is much easier.

5. REPORT QUALITY CONTROL.

As stated at the beginning of this chapter, the report is the only product of the investigation. If it does not adequately support the conclusions and recommendations of the investigative authority, it is likely to be rejected and nothing will happen that might prevent a future accident. The investigators will have wasted their time, not to mention a lot of money.

The accident records of the world are full of flawed reports. Either the investigative team was not capable of writing a good report or somehow the reviewing bureaucracy changed a good one to one that no longer made sense. Before publication, someone has to critically examine the report in terms of whether it will stand alone and weather the inevitable attacks on its conclusions. It is well to start with the certain knowledge that an aircraft accident report--any report-- is not going to make everyone happy. An aircraft accident is not a popular event in the first place and it is well to assume that someone is not going to like the report of it. If the report can be attacked and refuted based on its logic and content, it will be. Some basic guidelines.

A. CONSIDER THE READER. The person reading your report wasn't there. He did not see the accident scene as you saw it and may not have your technical background. You want to bring him along in series of logical steps so that at the end, he will understand it as you do. If it is necessary to include a glossary of the terms and acronyms you must use, fine. Put it in an appendix and tell the reader that he will find it easier to master the report if he keeps one finger inserted in the glossary. The glossary is for his benefit; not yours.

Use the SYNOPSIS wisely. This is your opportunity to give the reader the basic circumstances and results of the accident so he knows what kind of accident he is reading about. Likewise, spend some time on the HISTORY OF THE FLIGHT. All subsequent sections build on the knowledge gained by the reader of this section.

As the reader goes through the ANALYSIS part, he should be developing an idea of what did and did not cause the accident. This is the advantage of using the one-sentence conclusion suggested above in the F.A.C. method. By the time the reader has finished the ANALYSIS part, he should pretty well know what is going to be in the CONCLUSIONS part. If he finds it there, he's happy. If not, there is, in his view, something wrong with the report. He shouldn't find anything in the CONCLUSIONS that wasn't addressed and substantiated in the ANALYSIS. Likewise, he expects all the positive factors in the ANALYSIS part to show up in the CONCLUSIONS.

After he has finished reading the CONCLUSIONS, he will have formed a good idea of what the SAFETY RECOMMENDATIONS will be. Perhaps not the specific recommendations, but at least the general subjects to be covered. Same rules apply. If he sees a SAFETY RECOMMENDATION that is not supported by the CONCLUSIONS, he is not going to like the report.

B. AUDITING THE REPORT. One method of auditing the quality of an aircraft accident report is to review it backwards. Start with the SAFETY RECOMMENDATIONS and make sure they are supported by the CONCLUSIONS. Then make sure each CONCLUSION is supported by the ANALYSIS. Finally, make sure the ANALYSIS deals with the actual facts of the accident or the supportable reasoning and logic of the investigators. As a last step, make sure that there are no gross omissions in the report that will generate questions. In many investigations, it is obvious to the investigators that a particular subject (weather, for example) played no part in the accident and it is omitted from the report. This may be a mistake as it is not necessarily obvious to future readers who were not there. From their point of view, it is a case of stupid investigators forgetting to consider the weather. A simple statement to the effect that weather conditions were examined and not considered to be a factor can forestall a lot of questions.

6. DATA REPORTS.

The format for data reports vary with the country and the investigation authority. Some are designed by investigators and are meant to facilitate data collection. Some are designed by computer programmers and are meant to facilitate data entry into the computer. These forms are sometimes very difficult for the investigator to use. The best forms are probably somewhere between those two extremes. The investigators can collect the data without much trouble and the computer folks can enter it accurately.

There is a tendency in our computer-driven world to want to collect all possible data on every accident. This is just not possible and there has to be a meeting of the minds somewhere on what data is really needed. As an example, collecting a lot of engine data when the accident was a mid-air collision is a waste of someone's time. Moreover, it clutters up the data system with worthless data and may bias future analysis.

The data collection forms used by the National Transportation Safety Board were developed in 1982 by computer people. They were revised in 1983 and again in 1984 to make them more "user (investigator) friendly." Now, the NTSB system consists of a Preliminary Report (4 pages), a Factual Report (10 pages) and a number of supplements. The supplements are used as needed. Currently, there is a supplementary report for these subjects.

Wreckage Documentation
Cockpit Documentation
Second Pilot Information
Occupant Survival and Injury Information
Fire/Explosion
Airport/Airstrip
Aircraft Occupant and Injured Ground Personnel

At one time, the NTSB also used supplementary reports on these subjects. These are listed here as they might be useful in a different reporting system.

Rotorcraft
Aerial Application
Crash Kinematics
Water Contact/Ditching
Search/Rescue/Firefighting/Medical Treatment

In-Flight Collision
Air Traffic Control
Meteorology
Flight Data Recorder/Cockpit Voice Recorder

Attachments to the data reports are encouraged, but not mandatory. These include:

Photographs
Witness Statements
Test/Lab Reports
ATC Transcripts
Other Pertinent Factual Information

In addition to that, the NTSB has a 6-page form (plus instructions) to be completed by the pilot or operator. These are mailed to the NTSB and generally used for minor accidents when the NTSB will conduct only a limited investigation.

Under the NTSB system, these data reports are the only ones submitted for the majority of the aircraft accidents occurring in the United States. In the Factual Report, there is a small section for a narrative statement of facts, but the rest of the forms are of the "fill in the numbers" or "check the box" type. Keep in mind that the findings, conclusions and recommendations are developed by the NTSB staff and (sometimes) the Board members; not by the field investigator. In addition to the data reports, some accidents also get a full narrative report. In these cases, the format used is the ICAO format described above.

7. SUMMARY.

If this chapter seems a little preachy, it is because the authors have seen too many good investigations fail because of a poor report. When this happens, it is an embarrassment not only to the investigators but to the investigation system. Worse, the opportunity to correct a problem and prevent a future accident may have been missed. That's inexcusable.

39

Investigation Management

1. INTRODUCTION.

This chapter could have been covered in Part I, perhaps following Chapter 2. We decided to put it here because it seems more closely related to construction of the report than the description of investigation systems. The development of the report actually starts with the organization and management of the investigation.

Managing a one-person investigation shouldn't be a problem, although some people have difficulty with it. If there is more than one person involved, management creeps in. There must be some organization and division of labor. If a lot of people are involved, management becomes a real problem and someone's full time job. Consider that some investigations may involve over 100 people. We want to conduct the investigation thoroughly, but efficiently and at minimum cost. Further, we want to avoid having two people duplicating each other on the same subject, while another important aspect is not being investigated at all.

2. NTSB MANAGEMENT SYSTEM.

The NTSB will organize a large investigation into groups (propulsion, structures, systems, etc., see Chapter 2.) and put an NTSB investigator in charge of each group. This is a good first step toward management. Keep all the people with similar interests and skills together; give them a broad subject to cover and put someone in charge of them.

Even small investigations can benefit from that procedure. A small NTSB investigation is usually a single NTSB field investigator (who is the IIC) and a few parties to the investigation. Typically, these will represent the FAA, the airframe manufacturer and the engine manufacturer. In effect, the IIC has himself and three one-person groups to manage.

On big investigations, the NTSB starts with a meeting of all participants. Their procedures for conducting this initial meeting are instructive as they illustrate the NTSB's general management technique. The procedures are quoted here.

INITIAL MEETING
MAJOR ACCIDENT INVESTIGATION

1. Opening the Meeting

A. Call the meeting to order.

B. Introduce the IIC.

C. Excuse all members of the news media, attorneys, insurers, and persons presently representing claimants.

D. Distribute attendance roster.

2. Introductions

A. Board Member and Special Assistant

B. Public Affairs Officer

C. NTSB Personnel

D. U.S. government representatives (non-NTSB)

E. Foreign government representatives

F. State and local government representatives.

3. Opening Speech

We have available a paper containing guidance information regarding some of the Safety Board's more pertinent procedures regarding aircraft accident investigation. We ask that each person designated to participate in the investigation read this paper sometime today or tomorrow and adhere to these printed instructions for the duration of the on-scene phase of this investigation. Please keep in mind though that these guidelines are not intended to be all-encompassing. If procedural questions arise during the investigation, the IIC is your best source of information. By signing the attendance roster you are attesting that all NTSB guidelines will be followed.

Please keep in mind that our procedures, with periodic minor changes in the interest of efficiency, have stood the test of time. Without a doubt, we are the most experienced governmental aviation accident investigation body in the world. We investigate hundreds of catastrophic accidents every year and although our investigative protocol may seem strange and perhaps strict, it does work.

The Party System.

The NTSB allows participation by various interested parties in our investigations for two main reasons. First, parties, whether they be operators, manufacturers, unions, or other groups, assist the Safety Board by offering technical expertise that the Board may lack. We are a small organization and are not expected to be knowledgeable in every facet of every phase of the aviation industry. All persons participating in this investigation must be in a position to contribute specific factual information or skills which would not otherwise be supplied. NTSB Rules of Practice (49 CFR, Part 831.9) limit party status in aircraft accident investigations to only those organizations or agencies "whose employees, functions, activities, or products were involved in the accident or incident and who can provide suitable qualified technical personnel to actively assist in the field investigation." If your organization desires party status, please be prepared to describe the qualifications of the individuals you want to participate in this investigation when we ask for people to serve on the specialized groups.

Second, party participation is solicited by the Safety Board because it enables a company or organization to have immediate access to facts concerning the accident from which they may immediately initiate preventive or corrective action should a product or procedure be found lacking. We do not want companies to wait for NTSB recommendations. If a problem is discovered, we hope that it is remedied immediately by the company. This does not always happen, however, and this is the reason that the Federal Aviation Administration is, by law, always a party to our investigations. They can mandate immediate fixes.

The conduct of each party involved in this investigation will be managed by an organization or company party coordinator. This individual should be of enough stature within your company to be able to make decisions on behalf of the company during this on-scene phase. The party coordinator will be the NTSB's direct and official point-of-contact for your company, and therefore, should be immediately available to me at all times while we are here. We also desire that during our nightly progress meetings, the party coordinator be the single individual that speaks for that party. In addition, there is no such thing as an assistant or co-coordinator. A single individual speaks for the party.

Lastly, no participating organization will be allowed to be represented by an individual whose interests lie beyond the basic safety objectives of this investigation. The Safety Board has the ability to remove from the investigation any individual who fits this description. You will find that we're flexible by necessity, but a basic set of standards must be adhered to.

The Group System.

Participants from the selected parties, excluding the party coordinators, will be assigned to working groups lead by NTSB specialists for the duration of the on-scene investigation. As a rule, we do not solicit the help of anyone that is unwilling or unable to remain with their assigned groups for the entire on-scene phase of the investigation. Also, if you are assigned to a specific group, you will be required to remain with that specific group for the duration.

One of our goals will be to investigate this accident efficiently. This means that the working groups are going to be as small as possible and still be able to accomplish their assigned tasks. It is not necessary or desirable, for instance, for a flight attendant to be on the Power Plants working group. It is not necessary for an engine specialist or a pilot to be represented on the Survival Factors working group. It may not be necessary for anyone other than the NTSB specialist to be working on a group such as Weather. Decisions as to who will be on the working groups will be made by the NTSB group chairman and the

IIC. These decisions, of course, are not cast in stone simply because at this stage we do not have enough facts to positively determine how much help we will need. We believe, though, that is it better to start small and add people to the groups at a later date. Do not feel that you will miss out on information because you, as a party, are not involved in every working group. All pertinent information from all groups will be disseminated at the nightly progress meeting where the Party Coordinators will take copious notes.

Also, concerning participation by organizations or companies, we cannot tolerate individuals considered to be "floaters" being directly involved in the on-scene investigation. In other words, if you are not a party coordinator or assigned to a working group, you will not be issued an entry badge and will not be allowed to attend the nightly progress meetings. Exceptions to this rule will be routinely made for representatives of pertinent domestic and foreign government investigative organizations and rarely made in other instances.

Lastly, and this is most important, we desire that all contact with the news media concerning the activity of this investigation team, be made by the Board Member, NTSB Public Affairs Officer or IIC.

We will now identify party coordinators and assign individuals to working groups.

That's how the NTSB does it. They've had a lot of practice and you can tell by reading the opening speech that they have already encountered most of the problems that can occur.

As part of the initial meeting, each participant is given a document covering information for parties to the investigation. Here is a list of the topics covered in that document.

1. Responsibilities of the Board and Designated Parties to the Investigation.

2. Role of Parties to the Investigation.

3. Public Hearings.

4. Recovery and Security of Wreckage.

5. Handling of Factual Information.

6. Dissemination of Public Information.

7. Assignment and Duties of Group Members.

8. Safety Precautions During Accident Investigation.

9. Signing of Attendance Roster.

10. Observers and Accredited Representatives.

11. Party Recommendations as to Findings and Conclusions.

3. BASIC PRINCIPLES OF INVESTIGATION MANAGEMENT.

Once the NTSB investigation is split up into working groups, the management of each group is left to each group chairman. Some, as you might suspect, are better than others. Here are some basic principles that will work for a small group or for the investigation as a whole.

A. DIVIDE THE WORK. The NTSB does this by first dividing the investigation as a whole into working groups. This principle can be carried further as the activities of any working group can be subdivided into smaller subjects. As an example, the Operations Working Group will certainly want to look at flightcrew qualifications, training and operational procedures. Those are three separate subjects to be investigated and we don't yet know what kind of an accident we are talking about! As a rule, the smaller the subject break-down, the easier they are to assign and monitor. Here is where the report outline developed in Chapter 38 could be handy. It already provides a break-down of subjects that must be investigated.

B. MAKE ASSIGNMENTS. Assign investigators by name to each subject. Make sure all investigators know who is assigned to which subject. This insures that someone is responsible for each aspect of the investigation. It also let's everyone know who is doing what so there will be a minimum of duplication. If there is some overlap or common interests, each investigator knows who has that common interest. Psychologically, the investigator can now see how this is all going to work. At first, the investigation of the accident looks like a hopelessly massive project. Now that we have broken it down into small pieces, he realizes that he only has to do a small part of it.

C. MONITOR STATUS. The NTSB holds nightly progress meetings. The ideal way to run these meetings is to have each Group Chairman stand up and tell the IIC and the rest of the investigators what his group has done and learned so far and what is still to be done. This allows the IIC and the Group Chairmen to keep track of progress. They want all aspects of the investigation to proceed more or less simultaneously. If there is a problem with the investigation of one aspect, they react to it and get it back on track. This progress meeting serves another purpose. It is absolutely essential that each investigator know what other investigators have learned. No single investigator can know everything about an accident. Investigation is a synergistic effort. The total knowledge gained by the investigation team is greater than the sum of the knowledge of its individual investigators.

These same general techniques can be applied within each specialized group. Prior to the main meeting, each Group Chairman can ask his group members to state what they have done so far and what is still to be done. This becomes the basis for the Group Chairman's report to the IIC.

D. DEMAND EARLY DRAFT REPORTS. As soon as an aspect of the investigation is complete, it should be followed very shortly by a draft report of that aspect. If this is not done, the final report is likely to be flawed. See Chapter 38.

4. SUMMARY.

As mentioned, the report outline is a useful tool for both organizing the report and managing the investigation. The outline can be expanded and used for both investigation assignments and status.

Basically, investigation management involves three elements:

1. Have a plan for organizing the investigation. Let everyone know what it is.

2. Know what the final product (report) is going to look like. Let everyone know the format.

3. Monitor progress. Let everyone know what is going on.

40

Conclusion

If we look at aircraft accident rates in almost any definable sector of aviation, they are going down. Flying is becoming safer. Although aircraft accident investigation isn't the entire reason for progress in safety, it is certainly one of the main reasons. On the whole, we have done a good job and written some good reports. Action has resulted from those reports and accidents have been prevented. We are putting ourselves out of business, right?

Wrong.

Aviation is in a period of astonishing growth and that growth is likely to continue. The utility of the airplane is such that it is the mode of choice for transportation of mail, people and anything but bulk cargo; and they are working on that.

While the accident rates are going down, the total number of aircraft accidents is going up because of this growth factor. Indications are that the total number of accidents will continue to go up.

Also, as we get smarter, we will begin to focus our investigative skills on incidents instead of accidents. This is the ideal situation where we can enjoy the luxury of investigating something that happened, but didn't cause a lot of damage or injury. That's when we will be fully into the prevention mode and not just the correction mode. All things considered, there will be a continuing need for aircraft accident investigators.

Where do we go from here? Looking into our crystal ball, we see some changes coming.

MORE RECORDING DEVICES. It will become obvious that we cannot do without them. The use of solid state memory chips in avionics will be expanded. The recording of more FDR parameters will be mandated. The recording time of CVRs will be expanded to at least two hours.

GREATER USE OF COMPUTERS. At this time, aircraft accident investigation is very labor intensive. Much of what we do can be computerized. Among the equipment carried by the investigator of tomorrow will be a notebook computer. In it, in addition to all the basic formulas and conversion factors he could need, will be forms for collection of data. Through a modem or a built in Fax, he can transmit this data back to the home office. The computer will have a CD-ROM capability and the investigator will bring with him a CD with the maintenance manual, the parts catalog and pilot's handbook for the aircraft involved. With his modem, he has access to data files in both the NTSB and the FAA. Unlikely? Absolutely not. All that technology already exists and some investigators are doing that right now!

INCREASED EMPHASIS ON HUMAN FACTORS AND HUMAN PERFORMANCE. That's the growth part of our industry. We are going to look more and more to the human factors specialists for answers on why people acted as they did.

INCREASED EMPHASIS ON SAFETY MANAGEMENT AND ACCIDENT PREVENTION PROGRAMS. Aviation safety programs have long been optional, but that's changing. In the future, the operator's safety program may receive post-accident scrutiny along with his management.

BETTER ANALYTICAL TECHNIQUES. Most existing analysis techniques were borrowed and they are not entirely compatible with each other. We can do better. The computer will help.

BETTER INVESTIGATION TRAINING. We have passed the point where all investigative techniques can be learned in the field. Classroom training is becoming essential. It is most likely going to come down to a basic course in investigation followed by a number of short specialty courses.

How can you keep up with all this? A few suggestions:

PREPARE. If you intend to get into this business, start collecting the equipment and reference information you will need before there is an accident. After the accident, all those things become remarkably hard to find.

SEEK HELP AND ADVICE. This is an enormous field. This book covered a lot of different subjects, but barely scratched the surface. There will be times when you must rely on the knowledge of others.

DEVELOP COMPUTER SKILLS. You'll need them.

KEEP CURRENT. Like any other profession, things change. Not only are new pieces of hardware being hung on airplanes, new techniques for flying them are being developed. Every so often, someone comes up with a better way to investigate something. If you're not keeping track of these, you're obsolete. You need to join the International Society of Air Safety Investigators, attend their seminars and regularly read the aviation trade journals.

BUILD A TECHNICAL REFERENCE LIBRARY. There is a lot of information out there that is applicable to aircraft accident investigation, but no one is cataloging it or indexing it. There is no library. You either start collecting and saving it or you learn how to do document and literature searches or both. The bibliography in this book is a starting point. Buy a box of file folders and start collecting information of future value to you as an investigator. File these items alphabetically by subject so you can find them when you need them.

MAKE A CONTRIBUTION. The fraternity of professional aircraft accident investigators is fairly small and there are no secrets among us. Ideas are freely exchanged and anyone can have a good one. There is always a market for professional papers or articles on new investigation techniques and investigation experiences.

IN SUMMARY, aircraft accident investigation can be messy, but it can also be very rewarding. When you finish an investigation, no one has to tell you whether it was a good one or not. You know. If, as a result of your investigation, something is changed for the better and a future accident is prevented, you can take a lot of pride in that accomplishment. That's the whole idea. Good luck!

Appendix A

=========================

Bibliography and Sources of Information

1. INTRODUCTION.

As it turns out, there is a great deal of reference material available on the subject of aircraft accident investigation. Unfortunately, much of it is in the form of technical papers, articles or monographs and very little of it is cataloged or archived in a central location.

While some of the information deals specifically with aircraft accident investigation, much of it does not. An article dealing with the wear pattern and failure of bearings, for example, was not written specifically for the accident investigator.

The authors of this book have been collecting and filing information on this subject for years. Attempting to list all of the files in a bibliography would be futile as many of the documents are no longer available.

This bibliography is meant to be useful to the reader. We have limited the entries to those sources that the reader could reasonably obtain. In some cases, we have listed out-of-print documents which the reader might find with persistence.

The bibliography starts with reference material on the general subject of aircraft accident investigation and moves to references on specific topics. These are organized in the same order as the book chapters. Editorial comments are included where appropriate.

2. GENERAL INVESTIGATION

A. REFERENCES

International Civil Aviation Organization. <u>Manual of Aircraft Accident Investigation</u>. (Doc 6920-AN/855/4.) Montreal, Canada 1986.

Comment: For years, this was the standard reference work on aircraft accident investigation. Although it has not been technically revised since 1977, portions of it are still excellent and fully applicable.

International Civil Aviation Organization. <u>Aircraft Accident and Incident Investigation - 8th ed.</u> (Annex 13.) Montreal, Canada. 1994.

Comment: This contains the accepted definition of an aircraft accident and the ground rules on how accidents and incidents will be investigated and reported internationally. ICAO documents may be ordered from:

> Document Sales Unit
> International Civil Aviation Organization
> 1000 Sherbrooke Street West, Suite 400
> Montreal, Quebec
> Canada H3A 2R2   514-285-8219, Fax 514-288-4772

United States Air Force. Reporting and Investigating U. S. Air Force Mishaps. (AFR 127-4.) Washington, DC. GPO. 1990. (Replaced in 1994 by AFI 91-204.)

Comment: This provides United States military aircraft accident (mishap) definitions and methodology for determining causes.

United States Air Force Guide to Mishap Investigation, Vol. II, Investigation Techniques, (AFP 127-1.) Washington, DC. GPO. 1987.

Comment: This was the USAF investigation manual. Volume II was the technical part and each chapter was written by an expert in that subject. The authors of this book contributed five of the chapters. In 1994, the Air Force rescinded this volume, but there are many copies still available. For years it was used as a textbook by both Southern California Safety Institute and the University of Southern California. The thrust of the manual is primarily toward accidents involving turbine-powered high performance aircraft.

Ferry, Ted S., Readings in Accident Investigation. Thomas Books, Springfield, IL, 1984.

Comment: This is a collection of 28 technical articles covering a wide range of investigative topics. Although only a few of them deal directly with aircraft, most have some applicability to aircraft accidents.

Ferry, Ted S., Modern Accident Investigation and Analysis. 2nd ed., John Wiley & Sons, New York, 1988.

Comment: This is primarily a book on the process of investigation with a few chapters on the technical aspects.

Ellis, Glen, Air Crash Investigation of General Aviation Aircraft, Capstan Publications, Inc., Graybull, Wyoming, 1988.

Comment: This is a general treatment of the subject with primary emphasis on the legal aspects.

B. INFORMATION SOURCES

Transportation Safety Institute, Aviation Safety Division. TSI is the safety training organization within the United States Department of Transportation. It provides aircraft accident investigation courses for government investigators of the United States and other countries and private investigators who would otherwise be accorded status as a "party" to an NTSB investigation. In the past, it has published bound investigation manuals. At the present time, it distributes notebook and reference material in conjunction with its courses. It has a good technical library of reference material.

> Transportation Safety Institute
> 6500 S. MacArthur Blvd.
> P.O. Box 25082
> Oklahoma City, OK 73125   405-954-3614, Fax 405-954-3431

National Transportation Safety Board. The <u>NTSB Investigator's Manual</u> is primarily in internal procedural guide for the individual NTSB investigator. Presently, it is maintained in loose-leaf notebook form although it may be available on CD-ROM in the future. NTSB also publishes aircraft accident reports and special studies, many of which have direct applicability to aircraft accident investigation. All NTSB documents are available through the National Technical Information Service (NTIS).

> National Transportation Safety Board
> Public Inquiries
> 490 L'Enfant Plaza, East, SW
> Washington, DC 20594   202-382-6735

> National Technical Information Service
> Springfield, VA 22161

International Society of Air Safety Investigators (ISASI). This society publishes a quarterly journal, <u>FORUM</u>, which is a gold mine of technical papers and articles on aircraft accident investigation topics. They have and are expanding a technical reference library at their headquarters near Washington, DC.

> International Society of Air Safety Investigators
> Technology Trading Park
> Five Export Drive
> Sterling, VA 22170-4421   703-430-9668, Fax 703-450-1745

3. INVESTIGATION TECHNIQUES

A. INVESTIGATION AND PREVENTION THEORY

Wood, Richard H., <u>Aviation Safety Programs-A Management Handbook</u>, Englewood, Colorado, Jeppesen Sanderson, Inc. 1991.

B. FIELD INVESTIGATION.

Occupational Safety and Health Administration (OSHA) Standard 1030, <u>Bloodborne Pathogens</u>. (29 CFR 1910.1030.)

> **Comment:** This standard applies to aircraft accident scenes in the United States and establishes requirements for protection of investigators from biological hazards.

C. FIRE INVESTIGATION

Coordinating Research Council, Inc., <u>Handbook of Aviation Fuel Properties.</u> Society of Automotive Engineers, Warrendale, Pennsylvania, 1983.

DeHaan, John B., <u>Kirk's Fire Investigation.</u> 2nd ed., New York: John Wiley & Sons, 1983.

Dennett, Michael F., <u>Fire Investigation.</u> New Your: Pergamon Press, 1980.

Kuchta, J.M., Clodfelter, R.G., <u>Aircraft Mishap Fire Pattern Investigations</u>, Aero Propulsion Laboratory, Air Force Systems Command, Wright-Patterson Air Force Base, Ohio, 1985.

National Fire Protection Association (NFPA). The NFPA publishes a number of standards and guides covering all aspects of fire. Particulary useful are <u>#422 Guide for Aircraft Accident Response</u> and <u>#921 Fire and Explosion Investigations.</u> Write for a copy of their catalog.

National Fire Protection Association
1 Batterymarch Park
P.O. Box 9101
Quincy, MA 02269-9101   1-800-344-3555

D. TURBINE ENGINES.

General Electric, Guide, Mishap Investigation F110-GE-100, Evandale, Ohio, 1981.

LaChapelle, H. Vincent, J-79 Accident Investigation Training Manual, General Electric, Evandale, Ohio, 1984

Pratt and Whitney, F-100 Engine Investigation, (Training Manual.)

Pratt and Whitney, The Aircraft Gas Turbine Engine and its Operation. PWA Operating Instruction #200, Pratt and Whitney Aircraft Group, West Palm Beach, Florida, 1974.

Myers, C. W., Wagner, G. B., Nicol, F., Jet Engine Accident Investigation. General Electric Company, Cincinnati, Ohio, 1959.

Comment: This book is out of print, but there are numerous copies available in various technical libraries. It is a classic.

E. AIRCRAFT SYSTEMS. One way to accumulate knowledge on specific aircraft systems (propellers, hydraulic systems, etc.) is to purchase texts used to teach those systems to aircraft mechanics. It is worth ordering the catalogs of publishers of aviation books just to see what is available.

Jeppesen Sanderson, Inc.
55 Inverness Drive East
Englewood, CO 80112-5948   1-800-525-7379

Aviation Book Company
25133 Anza Drive
Santa Clarita, CA 91355-3412   1-800-423-2708

TAB Aviation Books
TAB/McGraw/Hill
Blue Ridge Summit, PA 17294-0850   1-800-822-8138

F. LIGHT BULB ANALYSIS.

Baker, J. Stannard and Thomas Lindquist, Lamp Examination for On or Off in Traffic Accidents, 3rd ed., Traffic Institute, Northwestern University, 1981.

Transportation Safety Board of Canada, A Guide to Light Bulb Analysis, TP 6255E, Transportation Development Center, Transport Canada, 1985, revised 1991.

Transportation Safety Board of Canada (formerly Canadian Aviation Safety Board), Light Bulb Filament Impact Dynamics Study, TP 6254E, Transportation Development Center, Transport Canada, 1985.

G. TIRES AND RUNWAY ACCIDENTS.

Aircraft Tire Manual, The Goodyear Tire & Rubber Company, Akron, Ohio, (undated).

Fricke, Lynn B., <u>Traffic Accident Reconstruction</u> (Vol. 2 of <u>The Traffic Accident Investigation Manual)</u>, Northwestern University Traffic Institute, Evanston, IL, 1990.

> **Comment:** This is the best reference text for evaluation of skid marks, calculation of speed and examination of the dynamics of a vehicle (aircraft or automobile) on the ground.

NASA TN D-6089, <u>A Comparison of Aircraft Ground Vehicle Stopping Performance on Dry, Wet, Flooded, Slush-, Snow-, and Ice Covered Runways</u>, National Aeronautics and Space Administration, Washington, D.C., 1970.

H. MID-AIR COLLISIONS.

Nelmes, Edwin V., <u>Evaluation of Scratch Marks in Mid-Air Collisions</u>., National Transportation Safety Board, Washington, DC, Undated.

> **Comment:** This book (pamphlet, actually) was the definitive study on the geometry of mid-air collisions. It was written sometime in the mid-1960s and its ideas have appeared in almost all subsequent accident investigation manuals including this one. It may be difficult to find, but worth the search to the investigator specializing in mid-air collisions.

I. WITNESSES.

Loftus, E. F., <u>Eye Witness Testimony</u>. Cambridge, MA. Harvard University Press. 1979.

> **Comment:** This book is not primarily on how to conduct a witness interview, but on the reliability of eye witnesses.

J. HUMAN FACTORS.

Hawkins, Frank H., <u>Human Factors in Flight</u>. 2nd ed., Ashgate Publishing Co., Brookfield, VT, 1993.

> **Comment:** This is a good basic text on human factors and it has an excellent bibliography.

Skjenna, Olaf, <u>Cause Factor: Human</u>. Minister of National Health and Welfare, Canada, 1988.

> **Comment:** Dr. Olaf Skjenna was an RCAF Flight Surgeon and Deputy Director of Civil Aviation Medicine. This book deals primarily with rotary wing human factors, but it contains a great deal of investigation information. The book is available from
>
>> Canadian Government Publishing Center
>> Supply and Services Canada
>> Ottawa, Canada K1A 0S9

K. HELICOPTER ACCIDENT INVESTIGATION.

<u>Helicopter Safety and Accident Investigation Manual, Techniques and Procedures</u>. Transportation Safety Institute, Oklahoma City, OK, 1986.

> **Comment:** This loose-leaf manual was prepared by Bell Helicopter Textron for use in the Transportation Safety Institute (TSI) investigation courses.

4. TECHNOLOGY

A. MATHEMATICS AND PHYSICS

Dole, Charles E., <u>Mathematics and Physics for Aviation Personnel</u>. Jeppesen Sanderson, Inc. Englewood, Colorado, 1991.

Comment: This is a programmed learning text and a fairly painless way of reviewing basic math and physics.

B. AERODYNAMICS

Anderson, John D. Jr., <u>Introduction to Flight</u>, 3rd ed, McGraw-Hill Inc., 1988.

Dole, Charles E., <u>Flight Theory for Pilots</u>. 4th ed., Jeppesen Sanderson, Inc., Englewood, Colorado, 1994.

Comment: This book complements <u>Aerodynamics for Naval Aviators</u>. It is not technically as deep, but it is better illustrated and more current.

Dole, Charles E., <u>Flight Theory and Aerodynamics</u>. John Wiley & Sons, New York, 1981.

Hurt, H. H., Jr., <u>Aerodynamics for Naval Aviators</u>., United States Navy, 1960.

Comment: This was the basic text on aerodynamics used by most military and many civilian pilots throughout the world. Although it is over 35 years old, basic aerodynamic principles haven't changed much. Originally, it was a government publication, but it has been widely reprinted and is now available through the Aviation Book Company (see earlier listing.)

Kershner, William K., <u>The Flight Instructor's Manual</u>, 3rd ed., Iowa State University Press, Ames, Iowa, 1993.

Comment: A pilot's eye view of aerodynamics and performance.

Mason, Sammy, <u>Stalls, Spins and Safety</u>, Macmillan Publishing Co., New York, N.Y., 1982.

Comment: Pilot-oriented. Non-technical.

Roed, Aage, <u>Flight Safety Aerodynamics</u>, Airlife Publications, Shrewbury, England, 1983.

Comment: Out of print, but worth the hunt. A well-illustrated review of the aerodynamics involved in aircraft crashes.

Roskam, Jan, <u>Airplane Design</u>, Roskam Aviation and Engineering Corp., Ottawa, Kansas, 1988.

Comment: This is a series of eight books used as aeronautical texts in many universities.

Roskam, Jan, and Chuan-Tau Edward Lan, <u>Airplane Aerodynamics and Performance</u>, Roskam Aviation and Engineering Corp., Ottawa, Kansas, 1982.

Saunders, George H., <u>Dynamics of Helicopter Flight</u>. John Wiley & Sons, New York, 1975.

Comment: This book is out of print, but it is an excellent basic text on the subject if it can be found.

C. STRUCTURAL FAILURE.

Advisory Circular 43-3, <u>Nondestructive Testing in Aircraft</u>, U.S. Department of Transportation, Federal Aviation Administration, Washington, D.C., 1973.

Advisory Circular 43-4, <u>Corrosion Control for Aircraft</u>, U.S. Department of Transportation, Federal Aviation Administration, Washington, D.C., 1975.

Ashby, Michael F. and Jones, David R. H., <u>Engineering Materials 1. An Introduction to their Properties and Applications</u>, Pergamon Press, Oxford, England, 1980.

Barson, John M. and Rolfe, Stanley T., <u>Fracture and Fatigue Control in Structures</u>, 2nd ed. Prentice-Hall, Inc., Englewood Cliffs, New Jersey, 1987.

Broek, David, <u>The Practical Use of Fracture Mechanics</u>, Kluwer Academic Press, Boston, Massachusetts, 1989.

Dole, Charles E., <u>Fundamentals of Aircraft Material Factors</u>, Jeppesen Sanderson, Inc., Englewood, Colorado, 1989.

Engineering Division Bulletin N. 63-1, <u>Metal Fatigue and its Recognition</u>, Civil Aeronautics Board, Bureau of Safety, 1963.

    Comment: This document is a useful source of information (text and graphics) on recognition of fatigue in aircraft structures.

Hurt, H. H. Jr., <u>Fundamentals of Helicopter Structures</u>, University of Southern California, 1967.

    Comment: Although this is out of print, there are many copies in the system. It was the companion to <u>Aerodynamics for Naval Aviators</u> and it applies to all aircraft; not just helicopters.

Latia, Douglas C., <u>Nondestructive Testing for Aircraft</u>, Jeppeson Sanderson, Inc., Englewood, Colorado, 1993.

Morin, C.R., et. al., <u>Techniques and Procedures for Structural and Material Aircraft Failure Investigation</u>, Transportation Safety Institute, Aviation Safety Division, Oklahoma City, Oklahoma, 1990.

Morin, Charles R., (ed), <u>Structures and Metallurgy</u>, Transportation Safety Institute, Aviation Safety Division, Oklahoma City, Oklahoma, 1986.

    Comment: The two Morin books are used as texts in TSI's Aircraft Accident Investigation courses. TSI's address was listed earlier.

Niu, Michael C. Y., <u>Airframe Structural Design</u>, Conmilit Press, Hong Kong, 1988.

    Comment: A well illustrated basic text on structural design.

Petroski, Henry, <u>To Engineer is Human</u>, Vintage Books, New York, N.Y., 1982.

    Comment: An excellent review of design and engineering failures.

Reithmaier, Larry (ed.), <u>Standard Aircraft Handbook</u>, 5th ed., TAB Aviation Books (McGraw-Hill), Blue Ridge Summit, PA, 1991.

    Comment: This is a basic handbook on common aircraft hardware including sizes and AN, MS and NAS numbers. It also includes information on correct installation of fittings and plumbing. A very useful document for parts identification at the scene.

Southworth, John, <u>Aerospace Vehicle Structures. An Expanded Glossary of Terms, Principles, and Methods</u>, Pegleg Books, Burbank, California, 1986.

Comment: A basic guide to the vocabulary and technical expressions used in aircraft design. Well illustrated.

Wulpi, Donald J., <u>Understanding How Components Fail</u>, American Society for Metals, Metals Park, Ohio, 1985.

D. CRASH SURVIVABILITY.

NTSB/SR-83/01. Safety Report: General Aviation Crashworthiness Project, Phase One. Washington DC, 1983.

NTSB/SR-85/01. Safety Report: General Aviation Crashworthiness Project: Phase Two--Impact Severtiy and Potential Injury Prevention in General Aviation Accidents. Washington DC, 1985.

NTSB-SR-80-2. Safety Report--The Status of General Aviation Aircraft Crashworthiness. Washington DC, 1980.

Note: NTSB reports are available from the National Technical Information Service, Springfield, VA 22161.

United States Army, Aviation Applied Technology Directorate, <u>Aircraft Crash Survival Design Guide,</u> <u>USAAVSCOM TR 89-D-22B.</u>, Fort Eustis, Virginia, 1989.

5. ANALYSIS

A. DATA RESEARCH

Air Data Research
8745 Grissom Road, Ste 368
San Antonio, TX 78251   210-680-6995, Fax 210-680-6885

Comment: Air Data Research maintains access to National Transportation Safety Board (NTSB), Federal Aviation Administration (FAA) and National Aeronautics and Space Administration (NASA) aviation safety data bases. This includes Accident and Incident files and Service Difficulty Reports (SDRs). The data is available by subscription on CD-ROM which permits analysis by individual investigators. Investigators may also request research into specific data bases and hard copy reports.

B. ANALYTICAL METHODS.

Accident Investigation and Research, Inc.
12-5480 Canotek Road
Ottawa-Gloucester
Ontario K1J 9H6, Canada   613-749-6711, Fax 613-749-2368

Comment: Among its many research projects, AIR, Inc. has developed a Radar Analysis Program (RAP) for radar-based flight path analysis and ELAN (Event Link Analysis Network), an anlytical method of correlating accident data. Both techniques are described in this book.

<u>System Safety Analysis Handbook</u>, New Mexico Chapter of the System Safety Society, P.O. Box 9524, Albuquerque, NM 87119-9524, 1993.

Comment: This is the most complete source of information on system safety concepts and their associated analytical techniques. Only available through the System Safety Society.

Vincoli, Jeffrey W., <u>Basic Guide to System Safety</u>. Van Nostrand Reinhold, New York, 1993.

Comment: A basic text and explanation of Fault Tree Analysis.

Appendix B

Technical Data

TRIGONOMETRIC FUNCTIONS					
DEG	SIN	COS	TAN	COT	DEG
0	0	1.000	0	+ ∞	90
1	.0175	.9999	.0175	57.29	89
2	.0349	.9994	.0349	28.64	88
3	.0523	.9986	.0524	19.08	87
4	.0698	.9976	.0699	14.30	86
5	.0872	.9962	.0875	11.43	85
6	.1045	.9945	.1051	9.514	84
7	.1219	.9926	.1228	8.144	83
8	.1392	.9903	.1405	7.115	82
9	.1564	.9877	.1584	6.314	81
10	.1737	.9848	.1763	5.671	80
11	.1908	.9816	.1944	5.145	79
12	.2079	.9782	.2126	4.705	78
13	.2250	.9744	.2309	4.331	77
14	.2419	.9703	.2493	4.011	76
15	.2588	.9659	.2680	3.732	75
16	.2756	.9613	.2868	3.487	74
17	.2924	.9563	.3057	3.271	73
18	.3090	.9511	.3249	3.078	72
19	.3256	.9644	.3443	2.904	71
20	.3420	.9397	.3640	2.747	70
21	.3584	.9336	.3839	2.605	69
22	.3746	.9272	.4040	2.475	68

DEG	SIN	COS	TAN	COT	DEG
23	.3907	.9205	.4245	2.356	67
24	.4067	.9136	.4452	2.246	66
25	.4226	.9063	.4663	2.145	65
26	.4384	.8988	.4877	2.050	64
27	.4540	.8910	.5095	1.963	63
28	.4695	.8830	.5317	1.881	62
29	.4848	.8746	.5543	1.804	61
30	.5000	.8660	.5774	1.732	60
31	.5150	.8572	.6009	1.664	59
32	.5299	.8481	.6249	1.600	58
33	.5446	.8387	.6494	1.540	57
34	.5592	.8290	.6745	1.483	56
35	.5736	.8192	.7002	1.428	55
36	.5878	.8090	.7265	1.376	54
37	.6018	.7986	.7536	1.327	53
38	.6157	.7880	.7813	1.280	52
39	.6293	.7772	.8098	1.235	51
40	.6428	.7660	.8391	1.192	50
41	.6561	.7547	.8693	1.150	49
42	.6691	.7431	.9004	1.111	48
43	.6820	.7314	.9325	1.072	47
44	.6947	.7123	.9657	1.036	46
45	.7071	.7071	1.000	1.000	45

DISTANCE		
MULTIPLY	BY	TO OBTAIN
Centimeters	3.2808×10^{-2}	Feet
(cm)	3.9370×10^{-1}	Inches
	1.0000×10^{-5}	Kilometers
	1.0000×10^{-2}	Meters
Feet	3.0480×10	Centimeters
(ft)	3.0480×10^{-1}	Meters
	3.0480×10^{-4}	Kilometers
	1.8939×10^{-4}	Miles
Inches (in)	2.5400	Centimeters
Meters (m)	1.0000×10^{2}	Centimeters
	3.2808	Feet
	6.2137×10^{-4}	Miles
Miles (Stat.)	5.2800×10^{3}	Feet
(sm)	1.6093	Kilometers
	1.6093×10^{3}	Meters
	8.6900×10^{-1}	Naut. Miles
Miles (Naut.)	6.0761×10^{3}	Feet
(nm)	1.8520×10^{3}	Meters
	1.1507	Stat. Miles

AREA		
MULTIPLY	BY	TO OBTAIN
Square Inches	6.4156	Cm^2
	6.9444×10^{-3}	$Feet^2$
	6.4516×10^{-4}	$Meters^2$
Square Meters	1.0764×10	$Feet^2$
Square Feet	1.4400×10^{2}	$Inches^2$
	9.2903×10^{-2}	$Meters^2$

PRESSURE		
MULTIPLY	BY	TO OBTAIN
Atmospheres	2.9921×10	Inches HG
	1.4696×10	PSI
	1.0132	Bars
Bars	9.8692×10^{-1}	ATM
	1.4504×10	PSI
	2.5906×10	Inches HG
Inches HG	3.3864×10^{-2}	Bars
	3.3864×10	Millibars
	4.9116×10^{-1}	PSI
	3.3421×10^{-2}	ATM
PSI	2.0360	Inches HG
	6.8046×10^{-2}	ATM
	6.8948×10^{3}	Pascals
PSF	4.7880×10	Pascals

VOLUME		
MULTIPLY	BY	TO OBTAIN
Gallons (U.S.)	3.7854	Liters
	1.3368×10^{-1}	$Feet^3$
	8.3270×10^{-1}	Imp. Gallons
Gallons (Imp)	1.2009	U.S. Gallons
	4.5460	Liters
Liters	3.5315×10^{-2}	$Feet^3$
	2.6417×10^{-1}	U.S. Gallons
Cubic Feet	2.8317×10^{-2}	$Meters^3$
Cubic Meters	3.5315×10	$Feet^3$

VELOCITY		
MULTIPLY	BY	TO OBTAIN
Cent/Second	3.2808×10^{-2}	Feet/Second
Feet/Second	3.0480×10	Cent/Second
	5.9248×10^{-1}	Knots
	3.0480×10^{-1}	Meters/Second
	6.8182×10^{-1}	Miles/Hour
Knots	1.6890	Feet/Second
	1.1516	Miles/Hour
	1.0130×10^{2}	Feet/Minute
	1.8532	Km/Hour
	5.1480×10^{-1}	Meters/Second
Miles/Hour	1.4667	Feet/Second
	1.6093	Km/Hour
	8.6898×10^{-1}	Knots
	8.8002×10	Feet/Minute

POWER		
MULTIPLY	BY	TO OBTAIN
Ft-Lbs/Second	1.818×10^{-3}	Horsepower
	1.356×10^{-3}	Kilowatts
	7.712×10^{-2}	BTU/Minute
Horsepower	4.241×10	BTU/Minute
	550	Ft-Lbs/Second
	7.457×10	Kilowatts
Kilowatts	5.689×10	BTU/Minute
	7.376×10^{2}	Ft-Lbs/Second
	1.341	Horsepower
BTU/Minute	1.297×10	Ft-Lbs/Second
	2.357×10^{-2}	Horsepower
	1.758×10^{-2}	Kilowatts

ENERGY		
Foot-Pounds	1.3558	Joules
	1.2851×10^{-3}	BTU
	1.3558×10^{7}	Ergs
British Thermal Units (BTU)	7.7817×10^{2}	Foot-Pounds
	1.0551×10^{3}	Joules
Joules	9.4771×10^{-4}	BTU
	7.3756×10^{-1}	Foot-Pounds

RADIAL MEASURE		
MULTIPLY	BY	TO OBTAIN
Degrees	1.7453×10^{-2}	Radians
	2.7778×10^{-3}	Revolutions
Radians	5.7296×10	Degrees
	1.5916×10^{-1}	Revolutions
Revolutions	3.6000×10^{2}	Degrees
	6.2832	Radians

FORCE		
MULTIPLY	BY	TO OBTAIN
Pounds	4.4482	Newtons
Newtons	2.2481×10^{-1}	Pounds

WEIGHT		
MULTIPLY	BY	TO OBTAIN
Pounds	4.5359×10^{-1}	Kilograms
Kilograms	2.2046	Pounds

MASS		
MULTIPLY	BY	TO OBTAIN
Slugs	1.5125×10	Kilograms
Kilograms	6.3592×10^{-2}	Slugs

WEIGHTS OF LIQUIDS (lb/U.S gal)	
Aviation Gasoline	5.87
Jet A/A1	6.74
Kerosene	6.7
JP-4	6.55
Aviation Oil	7.74
Fresh Water	8.35
Sea Water	8.55

TEMPERATURE CONVERSION	
°FAHRENHEIT =	9/5 (°C + 40) -40
-or-	1.8 (°C) + 32
°CELCIUS =	5/9 (°F + 40) -40
-or-	(°F - 32)/1.8
1° C = 1.8° F	See Chart

TEMPERATURE CONVERSION (Approximate) 1° C = 1.8° F

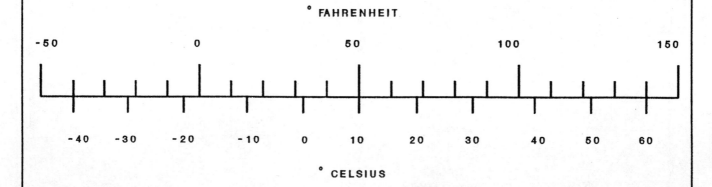

° FAHRENHEIT

° CELSIUS

Appendix C

============================

Code of Ethics and Conduct

The International Society of Air Safety Investigators (ISASI) is a society specifically dedicated to the development and improvement of aircraft accident or incident investigation. Their membership comes from over 100 different countries. Among their other activities, they have developed a Code of Ethics and Conduct for aircraft accident investigators. The Code is reprinted here with their permission.

International Society of Air Safety Investigators
Technology Trading Park
Five Export Drive
Sterling, VA 20164-4421, USA   703-430-9668, Fax 703-450-1745

INTERNATIONAL SOCIETY OF AIR SAFETY INVESTIGATORS CODE OF ETHICS AND CONDUCT

PREAMBLE. As noted in the ISASI Bylaws, the purpose of the Society is, "To promote the development and improvement of aviation accident or incident investigation." Implicit therein is a requirement for a baseline of agreement between the Members and the Society as to what constitutes professional behavior of the Members. Indeed, under the Bylaws, the Member covenants to support provisions of the Bylaws as a prerequisite to membership in the Society.

Therefore, as an Appendix to the Bylaws, this Code of Ethics and Conduct reflects behavior expected of ISASI Members. It has been prepared and adopted with the full realization that determination of the adherence of lack of adherence to these principles is a matter of judgement which can only be effected reasonably by peer review. Procedures governing adjudication of alleged violations of this Code are the responsibility of the Ethics and Conduct Committee as approved by the Executive Committee of the Society.

The Code has distinguished five Ethics and numerous related items of Conduct contained thereunder. Ethics are the axiomatic and aspirational major principles shown both under the heading of Ethics and as general headings in the Code of Conduct. They are broad goals towards which accident investigators "should" strive.

The Code of conduct is phrased in "shall" terms of expected Member behavior. The items constitute minimum levels of conduct which, if violated, constitute potential grounds for disciplinary action by the Society. Such disciplinary action can include expulsion from the Society.

It is recognized that provisions of this code will not apply to all Members during the totality of their work activities. However, insofar as investigations are conducted for safety purposes, and this Code does not conflict with other codes of professional behavior, Members are expected to adhere to the ISASI Code.

It is also recognized that operative words or phrases describing expected Member conduct are appropriate only if feasibility is assumed under the existent circumstances. Such an interpretation should be applied throughout this Code.

This Code has been adopted by the International Council. Recognizing the desirability of membership input to this Code, the Ethics and Conduct Committee shall report to the International Council annually the receipt of any suggestions for modification of the Code and their recommendations therefor. Thus, the membership is encouraged to communicate with the Ethics and Conduct Committee in these matters.

CODE OF ETHICS

1. INTEGRITY. *Each Member should at all times conduct his activities in accordance with the high standards of integrity required of his profession.*

2. PRINCIPLES. *Each Member should respect and adhere to the principles on which ISASI was founded and developed, as illustrated by the Society's Bylaws.*

3. OBJECTIVITY. *Each Member should lend emphasis to objection determination of facts during investigations.*

4. LOGIC. *Each Member should develop all accident cause-effect relationships meaningful to air safety based upon logical application of facts.*

5. ACCIDENT PREVENTION. *Each member should apply facts and analyses to develop findings and recommendations that will improve aviation safety.*

CODE OF CONDUCT

1. INTEGRITY. *Each Member Shall:*

1.1 Not attempt, or assist others to attempt, to falsify, conceal or destroy any facts or evidence which may relate to an accident.

1.2 Not make any misrepresentation of fact to obtain information that would otherwise be denied to him.

1.3 Be responsive to the feelings, sensibilities and emotions of involved persons, and shall avoid actions which might aggravate what may already be a delicate situation.

1.4 Not divulge fragmentary or unsupported information concerning the accident to parties external to the investigation no matter how publicly important such parties may appear to be.

1.5 Avoid actions or comments which might be reasonably perceived during the factfinding phase of the investigation as favoring one party or another.

1.6 Establish and adhere to the chain of authority with attendant responsibilities throughout the course of the investigation.

1.7 Not attempt to profit, nor accept profit, other than by normal processes of remuneration for professional services. (Note: Fee-splitting in the absence of actual work performed or acceptance of contingency fees for investigative activity are not acceptable conduct.)

1.8 Remain open-minded to the introduction of new evidence or opinions as to interpretation of facts as determined through analysis, and be willing to revise one's own findings accordingly.

1.9 Avoid any implication of professional impropriety by continuously applying the foregoing principles to one's own endeavors, and encouraging the application of these same principles to others associated with air safety investigation.

2. PRINCIPLES. *Each Member shall:*

2.1 Promote accident investigation as a fundamental element in accident prevention and encourage others to do the same.

2.2 Assist other Members to carry out their accident investigation tasks.

2.3 Not use membership status to effect personal gain or favor beyond signifying qualification to published membership criteria.

2.4 Not represent the Society or imply a position of the Society in public utterances on any issue unless prior written authority has been received from the Society President.

2.5 Seek advice of the International Council, via the Secretary, in the vent a situation arises where contemplated conduct by the Member may violate the Bylaws or the Code of Ethics and Conduct of the Society.

2.6 Submit evidence of violations of the ISASI Bylaws or this Code to the Society's Ethics and Conduct Committee in accordance with procedures approved by the International Council, and refrain from public discussion of the alleged violation until the committee findings have become a matter of appropriate record.

2.7 Encourage uninhibited, informal interchange of views among Members; however, any sensitive information thus gained shall not be made public or transmitted to others without clear approval of the person from whom the information was gained.

2.8 Have an obligation to improve the professional image of the Society; however, Members shall:

2.8.1 Refrain from unfounded criticism of officers of the Society either publicly or privately unless the matter is investigated thoroughly and brought to the attention of the President with reasonable time being allocated to review the situation and act accordingly.

2.8.2 Refrain from public criticism of any fellow Member unless that individual has first been apprised of the alleged basis for that criticism and given an opportunity for rebuttal.

2.9 Encourage and participate in the education, training and indoctrination of personnel likely to become involved actively in accident investigation.

2.10 Develop and implement a personal program for a continually improving level of professional knowledge applicable to air safety investigation.

2.11 Transfer promptly to the Treasurer of the Society and Society funds or property coming into the Member's possession unless specific use thereof has been authorized by the Bylaws.

3. OBJECTIVITY. Each Member shall:

3.1 Ensure that all items presented as facts reflect honest perceptions or physical evidence that have been checked insofar as practicable for accuracy.

3.2 Ensure that each item of information leading to fact determination be documented or otherwise identified for a reasonable time for possible follow-up by others.

3.3 Use the best available expertise and equipment in determining the validity of information.

3.4 Pursue fact determination expeditiously.

3.5 Following all avenues of fact determination which appear to have practical value towards achieving accident prevention action.

3.6 Avoid speculation except in the sense of presenting a hypothesis for testing during the fact-finding and analysis process.

3.7 Refrain from release of factual information publicly except to authorized persons, by authorized methods and then only when it does not jeopardize the overall investigation.

3.8 Handle with discretion any information reflecting adversely on persons or organizations and, when the information is reasonably established, notify such persons or organizations of potential criticism before it becomes a matter of public record.

4. LOGIC. Each Member shall:

4.1 Begin sufficiently upstream in each sequence of events so as to ascertain practicable accident prevention information.

4.2 Continue downstream in a sequence of events sufficiently to include not only accident prevention information but also crash injury prevention, search and survival information.

4.3 Ensure that all safety-meaningful facts, however small, are related to all sequences of events.

4.4 Delineate those major facts deemed not to be safety-related, explaining why they should not be considered as critical in the sequences of events.

4.5 Be particularly alert to value judgements based upon personal experiences which may influence the analysis; and where suspect, turn to colleagues for independent assessment of the facts.

4.6 Express the sequences in simple, clear terms which may be understood by persons not specializing in a particular discipline.

4.7 Include specialist material supporting the analysis either in an appendix or as references clearly identified as to source and availability.

4.8. Prepare illustrative material and select photographs so as not to present misleading significance of the data or facts thus portrayed.

4.9 List all documents examined or otherwise associated with the analysis and include an index thereof.

5. ACCIDENT PREVENTION. Each Member shall:

5.1 Identify from the investigation those cause-effect relationships about which something can be done reasonably to prevent similar accidents.

5.2 Document those aviation system shortcomings learned during an investigation which, while not causative in the accident in question, are hazards requiring further study and/or remedial action.

5.3 Communicate facts, analyses and findings to those people or organizations which may use such information effectively; such communications to be constrained only by established policies and procedures of the employer of the Member.

5.4 Provide specific, practical recommendations for remedial action when supported by the findings of the accident having been investigated singly or as supported by other cases.

5.5 Communicate the above noted information in writing, properly identified as a matter of record.

5.6 Encourage retention of relevant investigation evidence within the aviation system in such a manner as to form an effective baseline for further investigation of the given accident and/or facilitate analysis in connection with future accidents.

5.7 Demonstrate a respect for interpretation of facts by others when developing conclusions regarding a given accident and provide reasonable opportunity for such views to be made known during the course of the investigation.

Appendix D

===============

Glossary

Every effort has been made to define technical terms at the time they are used in the text. Those definitions will not be repeated here.

This Appendix contains a list of accepted descriptive terms commonly used in aircraft accident investigation. It also contains a convenient list of acronyms used in the text and a list of "V" speeds used in aerodynamic notation.

1. DESCRIPTIVE TECHNICAL TERMS.

When we describe what has happened to an aircraft part during or following an accident, we sometimes use terms that are technically inaccurate or terms that have different meanings to different readers.

Here is a list of common descriptive technical terms and their generally accepted meaning. This list was developed by the United States Air Force with the assistance of General Electric Company, General Motors Corporation and Pratt and Whitney Aircraft Group of United Technologies.

ARCED. Visible effects (burn spots, fused metal) of an undesired electrical discharge between two electrical connections. ALSO: FLASHED OVER.

BATTERED. Damaged by repeated blows or impacts.

BENT. Sharp deviation from original line or plane usually caused by lateral force. ALSO: CREASED, FOLDED, KINKED.

BINDING. Restricted movement such as tightened or sticking condition resulting from high or low temperature, foreign object jammed in mechanism, etc. ALSO: STICKING, TIGHT.

BOWED. Curved or gradual deviation from original line or plane usually caused by lateral force or heat.

BRINELLING. Circular surface indentations on bearing races usually caused by repeated shock loading of the bearing. FALSE BRINELLING is actually wear caused by bearing rollers sliding back and forth across a stationary race while true brinelling is plastic displacement of material.

BROKEN. Separated by force into two or more pieces. ALSO: FRACTURED.

BULGED. Localized outward or inward swelling usually caused by excessive local heating or differential pressure. ALSO: BALLOONED, SWELLING.

BURNED. Destructive oxidation usually caused by higher temperature than the parent material can withstand.

BURRS. A rough edge or a sharp projection on the edge or surface of the parent material.

CARBONED. Accumulation of carbon deposits. ALSO: CARBON-COVERED, CARBON-TRACKED, COKED.

CHAFFED. Frictional wear damage usually caused by two parts rubbing together with limited motion.

CHECKED. Surface cracks usually caused by heat.

CHIPPED. A breaking away of the edge, corner or surface of the parent material usually caused by heavy impact; not flaking.

COLLAPSED. Inward deformation of the original contour of a part usually due to high pressure differentials. ALSO: CRUSHED.

CORRODED. Gradual destruction of the parent material by chemical action. Often evidenced by oxide build-up on the surface of the parent material. ALSO: RUSTED, OXIDIZED.

CRACKED. Visible (not requiring special NDI techniques) partial separation of material which may progress to a complete break.

CROSSED. Material damage to parts (as in the case of crossed threads) or part rendered inoperative (as in the case of crossed wires) as a result of improper assembly.

CURLED. A condition where the tip(s) of compressor or turbine blades have been curled over due to rubbing against engine casings.

DENTED. A surface indention with rounded bottom usually caused by impact of a foreign object. Parent material is displaced, but seldom separated. ALSO: PEENED.

DEPOSITS. A build-up of material on a part either from foreign material or from another part not in direct contact. ALSO: METALIZING.

DISINTEGRATED. Separated or decomposed into fragments. Excessive degree of fracturing (breaking) as with disintegrated bearings. Complete loss of original form. ALSO: SHATTERED.

DISTORTED. Extensive deformation of the original contour of a part usually due to impact of a foreign object, structural stress, excessive localized heating or any combination of these. ALSO: BUCKLED, DEPRESSED, TWISTED, WARPED.

ECCENTRIC. Part(s) wherein the intended common center is displaced significantly. ALSO: NON-CON-CENTRIC.

ELECTRICAL CIRCUITS

GROUNDED. Current path to ground.

OPEN. Incomplete electrical circuit due to separation at or between electrical connections.

SHORTED. Undesired current path between leads or circuits that are normally at different electrical potentials.

ERODED. Material carried away by flow of fluids or gases; accelerated by heat or grit.

FLATTENED OUT. Permanent deformation beyond tolerance limits. Usually caused by compression.

FRAYED. Worn into shreds by rubbing action.

FUSED. Joining together of two materials. Usually caused by heat, friction or current flow.

GALLED. Chafing or sever fretting caused by slight relative movement of two surfaces under high contact pressure.

GLAZED. Undesirable development of a hard, glossy surface due to rubbing action, heat or varnish.

GOUGED. Scooping out of material usually caused by a foreign object. ALSO: FURROWED.

GROOVED. Smooth, rounded furrow or furrows of wear, usually wider than scoring, with rounded corners and a smooth groove bottom.

HOT-SPOT. Subject to excessive temperature usually evidenced by change in color and appearance of part. ALSO: HEAT DISCOLORED, OVERHEATED.

MELTED. Deformation from the original configuration due to heat, friction or pressure.

MIS-MATCHED. Improper association of two or more parts.

MIS-POSITIONED. Improper installation of a part resulting is damage to the installed part or to associated parts. ALSO: MIS-ALIGNED, REVERSED.

NICKED. A sharp surface indention caused by impact of a foreign object. Parent material is displaced; seldom separated.

OUT-OF-ROUND. Diameters of part not consistent.

OUT-OF-SQUARE. Deformation of right angle relationship of part surfaces.

PEELED. A breaking away of surface finishes such as coatings, platings, etc. Peeling would be flaking of very large pieces. A blistered condition usually precedes or accompanies peeling. ALSO: BLISTERED, FLAKED.

PICK-UP. Transfer of metal from one surface to another. Usually the result of rubbing two surfaces together without sufficient lubrication.

PITTED. Small irregular shaped cavities in the surface of the parent material usually caused by corrosion, chipping or heavy electrical discharge.

PLUGGED. Total or partial blocking of pipe, hoses, tubing, channeling or internal passages. ALSO: CLOGGED, OBSTRUCTED, RESTRICTED.

POROUS. Voids in a material located internally. Usually applied to cast material or welds. ALSO: POCK-MARKED, PERFORATED.

ROLLED-OVER. Lipping or rounding of a metal edge. ALSO: LIPPED, TURNED.

RUBBED. To move with pressure or friction against another part.

RUPTURED. Excessive breaking apart of material usually caused by high stresses, differential pressure, locally applied force or any combination of these. ALSO: BLOWN, BURST, SPLIT.

SCORED. Deep scratch or scratches made during part operation by sharp edges of foreign particles.

SCRATCHED. Light narrow, shallow mark or marks caused by movement of a sharp object or particle across a surface. Material is displaced; not removed.

SEIZED. Parts bound together because of expansion or contraction due to high or low temperature, foreign object jammed in mechanism, etc. ALSO: FROZEN, JAMMED, STUCK.

SHEARED. Dividing a body by cutting action, i.e., division of a body so as to cause its parts to slide relative to each other in a direction parallel to their plane of contact. ALSO: CUT.

SPALLED. Sharply roughened area characterized by progressive chipping-away of surface material. Not to be confused with flaking.

STRETCHED. Enlargement of a part as a result of exposure to operating conditions. ALSO: GROWTH.

STRIPPED. A condition usually associated with fastener threads or electrical insulation. Involves removal of material by force.

TORN. Separated by pulling apart.

WORN. Material of part is consumed as a result of operation.

2. ACRONYMS USED IN THE TEXT.

AAIB	Aircraft Accidents Investigation Branch (Britain)
AC	Alternating Current
ACARS	Arinc Communications Addressing & Reporting System
AD	Airworthiness Directive
ADI	Attitude Director Indicator
ADREP	Accident Data Reporting System (ICAO)
AG	Air-to-Ground
AGL	Above Ground Level
ALPA	Air Line Pilots Association
AOA	Angle of Attack
ARINC	Aeronautical Radio Incorporated
ASRS	Aviation Safety Reporting System (USA)
ATC	Air Traffic Control
BASI	Bureau of Air Safety Investigation (Australia)
CAB	Civil Aeronautics Board
CADC	Central Air Data Computer
CAM	Cockpit Area Microphone
CB	Circuit Breaker
CFI	Certified Flight Instructor (USA)
CFR	Code of Federal Regulations (USA)
CG	Center of Gravity
CRT	Cathode Ray Tube
CSD	Constant Speed Drive
CTSB	Canadian Transportation Safety Board

CVR	Cockpit Voice Recorder
DBV	Diagonally Braked Vehicle
DC	Direct Current
DFDR	Digital Flight Data Recorder
DOT	Department of Transportation (USA)
EFIS	Electronic Flight Instrument System
ELAN	Event Link Analysis Network
EPA	Environmental Protection Administration (USA)
EPR	Engine Pressure Ratio
FAA	Federal Aviation Administration (USA)
FAR	Federal Aviation Regulation (USA)
FBI	Federal Bureau of Investigation (USA)
FDR	Flight Data Recorder
FM	Flight Manual
FMS	Flight Management System
FMEA	Failure Mode and Effects Analysis
FOD	Foreign Object Damage (also the foreign object)
FPD	Freezing Point Depressant
FPS	Feet Per Second
FSF	Flight Safety Foundation
FTA	Fault Tree Analysis
G-LOC	G-induced Loss of Consciousness
GPS	Global Positioning System
GPWC	Ground Proximity Warning Computer
GPWS	Ground Proximity Warning System
GW	Gross Weight
HSI	Horizontal Situation Indicator
HUD	Head Up Display
ICAO	International Civil Aviation Organization
IFR	Instrument Flight Rules
IGV	Inlet Guide Vanes
IIC	Investigator In Charge
ILS	Instrument Landing System
IMC	Instrument Meteorological Conditions
INS	Inertial Navigation System
ISASI	International Society of Air Safety Investigators
ISO	International Standards Organization
KCAS	Knots Calibrated Airspeed
KIAS	Knots Indicated Airspeed
KEAS	Knots Equivalent Airspeed
KTAS	Knots True Airspeed
LLF	Limit Load Factor
LLWAS	Low Level Windshear Alerting System
LLWS	Low Level Wind Shear
LTA	Less Than Adequate
MAC	Mean Aerodynamic Chord
MORT	Management Oversight Risk Tree
MPH	Miles Per Hour
MSL	Mean Sea Level
NASA	National Aeronautics and Space Administration (USA)
NATO	North Atlantic Treaty Organization
NDE	Non Destructive Evaluation
NDI	Non Destructive Inspection

NDT	Non Destructive Testing
NMAC	Near Mid Air Collision
NOTAM	Notice To Airmen
NTSB	National Transportation Safety Board (USA)
OSHA	Occupational Safety and Health Administration (USA)
POH	Pilots Operating Handbook
PSI	Pounds Per Square Inch
QAR	Quick Access Readout (Recorder)
RPM	Revolutions Per Minute
SAE	Society of Automotive Engineers (USA)
SAS	Stability Augmentation System
SB	Service Bulletin
SDR	Service Difficulty Report (USA)
SEM	Scanning Electron Microscope
SOFA	Status of Forces Agreement
STANAG	Standard NATO Agreement
STEP	Simultaneous Time and Events Processing
TCAS	Traffic Alert and Collision Avoidance System
ULF	Ultimate Load Factor
USAF	United States Air Force
USCS	United States Customs Service
USGS	United States Geological Survey (Map)
USN	United States Navy
VFR	Visual Flight Rules
VMC	Visual Meteorological Conditions
VVI	Vertical Velocity Indicator
WYSIWYG	What You See Is What You Get

3. "V" SPEEDS.

V_A	Design Maneuvering Speed
V_B	Design Speed for Maximum Gust Intensity
V_C	Design Cruising Speed
V_D	Design Diving Speed
V_{DF}	Demonstrated Flight Diving Speed
V_F	Design Flap Speed
V_{FC}	Maximum Speed for Stability Characteristics
V_{FE}	Maximum Flap Extended Speed
V_H	Maximum Level Flight Speed With Maximum Continuous Power
V_{LE}	Maximum Landing Gear Extended Speed
V_{LO}	Maximum Landing Gear Operating Speed
V_{LOF}	Lift-Off Speed
V_{MC}	Minimum Control Speed With Critical Engine Inoperative
V_{MO}	Maximum Operating Limit Speed
V_{MU}	Minimum Unstick Speed
V_{NE}	Never-Exceed Speed
V_{NO}	Maximum Structural Cruising Speed
V_R	Rotation Speed
V_S	Stalling Speed
V_X	Best Angle of Climb Speed
V_Y	Best Rate of Climb Speed
V_1	Takeoff Decision Speed (Formerly Critical Engine Failure Speed)
V_2	Takeoff Safety Speed

Index